International Nuclear Science Technology and Engineering Conference (iNuSTEC 2019)

Sustainable Nuclear Economy for the Nation Progress

IOP Conference Series: Materials Science and Engineering
Volume 785

Bandar Baru Bangi, Malaysia
29 - 31 October 2019

Editors:

Faridah Mohamad Idris
Julie Andrianny Murshid
Aznan Fazli Ismail
Syazwani Mohd Fadzil
Mohd Idzat Idris
Imran Yusuff
Norsyahidah Mohd Hidzir
Irman Abdul Rahman
Khoo Kok Siong
Faizal Mohamed
Mohd Syukri Yahya

ISBN: 978-1-7138-3512-7
ISSN: 1757-8981

Printed from e-media with permission by:

Curran Associates, Inc.
57 Morehouse Lane
Red Hook, NY 12571

Some format issues inherent in the e-media version may also appear in this print version.

Institute of Physics
Dirac House, Temple Back
Bristol BS1 6BE UK

Phone: 44 1 17 929 7481
Fax: 44 1 17 920 0979

techtracking@iop.org

Additional copies of this publication are available from:

Curran Associates, Inc.
57 Morehouse Lane
Red Hook, NY 12571 USA
Phone: 845-758-0400
Fax: 845-758-2633
Email: curran@proceedings.com
Web: www.proceedings.com

TABLE OF CONTENTS

ENERGY

ENVIRONMENT

INFRASTRUCTURE

SUSTAINABILITY

IOP Conf. Series: Materials Science and Engineering **785** (2020) 011001 doi:10.1088/1757-899X/785/1/011001

IOP Conference Series: Materials Science and Engineering

International Nuclear Science, Technology and Engineering Conference 2019 (iNuSTEC2019)

"Sustainable Nuclear Economy For The Nation Progress"

Universiti Kebangsaan Malaysia
Bangi, Selangor, Malaysia
29-31 October 2019

Editors

Faridah Mohamad Idris
Malaysian Nuclear Agency (Nuclear Malaysia)
Julie Andrianny Murshid
Malaysian Nuclear Agency (Nuclear Malaysia)
Aznan Fadzli Ismail
Universiti Kebangsaan Malaysia (UKM)
Syazwani Mohd Fadzil
Universiti Kebangsaan Malaysia (UKM)
Mohd Idzat Idris
Universiti Kebangsaan Malaysia (UKM)
Imran Yusuff
Universiti Kebangsaan Malaysia (UKM)
Norsyahidah Mohd Hidzir Universiti Kebangsaan Malaysia (UKM)
Irman Abdul Rahman Universiti Kebangsaan Malaysia (UKM)
Khoo Kok Siong
Universiti Kebangsaan Malaysia (UKM)
Faizal Mohamed
Universiti Kebangsaan Malaysia (UKM)
Mohd Syukri Yahya
Universiti Tenaga Nasional Malaysia (UNITEN)

All papers have been peer-reviewed

Supporting/Sponsoring Organisations

Ministry of Education Malaysia; Ministry of Energy, Science, Technology, Environment & Climate Change; Malaysian Nuclear Agency; ASME Malaysia Section; Malaysian Welding and Joining Society; Reactor Interest Group; Asia-Oceania Neutron Scattering Association, Japan Atomic Energy Society.

IOP Conf. Series: Materials Science and Engineering **785** (2020) 011001 doi:10.1088/1757-899X/785/1/011001

PREFACE

All praises to God the Al-Mighty for giving us the strengths and the His blessing in making Nuclear Science, Technology and Engineering Conference 2019 (iNuSTEC2019) a success.

The history in organizing Nuclear Science, Technology and Engineering Conference started back in 2010. During that year Then, it was called NuSTEC2010 and was organized only at national and regional levels. From that year onwards, we succeeded in organizing our second (iNuSTEC2011), third (iNuSTEC2012), fourth (international iNUSTEC2013), fifth (NuSTEC2014), sixth (international iNuSTEC2015), seventh (international iNuSTEC2016), eighth (international iNuSTEC2017) conferences, ninth (iNuSTEC2018) and tenth (iNuSTEC2019). Based on these achievements, we have decided to reconvene the conference sustainability by organizing regional/international iNuSTEC on annual basis.

The iNuSTEC2019 was held from 29-31 October 2019 at Universiti Kebangsaan Malaysia (UKM) in Bangi, Selangor, Malaysia. The conference continued the tradition of continuous effort by Malaysian Nuclear society to bring together the nuclear communities in an annual event to discuss and to discourse on progress in nuclear researches and related fields. The conference particularly encouraged the interaction of research students, research institutions and academics with the more established community. iNuSTEC2019 also played an essential role in introducing and fostering new research talents and in advancing the future profile of nuclear science, technology and engineering regionally and globally.

iNuSTEC2019 comprised of the conference parallel sessions, Zarah dan Daya 2019, Women in Nuclear Conference 2019 and Nuclear Youth 2019 competition. At the national forefront of science and energy development, we were honoured to be able to highlight the usage of science and engineering in the nuclear fields and to promote the use of nuclear science and technology for peaceful use through these conference, workshop and side-meetings.

We would like to acknowledge the Ministry of Energy, Science, Technology, Environment and Climate Change Malaysia (MESTECC), Ministry of Education Malaysia (MOE), Malaysian Nuclear Agency (Nuklear Malaysia), International Atomic Energy Agency (IAEA), Japan Atomic Energy Society (JAES), China Nuclear Society (CNS), SOKENDAI (The Graduate University For Advanced Studies), Japan, Alex Corporation (M) Sdn Bhd, Sinaran Utama Teknologi Sdn Bhd, Reltech Lab Sdn Bhd, Cerenkov Scientific Sdn Bhd, Intech Scientific Sdn Bhd, Exprojaya Sdn Bhd, Arey Solutions Sdn Bhd and Rem Consultants Sdn Bhd, Asia-Oceania Neutron Scattering Association (AONSA), ASME Section Malaysia, Malaysian Welding and Joining society, Institute of Nuclear Engineers (INE MNS), Kelab Sains Nuklear (KESAN), Nuclear Engineering Student Society (NESS), NGOs, universities and companies for their moral and financial support. Special thanks to iNuSTEC2018 organizing committees for their effort and commitment during this conference. We look forward to meeting all of you in 2019 – iNuSTEC2020 @ UNITEN.

Assoc Prof Dr Faizal Mohamed (UKM)
For Chairs of iNuSTEC2019
Bangi, Selangor
Malaysia

IOP Conf. Series: Materials Science and Engineering **785** (2020) 011001 doi:10.1088/1757-899X/785/1/011001

MESSAGE FROM:
PRESIDENT, MALAYSIAN NUCLEAR SOCIETY (MNS)

On behalf of the Malaysian Nuclear Society (MNS), I would like to express our thanks to all participants of the International Nuclear Science, Technology and Engineering Conference 2019 (iNuSTEC2019). MNS is honoured to receive supports from distinguished international and national experts in the field of nuclear (neutron) science and engineering as International Advisory and National Advisory Panels for iNuSTEC2019. We are indeed pleased to have Organizing Committees of iNuSTEC2019 and the Secretariats working with along to ensure the success of iNuSTEC2019. At the forefront of nuclear science and engineering MNS collaborates with universities and related NGOs to bring the success of iNuSTEC2019. MNS members shall strive to make it better and stronger in future.

In conjunction with this conference, MNS is also honoured to organise Zarah dan Daya 2019, Women in Nuclear Conference 2019 and 4th Nuclear Youth Competition 2019 (NYC2019) to help bring relevant sector and the young Z nuclear generation into the main stream of nuclear development and issues.

This conference brings together experienced and young researchers to discuss and debate latest development in all areas of nuclear science, engineering and technology applications for friendly purposes at international level. This would spur and catalyse further research and development leading to the sustainable and improvement of the quality of life without compromising the quality of the environment in Malaysia and in the world.

We would like to express our gratitude to UKM for hosting iNuSTEC2019, Zarah dan Daya 2019, WiN Conf 2019 and NYC2019 and, to all the personnel involved for their unwavering efforts to make this event a reality. I wish to also record my appreciations and thanks to all sponsors for their generous support. May all these efforts bring benefits to all of us and I congratulate everyone involved in making this conference successful event.

Thank you.

Assoc Prof Dr Abdul Aziz Mohamed
President, MNS

IOP Conf. Series: Materials Science and Engineering **785** (2020) 011001 doi:10.1088/1757-899X/785/1/011001

CONTENTS

PAPERS

IOP Conf. Series: Materials Science and Engineering **785** (2020) 011001 doi:10.1088/1757-899X/785/1/011001

PROCEEDING EDITORIAL COMMITTEE

IOP Conf. Series: Materials Science and Engineering **785** (2020) 011001 doi:10.1088/1757-899X/785/1/011001

iNuSTEC2019 ORGANISING COMMITTEE

Patrons:
Prof. Ir. Dr. Mohd Hamdi Abd Shukor
Vice Chancellor, Universiti Kebangsaan Malaysia (UKM)

Advisors:
Professor Dr. Mohammad B. Kassim,
Dean, Faculty of Science and Technology, Universiti Kebangsaan Malaysia
Prof Dr Nahrul Khair Alang Md Rashid (Universiti Teknologi Malaysia)
Advisor, Malaysian Nuclear Society

Chairman:
Assoc. Prof. Dr Faizal Mohamed
Deputy Dean, Faculty of Science and Technology, UKM

Co-chairman:
Assoc Prof Abdul Aziz Mohamed
President, Malaysian Nuclear Society (MNS)

Deputy Chairman:
Assoc. Prof. Dr Irman Abdul Rahman
Head, Nuclear Science Programme, UKM

Co-deputy Chairman:
Ts. Dr. Siti Aiasah Hashim, Win Malaysia/Nuklear Malaysia

Secretary:
Dr Syazwani Mohd Fadzil, UKM
Dr Faridah Mohamad Idris, MNS/ Nuklear Malaysia

Treasurer:
Dr Julie Andrianny Murshidi, MNS Treasurer

Deputy Treasurer:
Dr Mohd Idzat Idris, UKM

IOP Conf. Series: Materials Science and Engineering **785** (2020) 011001 doi:10.1088/1757-899X/785/1/011001

INTERNATIONAL ADVISORY COMMITTEE

Prof. Emiretus Dr. Noramly Muslim, MNS Honorary Fellow
Prof. Emiretus Dr. Misaki Saito, JAERI, Japan.
Dato' Raja Aziz Raja Adnan, IAEA, Vienna, Austria
Dr Yamashita, JICC, Jaica
Prof Dr Il Hwang Soon, Seoul National University, Korea
Prof. Sung-Min Choi, KAIST, Republic of Korea
Prof. Dr. Edy Giri Rachman Putra, BATAN, Indonesia
Dr Mohd Abd Wahab bin Yusof, Nuklear Malaysia
Assoc. Prof. Dr. Mokhtar Awang, ASME Malaysia Section

NATIONAL ADVISORY COMMITTEE

Prof. Dr. Bachok M.Taib, Universiti Sains Islam Malaysia.
Prof. Dr. Hj. Khudzir Hj Ismail, Universiti Teknologi MARA.
Prof. Emeritus Dato' Wira Ir. Dr. Md Noor bin Salleh, International Islamic Unviversity Malaysia.
Prof Dr. Zanariah Abdullah, Universiti Malaya.
Prof Dr Wan Ahmad Tajuddin Wan Abdullah, Universiti Malaya
Prof Madya Dr Izham bin Datuk Zainal Abidin, UNITEN
Dr Khaidzir Hamzah, Universiti Teknologi Malaysia.
Assoc. Prof. Dr. Abu Hassan, Universiti Teknologi MARA.
Dr Mokhtar Awang, Universiti Teknologi PETRONAS /ASME Malaysia Section.
Assoc Prof Dr Samsudin Bani, President of Malaysian Welding and Joining Society (MWJS).

IOP Conf. Series: Materials Science and Engineering **785** (2020) 011001 doi:10.1088/1757-899X/785/1/011001

TECHNICAL COMMITTEE

Universiti Kebangsaan Malaysia:

Technical and Logistics Committee
Dr Imran Yusuff
Dr Mohd. Amir Radhi bin Othman
Wan Mohd Aimran bin Wan Mohd Kamil
Azlan Shah Mustafa

Scientific Committee
Dr Aznan Fazli Ismail
Assoc Prof Dr Yap Chi Chin
Dr Chen Ruey Shan
Dr Nur Adlyka binti Ainul Annuar
Assoc Prof Dr Irman Abdul Rahman
Dr Idris Zulkifle

Protocol And Publicity Committee
Assoc Prof Dr Khoo Kok Siong
Dr Nurul Shazana binti Abdul Hamid
Munira Hasan
Diha Ariane Salihuddin
Che Mohd Zaidi bin Che Daud
Tazidi Taib

Banquets And Souvenirs Committee
Dr Norsyahidah binti Mohd Hidzir
Dr Siti Aisyah binti Shamsudin
Dr Zalita binti Zainuddin
Dr Rozidawati Awang
Farahin Othman
Takim Saring

IOP Conf. Series: Materials Science and Engineering **785** (2020) 011001 doi:10.1088/1757-899X/785/1/011001

MALAYSIAN NUCLEAR SOCIETY

Khair'ah Binti Yazid@Khalid, Nuklear Malaysia
Anas Bin Mohamad Fauzi, Uniten
Dr Hassan Bin Mohamed, Uniten
Dr Julia Bt. Abdul , Nuklear Malaysia
Dr Shukri Mohd, Nuklear Malaysia
Dr Muhammad Rawi B. Mohamed Zin, Nuklear Malaysia
Dr Mohd Syukri Yahya, Uniten
Khairul Azhar Bin Abdul Halim, Nuklear Malaysia
Ikmal Azrai Bin Abdul Aziz, Arey Solution
Mohd Rizal B. Mamat @ Ibrahim, Nuklear Malaysia
Anwar B. Abdul Rahman, Nuklear Malaysia
Dr. Hafizal Yazid, Nuklear Malaysia
Azraf Bin Azman, Nuklear Malaysia
Shalina Bt. Sheik Muhamad, Nuklear Malaysia
Nazaratul Ashifa Bt. Abdullah Salim, Nuklear Malaysia
Dr Mazleha Maskin, Nuklear Malaysia

Universiti Teknologi Malaysia:

Dr Khaidzir Hamzah
Dr Jasman Zainal
Dr Zulhairun Bin Abdul Karim

Universiti Tenaga Nasional:

Assoc Prof Dr Nasri A. Hamid
Dr Mohd Syukri Yahya,
Dr Hasan Mohamed

Universiti Teknologi MARA:

Assoc Prof Dr Abu Hassan Husin

Universiti Sains Islam Malaysia:

Dr Mohd Hafez bin Mohd Isa

Universiti Malaya:

Prof Hasan Abu Kassim,
Assoc Prof Dr Norhasliza Yusof

Universiti Tun Hussien Onn Malaysia:

Assoc Prof Dr Zaidi Embong

Institute of Nuclear Engineers:

Safwan Shalbi
Muhammad Aliff Ashraff Rosdi
Asyraf Arif Bin Abu Bakar
Abdul Hannan Damahuri

Zarah dan Daya 2019 Committee

Advisors
Prof Dr Wan Ahmad Tajuddin Wan Abdulllah, UM/NCPP
Prof Shahidan Radiman, UKM/NCPP
Dr Izyan Hazwani Hashim, UTM/NCPP
Dr Faridah Mohamad Idris, Nuklear Malaysia/NCPP

Chairman
Dr Nurfikri Norjoharuddeen, UM/NCPP

Deputy Chairman
Dr Imran Yusuff, UKM/NCPP

Secretary
Nurul Hidayah Binti Mohamad Nor, UM/NCPP

Treasurer
Abu Ubaidah Amir bin Ab Maalek, UM/NCPP

Logistics
Zukhairmira Zolkapli, UM/NCPP
Khasmidatul Akma bt Mohamed Kamal Azmi UM/NCPP
Muhammad Hafizuddin Al-helmy bin Nouxman UM/NCPP

Technical
'Atiqah Mohamad Jainal UM/NCPP
Fahmi Maula UM/NCPP
Shafiq Rasulan UM/NCPP

WinConf Committee:

Chairperson
Dr. Siti A'iasah Hashim

Secretary
Ruzalina Baharin

Technical committee
Dr Hasni Hasan (Chair)
Dr Noor Hasnah Mohd Khairullah
Datin Zarina Masood
Puan Rokiah Mohd Sabri
Naurah Mat Isa
Roha Tukimin
Hasfazilah Hassan

KELAB SAINS NUKLEAR (KESAN) SECRETARIAT

iNuSTEC:

1. Nursyasya Farzana Binti Norjamalullail
2. Muhammad Aiman Bin Mohd Yunus
3. Sheldon Thony
4 Batari Toja Binti Iskandar
5. Nurul Fathin Binti Samsor
6. Jayasuria A/P Ravinthiran
7. Nur Hanani Binti Zulkifli
8. Nur Amira Hidayah
9. Anis Arisya Binti Abdul Malek
10. Izzah Husna Binti Abdul Jabar
11. Nur Azira Binti Azana Azhar
12. Ahmad Harizuddin Bin Mohd Sani
13. Nur Aziera Binti Zamshahari
14. Ahmad Idham Bin Ibrahim
15. Sharendravarman A/L Balakrisnan
16. Nurul Muyassarah Binti Ibrahim
17. Nur Aina Najwa Bin Zolkiflee
18. Lim Teng Sheng
19. Aina Aqilah Binti Mohd Asmadi
20. Revathy A/P Suresh

Nuclear Youth Competition

1. Muhammad Ariff Baihaqi Bin Jamaludin
2. Nur Sabrina Binti Tajuddin
3. Nashihah Binti Ab Karim
4. Nurul Najihah Xiwei Binti Mohd Kamal
5 Nur Nadhilah Faqihah Binti Mohd Ferdawos
6. Dayang Siti Mahani Binti Mohd Sukarni
7. Nuresa Binti Othman
8. Siti Khatijah Binti Marakar
9. Nur Shaihira Binti Abdul Halim
10. Nur Aina Binti Aminuddin
11. Nur Afiqah Binti Nasruddin
12. Nurul Atiqah Binti Faizul
13. Seri Zurina Binti Abd Kayum
14. Anis Balqis Binti Mohd Azmi
15. Muhammad Hazim Bin Yahya
16. Mohamad Danial Bin Sakaria
17. Nurul Farihah Binti Mohd Nazib
18. Nooramira Binti Mohd Saidin
19. Muhammad Haziq Bin Mohammad Zainuri
20. Nur Syafinaz Binti Mohd Amirul Manja
21. Kavipriya A/P Asokan
22. Nurul Syakirah Binti Zainal
23. Farah Ainin Sofiya Binti Muhd Syazli
24. Nur Fatin Hanani Binti Razali
25. Aqilah Iman Binti Rahim
26. Nur Fathin Kamila Binti Aminudin
27. Nur Azira Binti Jasman
28. Shazmin Hamimi Binti Hafiz
29. Auni Syamimi Amni Binti Ahmad Abdul Murad
30. Rabiatun Nisrin Binti Yusop
31. Nur Fyrzanah Hasynah Binti Farzeli
32. Ainur Fatihah Binti Alpenzy
33. Nurul Nadia Binti Salim
34. Ahmad Daniel Bin Ahmad
38. Amir Aiman Bin Adnan
39. Zulhafizan Saad Bin Jamaluddin
40. Mohamad Ehsan Bin Songip
41. Zainul Ariffin Bin Zawawi
42. Husna Nabeela Binti Hussain
43. Afifah Mardhiah Binti Mohamed Radzi
44. Muhammad Ammar Bin Mohd Suhaimie
45. Mohamad Firdaus Bin Noor Azman
46. Nurul Iffah Zahirah Binti Ahmad Zabari
47. Farra Aidura Binti Azli
48. Shahrul Sazuan Bin Ariffin
49. Megat Helmi Salihin Bin Megat Nirman Bukhari
50. Muhd Amir Bin Mohd Abd Nasir
51. Muhammad Nur Izzat Bin Mamat Saidi
52. Muhamad Arsyad Bin Aziz
53. Nur Farahatun Hawa Binti Ramlee
54. Saidatul Aisyah Binti Kamarulzaman
55. Muhammad Hisyam Bin Ibrahim
56. Haziq Bin Hassan
57. Mimi Nur Umaira Binti Idham Wazir
58. Nadhirah Binti Abd Ajed
59. Zulhazrin Azahra Binti Alias
60. Athina Binti Atiman
61. Farah Arissa Binti Halim
62. Mohamad Shatifi Bin Abdullah
63. Muhammad Fawwaz Bin Shaharudin
64. Nurfitri Syahada Binti Mohd Yusof
65. Nur Anis Zulaikha Binti Kamarudin
66. Nurul Syafiqah Binti Mulldi
67. Muhammad Fikri Bin Mohd Fahrul
68. Muhammad Amirul Hafiz Bin Sabri

ACKNOWLEDGEMENT

iNuSTEC2019 is organized by Malaysian Nuclear Society (MNS)
and Universiti Kebangsaan Malaysia (UKM)

In Collaboration with

Malaysian Nuclear Agency (Nuclear Malaysia)
Universiti Tenaga Nasional (UNITEN)
Universiti Teknologi MARA (UiTM)
Universiti Islam Antarabangsa Malaysia (IIUM)
Universiti Teknologi Malaysia (UTM)
Universiti Malaya (UM)
Universiti Sains Islam Malaysia (USIM)
Universiti Tun Hussien Onn Malaysia (UTHM)

Co-organized by
National Centre for Particle Physics (NCPP)
Institute of Nuclear Engineers (INE-MNS)
Kelab Sains Nuklear (KESAN)
Nuclear Engineering Student Society (NESS)

Supported by

ASME Malaysia Section
Malaysian Welding and Joining Society

E MNS COUNCIL MEMBERS 2019/2020

Advisor / Past President	Dr Nahrul Khair Bin Alang Md Rashid (UTM)
President	Dr Abdul Aziz Bin Mohamed (UNITEN)
Vice President	Dr Faizal Bin K.P. Kunchi Mohammed (UKM)
Secretary	Dr Faridah Bt Mohamad Idris (Nuklear Malaysia)
Deputy Secretary	Anas Bin Mohamad Fauzi (Uniten)
Treasurer	Dr Julie Andrianny Bt. Murshidi (Nuklear Malaysia)
Deputy Treasurer	Khair'ah Binti Yazid@Khalid (Nuclear Malaysia)

Members

Dr Hassan Bin Mohamed (Uniten)
Dr Julia Bt. Abdul Karim (Nuklear Malaysia)
Dr Shukri Mohd (Nuklear Malaysia)
Dr Muhammad Rawi B. Mohamed Zin (Nuklear Malaysia)
Dr Mohd Syukri Yahya (Uniten)

Auditors

Khairul Azhar Bin Abdul Halim (Nuklear Malaysia)
Ikmal Azrai Bin Abdul Aziz (Arey Solution)

MNS Chapters

RadTech (Radiation Tech)
XApp (X-ray App)
NuEnergy (Nuclear Energy)
Neutron Photon (Neutron- Photon)
Student Chapter (UKM, UiTM, UNITEN, UTM - NESS)
Women in Nuclear Malaysia (WiN Malaysia)
Institute of Nuclear Engineers (INE MNS)

Visit us at http://www.nuklearmalaysia.org

IOP Conf. Series: Materials Science and Engineering **785** (2020) 011001 doi:10.1088/1757-899X/785/1/011001

With Best Compliments

SOKENDAI
(THE GRADUATE UNIVERSITY FOR ADVANCED STUDIES), JAPAN

ALEX CORPORATION (M) SDN BHD, MALAYSIA

SINARAN UTAMA TEKNOLOGI SDN BHD, MALAYSIA

RELTECH LAB SDN BHD, MALAYSIA

CERENKOV SCIENTIFIC SDN BHD, MALAYSIA

INTECH SCIENTIFIC SDN BHD, MALAYSIA

CREST LAB SDN BHD / OLYMPUS, MALAYSIA

EXPROJAYA SDN BHD, MALAYSIA

REM CONSULTANTS SDN BHD, MALAYSIA

AREY SOLUTIONS SDN BHD, MALAYSIA

UNIVERSITI
KEBANGSAAN
MALAYSIA
The National University
of Malaysia

IOP Conf. Series: Materials Science and Engineering **785** (2020) 011003 doi:10.1088/1757-899X/785/1/011003

Peer review statement

All papers published in this volume of *IOP Conference Series: Materials Science and Engineering* have been peer reviewed through processes administered by the proceedings Editors. Reviews were conducted by expert referees to the professional and scientific standards expected of a proceedings journal published by IOP Publishing.

IOP Conf. Series: Materials Science and Engineering **785** (2020) 012001 doi:10.1088/1757-899X/785/1/012001

Gamma and neutron source term calculation for irradiated TRIGA Mark II fuel using ORIGEN2

Nurlaila Syamsul Bahri[1], Jasman Zainal[1, a)], Abi Muttaqin Jalal Bayar[2], Mohamad Hairie Rabir[2]

[1]Department of Nuclear Engineering, School of Chemical and Energy Engineering, Faculty of Engineering, Universiti Teknologi Malaysia, 81310, Johor, Malaysia.
[2]Nuclear & Reactor Physics Section, Reactor Technology Centre, Technical Support Division, Malaysian Nuclear Agency, Bangi, 43000 Kajang, Selangor, Malaysia.

[a]jasman@utm.my

Abstract. TRIGA Mark II has served as nuclear research reactor for decades, which is to provide research facility towards nuclear development as well as for trainings in the aspect of education and industry. Standard TRIGA fuel, $UZrH_{1.6}$ is used in this reactor which can be found in two types; aluminium or stainless steel cladding. In this study, stainless steel cladded fuel used in Reactor TRIGA PUSPATI (RTP) that consists of fissile U-235 with 19.9% enrichment and 20% weight of Uranium is used as reference for input data. The irradiated fuel inventories and fuel burnup were determined using ORIGEN2 code by referring to PWR data library, consists of gamma and neutron source term comes from activation products, actinides and fission products after 18 years of irradiation and 5 years of cooling period with 200 days of time step. The outcomes from this study can be concluded with the inventories from fission products to be the highest in all categories, in which the U-235 depleted up to 52.9%. The results from this simulation is crucial as it is then carried forward for radiation shielding for interim storage facility.

1.0 Introduction

TRIGA Mark II research reactor was installed at Malaysian Nuclear Agency in the year of 1982, named as Reactor TRIGA PUSPATI (RTP) and went critical for the first time on 28 June of the same year with excess reactivity of $0.15 [1]. RTP is the one and only research reactor available in Malaysia and it has been operating for 37 years with maximum thermal power output of 1 MW. The core consists of standard TRIGA fuel $UZrH_{1.6}$ which is a mixed of 8.5, 12, and 20 wt% of uranium content with 19.9% enrichment of U-235. All the fuels are stainless-steel cladded (type-104). Through its operation years, RTP core have been reshuffled for few times in order to utilize the fuel to control the criticality of the core. As in the year 2017, the core operates with its latest 15th configuration of fuel assembly as shown in Figure 1 where the fuels together with control rods and dummy elements are arranged in all seven circular rings.

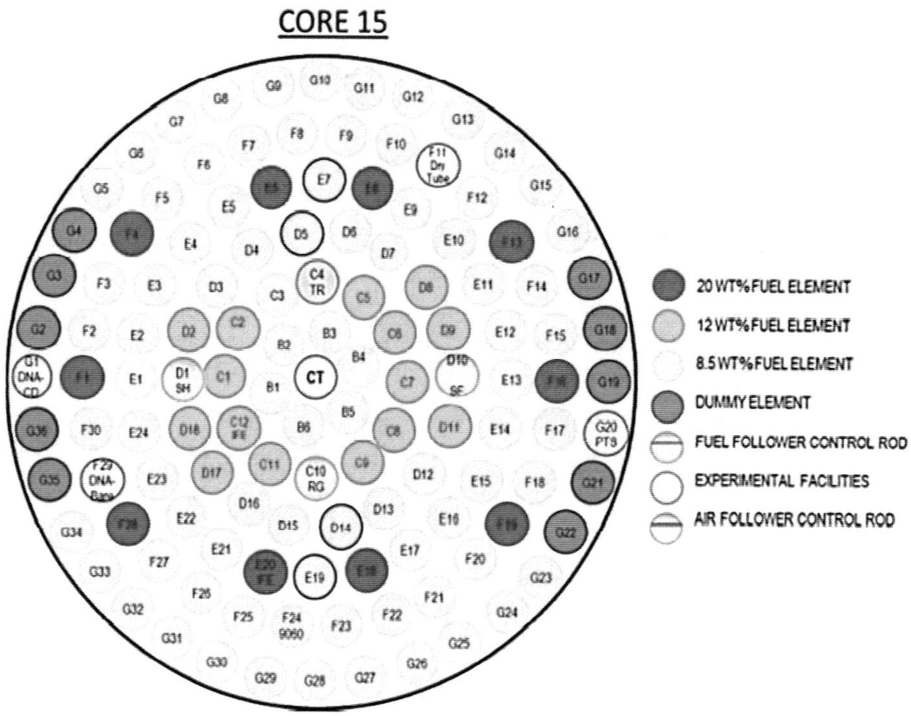

Figure 1. The Core-15 configuration, the latest configuration of RTP core [1].

In this work, source term estimation in terms of neutron and gamma of irradiated fuel was simulated by using ORIGEN2 by taking the data of fresh 20 wt% uranium content fuel in RTP as the target of reference. The use of ORIGEN2 is based on its capability to calculate buildup, decay and processing of various selection of radioactive material. This paper describes the methods used in order to get the fuel burnup, photon spectrum and also the neutron source of a single fuel after irradiation.

2.0 Fuel element source term estimation

The estimation was made based on the theoretical limit for U-235 depletion in TRIGA fuel that is around 25% to 50% reduced from its original value, and therefore leads into the situation where the fuel should be removed out from reactor operation [2]. For the burnup calculation, few assumptions were made as initial conditions. Since the ORIGEN2 code simulates for one fuel assembly only, an average power for one fuel is required to be determined. The TRIGA reactor is fixed to produce 1 MW thermal power in the first six years and reduced to 750 kW for the rest of irradiation period.

A comparison was made by referring to the irradiation history of RTP during its first criticality in which the situation of 66 fuels of 8.5 wt% fuels is substituted with 20 wt% fuels in the core to produce 1 MW thermal power. Thus, a total of 23 fuels of 20 wt% uranium content fuel is determined to be in the hypothetical core, with an average power of 0.04348 MW (1 MW) and 0.032607 MW (750 kW) is produced in one fuel.

The total irradiation time is one of the variable in this work, which is depending to the output where the depletion limit of U-235 is reached. Considering the total of six hours operation per day, five days a week, the total time were then standardized according to unit day with 200 days of time step. The fuel is also allowed to cool, assuming the period of decay within 5 years to be considered, where the output data were distinguished for every 300 days interval.

There are several libraries to choose from, but for this work, the cross section libraries for PWR is used which correspond to the irradiation power command (IRP). The libraries are identified by NLIB numbers, which for PWR case, NLIB=204 (activation products), NLIB=205 (actinides) and NLIB=206 (fission products) are used, and thus, resulting three separate and distinct nuclide lists as the output. Photon data yield library also included, which gives the photon per disintegration in eighteen energy groups. The data of fuel includes the weight and atomic mass for every isotopes in fuel meat (active region in the fuel where energy is produced by fission process) and cladding materials which are as shown in Table 1 and Table 2.

Table 1. Composition of fuel meat of TRIGA fuel.

Isotope	Mass (g)
U-235	99
U-238	396
Zr	1955.38
H	34.57

Table 2. Composition of cladding for SS-clad TRIGA reactor fuel.

Isotope	Mass (g)
C	0.66
N	1.07
Si	8.19
P	0.37
S	0.25
Cr	155.69
Mn	16.39
Fe	564.12
Co	0.66
Ni	73.09

The composition data for the cladding materials were taken by varying the total composition used in the ORIGEN2 run for TRIGA reactor with SS-clad fuel. These composition were described which is by referring to midrange burnup of 19,492 MWd/MTHM, that equates to 3.8 MWd for a fuel element after irradiated for 3.469 years at a constant power of 0.003 MW [3].

3.0 Results and discussions

3.1 Burnup calculation

Burnup calculation were performed as it is one of the outputs provided in ORIGEN2 together with the summarized information of fluxes, specific power and infinite multiplication factor data for the simulation, at which the core being operated at 1 MW (for the first six years) and 750 kW (the rest of the operation period). After cessation of fuel irradiation (continuous 1200 days of operation), the results obtained regarding burnup of individual TRIGA fuel is as shown in Table 3. The mass of U-235 in the irradiated fuel is 52.9% depleted from its original value.

Table 3. Burnup data of the irradiated TRIGA fuel.

Average power produced by one fuel	0.0362319 MW
Fuel burnup	43.4782 MWd
Average neutron flux	1.26E+14 n/cm^2.s
$\mathbf{K_{inf}}$	1.28430

The irradiation was stopped after continuous 1200 days (18 years) of operation as the depletion limit is reached. This action was taken by referring to the change of mass of U-235 in the fuel, in the same time the mass of U-238 was ensured to have no significant change innot change its value as it is supposed to experience scattering with fast neutron, which resemble the realistic fission process occur in RTP. The relevance of this results were referred to the study done by Rabir (2017) which showing the change of the mass of major radioisotope mass over its burnup value in an irradiated TRIGA fuel [1]. Another action was taken with the aim to ensure the output's reliability, which is by comparing them with a data of nuclear mass inventories of spent research reactor fuel assemblies, which applied the use of ORIGEN and WIMS code in determining irradiated fuel inventories, including TRIGA type (specifically for 20wt% uranium content fuel). The comparison can be referred in Table 4.

Table 4. Comparison between the mass results obtained for RTP fuel and data of TRIGA in [4].

Criteria	**Results based on RTP TRIGA fuel**		**Data of TRIGA fuel provided in [4]**	
	Before irradiation	After irradiation	Before irradiation	After irradiation
Mass of U-235	99 g	46.65 g	98 g	45.65
Mass of U-238	396 g	390.8 g	392 g	383.5 g

3.2 Gamma and neutron source term calculation

The neutron production rate simulated from ORIGEN2 are relatively compact and straightforward, which consists of neutron production rate from (alpha, n) reactions and spontaneous fission for each radioisotope involved. The results are as shown in Figure 2 and Figure 3, where a comparison was made between the values after irradiation ends and after 5 years of cooling.

Table 5. Summary of total neutron production rate.

Neutron source	**Total neutron production rate (neutron/second)**	
	After irradiation	After 5 years of cooling
(alpha, n) reaction	6.61E+03	1.50E+03
Spontaneous fission	3.07E+04	4.54E+03

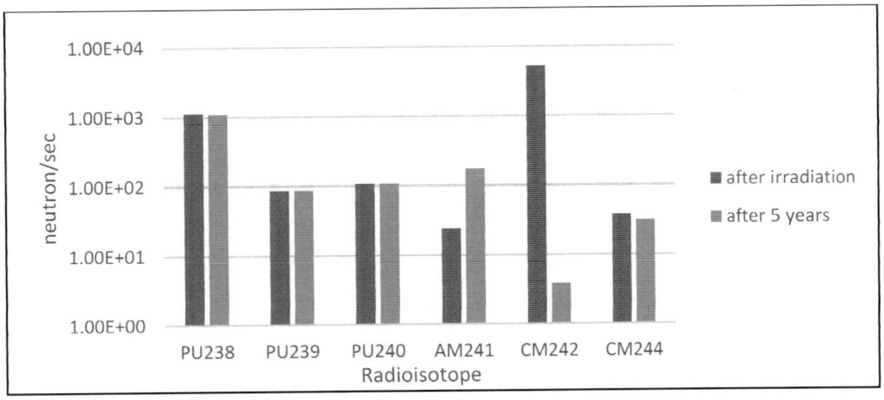

Figure 2. Neutron source from (alpha, n) reaction.

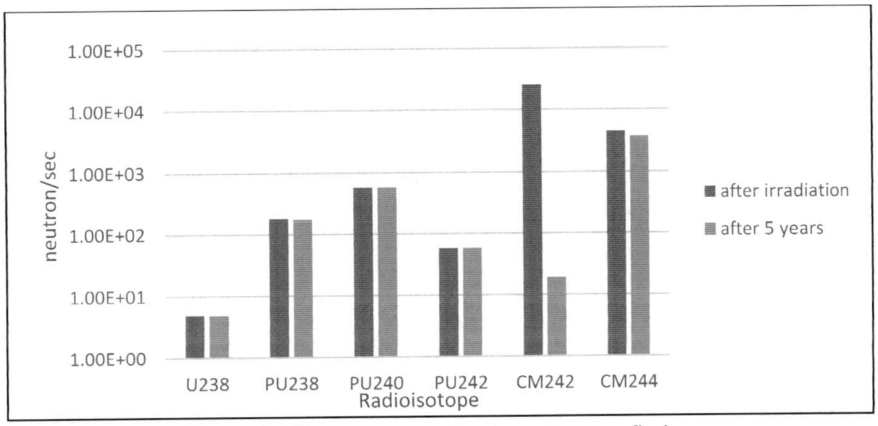

Figure 3. Neutron source from spontaneous fission.

As for photon emission rate, the output data were separated into three segments according to the type of radioisotopes. The activation products includes the low proton number (Z) and radioisotopes that is a product from structural materials of the fuel. The actinides includes heavy isotopes and their daughters, and finally the fission products that includes a significant fission product yield. Gamma energy and the change of gamma emission rate from the point of irradiation ends and after 5 years of cooling for all activation products, actinides, and fission products for eighteen groups of gamma are as shown in Figure 4, Figure 5, and Figure 6. The summary of the total photo emission rate from all 18 energy groups as shown in Table 6.

Table 6. Summary table of the total photo emission rate from all 18 energy groups.

Type of radioisotope	Total photo emission rate from all 18 energy groups (photon/second)	
	After irradiation	After 5 years of cooling
Activation products	2.460E+13	1.516E+06
Actinides	2.513E+14	1.257E+10
Fission products	1.301E+15	1.519E+13

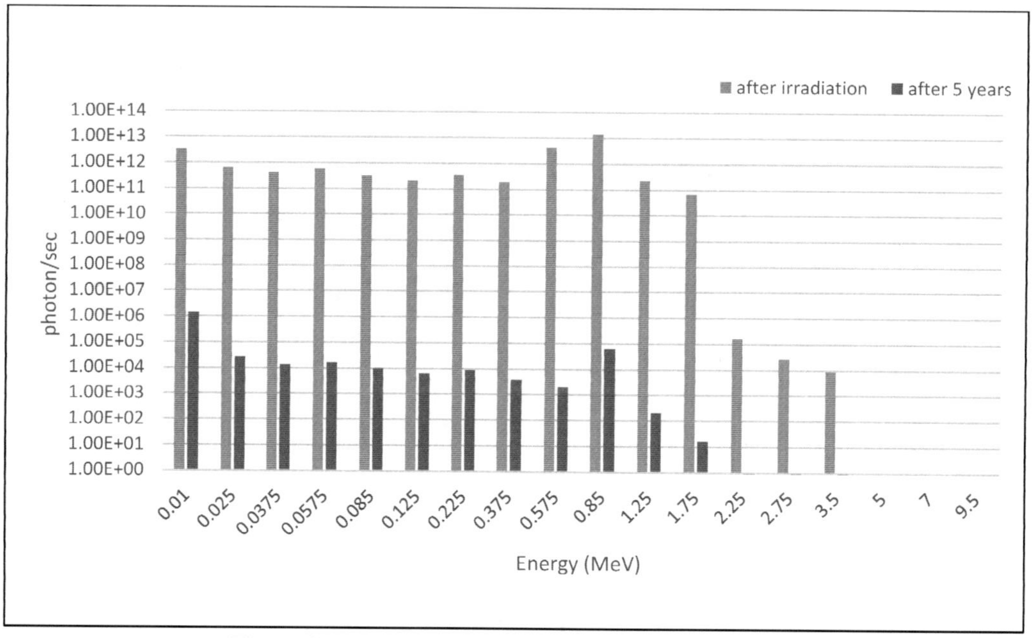

Figure 4. Photon emission rate from activation products.

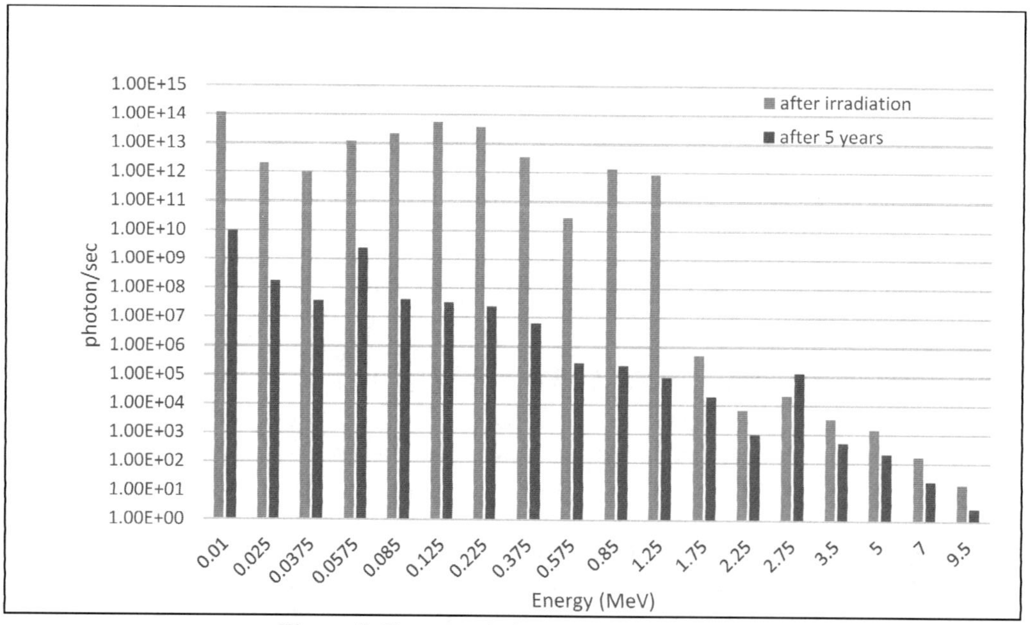

Figure 5. Photon emission rate from actinides.

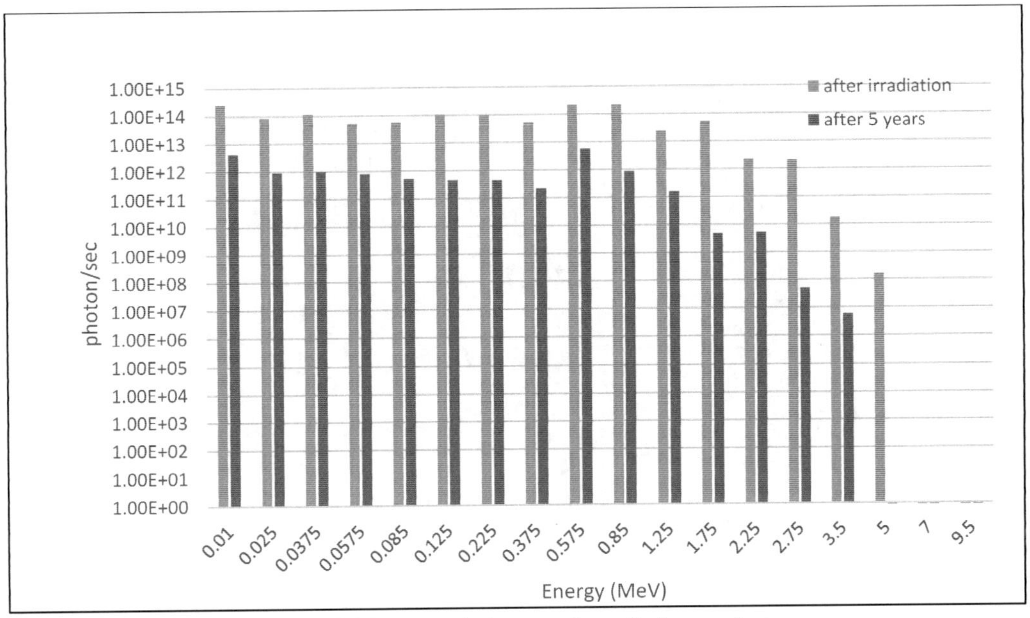

Figure 6. Photon emission rate from fission products.

From the results above, it can be seen that radioisotope from fission products to be the main contributor the photon emission rate compared to other two, followed by actinides in both after irradiation and after 5 years of cooling. This is probably because of the longer half-life radioisotopes which emits gamma in fission products that can be significantly found, such as Cs-137 ($t_{1/2}$=30.17 years). As for activation products, the radioisotopes involved are mainly those produced from the structural material (cladding) which have shorter half-life, and this can be seen from the drastic change of photon emission rate in activation products after irradiation until after 5 years the fuel allowed to cool down and decay.

4.0 Conclusion

The results of the gamma and neutron source term from irradiated TRIGA Mark II fuel were calculated using ORIGEN2, which resulting fission products to emits highest value of photon emission rate. The source of neutron can be found from two; (alpha, n) reaction and spontaneous fission occurs inside of the irradiated fuel in both event; after irradiation and after 5 years of cooling, with the total of 3.73E+04 neutron/second and 6.04E+03 neutron/second respectively. This work is done based on the photon and PWR data libraries on 20 wt% uranium content fuel only. The data provided are respected to one fuel assembly only. This results can be used as the reference for further studies involving fuel inventories of TRIGA fuel, such as to estimate surface dose calculation on interim storage for TRIGA spent nuclear fuel.

References

[1] Rabir M H B, Zin M R B M, Karim J B A, Bayar A M B J, Usang M D A, Mustafa M K A B, Jalil M H B 2017 Neutronics calculation of RTP core *AIP Conf. Proc.* **1799.**

[2] International Atomic Energy Agency 2013 Management and storage of research reactor spent nuclear fuel: proceedings of a technical meeting held in Thurso, United Kingdom. (Vienna: International Atomic Energy Agency) *ISBN 978-92-0-138210-8.*

[3] Sterbentz J W 1997 Radionuclide mass inventory, activity, decay heat, and dose rate parametric data for TRIGA spent nuclear fuels INEL-96/0482 (Idaho Falls, Idaho: Lockheed Martin Idaho Technologies Company, Idaho National Engineering and Environmental Laboratory).
[4] Pond R B, Mato J E 2000 Nuclear mass inventory, photon dose rate and thermal decay heat of spent research reactor fuel assemblies rev. 2 ((United States: Argonne National Lab.(ANL), Argonne, IL) ANL/RERTR/TM-26 .

IOP Conf. Series: Materials Science and Engineering **785** (2020) 012002 doi:10.1088/1757-899X/785/1/012002

Percentage depth dose measurement in high energy photons and electrons by using the Al_2O_3 optically stimulated luminescent (OSL) dosimeter

Mohd Fahmi Mohd Yusof[1,a)], Nur Natasha Md Othman[1], Reduan Abdullah[1], Ahmad Bazlie Abdul Kadir[2], Husaini Salleh[2]

[1]School of Health Sciences, Universiti Sains Malaysia, 16150 Kelantan, Malaysia
[2]Health Physics and Secondary Standard Dosimetric Laboratory, Malaysian Nuclear Agency, 43000 Bangi, Selangor, Malaysia.

[a]mfahmi@usm.my

Abstract. The study focused on the measurement of percentage depth dose (PDD) in high energy photons and electrons by using the optically stimulated luminescent (OSL) dosimeters. The beam output calibration was performed on 6 MV photons and 6 MeV electrons and compared to that in the ionization chamber. The PDD was measured in 6 MV photons and 6 MeV electrons and the PDD curves were plotted by using the OSL dosimeters. The PDD curves were compared to that in other common dosimeters including ionization chamber, Gafchromic® EBT2 film dosimeter and the treatment planning software (TPS). The results showed that the beam output calibration measured by using the OSL dosimeters was in good agreement to the ionization chamber within 1.6 and 5.9% percentage of discrepancy for both 6 MV photons and 6 MeV electrons respectively. The measured PDD by using the OSL dosimeters were also in in good agreement to the ionization chamber, EBT2 film and TPS software within percentage 13.4 and 10.4% of discrepancies for 6 MV photons and 6 MeV electrons respectively. The overall results indicated the suitability of OSL dosimeters to be used for direct dosimetry measurements in high energy photons and electrons.

1. Introduction

The use of *in-vivo* dosimeters for dosimetry studies in clinical radiotherapy would help to provide comparison of delivered dose in the patients between the planned and the actual delivered dose. The treatment planning software (TPS) provided a simulation of dose to the target volume taking into account the density and morphology of the tissue volume surrounding it provided by the inputs from the computed tomography (CT) images. The delivered dose to the patients however cannot be measured on the target volume as *in-vivo* detection is not possible in patients. Therefore several indirect dosimeters such as radiochromic film dosimeter and thermoluminescent dosimeter (TLD) are introduced for *in-vivo* dose measurements. This is achieved by using the solid water phantoms and anthropomorphic phantoms to simulate the patient. These dosimeters are placed in the region of interest inside the phantom volume to measure the dose delivered from the treatment units.

For the recent years, the use of optically stimulated luminescent (OSL) dosimeter for dosimetry had been increased in the application of dose assessment resulted from the ionizing radiation absorbed by the human body. In radiotherapy, OSL had been used as an alternative dosimetry technique when rigorous measurement protocols are established especially in radiotherapy and brachytherapy [1-7].

Previous studies suggested the good reproducibility and linear dose response of the optically stimulated luminescent (OSL) dosimeters for standard clinical dose measurements and are suitable for the remoted dosimetry works involving high energy photons and [8][9]. It can be also used in variety of clinical applications such as surface dosimetry, entrance and exit dose measurements, dose mapping and brachytherapy [10-12]. OSL dosimeter utilizes aluminium oxide crystal to include a latticework of dosimetric traps. The typical technical specification and the useful dose range for OSL nanoDots are between 5 and 1500 cGy for general applications. As for medical dosimetry applications, the linear dose response is up to 300 cGy. It also also mentioned that the OSL nanoDots having a very wide operating energy range which is 5 keV to 20 MeV makes it an ideal dosimetry device to use in many dosimetric works application in medical physics including diagnostic imaging X-rays, nuclear medicine imaging dosimetry and radiotherapy dosimetry. It has high sensitivity toward radiation on which it can detect as low as 10 μSv of ionizing radiations.

Percentage depth dose (PDD) is one of the essential parameters to evaluate the attenuation property of a medium and the detection characteristic of a dosimeter [13]. The PDD measurement is commonly measured by using ionization chamber in a water phantom as suggested by many available international dosimetric protocols. Besides the measurement by using the physical dosimeters, the PDD can also being measured by the mean of computer simulations such as the Monte Carlo and treatment planning software (TPS) simulations. This study measured the PDD in 6 MV photons and 6 MeV electrons by using the nanoDot OSL dosimeters based on the proposed method by the international atomic energy agency [14]. The measured PDD using the nanoDot OSL dosimeters were compared to those in the ionization chamber, gafchromic EBT2 film dosimeter and the TPS.

2. Methodology

2.1. Dose Linearity of the OSL Dosimeters

The linearity of the OSL dosimeter to the dose from the photons and electron was determined to measure the behavior of the OSL signals at increased dose of photons and electrons. The OSL dosimeters were irradiated in high energy photon and electron by using Primus linear accelerator at 6 MV photons and 6 MeV electrons. The OSL dosimeters was placed at the central axis of the beam on the solid water phantom by using the source to surface distance (SSD) of 100 cm with the depth of measurement at the depth of maximum dose, d_{max} (approximately 1.5 cm) based on the calibration condition of the available dosimetric protocols (IAEA TRS 398, 2000). Bolus material with approximate thickness of 1.5 cm was placed on the solid water phantom and the SSD was set at 100 cm to the surface of the bolus to ensure that the OSL dosimeters were at the d_{max} as shown in Figure 1. The field size of 10 × 10 cm was used for photons while the applicator of 10 ×10 cm size was used for electrons to provide the field size of the calibration condition. These OSL dosimeters were irradiated at increased dose between 50 and 1000 cGy based on the monitor unit (MU) selection on the control panel (1 cGy = 1 MU).

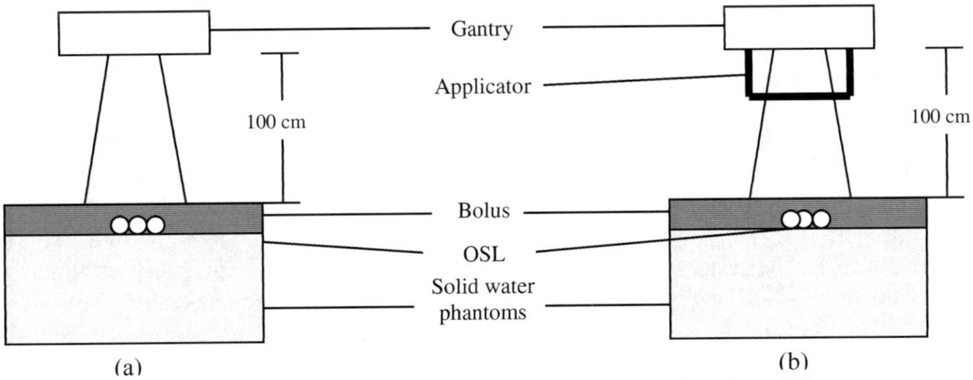

Figure 1. The experimental set up for the measurement of maximum dose in solid water phantom for (a) photons and, (b) electrons.

2.2. Measurement of Percentage Depth Dose in High Energy Photons and Electrons

The PDD in solid water phantoms were measured in 6 MV photons and 6 MV electrons by using the OSL dosimeters. The OSL dosimeters were first placed the surface of the solid water phantoms to measure the surface dose. The dose at depth was measured by placing the bolus material on the depth above the OSL dosimeters to eliminate the air gaps between the solid water phantoms when the OSL dosimeters were present. The SSD was maintained at 100 cm every time the depth was increased. The signals from the exposed OSL dosimeters were sent for readout by using the OSL reader model MicroStar InLight reader (Landauer Inc.). The PDD in OSL dosimeters were calculated by using the equation of

$$PDD_{OSL} = \frac{R}{Rx} \times 100 \quad (1)$$

with R is the OSL reading at any depth and Rx is OSL reading of maximum dose (d_{max}). The PDD curves were plotted for photons and electrons and compared to that in ionization chamber, EBT2 film dosimeter and the TPS simulations. The percentage difference of the PDD at each depth of measurement was calculated by using the equation:

$$Percentage\ difference = \frac{dose\ with\ bolus - dose\ without\ bolus}{dose\ without\ bolus} \times 100\% \quad (2)$$

2.3. Measurement of Percentage Depth Dose by Using Ionisation Chamber and Treatment Planning Software

The PDD measurement by using the ionisation chamber was performed based on the department's quality assurance (QA) method [14]. The PDD was measured by using the solid water phantoms and Markus plane parallel ionisation chamber at 100 cm SSD and 10 x 10 cm^2 field size on the surface. The charges at depths were measured by adding the solid water phantom on the surface while maintaining the SSD. The percentage ionisation depth (PDI) was measured by using the equation:

$$PDD = \frac{D}{D_{max}} \times 100\% \quad (3)$$

with D and D_{max} is the dose at depth and maximum dose respectively.

The measurement of PDD by using the treatment planning software (TPS) was carried out by scanning the solid water phantoms by using the computed tomography (CT) scanner. The CT images of the phantoms were acquiesced by using the TPS software [15]. The 6 MV photons and 6 MeV electrons were simulated similar to that in the measurement by using ionisation chamber. The dose ast depth were calculated based on the CT information provided. The PDD curves in ionisation chamber, and TPS were plotted and compared to that in the OSL nanoDots.

3. Results and Discussion

The linearity of OSL signal measured at the increased dose for 6 MV photons and 6 MeV electrons is illustrated in Figure 2. The results showed that the OSL signals were in good linearity with the increased dose (MU) for both 6 MV photons and 6 MeV electrons shown by the linear regression (R^2) values of close to 1. The results were in good agreement with the previous work on the dose linearity of OSL dosimeters in high energy photons and electrons [9]. This indicated that the OSL dosimeters are suitable to be used for the dosimetric studies for photons and electrons doses between 50 and 1000 cGy.

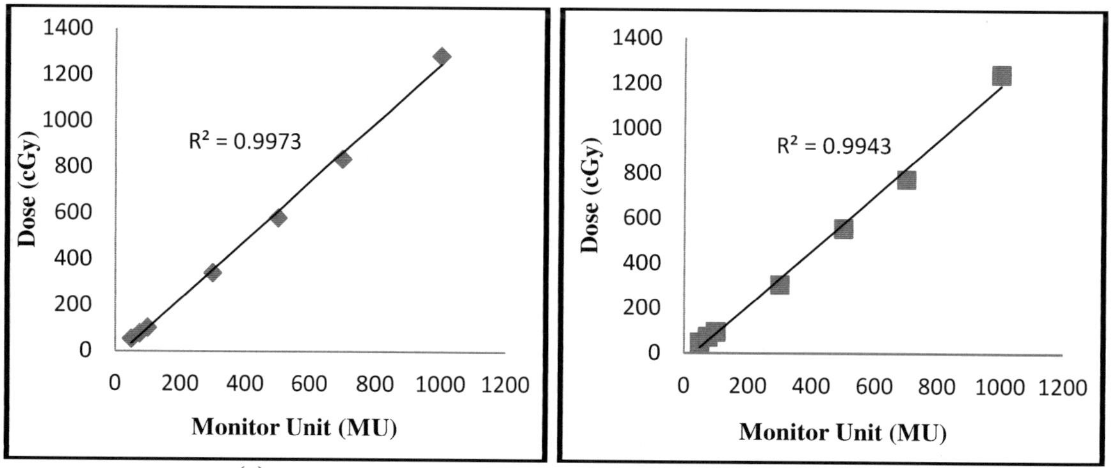

(a) (b)

Figure 2. The OSL signals in cGy at the increased dose (MU) for (a) photons and, (b) electrons.

The measured PDD and the PDD curve of 6 MV photons and 6 MeV electrons by using the OSL dosimeters in comparison to other dosimeters are illustrated in Figure 3. The results showed that the PDD measured by using the OSL dosimeters were in good agreement to that in the other dosimeters of ionisation chamber, EBT2 film dosimeter [16] and the TPS simulation. The comparison of PDD in OSL dosimeter to ionisation chamber as the standard dosimeters by many available international protocols showed an agreement within maximum of 13.4 and 10.4% percentage of discrepancies to other dosimeters in 6 MV photons and 6 MeV electrons respectively [5,17]. The depth of maximum dose D_{max} measured in OSL dosimeters also showed good agreement to the ionisation chamber, EBT2 film dosimeter and TPS at approximate depth of 1.5 cm for both photons and electrons. The surface dose measured in OSL dosimeter however was lower than all other dosimeters with percentage of discrepancies between 25 and 29% for 6 MV photons and 9.5 and 13.1% for 6 MeV electrons [5,17].

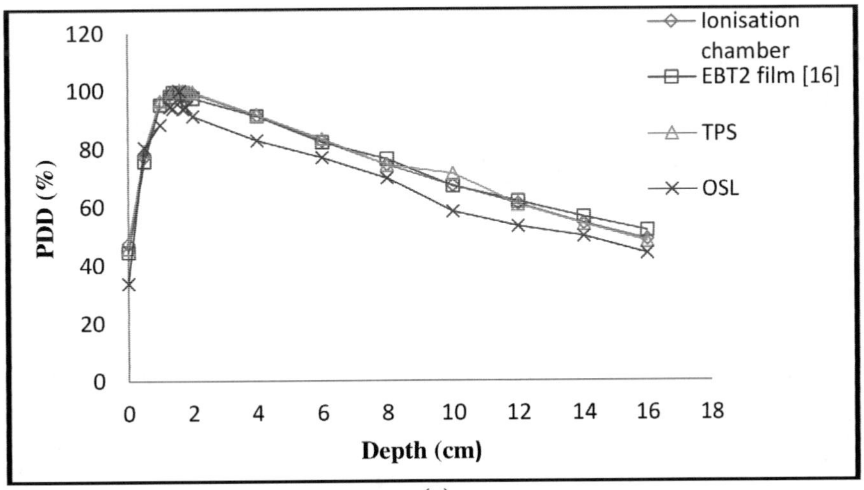

(a)

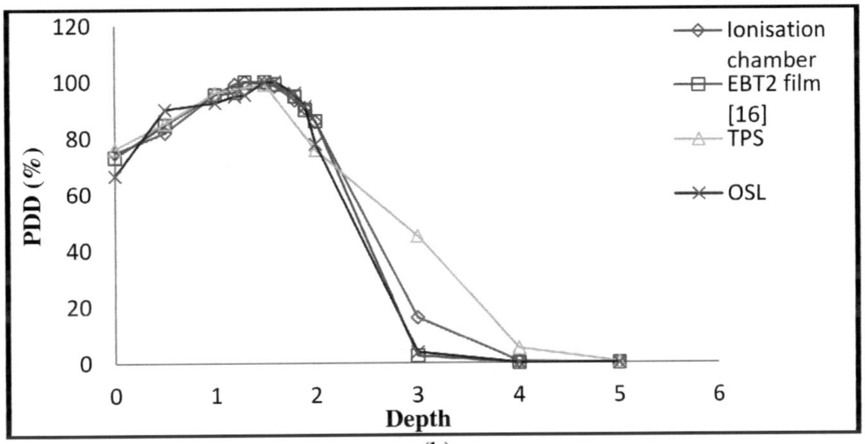

(b)

Figure 3. The PDD curves in (a) 6 MV photons and (b) 6 MeV electrons, measured by using OSL dosimeters in comparison to the ionization chamber, EBT2 film and treatment planning system (TPS).

The measurement of dose at the depth of maximum dose, D_{max} by using the OSL dosimeters in comparison to the output calibration of 6 MV photons and 6 MeV electrons by using the ionisation chamber is presented in Table 1. The results showed that the dose measured by using the OSL dosimeters were in excellent agreement with the ionization chamber with 1.02 and 0.17% percentage of discrepancies for 6 MV photons and 6 MeV electrons respectively. The results indicated the suitability of the OSL dosimeters to be used in the dosimetric studies in high energy photons and electrons.

B

Table 1. The measured dose at depth of maximum dose, d_{max} by using OSL dosimeters in comparison to the output calibration by using ionisation chamber at 6 MV photons and 6 MeV electrons.

Energy	**Dose (cGy)**		**Percentage of discrepancy (%)**
	OSL	**Ionisation chamber**	
6MV	101.63	100.6	1.02
6MeV	94.05	93.89	0.17

4. Conclusion

The nanoDot OSL dosimeters showed linear dependency to the increased dose between 50 and 1000 cGy in both 6 MV photons and 6 MeV electrons. The measurement of PDD showed good agreement of PDD curve of the OSL dosimeters to other common dosimetry tools of ionisation chamber, EBT2 film dosimeter and the treatment planning system (TPS) in both photons and electrons. The measured dose at d_{max} by using the OSL dosimeters showed good agreement to the output calibration of 6 MV photons and 6 MeV electrons by using ionisation chamber. The overall results indicated the suitability of OSL dosimeter for dosimetry works involving clinical photons and electrons.

5. References

[1] Hu B, Wang Y, and Zealy W 2009 Performance of Al2O3: Optically stimulatedluminescence dosimeters for clinical radiation therapy applications. *Australas Phys. Sci. Eng. Med.* **32(4)** p 226-232.

[2] A. S. Pradhan, Lee J I and Kim J L. 2008 Recent developments of optically stimulated luminescence materials and techniques for radiation dosimetry and clinical applications. *J Med. Phys.* **33**(3) p 85-99.

[3] Tien C J, Ebeling R, Hiatt J R, Curran B and Sternick E 2012 Optically stimulated luminescent dosimetry for high dose rate brachytherapy, *Front. Oncol.* **91**(2) p 1-7.

[4] Bäck S J, Mattsson, Kjaer-Kristoffersen F and Medin J 2004 Real-time optical-fibre luminescence dosimetry for radiotherapy: physical characteristics and applications in photon beams, *Phys. Med. Biol.* **49**(9) p 1655-1699.

[5] Jursinic P A 2007 Characterization of optically stimulated luminescent dosimeters, OSLDs, for clinical dosimetric measurements, *Med. Phys.* **34**(12) p 4594-4604.

[6] Dunn L, Lye J, Kenny J, Lehmann J, Williams I and Kron T 2013 Commissioning of optically stimulated luminescence dosimeters for use in radiotherapy, *Radiat.Meas.* **52** p 31-39.

[7] Knežević Ž, Stolarczyk L, Bessieres I, Bordy J M, Miljanić S and Olko P 2013 Photon dosimetry methods outside the target volume in radiation therapy: Optically stimulated luminescence (OSL), thermoluminescence (TL) and radiophotoluminescence (RPL) dosimetry, *Radiat Meas.* **57** p 9–18.

[8] Yusof M F M, Yahya M H, Rosnan M S, Abdullah R and Kadir A B A. 2017 Dose measurement using AL_2O_3 dosimeter in comparison to LiF:Mg,Ti dosimeter and ionization chamber at low angh energy x-ray *AIP Conf. Proc.* **1799** p 040071-0400077.

[9] Yusof M F M, Joohari N A, Abdullah R, Shukor N S A, Kadir A B A and Isa N M. 2018. Dosimetry of Al_2O_3 optically stimulated luminescence dosimeter at high energy photons and electrons. *IOP Conf. Ser.: Mater. Sci. Eng.* **298** p 012023.

[10] Viamonte L, da Rosa A R, Buckley L A, Cherpark A, and Cygler J E 2008 Radiotherapy dosimetry using a commercial OSL system, *Med. Phys.* **35(4) p** 1261-1266.

[11] Kurokawa C and Urushiyama A. 2013 Measurements for dose distributions with a photo-stimulable luminescence dosimeter sheet. *J. Phys.: Conf. Ser.* 444 p 012035.

[12] Wouter C, Dirk V, Paul L and Tom D 2017 A reusable OSL-film for 2D radiotherapy dosimetry. Phys Med Biol. **62** p 8141-8154.

[13] Khan M F 2010 The physics of radiation therapy (Philadelphia: Lippincott William & Wilkins)
[14] Podgorsak E B 2005 Radiation Oncology Physics: A handbook for teachers and students (Vienna: International Atomic Energy Agency)
[15] Yusof M F M, Abdullah R, Tajuddin A A, Hashim R and Bauk S 2016 Attenuation properties and percentage depth dose of tannin-based *Rhizophora* spp. particleboard phantoms using computed tomography (CT) and treatment planning system (TPS) at high energy x-ray beams. *AIP Conf. Proc.* **1704** p 040003.
[16] Yusof M F M, Hamid P N K A, Tajuddin A A, Abdullah R, Hashim R, Bauk S, Isa N M, Isa M J M 2017 Characterization and attenuation study on Tannin-added *Rhizophora* spp. particleboard at high energy photon and electron *AIP Conf. Proc.* **1799** p 040002.
[17] Akbas U, Kesen N D, Koksal C and Bilge H 2018 A comparison of different dosimetry systems for surface and builup region dose measurements *Phys. Medica* **52**(1) p143

Acknowledgement
The author would like to acknowledge the financial support of USM Bridging Grant scheme no. 304.PPSK.6316324. The author would also like to acknowledge the Department of Nuclear Medicine, Radiotherapy and Oncology, Hospital University Sains Malaysia (HUSM) for the use of the linear accelerator facility in this study.

IOP Conf. Series: Materials Science and Engineering **785** (2020) 012003 doi:10.1088/1757-899X/785/1/012003

Electron range after titanium foil window from low energy electron accelerator

Mohd Rizal Md Chulan[1,a)], Mohd Faiz Mohd Zin[1], Leo Kwee Wah[1], Mukhlis Mokhtar[1], Mohd Azhar Ahmad[1], Abd Halim Baijan[1], Rokiah Mohd Sabri[1], Khaidawaton Abd Malik[1]

[1]Accelerator Development Center, Technical Support Division, Malaysian Nuclear Agency, Bangi 43000 Kajang, Selangor, Malaysia.

[a]rizal@nuclearmalaysia.gov.my

Abstract. Currently, the low energy electron accelerator machine in Malaysian Nuclear Agency is at the commissioning stage. Various parameters are tested and studied. One of the parameters is the electron range in the air through titanium foil window. This parameter is important for the process of sample irradiation. There are two methods to determine this parameter. The first is through manual calculations and second is through Monte Carlo code simulations using PHITS. This paper describes the results of both methods and attempts to make comparisons between them.

1. Introduction

Malaysian Nuclear Agency has successfully developed a low-energy electron accelerator with the maximum energy of 250 keV and beam current output of 30 mA. For this device to work properly, it needs to go into the commissioning process to ensure the reliability of the resulting radiation dose. There are several parameters to be tested and studied and one of them is the range of the electrons in the air after passing through a titanium foil.

The electron range in the matter is the maximum penetration distance of an electron in the matter (air) measured in a straight line [5]. The unit of electron range is g/cm^2. The range depends on its initial energy and on the material through which it passes [3]. By knowing the density of a substance, the electron range within a material can be calculated.

In this paper, two methods have been used and shown. The first is a well-known method to estimate electron range as shown by Katz and Penhold (1952) for electron energy less than 2.5 MeV. It is a method where the electron range is manually calculated. The second method is a simulation of the electron transport using PHITS.

PHITS, which stands for **P**article and **H**eavy **I**on **T**ransport Code **S**ystem, has been developed by JAEA. It is a general purpose Monte Carlo particle transport simulation code and is used in many studies and one of them is on accelerator technology studies [4].

This study is important because by calculating the electron range in the air after the titanium foil window, the positioning of the sample for irradiation purposes can be determined. The sample for irradiation purpose will be positioned in between the range of electron in the air.

IOP Conf. Series: Materials Science and Engineering **785** (2020) 012003 doi:10.1088/1757-899X/785/1/012003

2. Methodology

2.1. Study case

For both calculation and simulation, the consideration is when an electron is accelerated in an accelerating tube (vacuum) through a 25 μm thick titanium window before it goes out into the air (diagram 1). The distance traveled electrons in a straight line through the air will be calculated by manual calculation and by simulation via PHITS 3.10 for the initial electron energies of 100 keV, 150 keV, 200 keV and 250 keV.

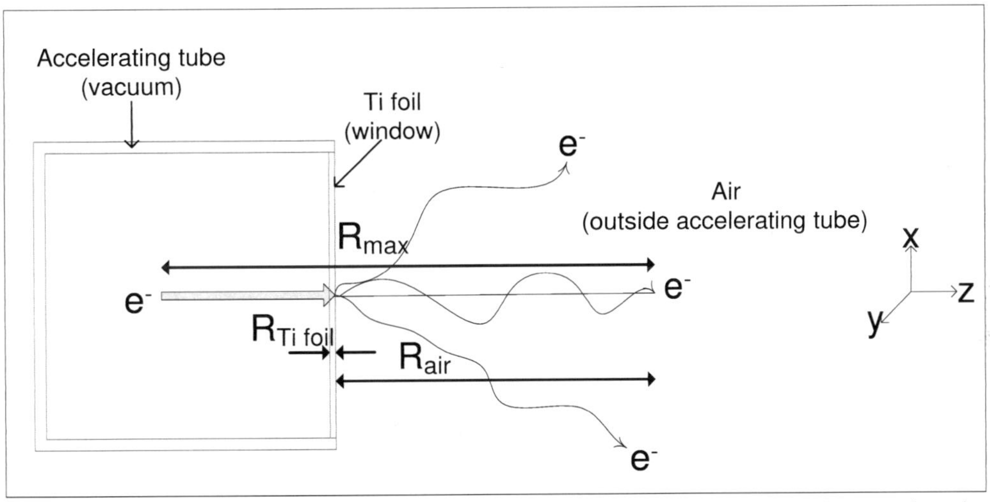

Diagram 1. The event for calculation and simulation of electron range in air after passing through a titanium foil.

2.2. Calculation of Electron Range Through Air After Passing a Titanium Foil

The maximum range, R_{max}, of an electron at energy less than 2.5 MeV can be computed from an empirical formula given by Katz and Penfold [2]:

$$R_{max}\ [g/cm^2] = 0.412\ E_\beta^{1.265-0.0954\ ln(E_\beta)} \tag{1}$$

where E_β is the maximum electron energy in MeV.

The range can also be expressed as density thickness (g/cm²) of the material given by [1], as follows:

$$R_{material}(g/cm^2) = \rho_{material}\ (g/cm^3) \times t_{material}\ (cm) \tag{2}$$

2.3. PHITS Simulation

PHITS ver. 3.10 was used for this study. In this simulation, the source in the form of a pencil-shaped electron beam was used. A total of 100,000 particles per batch (5 batches) was simulated at 10 cm from window inside vacuum cylinder while the outside cylinder is consisting of air. Simulation error is set to 5% and any values above this limit was ignored. Initial electron energies of 100 keV, 150 keV, 200 keV and 250 keV were used in this simulation.

IOP Conf. Series: Materials Science and Engineering **785** (2020) 012003 doi:10.1088/1757-899X/785/1/012003

3. Results and Discussion

From the calculations, we found that the values of the electron range increases along with the electron energy. The same goes for PHITS simulations (Figure 1). This shows that the PHITS simulation results adheres to the equations provided by Katz and Penfold.

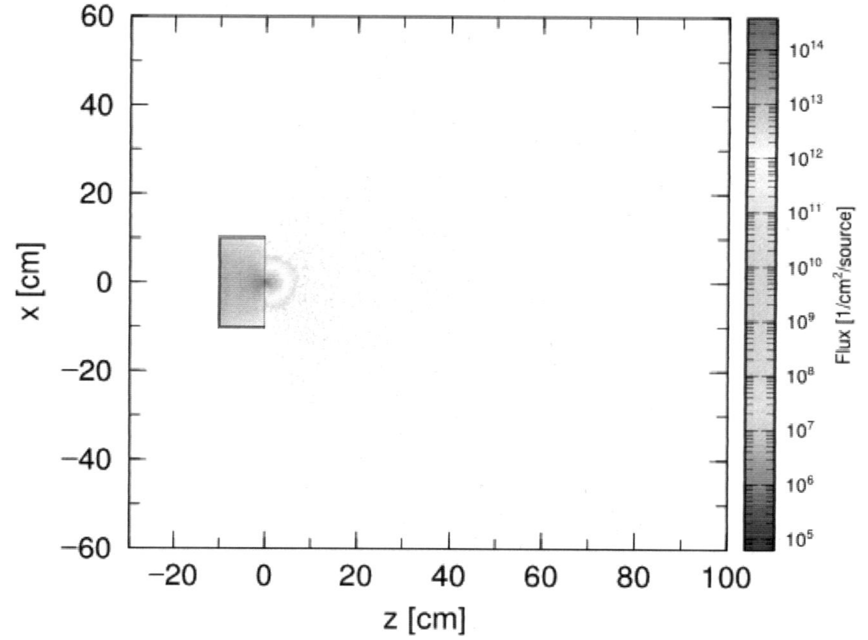

(a) Electron *track* with initial energy of 100 keV

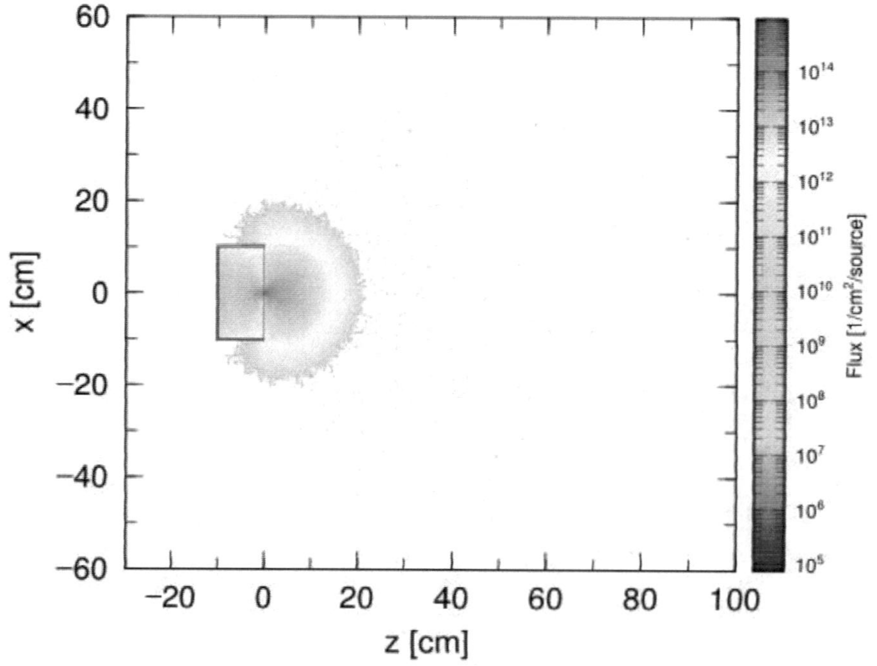

(b) Electron *track* with initial energy of 150 keV

IOP Conf. Series: Materials Science and Engineering **785** (2020) 012003 doi:10.1088/1757-899X/785/1/012003

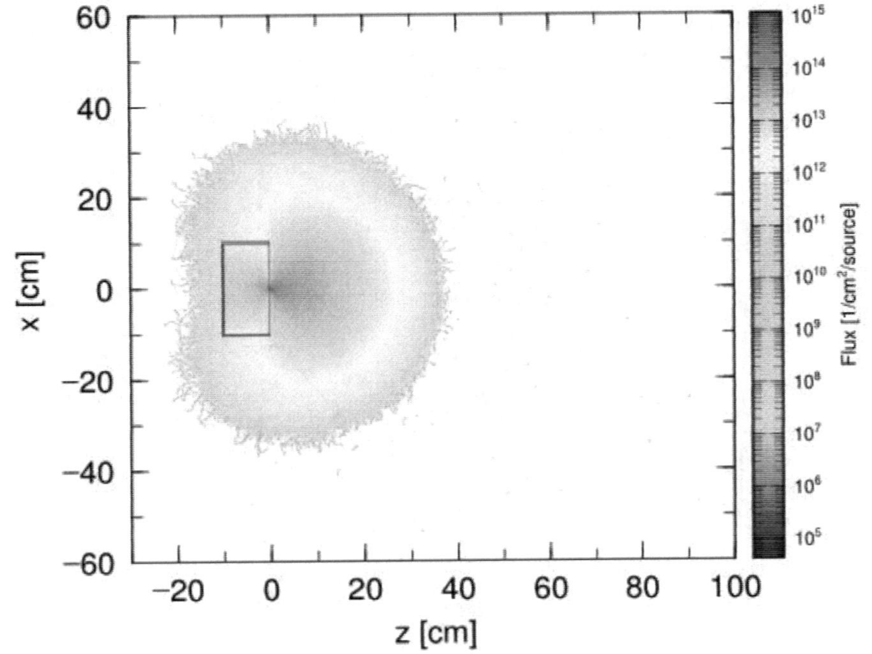

(c) Electron *track* with initial energy of 200 keV

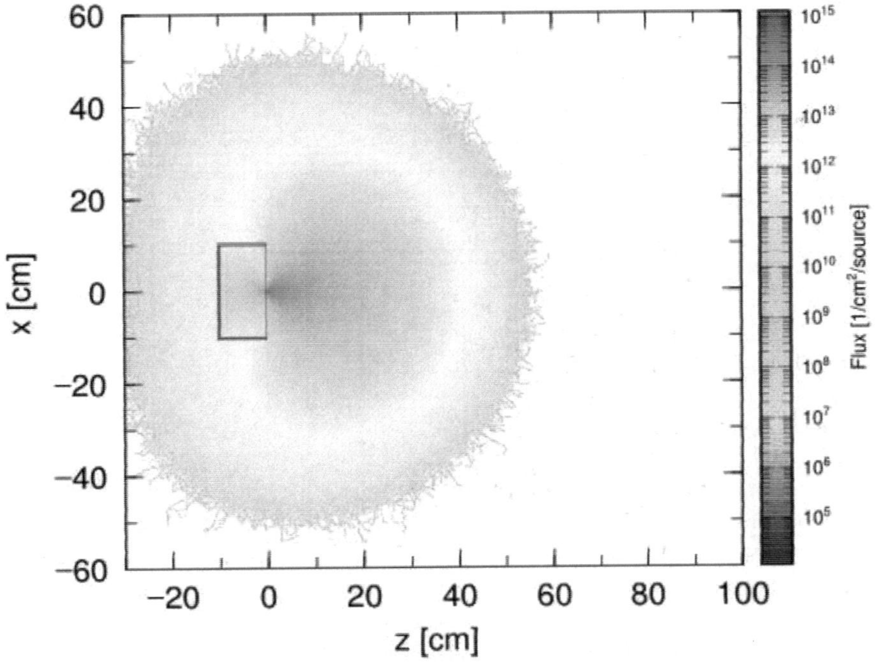

(d) Electron *track* with initial energy of 250 keV

Figure 1: PHITS simulation results for the electron *track* of (a) 100 keV, (b) 150 keV, (c) 200 keV and (d) 250 keV.

Electron *track* means the value of electron fluence in any specific region. From the PHITS simulation results, it is found that the electron *track* length through the air increases with the increase of initial electron energy. The value of the electron flux ((1/cm^2/source) decreases at larger electron distance in the air for each initial energy considered. This indicates that if the initial electron energy is high then the penetration distance is far. Its flux value decreases as more electrons are absorbed by air.

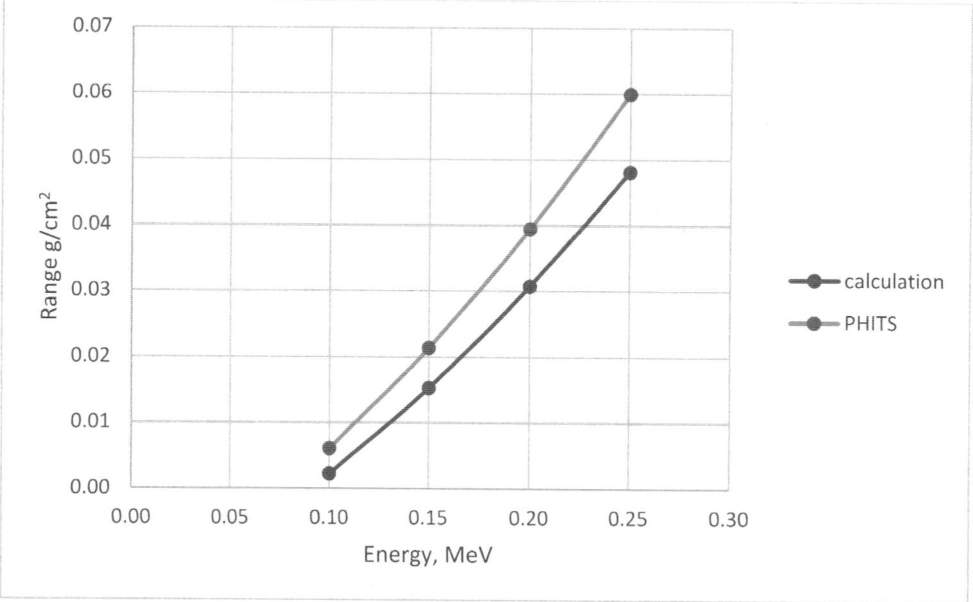

Figure 2: Plot of electron ranges in air vs energy values from manual calculation and PHITS simulation

From Figure 2, it is found that there is a difference in the values of electron range between manual calculation and PHITS simulation. This difference may be due to the parameters used in the calculation and simulation by PHITS. From diagram 1, the electron range can be expressed as:

$$R_{max} = R_{Ti\,foil} + R_{air} \tag{3}$$

$$R_{Ti\,foil} = \rho_{Ti\,foil} \times t_{Ti\,foil} \tag{4}$$

$$R_{air} = \rho_{air} \times t_{air} \tag{5}$$

where,

$R_{Ti\,foil}$ is range of electron in Ti foil,
$r_{Ti\,foil}$ is density of Titanium foil,
$t_{Ti\,foil}$ is thickness of titanium foil,
R_{air} is range of electron in air,
r_{air} is density of air and,
t_{air} is thickness of air.

Hence, the actual thickness (distance), t_{air} electron in air can be computed:

$$t_{air} = \frac{R_{max} - R_{Ti\,foil}}{\rho_{air}} \tag{6}$$

From equations (3) to (6), the linear electron distance (t_{air}) in the air after the titanium window can be estimated for the initial electron energy values of 100 keV, 150 keV, 200 keV and 250 keV. Similarly, in the PHITS simulation, the electron distances in the air can be obtained by taking a flux value of less than 0.05 (tolerance error). From these values (Fig. 3), it is possible to determine the optimal position of the sample for the purpose of irradiation for specific electron energy. The sample will be placed in between the electron range. Exceeding the electron range value means the absence of electron.

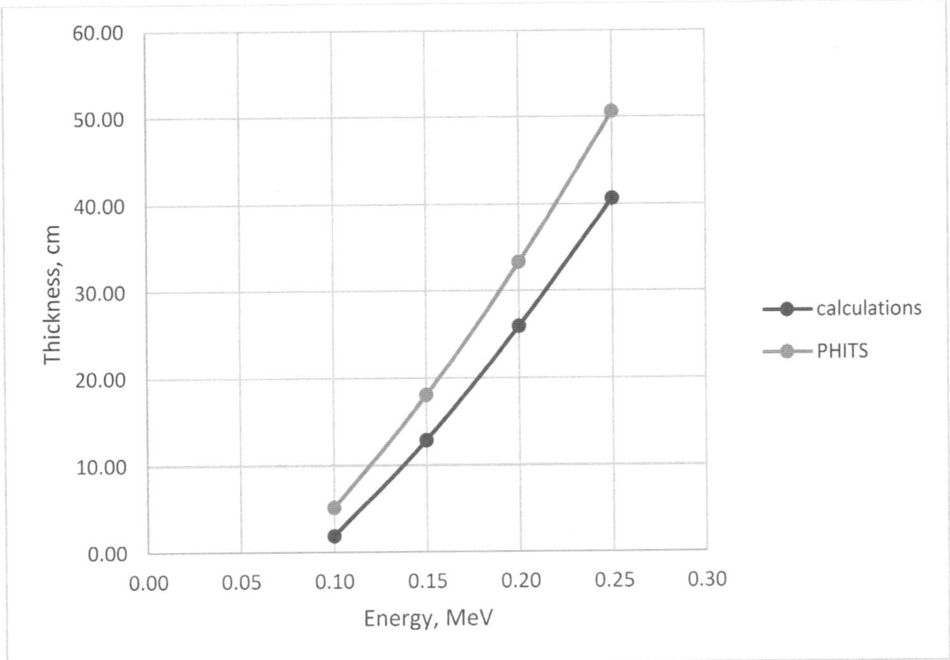

Figure 3: Plot of the electron distance in the air vs energy after titanium window from manual calculation and simulation of PHITS.

4. Conclusions

In comparison between the manual calculations and PHITS simulations, the patterns as shown in Figure 3 are similar despite the differences in values. This indicates that the PHITS simulation results is acceptable.

5. References

[1] Holbert K E 2012 Charged Particle Ionization and Range (EEE460-Handout).
[2] Katz L and Penfold A S 1952 *Rev. Mod. Phys.* **24** p 28.
[3] National Council on Radiation Protection and Measurement (NCRP) 1977 radiation protection design guidelines for 0.1 – 100 MeV particle accelerator facilities *NCRP rep. no. 51* p.28
[4] Sato T, Niita K, Matsuda N, Hashimoto S, Iwamoto Y, Noda S, Ogawa T, Iwase H, Nakashima H,Fukahori T, Okumura K, Kai T, Chiba S, Furuta T and Sihver L 2013 Particle and Heavy Ion Transport code System, PHITS version 2.52 J. of Nucl. Sci. and Tech. **50** p 913-923.
[5] Everhart T E and Hoff P H 1971 Determination of kilovolt electron energy dissipation vs penetration distance in solid materials *J. Appl. Phys.* **42** p 5837.

IOP Conf. Series: Materials Science and Engineering **785** (2020) 012003 doi:10.1088/1757-899X/785/1/012003

Acknowledgements
We extend gratitude to Dr. T. Furuta (Japan Atomic Energy Agency, JAEA) and Dr. Hiroshi Iwase (High Energy Accelerator Research Organization, KEK), Japan, for their guidance and willingness to conduct a workshop namely NM-KEK PHITS Course 2019 at Nuklear Malaysia in which both contributed in-depth on this subject.

Determination of the combined effect of boron and gadolinia on the reactivity and safety parameters of (U, Th)O_2 fuel

E H Uguru[1,2,a)], S F A Sani[1,b)], M U Khandaker[1,3], M H Rabir[4] and J A Karim[4]

[1] Department of Physics, University of Malaya, 50603 Kuala Lumpur, Malaysia
[2] Department of Industrial Physics, Ebonyi State University, P.M.B. 053, Abakaliki, Nigeria
[3]Center for Biomedical Physics, School of Healthcare and Medical Sciences, Sunway University, 47500 Bandar Sunway, Selangor Darul Ehsan, Malaysia
[4]Nuclear and Reactor Physics Section, Reactor Technology Center, Technical Support Division, Malaysian Nuclear Agency, Bangi, 43000 Kajang, Selangor, Malaysia

[a]ugurueddy2008@yahoo.com
[b]s.fairus@um.edu.my

Abstract. This study investigated the combined effect of boron in light water moderator and gadolinia in fuel on the neutronic properties of (U, Th)O_2 fuel in Westinghouse small modular reactor. The reactivity control of boron and gadolinia, and the potential of gadolinia in increasing thermal conductivity of its mixed fuel pellet motivated this study. Monte Carlo N-particle code integrated with CINDER90 was used for the calculations. Four fuel assemblies were used: one without boron and gadolinia, one with borated water only while the other two contained borated water plus 16 and 24 gadolinia fuel rods in each of the FAs. The BOC safety parameters was calculated at hot zero power (293.6 K) and hot full power (600 K). The results showed that the moderator temperature coefficient of reactivity was highest in FA without boron and gadolinia and smallest in FA containing boron plus 24 gadolinia fuel rods, while a reverse phenomenon was observed for the fuel temperature coefficient of reactivity. The effective delayed neutron fraction was slightly larger in FA without neutron absorber. The results show that boron and gadolinia decreased significantly the MTC but increased FTC slightly in addition to their initial neutron multiplication suppression role.

1. Introduction

During normal reactor operation, starting from the beginning of cycle (BOC) and to the end of cycle (EOC), the state of nuclear reactor as a multiplying medium requires a balanced neutron multiplication [1]. Although, the desire of every reactor operator is to maintain criticality (where the neutron multiplication factor is unity), some level of excess reactivity (supercritical) is needed to maintain normal reactor power over a long period of fuel burn-up [2]. This balanced reactivity is achieved using reactivity control materials. Indeed, mechanical control rod was the only reactivity control measure used in the early stage of nuclear reactor especially in light water designs (PWR and BWR) while innovative soluble boron chemical shim and burnable neutron absorber materials were introduced in the later stage of nuclear technology [3]. Presently, reactivity control mechanism, especially in large light water reactors, involves the use of integral fuel burnable absorber (IFBA) in fuel and soluble boron in coolant water for the BOC reactivity control and normal operational flux flattening, respectively.

Today, reliable data and experience have been gained on the use of UO_2 fuel for different nuclear reactor operations, but such cannot be said for thorium fuel that is undergoing series of research and development processes. The Westinghouse small modular reactor (W-SMR) core design with UO_2 reference fuel contains IFBA and pyrex for reactivity control. However, this study group is proposing a W-SMR thorium core utilising IFBA only for BOC reactivity control owing to the large thermal neutron absorption cross section of ^{232}Th. It is therefore, pertinent to determine the combined reactivity effect of boron and gadolinia burnable absorber and their potential impact on reactor safety parameter in $(U, Th)O_2$ as a potential UO_2 fuel alternative in the nearest future. This is very important for SMR and $(U, Th)O_2$ owing to the role of safety parameters as a reactor design and fuel option deciding factor.

This study was to determine the impact of soluble boron and gadolinia on moderator temperature coefficient of reactivity (MTC), fuel temperature coefficient of reactivity (FTC), delayed neutron fraction and average neutron generation time in $(U, Th)O_2$ fuel. The W-SMR was used because of its design and passive safety system similarity with AP1000 core and the existing technical know-how on the operation of PWRs. The benchmarked MCNPX integrated with CINDER90 burn-up code was used for the calculations [4].

2. Material and method

Four robust fuel assemblies with boron and gadolinia used in this study are shown in Figure 1. The detailed description of the dimensional values is described in ref. [5]. Table 1 contains $(U, Th)O_2$ fuel and other material specifications used. The four sets of Fuel Assemblies (FAs) filled with the same $(U, Th)O_2$ fuel were independently studied: one without boron and gadolinia, one with 3600 ppm concentration of soluble boron in the moderator while the other two contained the same boron concentration in moderator plus 16 and 24 gadolinia fuel rods in each of them. MCNPX 2.7 version with ENDF/B-VII.0 evaluated nuclear data library was used to model the FA geometry as shown in figure 1. As a general-purpose, time-dependent and generalised-geometry simulation code, MCNP uses a continuous energy cross-section library to solve particle (e.g. electron, photon, neutron and coupled system) transport problems.

The beginning of cycle (BOC) safety parameters were calculated at 293.6 K hot zero power and 600 K hot full power. For the calculation of MTC and FTC, the temperature cross sections 70c, 71c and TMP card were adequately used on the fuel, moderator and other related materials at the material and cell cards. For the MTC, the light water temperature was lowered from the 600 K to 293.6 K while the fuel temperature remained at hot full power. This was achieved in MCNP by changing the moderator cross section libraries from 71c to 70c and its density from operating density to room density of 99.9 g/cc. For FTC, the moderator remained at its room temperature and density while the fuel temperature was lowered to 293.6 K without change in density since temperature change has a negligible effect on the fuel density. The prompt neutron multiplication factor (k_{prompt}) was calculated by setting TOTNU card to "NO". In each case, MCNPX performed KCODE calculation to determine the corresponding infinite neutron multiplication factors k_{hzp}, k_{hfp} and k_{prompt}. The binding energy effect of hydrogen in light water was accounted for by using thermal treatment ($S(\alpha, \beta)$) capability of MCNPX. The KCODE calculation was performed in each of the four FAs to determine the infinite neutron multiplication factor (k_{inf}) at normal reactor operating conditions. Use was made of 10,000 initial neutrons source, 50 ineffective and 300 effective cycles and 1.0 initial guess of effective criticality with WIMS 69 energy group at 9 MWth fuel assembly thermal power. The MTC, FTC and delayed neutron fraction were calculated by using (1), (2), (3) and (4) accordingly. The average neutron generation time was calculated by using (5) which is equal to the prompt neutron lifetime at criticality. The calculation of prompt neutron removal lifetime was done by MNCPX.

IOP Conf. Series: Materials Science and Engineering **785** (2020) 012004 doi:10.1088/1757-899X/785/1/012004

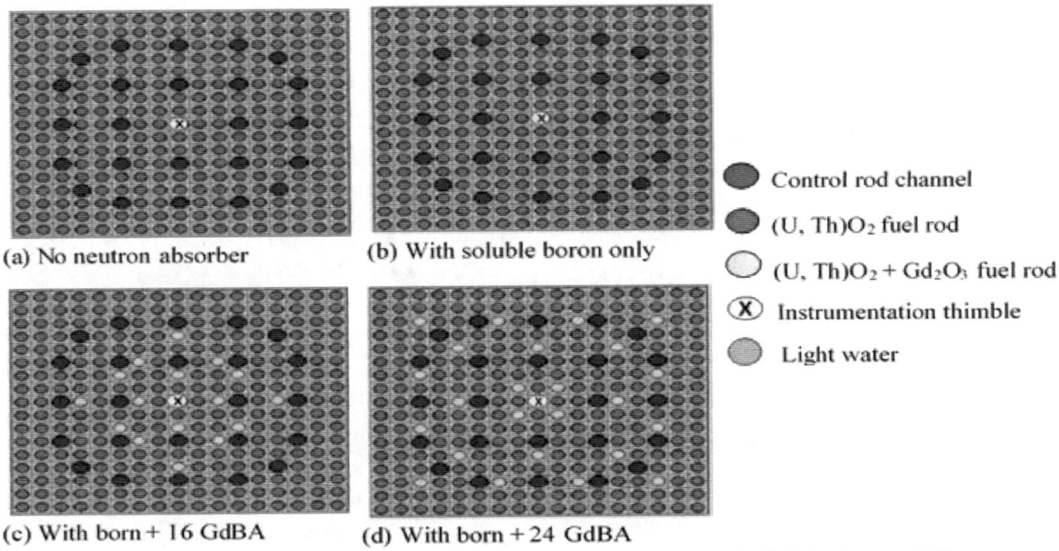

Figure 1. Horizontal cross-sectional view of the modelled fuel assemblies.

Table 1. Composition of fuel and other materials.

Material/Fuel	Heavy metal weight percent (w/o)	Density (g/cm^3)
Fuel 1: $(U, Th)O_2$	^{235}U: 13.8, ^{238}U: 11.2, Th:75	11.5
Fuel 2: $(U, Th)O_2 + Gd_2O_3$	^{235}U: 13.8, ^{238}U: 11.2, Th:66.4 Gd: 8.6	11.012
Cladding: Zircaloy-4	Sn: 1.3 Fe: 0.23 Cr: 0.27 Zr: 98.20	6.56
Reflector	Light water (H_2O)	0.756
Gap	Helium gas (He)	0.00164
Gadolinium isotopes	^{152}Gd: 0.20, ^{154}Gd: 2.18, ^{155}Gd: 14.80, ^{156}Gd: 20.47, ^{157}Gd:15.65, ^{158}Gd:24.84, ^{160}Gd:21.86	7.9

$$MTC = \frac{\delta\rho}{\delta T} = \delta\rho\left(\frac{1}{T_{hfp} - T_{hzp}}\right) \tag{1}$$

$$\delta\rho = \frac{k_{inf}^{hfp} - \kappa_{inf}^{hzp}}{\kappa_{inf}^{hfp} \times \kappa_{inf}^{hzp}} \tag{2}$$

$$FTC = \frac{\delta\rho}{\delta T} = \delta\rho\left(\frac{1}{T_{hfp} - T_{hzp}}\right) \tag{3}$$

$$\beta_{eff} = \frac{k_{inf} - k_{prompt}}{k_{inf}} \tag{4}$$

$$\Lambda = \frac{\ell_{prompt}}{k_{inf}} \tag{5}$$

where,

$\delta\rho$ = Change in reactivity due to change in either moderator or fuel temperature,
δT = Change in fuel or moderator temperature (K),
k_{inf} = Infinite neutron multiplication factor before any change,
κ_{inf}^{hzp} = Infinite neutron multiplication factor before change in moderator temperature and density or in fuel temperature,

κ_{inf}^{hfp} = Infinite neutron multiplication factor after change in moderator temperature and density or in fuel temperature,
β_{eff} = Effective delayed neutron fraction,
k_{prompt} = Multiplication factor for prompt neutron only,
Λ = Average neutron generation time and,
ℓ_{prompt} = Prompt neutron lifetime or prompt neutron removal time.

3. Results and discussion

3.1. Beginning of cycle parameters

The beginning of cycle events is paramount to reactor operational safety and efficiency because of how quick events happen within short time scale. In view of this, the initial k_{inf} and effective delayed neutron fraction were analysed as shown in Figure 2. From Figure 2 (a), the initial k_{inf} value in FA without any absorber was much higher than others owing to higher fission reactions while that of FA with boron + 24 gadolinium burnable absorber (GdBA) rods was the least because of the larger number of absorber rods. By the design specification of AP1000 for UO_2 fuel with initial k_{eff} of 1.205 [6] for a fresh core without external control mechanism, the values are within acceptable range for a fuel assembly study. This shows that increasing the GdBA rods above 24 with 3600 ppm boron concentration in such FA arrangement will not be efficient. As shown in figure. 2 (b), all the FAs has effective delayed neutron fraction value that are respectively greater than the 0.0065 value for fissile ^{235}U but less than 0.0075 designed value for UO_2 fuelled AP1000 reactor. This difference was expected due to smaller delayed neutron fraction value for fissile ^{233}U compared to ^{239}Pu counterpart from ^{238}U. Because of the fewer delayed neutron fraction and regardless of the higher thermal neutron absorption cross section of ^{232}Th, thorium fuelled reactor requires tighter control measures at the BOC.

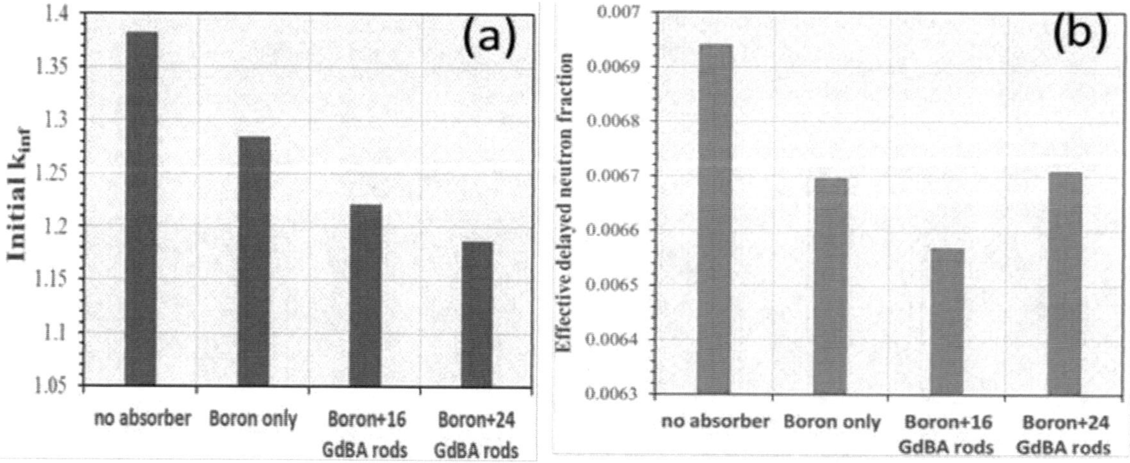

Figure 2. The variation of **(a)** initial k_{inf} **(b)** effective delayed neutron fraction.

The moderator and fuel temperature coefficients of reactivity are amongst safety parameters of interest which were calculated using (1) and (3) respectively. For optimised and inherent reactor safety, MTC and FTC of the reactor core are designed to be negative throughout the cycle operation. The MTC and FTC provide a measure of change in core reactivity caused by temperature change in moderator and fuel respectively. A negative MTC represents reduction in reactivity in an event of moderator temperature increase and the reverse is the case for a positive MTC. In this way, a negative MTC reactor core is self-limiting in any sudden moderator temperature increase, and therefore, maintains stability in reactor power.

The variation of MTC and FTC was presented in Figure 3. It shows that MTC was significantly negative in FA without boron and GdBA rod but significantly less negative in FAs with either boron only or boron plus GdBA absorbers. Since the FAs contained the same fuel and fissile enrichment, difference in MTC was attributed to the presence of boron and GdBA which hardened the thermal neutron spectrum and decrement of fuel load by increasing GdBA fuel rods. On the other hand, the FTC was slightly less negative in FA without boron and GdBA rod while its negativity increased slightly as the number of the GdBA rod increases, therefore, showed a reverse occurrence to MTC. For the fact that FTC is a measure of the number of neutrons absorbed by fertile nuclide in the fuel at fuel temperature change, it therefore, means that boron and GdBA increased thermal neutron absorption in the system. Apart from the increased thermal neutron absorption by neutron absorbers, gadolinia raised the thermal motion of the fuel resulting to larger doppler broadening and therefore, decreases thermal neutron resonance escape probability. This led to the increased negative FTC in FAs with neutron absorber which increased as the number of GdBA rods increases. Generally, the MTC in all the FAs was respectively negative but below the -10 to -60 pcm/K MTC for UO_2 fuelled AP1000 reactor core [7]. This was attributed to the type of fuel studied since MTC depends on fast fission factor, fissile and fertile nuclides neutron absorption cross section at various energies, weight percent of fissile material and the reactor initial fuel load. The FTC in all the FAs was significantly above the -2 pcm/K its designed value for AP1000 reactor [8] due to large thermal neutron absorption cross section of ^{232}Th compared to ^{238}U and the enhanced neutron absorption by the reactivity control materials.

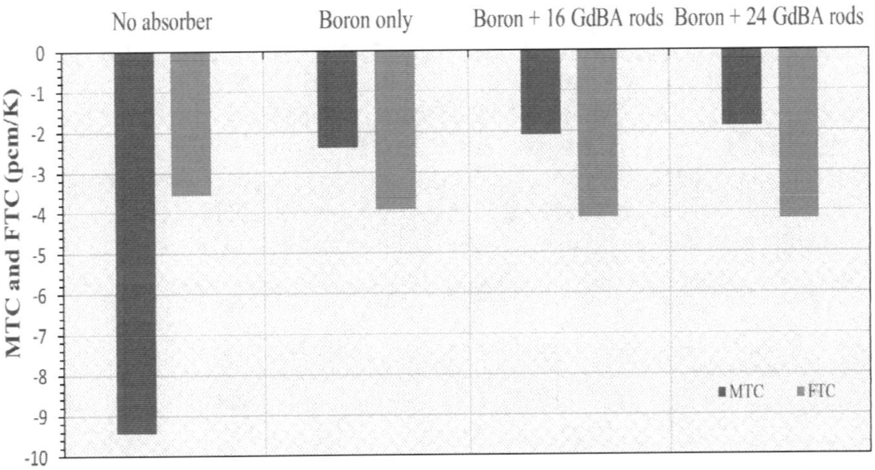

Figure 3. Moderator and Fuel temperature coefficients of reactivity.

3.2. In-cycle burn-up calculations

The performance of the fuel overtime was studied using Burn card of MCNPX and the integrated CINDER90 burn-up code. The variation of neutron multiplication factor with burn-up in all the FAs decreased sharply at BOC due to rapid and increased production of reactor poisons which removed more neutrons from the system. However, it quickly reached equilibrium and maintained a relatively stable and descending slope except for the FA with boron + 24 GdBA rods which experienced reactivity swing between 10 to 30 GWd/tHM. The variation of k_{inf} in all the FAs at BOC was discussed in subsection 3.1. The k_{inf} of the FAs without absorber and that with boron only had a smooth and linear variation as shown in Figure 4 (a). The k_{inf} of the FA with boron + 16 GdBA rods whose value was in agreement with the 1.205 designed k_{eff} of AP1000 for UO_2 varied linearly without noticeable positive reactivity swing which is desired for stable reactor power.

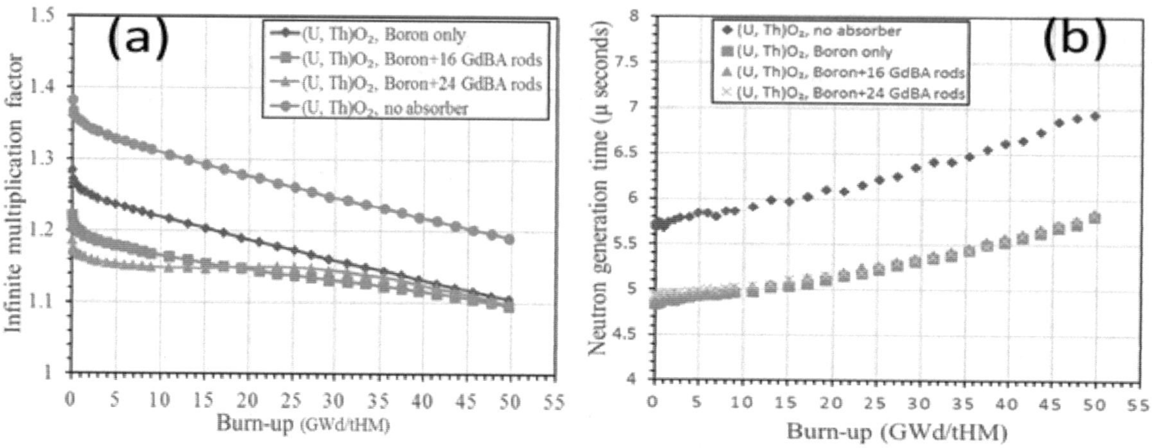

Figure 4. Variation of **(a)** neutron multiplication factor and **(b)** neutron generation time with burn-up.

The positive reactivity swing in FA with boron + 24 GdBA rods was noticeable but relatively small. The average neutron generation time (Λ) shown in Figure 4 (b) was larger in FA without any neutron absorber and slightly smaller in FAs with boron only between 0 - 20 GWd/tHM but matched and varies smoothly with the values in FAs with boron + GdBA rods when the gadolinia effect was almost zero. This phenomenal occurrence was due to the larger thermal neutron absorption cross section of boron and gadolinium compared to the fuel elements. This results to the reduction of prompt neutron lifetime due to faster removal of thermal neutron in the FAs with boron and GdBA rod leading to the observed shorter average neutron generation time. Therefore, the average neutron generation time increases with burn-up, and inversely proportional to the thermal neutron absorption cross sections of the nuclear fuel materials.

4. Conclusion

This work investigated the combined effect of soluble boron and gadolinia on reactivity and safety parameters with emphasis on their effect on the MTC, FTC, delayed neutron fractions and mean neutron generation time. The MTC in all the FAs was respectively negative showing a larger negativity in FA without boron and gadolinia and smaller negativity in FA with boron + 24 GdBA rods. On the other hand, the FTC was less negative in FA without boron and GdBA rods but most negative in FA with boron + 24 GdBA rods. Although, the MTC in all the FAs was respectively negative, it was below the -10 to -60 pcm/K value for UO_2 fuelled AP1000 reactor core but within acceptable value for a fuel assembly study. The FTC in all the FAs was significantly above the -2 pcm/K designed value for AP1000 reactor. The effective delayed neutron fraction in all the FAs were respectively greater than the 0.0065 value for fissile ^{235}U but less than 0.0075 designed value for UO_2 fuelled AP1000 reactor. The average neutron generation time increased with burn-up but inversely proportional to the thermal neutron absorption cross sections of the nuclear fuel materials. This shows that boron and gadolinia shortened the prompt neutron lifetime and therefore, decreased neutron generation time except at criticality region where prompt neutron lifetime equals the generation time.

5. References

[1] Larmier C, Zoia A, Malvagi F, Dumonteil E and Mazzolo A 2018 Neutron multiplication in random media: Reactivity and kinetics parameters *Ann. Nucl. Energy* **111** p 391-406.
[2] Wojciechowski A 2016 Criticality of the thorium burnup in equilibrium state *Prog. Nucl. Energy* **92** 81-90
[3] Yahya M S and Kim Y 2016 Burnable absorber-integrated guide thimble (BigT)–II: application to 3D PWR core design *J. Nucl. Sci. Technol.* **53** (10) p 1521-1527

[4] Kim S M and Kim M H 2014 A study on MCNPX-CINDER90 system for activation analysis *Transac. of the Korean Nuc. Soc.* (Pyeongchang, Korea) p 30-31.
[5] Uguru E H, Sani S F A, Khandaker M U and Rabir M H 2020 Investigation on the effect of ^{238}U replacement with ^{232}Th in small modular reactor (SMR) fuel matrix *Prog. Nucl. Energy* **118** p 103108.
[6] Compton W K 2015 Impact of configuration variations on small modular reactor core performance *Master thesis*.
[7] Housiadas C and Antonopoulos-Domis M 2001 Estimation of the moderator temperature coefficient of reactivity via noise analysis using closed-loop transfer functions *Ann. Nucl. Energy* **28** (10) p 983-991
[8] Thilagam L, Sunny C S, Subbaiah K, Devan K, Lee Y S and Jagannathan V 2007 Doppler coefficient of reactivity - Benchmark calculations for different enrichments of UO_2, *US Joint Int. Topical Meeting on Math. and Comp. and Supercomp. in Nucl. App. April 15-19 Monterey, California USA* (American Nuclear Society).

Acknowledgements
This work was supported by the University of Malaya, Malaysia through the Research University Grant – Faculty Programme with Grant No. GPF036B-2018.

IOP Conf. Series: Materials Science and Engineering **785** (2020) 012005 doi:10.1088/1757-899X/785/1/012005

Neutron attenuation and mechanical properties of polymer composites filled with boron carbide particles

M Z Nurazila[1,a], Y Hafiza[1] and M A Megat Harun[1]

[1]Malaysian Nuclear Agency, 43000 Kajang, Bangi Selangor, Malaysia

[a]nurazila@nuclearmalaysia.gov.my

Abstract. In this work, neutron attenuation and mechanical properties of thermoplastic natural rubber (TPNR) blend filled with boron carbide (B_4C) have been studied as a function of filler loading. Thermoplastic natural rubber of high density polyethylene/natural rubber (HDPE/NR) blend with different amounts of boron carbide (5–20 wt%) have been prepared via melt blending method. All samples were subjected to neutron transmission and tensile tests. The results showed that neutron shielding performance of the composites were improved significantly with the addition of B_4C into TPNR matrix. On the other hand, mechanical properties of the composites were found to decrease with increasing filler loading.

1. Introduction

Recently, polymer-based composites have gained much attention by the radiation protection industry due to their excellent properties such as lightweight, good processability, good flexibility and low manufacturing cost [1]. For neutron radiation, polymers served as an ideal neutron shielding material because of their high hydrogen content. Neutron shielding requires low atomic number elements such as hydrogen with high scattering cross sections that can effectively moderate or thermalize incident neutrons.

Polyethylene is one of the effective neutron shielding materials and displays excellent attenuation behavior because of its hydrogen content (14% by weight) and its density [2, 3]. It is inexpensive and resistant to many chemicals and has hydrophobic nature which reduces the possibility of a superficial contamination to a minimum. In polyethylene, the hydrogen capturing of thermal neutrons is through the $^1H(n, \gamma)^2H$ reaction which has a cross section of 0.33 barn for neutrons in thermal equilibrium at room temperature ($E_n = 0.027$ eV).

Elements with high absorption cross sections for thermal neutrons like boron, cadmium and gadolinium are often dispersed in the shielding material to absorb moderated/thermalized neutrons. Adding boron in polyethylene can reduce the dose from secondary gamma production from radiative capture (n,γ). Boron is an excellent choice to absorb neutrons due to its high neutrons cross section of boron-10 isotope which is 3840 barns [4]. The capturing of thermal neutrons by boron is through the $^{10}B(n,\alpha)^7Li$ reaction. Another advantage of the boron-10 isotope is that the neutron capture reaction is creating alpha radiation that is easily stopped by the material.

In this study, thermoplastic natural rubber (TPNR) composites consist of high density polyethylene/ natural rubber (HDPE/NR) blend were fabricated for neutron shielding materials. TPNR which has intermediate properties between rubber and plastic provide flexibility in shaping and making it attractive to be used as a matrix in a composite material. The unique properties of elastic and rigid of TPNR could provide wide area coverage that require radiation shielding. Boron carbide (B_4C) with high thermal neutron cross section was used as filler in order to provide shielding effect against thermal neutrons [5-7]. Then the attenuation and mechanical properties of the TPNR/B_4C composites were studied as a function of filler loading.

2. Materials and method

2.1. Sample preparation
Thermoplastic natural rubber (TPNR) blend was prepared via melt blending of natural rubber and high density polyethylene (NR/HDPE) with the ratio of 60/40. Different amounts of boron carbide (B_4C) powders were added to get composites with different concentrations (5, 10, 15 and 20 wt%). Liquid natural rubber (LNR) was used as a compatibilizer. Mixing was done using an internal mixer (Haake Rheomix 600P) with a processing mixing temperature at 135 °C, a rotor speed of 55 rpm and mixing time of 15 minutes. Then, the composite compound was moulded in a compression moulding machine at 135 °C to form a slab (15 cm x 15 cm).

2.2. Neutron transmission test
Neutron transmission tests were conducted in TRIGA reactor Malaysian Nuclear Agency. All samples with 1 mm thickness were exposed to thermal neutron for equal interval of time (2 minutes) using the thermal neutron beam. Composite samples were installed at the beam port located 719 cm from the source. The transmitted neutron intensity (I) for each composite was determined by recording the corresponding counts using a fission chamber detector (Model 3053, LND Inc.). The incident neutron intensity (I_o) was determined when no composite material is in place. Total neutron transmission (I/I_o) and macroscopic cross section (Σ) of the composites were calculated using Beer-Lambert Law (equation 1) [8] where x is the thickness of the samples. Half value layer (HVL) of the samples were also calculated using equation 2.

$$I = I_o e^{-\Sigma x} \tag{1}$$

$$HVL = \frac{ln\,2}{\Sigma} \tag{2}$$

2.3. Tensile test
Five dumbbell-shaped specimens for each sample type were cut according to ASTM D638-91a with 1 mm thickness for tensile test. Tensile strength and elongation at break were measured using Instron 8874 tensile test machine with cross-head speed of 50 mm/min at room temperature. The test was performed according to ASTM D638 with maximum load of 5000 N. The average of the values was taken.

3. Results and discussion

3.1. Neutron attenuation
The neutron shielding characteristics of TPNR/B_4C composites at various filler loadings (5-20 wt%) were determined. Figure 1 shows neutron transmission of TPNR/B_4C composites as a function of filler loading and the values are listed in Table 1. It is clear from the figure that neutron transmission value decreases as filler loading increases. TPNR matrix have neutron transmission value of 0.674. The incorporation of B_4C filler into TPNR matrix decreases the neutron transmission and the value reaches to 0.302 at 20 wt% of B_4C.

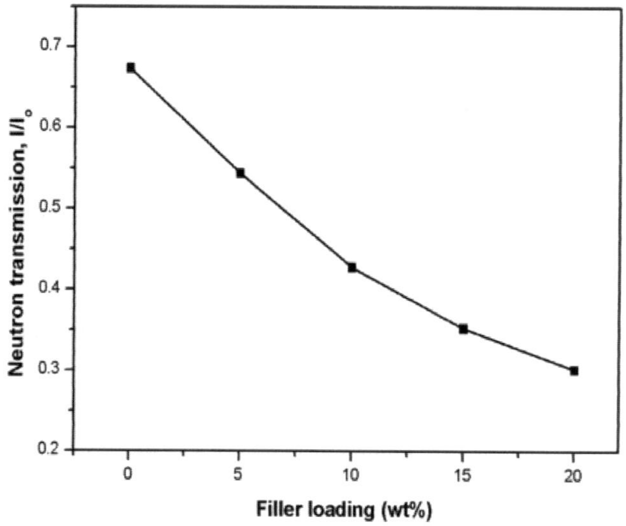

Figure 1. Neutron transmission versus filler loading for TPNR/B_4C composites.

The results for macroscopic cross section and the half value layer (HVL) are shown in Table 1. It can be seen that the highest value comes from a composite containing 20 wt% of B_4C. The attenuation of the 20 wt% B_4C composite is about 217% higher than the TPNR. A significant increase in macroscopic cross section as the B_4C filler loading increases means that the probability of interaction increases as B_4C content became higher. Therefore, more neutron can be absorbed by shielding material. Increasing of macroscopic cross section value of TPNR composites with B_4C loading is due to the high thermal neutron absorption cross section of boron-10 isotope in B_4C. Natural boron consists primarily of two stable isotopes, ^{11}B (80%) and ^{10}B (20%) and its (n, α) reaction cross section for thermal neutrons is about 3840 barns.

Increasing the percentage of B_4C filler also causes the half value layer (HVL) to decrease. Half value layer indicates the required thickness of an absorber to reduce the radiation level to half of its initial value [9]. In this work, HVL decreases from 0.207 cm to 0.065 cm with B_4C loading increase from 0 wt% to 20 wt% indicating an improvement of the attenuation properties of the composites. The concept of HVL is very useful in doing rapid, approximate shielding calculations [10].

Table 1. Neutron attenuation properties of TPNR composites at various B_4C loadings.

Filler loading (wt%)	Neutron transmission I/I_o	Macroscopic cross section (cm^{-1})	Half value layer (cm)
0	0.674	3.343	0.207
5	0.543	5.085	0.136
10	0.428	7.015	0.099
15	0.352	8.692	0.080
20	0.302	10.604	0.065

3.2. Mechanical properties

Mechanical properties of the radiation shielding materials are important to minimize degradation of the materials during usage. The tensile strength and elongation at break of TPNR/B_4C composites at different B_4C loading are shown in Figure 2 and Figure 3 respectively and the values are summarized in Table 2. From Figure 2, it can be seen that the tensile strength decreases by loading B_4C into TPNR matrix. The tensile strength decreases from 3.9 MPa to 2.74 MPa when TPNR is loaded with 5 wt% of B_4C. Only 1.82 MPa of tensile strength is observed at 20 wt% loaded samples.

Small amount of filler gives better strength to the composites might be because of good rubber-filler interaction occur as a result of good dispersion of the filler particles into the matrix. Wang and Chen [11] reported in their study that lower amounts of filler can easily dispersed in the composites thus produced a significant improvement in the filler-matrix interfacial bonding. Better dispersion of the filler in the matrix may increase the efficiency of the stress transfer from matrix to filler phase. At higher filler loading, filler particles are no longer wetted properly by the TPNR matrix resulting in poor interfacial adhesion. In addition, the agglomeration and the aggregation of the filler particles in the TPNR matrix may have formed at higher filler loading and it caused interruption to the filler-matrix bonding [12, 13]. Thus, it will produce stress concentration points leading to early failure.

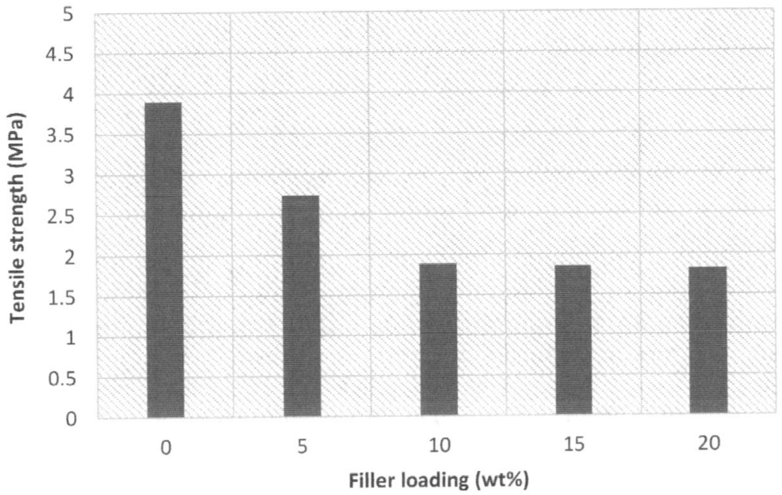

Figure 2. Tensile strength of the composites as a function of filler loading.

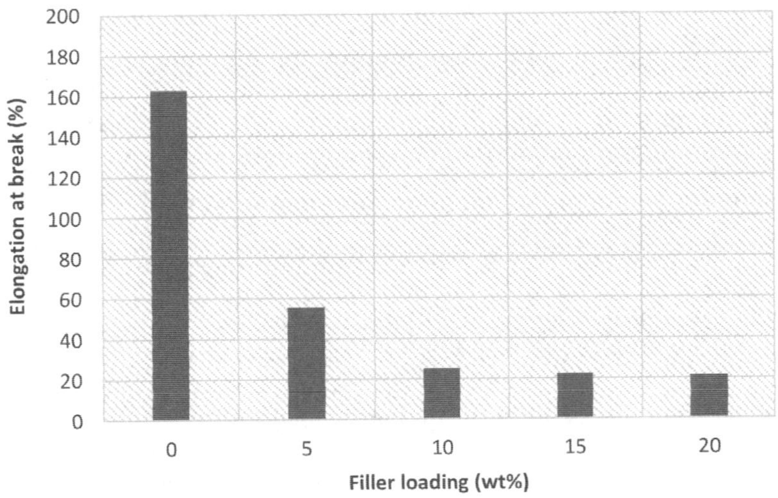

Figure 3. Elongation at break of the composites as a function of filler loading.

Elongation at break of the composites shows similar trend like tensile strength. As the filler loading increases, the elongation at break tend to decrease. The elongation at break value decreases from 162.96% to 55.55% by loading 5 wt% of B_4C. The lowest value of 21.29% can be observed at maximum filler loading of 20 wt%. The decreasing of elongation at break at higher filler loading is due to the enhancement in rigidity of the composites [14]. Azahari *et al.* [15] stated that the reduction in elongation at break may possibly affected by the attractive forces between the fillers and the polymer molecules. The forces restrict the free mobility of the polymer chains to form a cross-linked network. Thus, it will increase the resistance to stretch upon the application of strain.

Table 2. Tensile properties of TPNR/B_4C composites at different B_4C loadings.

Filler loading (wt%)	Tensile strength (MPa)	Elongation at break (%)
0	3.9	162.96
5	2.74	55.55
10	1.88	25.01
15	1.85	22.23
20	1.82	21.29

4. Conclusion

TPNR/B_4C composites with various B_4C loadings were successfully fabricated and studied. Neutron transmission and tensile tests were conducted to investigate the neutron attenuation and mechanical properties of TPNR/B_4C composites. It has been found that neutron transmission and half value layer of composites decreased while macroscopic cross section increased indicating an improvement of the attenuation properties of the composites with the increase of B_4C loading into TPNR matrix. On the other hand, tensile strength and elongation at break of the composites were decreased with increasing filler loading.

5. References

[1] El-Fiki S, El-Kameesy S U, Nashar D E E, Abou-Leila M A, El-Mansy M K, Ahmed M 2015 Influence of bismuth contents on mechanical and gamma ray attenuation properties of silicone rubber composite *Int. J. Adv. Res.* **3** p 1035–39.

[2] Mensah W O, Fletcher J J and Danso K A 2012 Assessment of radiation shielding properties of polyester steel composite using MCNP5 *Int. J. Sci. Technol,* **2** 445-461.

[3] Harrison C, Burgett E, Hertel N and Grulke E 2008 Polyethylene/boron composites for radiation shielding applications *AIP Conf. Proc.* **969** p 484–491.

[4] Tariq Y and Muhammad N K 2008 High density polyethylene/boron carbide composites for neutron shielding *e-polymers* **059** p 1-7.

[5] Chilton A B, Shultis J K and Faw R E 1984 *Principles of Radiation Shielding* (Englewood Cliffs, NJ: Prentice-Hall).

[6] Shultis J K and Faw R E 1996 *Radiation Shielding* (Englewood Cliffs, NJ: Prentice-Hall).

[7] Price B T, Horten C C and Spinnery K T 1957 *Radiation Shielding* (Elmsford, NY: Pergamon).

[8] Kipcak A S, Baysoy D Y, Moroydor D E and Piskin S 2011 Characterization of the kurnacovite mineral and its absorption behavior due to neutron radiation *Research Bulletin of the Australian Institute of High Energetic Materials* **1** p 38-46.

[9] Biswas R, Sahadath H and Mollah A S 2016 *J. Rad Res. and Appl. Sci.* **9** p 26-34.

[10] Basyigit C, Uysal V, Kilincarslan S, Mavi B, Gunoglu K, Akkurt I, Akkas A 2012 *AIP Conf. Proc.* **1400** p 232.

[11] Wang J and Chen D 2013 Mechanical properties of natural rubber nanocomposites filled with

thermally treated attapulgite *J. of Nanomaterials* **2013** p 1-11.
[12] Sirichai K, Piyaporn N, Ekachai W and Narongrit S 2012 Use of bagasse fiber ash as secondary filler in silica or carbon black filled natural rubber compound. *Mat. Des.* **41** p 74-82.
[13] Jaewoo K, Byung C L, Young R U and William H M 2014 Enhancement of thermal neutron attenuation of nano-B_4C, -BN dispersed neutron shielding polymer nanocomposites *J. Nucl. Mater.* **453** p 48-53.
[14] Ismail H, Salleh S Z and Ahmad Z 2013 Properties of halloysite nanotubes-filled natural rubber prepared using different mixing methods *Mat. Des.* **50** p 790-797.
[15] Azahari N A, Othman N and Ismail H 2012 Effect of attapulgite clay on biodegradability and tensile properties of polyvinyl alcohol/corn starch blend *Film. Inter. J. Polym. Mat.* **61** p 1065-1078.

Acknowledgments
The authors would like to express their gratitude to Malaysian Nuclear Agency for supporting this research and all parties that are contributed to this work.

IOP Conf. Series: Materials Science and Engineering **785** (2020) 012006 doi:10.1088/1757-899X/785/1/012006

Statistical Estimation of ideal and realistic muon interaction on Al, Fe, and Cu absorbers

M I H M Norizan[1], I H Hashim[1,2,3,a)], F Ibrahim[1], N N A M A Ghani[1]

[1]Department of Physics, Faculty of Science, Universiti Teknologi Malaysia, 81300, Johor Bahru, Johor.
[2]Research Center for Nuclear Physics, Osaka University, 567-0047 Ibaraki-shi, Osaka, Japan.
[3]National Centre Particle Physics, Universiti Malaya, 50603 Kuala Lumpur, Malaysia.

[a]izyan@utm.my

Abstract. Interaction of charged muons in various absorbing materials is almost similar to other charged particles. Their behaviour has been observed by various experimental work in material science and nuclear and particle physics. A statistical model is developed in order to study the behaviour of muons after interacting with absorbing material. In this study, ideal and realistic beam structure from high intensity muon facility around the world such as at Japan Accelerator Research Complex, Research Centre Nuclear Physics, Osaka University and Paul Scherrer Institute is assumed. Three absorbing materials (aluminium, iron and copper) with thickness range of 0.00 ~ 0.12 g/cm^2 are used to observe the momentum straggling of the beams. Performance for our model was checked by comparing the value of stopping power with others work for muons with initial kinetic energy of 10-40 MeV with and without including shell effect correction term. The effect of the energy losses towards the beam structure will be discussed.

1. Introduction

Interaction of particles with matter depends on their charged and energy of the incident particle. These interactions will determine the nuclear reaction occur in the nucleus and the electron at surrounding orbits. Muon interaction is very much similar to electron interaction where positive muons tends to behave as electron and interact at 1s and 2s electronic quantum levels in muonium. While for the negative muon which has 207 times of electron mass interact at the atomic level of a muonic atom. Both μ^+ and μ^- are subject to electromagnetic and weak interaction due to Coulomb interaction, their magnetic moments and lepton flavour conservation. Low momentum muons usually stop at short range for medium-heavy nuclei absorber and their stopping probability is very high. For medium to high momentum muons, thicker absorber is required for initiating reactions.

As the muon passes through matter, it will lose its energy by atomic collisions or deflections from the incident direction due to many interactions. Energy straggling and range straggling are introduced routing in large divergences of the particles in matter [2]. The mean energy losses, the straggling associated with the energy losses is crucial for understanding the energy losses by muon [3]. The incoming particle suffer an inelastic collision with the atomic electrons of the absorbing material and elastic scattering from the nuclei. In various calculations by Bethe-Bloch formula, the muon beam structure only concerns on 1 keV energy bin. However, experimental estimation requires certain amount of beam spread to evaluate realistic muon behaviour after interaction with the materials.

World-renowned muon facilities provide various distribution for the initial muon beam suitable for materials science, nuclear and particle physics experiment [1]. Two of them are located in Japan at the Japan Accelerator Research Complex (J-PARC) and the Research Center for Nuclear Physics, Osaka University, while another one located in Switzerland is the Paul Scherrer Institute (PSI). These facilities provide high intensity continuous or pulsed muon beams as shown in Table 1. All three facilities of our concerned shows different momentum width that affect the muon behaviour after passing through materials.

Table 1. Comparison of muons facilities around the world.

	PSI	J-PARC	MuSIC
Beam power	**1200 kW**	**1000kW**	**0.4 kW**
Beamline channel	**μE1 (Mode B)**	**D2-beamline, MLF**	**MuSIC-M1**
Muon Intensity [μ/sec]	**2×10^4**	**1×10^4**	**0.5×10^4**
Muon Structure	**Continuous**	**Pulse**	**Continuous**
Momentum width (FWHM)	**~1%**	**~15%**	**~5%**

The present work aims to study the muon behaviour on different absorbing materials. A muon energy loss model is developed based on Bethe-Bloch formula, idealistic and realistic muon beam width are tested on various absorbing materials and thickness range. The performance of the model is verified from previous work in reference [4].

2. Muon energy losses model

A useful model to predict the muon energy losses is by implementing the ideal and real cases for the incident beams. By using a C++ program, the statistical simulations generate the muon momentum distribution and thus shows impact of beam straggling after material absorber. Incident muon beams will be assumed to irradiate the absorbing material. Muon will suffer energy losses and energy straggling is introduced in the absorbing material. The energy losses of muons are known by calculating the stopping power given by the Bethe-Bloch formula in Eq. 1. In this study, muon beams are assumed to irradiate three different absorbing materials; Al, Fe and Cu. Due to the energy losses, S. Kumar et al. [3] states that the range of muon momentum becomes wider as the thickness of the absorbing material increase. In order to observe the momentum straggling, this model include calculations for the absorbing materials with the thickness of 0.04 g/cm^2, 0.08 g/cm^2 and 0.12 g/cm^2.

The mean energy loss per unit path length is given by quantity called stopping power or simply -dE/dx given by Bethe-Bloch formula [5]

$$-\frac{dE}{dx} = 2\pi N_a r_e^2 m_e c^2 \rho \frac{Z}{A}\frac{z^2}{\beta^2}\left[ln\left(\frac{2m_e\gamma^2 v^2 W_{max}}{I^2}\right) - 2\beta^2 - \delta - 2\frac{C}{Z}\right] \quad (1)$$

with $2\pi N_a r_e m_e c^2 = 0.1535\ cm^2/g$, r_e is defined as an electron radius which is 2.817×10^{-13} cm and the m_e is the rest mass of an electron. The constant Avogadro's number, N_a is $6.022 \times 10^{23}\ mol^{-1}$ and z is a charge of incident particle in unit of e. The Lorentz factor, γ is defined as $1/(\sqrt{1-\beta^2})$, where $\beta$is the ratio of velocity of the incident particle, v to the speed of light, c. The maximum energy transfer in a single collision is known as W_{max}. Here the parameter involving the absorbing material is the density of the material, ρ, the atomic number of the absorbing material, Z the atomic weight of the absorbing material, A, the mean excitation potential, I, and the density correction, δ.

Shell corrections, C accounts for the effect which arises when the velocity of the incident particle is comparable or smaller than the orbital velocity of the bound electron. At such energies, the assumption that the electron is stationary with respect to the incident particle is no longer valid and the Bethe-Bloch formula breaks down. The correction is generally small in number and give there an empirical formula for this correction given by [6]:

$$\begin{aligned} C(I,\eta) = {} & [(0.422377\eta^{-2} + 0.0304043\eta^{-4} - 0.00038106\eta^{-6}) \times 10^{-6}\, I^2] \\ & +[(3.850190\eta^{-2} - 0.1667989\eta^{-4} + 0.00157955\eta^{-6}) \times 10^{-9}\, I^3] \end{aligned} \tag{2}$$

where $\eta = \beta\gamma$ and I is the mean excitation potential in eV. Hence, in order to understand and illustrate the interactions of muons, development of the energy losses model is required to give a picture of the experiments. The energy losses of the incident beam include the effect of the material surrounding the experimental area which include elements that make up the set-up such as aluminium, iron and copper.

Comparison with calculation from Groom *et al.* [4] have been made and summarized in Table 2. In general, the agreement is regarded as adequate, but is worse for higher energy muons. In the second run, the simple shell corrections as given by Eq. 2 has been used, and under this condition, the value of stopping power somewhat overcorrects.

Table 2. Comparison of stopping power for muons (in MeV $g^{-1}\ cm^2$) including and excluding the shell correction, with those from Groom *et al.* [4].

Energy (MeV)	10 MeV	20 MeV	30 MeV	40 MeV
Aluminium (Z=13)				
This calculation				
Without shell correction	**6.40**	**3.98**	**3.13**	**2.70**
With shell corrections	**6.26**	**3.83**	**2.98**	**2.55**
Groom et al. [4]	**6.19**	**3.80**	**2.96**	**2.53**
Iron(Z=26)				
This calculation				
Without shell correction	**5.71**	**3.58**	**2.83**	**2.45**
With shell corrections	**5.56**	**3.44**	**2.68**	**2.30**
Groom et al. [4]	**5.49**	**3.40**	**2.65**	**2.27**
Copper(Z=29)				
This calculation				
Without shell correction	**5.49**	**3.45**	**2.73**	**2.36**
With shell corrections	**5.35**	**3.31**	**2.59**	**2.22**
Groom et al. [4]	**5.28**	**3.28**	**2.56**	**2.20**

Based on the comparison of different calculation with and without shell correction term for our statistical energy losses model, we conclude that the calculations for this model must include shell correction terms in order to best re-enact the real-life conditions. Compared to Groom et al. [4] calculations, the muon stopping power at 10 MeV are higher by 1.1%, 1.2% and 1.3% through Al, Fe and Cu respectively. Without the shell correction, the stopping powers for this case are higher by 3.4%,

4.0% and 3.9% respectively. Through lower Z material, the agreement is more precise with Groom et al. [4]. In all cases, the agreement improves with increasing energy.

In this model, the calculations are done on the low-energy muons at common momentum as used in the three muon facilities; J-PARC, PSI and MuSIC (30 MeV/c). Two cases of muon beams will be considered which is realistic and ideal muon beams. The distribution function of the real beam case is in the form of a Gaussian function. Meanwhile, ideal beam means no tail effect in the incident beams. These are standard that can be achieved at their highest efficiency and performance in these advanced muon facilities. Different from the practical muon beams, the true value has 100% probability of lying within the limits p-σ and p+σ.

The intensity *I* of the muon beams correspond to the integral of the spectra. In a real experiment, the range can be determined by passing a beam of particles at the desired energy through different thicknesses of the material under test and measuring the ratio of transmitted to incident particles [6]. In this model, the transmission ratio is obtained by taking the number of counts of the resulting muon beams normalized to the number of counts of the incident muon beams. From the number-distance curve, the mean range is the range where roughly half of the particle is absorbed. For the purpose of finding a well-defined number for mean range of the muons, the incident muon beams are assumed to irradiate the three-absorbing materials with the thickness starting from 0.02 g/cm^2 with increments of 0.02 g/cm^2 until the transmission ratio becomes zero.

3. Results and discussion

3.1 Muon energy losses in Al, Fe, and Cu of varying thickness

The energy losses spectra through varying thickness of absorbing materials in the range of 0.00 – 0.12 g/cm^2 were plotted. After the irradiation, for real and ideal beam cases the mean incident momentum is 30 MeV/c from J-PARC muons facilities through Al, Fe and Cu are presented in Figure 1 and 2 as a function of the projectile momentum respectively. From Figure 1, the momentum distribution of the incident muons through increasing thickness in the range of 0.00 – 0.12 $g/cm^2$is broadened widely from 3.77 to 10.93, 10.46, and 9.75 MeV/c for Al, Fe and Cu respectively. Then, for each absorbing material with thickness 0.12 g/cm^2, the incident muons loses their momentum by at most 16.66, 14.86, and 13.29 MeV/c respectively.

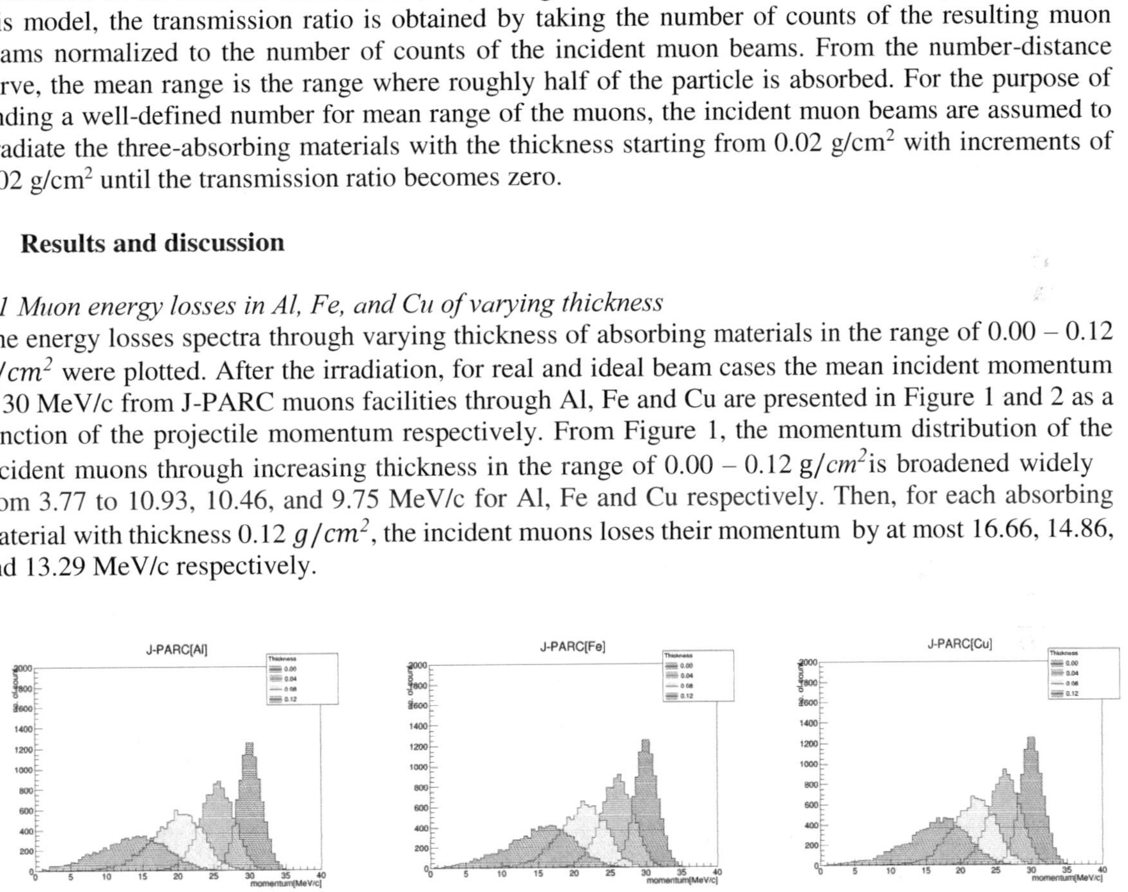

Figure 1. Momentum distribution for real cases of muon from J-PARC at various thickness after irradiation through different absorbing materials: (a) Al (b) Fe and (c) Cu.

From Figure 2, in Al, almost all muons did not lose all their energy except through the thickness of 0.12 g/cm^2 of aluminium where only 0.08% of muons stopped in the target. Meanwhile for Fe and Cu, both show non-stopped muons in the calculated beam at a given thickness. The momentum distribution of the incident muons through increasing thickness in the range of 0.00 – 0.12 g/cm^2 is broadened widely from 3.1 to 17.4, 11.5, and 9.9 MeV/c for Al, Fe and Cu respectively.

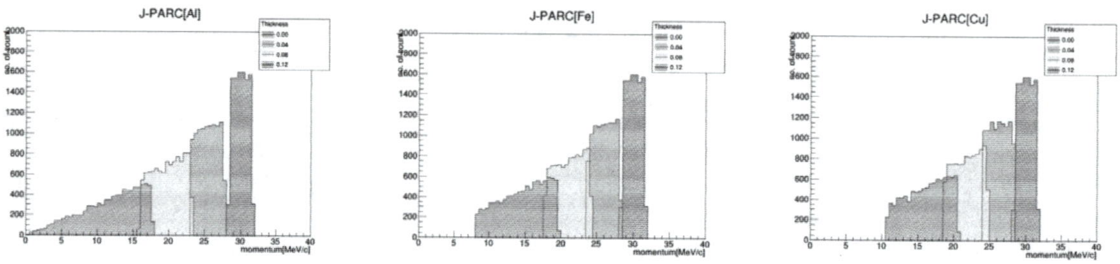

Figure 2. Momentum distribution for ideal cases muon from J-PARC at various thickness after irradiation through different absorbing materials: (a) Al (b) Fe and (c) Cu.

Figure 3 plots the width of the beams against the thicknesses of the absorbing materials to visualise the broadening pattern of muons through these materials. The width broadening of the beams shows a similar pattern as reported by Newhauser *et al.* [7] as they observed the beam width broadening of proton in water.

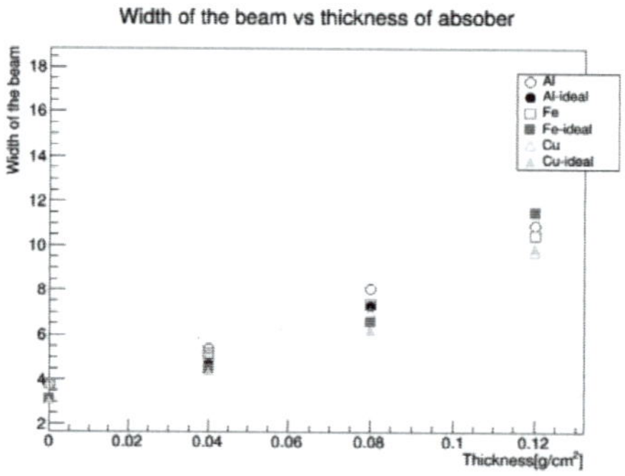

Figure 3. Broadening of the beam width in different material due to energy losses.

The analysis on the range of muons is done by calculating the muon transmission ratio. The result of our calculation for the J-PARC, MuSIC and PSI real and ideal muon beam cases are shown in Figure 4 (a) and (b) respectively. It is observed that the value of mean range constant despite different initial momentum distribution, is found to be 0.053, 0.021 and 0.019 cm through Al, Fe and Cu respectively. Besides, the cut-off in ideal beam cases observed to be sharper, which means that they show less pronounced straggling than in the real beam cases.

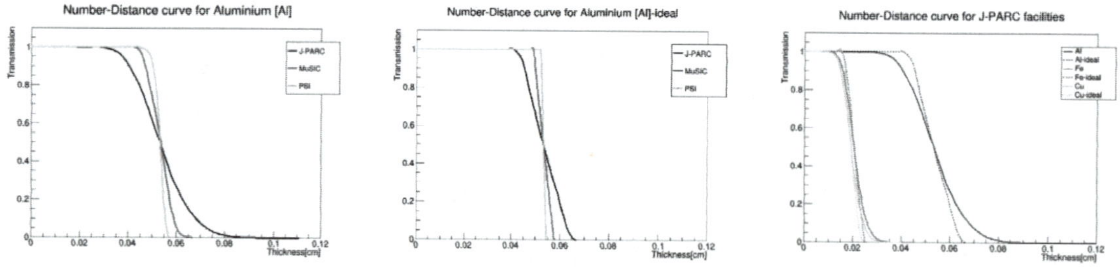

Figure 4. Number-distance curve; (a) for a real case in Al, (b) for an ideal case in Al, and (c) for J-PARC facility only in Al, Fe and Cu.

Figure 4(c) show that muons through aluminium has the highest mean range compared to iron and copper. This can be explained by the density of the material which affects the range of muons in material.

The stopping power increases with increasing density [8]. Hence, the range is inversely proportional to the density of the absorbing material.

4. Conclusions

This study successfully developed a statistical model to understand the muon behaviour when interacting with matter. After checking the performance of the statistical model, we conclude that shell corrections must be included into the stopping power formula in order to get a better result. The effect of the increasing path length on the width of beam distribution for muons passing through different absorbing materials at various thicknesses was clearly demonstrated where the width gets wider an absorbing material. The density of the absorbing material affects the mean range of the muon beam as the range is inversely proportional to the density. Finally, the ideal beam demonstrates smaller straggling effect as compared to the real beam cases.

5. References

[1] Hino Y, Kuno Y, Sato A, Sakamoto H, Matsumoto Y, Tran N H, Hashim I H, Fukuda M, Hayashida Y, Ogitsu T and Yamamoto A 2014 *Nucl. Phys. B- Proc. Suppl.* **253**, p 206-207.

[2] Medenwaldt R, Møller S P, Uggerhøj E, Worm T, Hvelplund P, Knudsen H, Elsener K and Morenzoni E 1991. *Nucl. Instr. and Meth. in Phys. Res. B: Beam Interact. with Mater. and Atoms* **58**(1) p 1-5.

[3] Kumar S, Diwan P K, and Kumar S, 2015 *Radiat. Phys. and Chem.* **106** p 21-25.

[4] Groom D E, Mokhov N V and Striganov S I 2001 *Atomic Data and Nucl. Data Tables*, **78**(2) p 183-356.

[5] Miyake Y, Nishiyama K, Kawamura N, Strasser P, Makimura S, Koda A, Shimomura K, Fujimori H, Nakahara K, Kadono R, and Kato M 2009 *Nucl. Instr. and Meth. in Phys. Res. A: Accel., Spectro., Det. and Assoc. Equip,* **600**(1), pp.22-24.

[6] Leo W R, *Tech. for Nucl. and Particle Phys. Exp.* 1987 (Berlin Heildelberg: Springer-Verlag).

[7] Newhauser W D and Zhang R 2015 The physics of proton therapy *Phys. in Med. and Biol.* ***60***(8), p R155.

[8] Knoll G F 2010 *Radiat. Det. and Meas.* (John Wiley & Sons).

IOP Conf. Series: Materials Science and Engineering **785** (2020) 012007 doi:10.1088/1757-899X/785/1/012007

Study on the shielding materials for low-energy gamma sources

A H Aminordin Sabri[1,a)], M Z Abdul Aziz[2], S F Olukotun[3], F Tabbakh[4], S M Tajudin[1]

[1]Faculty of Health Sciences, Universiti Sultan Zainal Abidin (UniSZA), Terengganu, Malaysia.
[2]Oncological and Radiological Science Cluster, Advance Medical and Dental Institute, Universiti Sains Mala[1]ysia, Penang, Malaysia.
[3]Department of Physics and Engineering Physics, Obafemi Awolowo University, Ile-Ife, Nigeria.
[4]Nuclear Science and Technology Research Institute, Tehran, Iran.

[a]adilahanim@unisza.edu.my

Abstract. Photon dosimetry is indispensable in designing an irradiation facilities shielding. Monte Carlo simulation was utilized to investigate the efficiency of relatively new developed clay and Gadolinum (Gd)-doped polymer as a radiation shielding material for low energy gamma sources (Am-241 and Co-57). The calculated linear attenuation coefficient (μ) of Am-241 and Co-57 for clay is higher within 6.6 % and 0.9 % compared to ordinary concrete, respectively. The μ value for Gd-doped polymer is higher by a factor of 9 and 5 compared to clay for Am-241 and Co-57, respectively. A thickness of 2 cm and 5 cm from both clay and concrete were adequate to attenuate almost 90 % incident photons from Am-241 and Co-57, respectively. The same thickness of 2 cm by Gd-doped polymer could attenuate almost 95 % of Co-57 photons. 3 cm thickness of clay and concrete could shield the gamma source dose rate of Am-241 (1 MBq) down to 0.05 μSv/hr, while almost 10 cm needed for Co-57 source. Gd-doped polymer with thickness of 2 cm could shield almost 94 % the dose rate from Co-57 source. For higher energy gamma sources, clay and ordinary concrete need to be doped with a higher Z element to ensure safety of the radiation.

1. Introduction

Photon dosimetry is indispensable in designing an irradiation facilities shielding. For example, lead (Pb-82) material and ordinary concrete in the order of millimeter (mm) and centimeter (cm) thickness, respectively had been widely used to attenuate the photons, either used as a shielding material for the wall thickness of an irradiation facility room [1][2], such as our radiotherapy room (150 keV) or to shield the emission of photons from a gamma source in a source pig.

The development and properties of non-lead and non-concrete material as a radiation-shielding material have been studied previously by many researchers [3-10]. Lead material already known for its toxicity, expense and limitation in abundance [11-13]. For example, Olukotun *et al.* proposed the use of clay and kaolin as an effective shielding material for photons [9], Tabbakh studied carbohydrate based compounds as new shielding materials [8] for gamma radiation shielding, while Akkurt and Canakci doped the clay with boron as a shielding material for higher-energy gamma sources [14]. Previous studies conducted by Kaura *et al.* and Agar *et al.* were proposed the use of metallic alloy as a potential gamma-shielding material by doping with a high atomic number (Z) element [15][16].

The primary concern in many cases of photon shielding studied was for the transmitted photons, both scattered and unscattered photons that penetrate the shielding materials. It was depends on the point at which the quantity of interest will be calculated or measured from the attenuating material. As example, Sayyed *et al.* had studied the attenuation properties for selected germanate glasses based shielding compound through theoretical calculation [17]. Another studies of radiation attenuation had been performed by theoretical calculation for various concretes [18] and by experimental work for glass based shielding compound [19][20]. In addition to attenuation coefficient (μ) and its related parameters, it is necessary to evaluate intensity transmitted or reflected due to the shielding material, particularly for a newly developed shielding material and photon energy (keV). As we need to ensure the radiation dose rate is as low as possible as per background radiation, such evaluation is important when designing a radiation shielding facility or for source storage. Thus, we conduct the study of our relatively new developed clay [10] and Gadolinum (Gd)-doped polymer [28] as well as ordinary concrete [27] as a radiation shielding material for low energy gamma sources (Am-241 and Co-57).

In this study, the transmitted photons and its ambient dose equivalent (Sv/photon) of the clay and Gadolinum (Gd)-doped polymer materials were evaluated by Monte Carlo simulation (EGS5 code [21][22]). These compound as a shielding material will be optimized by calculating for photon (gamma) energy below 150 keV as based on our irradiation facility to have an Am-241 gamma source at UniSZA, which has a maximum setting of 150 keV. In the first case, photon dosimetry using clay and Gadolinum (Gd)-doped polymer materials were calculated for gamma ray of low-energy source such as Am-241 (16.1 keV, 26.3 keV, and 59.5 keV photons). In the latter case, we include the calculation for gamma ray of Co-57 (14.4 keV, 122 keV, and 136 keV photons).

2. Calculated transmitted photons for multienergy gamma sources

Peviously in our study, clay had been successfully demonstrated to decrease transmitted photon compared to ordinary concrete for low-incident photon energy because of dominant photoelectric absorption [9]. A pencil beam of photons was incident to the center of cylinder clay. The photon's interaction, such as pair productions, Compton scattering (including Rayleigh scattering) and photoelectric absorption were considered in our calculation. However, Compton scattered photons that penetrate the clay-shielding material were ignored for comparison with the attenuation coefficients generated by XCOM [23]. Figure 1 shows the calculation of transmitted photons as a function of ball clay thickness for a pencil beam of 150 keV incident photon. The full square points were fitted with an exponential function to obtain linear attenuation coefficient (μ) values of 1.3892 cm^{-1}. The calculated μ value has agreed with the theory, within 1.06% to justify the subsequent calculations. The error of 0.23% shown in the figure for the calculated μ value comes from exponential fitting error.

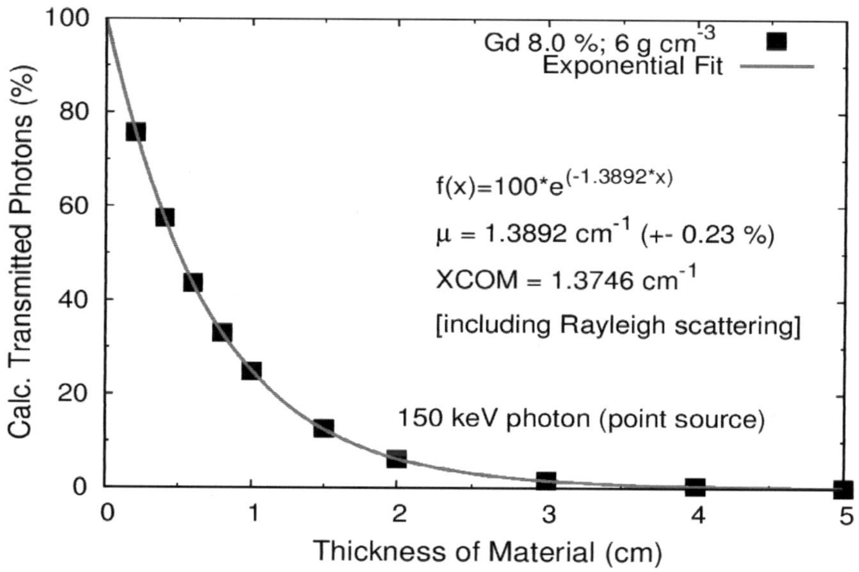

Figure 1. Calculated μ of Gadolinum (Gd)-doped polymer for single-photon energy compared to theoretical value. Calculated μ value is 1.3892 cm^{-1} for 150 keV incident photon to have a good agreement with the theory value within 1.06 %.

For the latter point, μ value was calculated for multi-energies gamma source of Am-241. The branching ratio of the gamma source was sampled according to JRIA data book [24]. Table 1 shows the gamma energies and its corresponding branching ratio adopted in the calculation.

From Table 2, the calculated linear attenuation coefficient (μ) of Am-241 and Co-57 for clay is higher within 6.6 % and 0.9 % compared to ordinary concrete, respectively. The μ value for Gd-doped polymer is higher by a factor of 9 and 5 compared to clay for Am-241 and Co-57, respectively. A thickness of 2 cm clay is adequate to attenuate up to 90% of the photons from Am-241 source. The calculated μ value is 0.7527 cm^{-1}, which is higher, within 5.47%, than the theoretical values of 59.5 keV (0.7126 cm^{-1}). The amount of difference was as expected, as the Am-241 source has lower gamma energies of 16.1 and 26.3 keV photons, which are not considered in the theoretical value.

Meanwhile,. a thickness of 2.0 cm Gadolinum (Gd)-doped polymer is adequate to attenuate up to 95% of the photons from Co-57 source. The calculated μ value is 1.5601 cm^{-1}, which is higher, within 13.67%, than the theoretical values for 122 keV. A thickness of 6 cm clay is adequate to attenuate up to 90% of the photons from Co-57 source. The calculated μ value is 0.3304 cm^{-1}, has a good agreement, within 1.59% with the theory.

Table 1. The list of gamma energies and its branching ratio [24].

Source	Energy (MeV)	Branching ratio (%)
Am-241	0.0263	2.4
	0.0595	35.9
	0.0161	37.9
Co-57	0.0144	9.2
	0.122	85.6
	0.136	10.7

Table 2. The calculated linear attenuation coefficient (μ) cm^{-1} for gamma sources.

Sources	Calculated Linear Attenuation Coefficient (μ) cm^{-1}		
	Ordinary concrete [27]	Clay [10]	Gd-Doped Polymer [28]
Am-241	0.7043	0.7527	6.9471
Co-57	0.3274	0.3304	1.5601

3. Calculated ambient dose equivalent for multi-energy gamma source

Calculation of ambient photon dose (Sv/photon) becomes important for the public and radiation workers, as the radiation must be estimated for the radiation safety area. It is necessary to ensure that the dose rate from the gamma source attenuated down to background radiation level for radiation safety, particularly if the storage location of the gamma source is easily assessable. As an example, the measured background dose rate at our X-ray laboratory is 0.055 ± 0.008 mSv/hr. Ambient dose equivalent is one of the dose unit as an operational quantity by ICRU [25] body for radiation protection. For example, an environmental dosimeter or ionization chamber survey meter was calibrated to measure ambient dose in the unit of Sievert (Sv) to read the photon dose equivalent at 1 cm depth (H*(10)). In such a case, the calculated and measured values of photon dose rate are comparable in absolute values [26].

Table 3. The calculated and theoretical dose rates for 1MBq gamma sources.

Source	1 MBq Dose Rate (μSv/hr) at 20 cm		% of difference
	Calculated	Theoretical	
Am-241	0.3796	0.3775	0.54
Co-57	0.5552	0.5650	1.72

Surface crossing was used to calculate the ambient dose on the surface of 2 cm x 2 cm scoring region at a distance of 20 cm from the gamma source in air. In this part of the calculation, the solid angle of the source photons was 4π that was isotropically incident to the clay. To obtain ambient dose equivalent, air absorbed dose (Gy) was calculated as a first step by using kerma approximation, which calculates collision kerma by multiplying mass energy absorption coefficient to the energy fluence. The following conversions were used to convert the unit of MeV g^{-1} to Gy;

$$1\ \text{MeV} = 1.602 \times 10^{-13}\ \text{J} \tag{1}$$

$$1\ \text{MeV/g} = 1.602 \times 10^{-13}\ (\text{J/MeV}) \times 1000(\text{g/kg}) = 1.602 \times 10^{-10}\ \text{Gy} \tag{2}$$

A function was adopted in the calculation to evaluate the ratio of ambient dose equivalent (Sv) to air absorbed dose. Then the ambient dose equivalent was calculated by multiplying the ratio of ambient dose to air absorbed dose. As a first step, the calculated ambient dose rate without the clay was compared to the theory 1 cm depth dose rate at a distance of 20 cm for Am-241. For the theoretical dose rate, the dose rate conversion factor (μSv.m^2.MBq^{-1}.hr^{-1}) for above 10 keV was obtained from radioisotope data book [24]. The constant value for Am-241 at 1-meter distance is 0.0151 μSv.m^2.MBq^{-1}.hr^{-1}. Table 3 shows the result of calculated dose rate compared to the theoretical dose rate (μSv/hr) at a distance of 20 cm.

Figure 2 (a) and (b) shows the calculated dose rate at a distance of 20 cm for 1 MBq of Am-241 and Co-57 sources as a function of clay, ordinary concrete and Gadolinum (Gd)-doped polymer thickness. From both graphs, a thickness of 2 cm and 5 cm from both clay and concrete were adequate to attenuate almost 90% incident photons from Am-241 and Co-57, respectively. The same thickness of 2 cm by Gd-doped polymer could attenuate almost 95 % of Co-57 photons. 3 cm thickness of clay and concrete could shield the gamma source dose rate of Am-241 (1 MBq) down to 0.05 μSv/hr, while almost 10 cm needed for Co-57 source. Gd-doped polymer with thickness of 2 cm could shield almost 94 % the dose rate from Co-57 source, which could be considered a safe value as per our background radiation value.

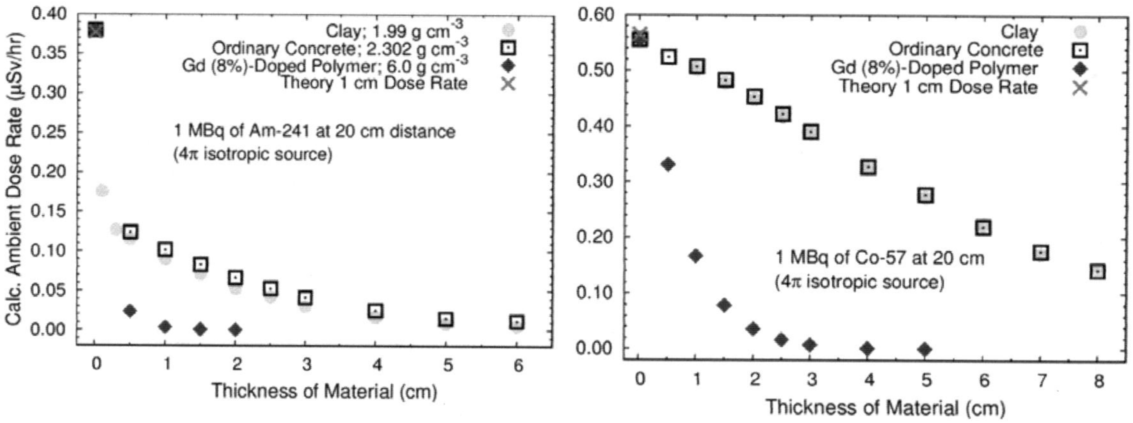

(a) (b)

Figure 2. The calculated dose rate (μSv/hr) at a distance of 20 cm for 1 MBq of **(a)** Am-241 and, **(b)** Co-57 sources. The open square points are the ordinary concrete, the green circle is the clay material, the blue square is the Gadolinum (Gd)-doped polymer, while the cross point at 0 cm (without clay) is the theoretical dose rate.

4. Conclusion

An approach of calculations for photon energy spectra and dose rate either it is transmitted or reflected one must be considered, particularly in exploring the new element or compound materials when designing a radiation shielding. For clay and ordinary concrete, a thickness of 2 and 6 cm is adequate to attenuate up to 90 % of the photons from Am-241 and Co-57 sources, respectively. The clay μ is higher within 6.6 % and 0.9 % compared to ordinary concrete for Am-241 and Co-57 sources, respectively. While the μ values for the proposed Gd (8%)-doped polymer is higher by a factor of 9 and 5 for the same sources, respectively. A thickness of 3 cm clay and concrete were adequate to shield dose rate of 1 MBq Am-241 down to 0.05 μSv/hr at 20 cm distance, while almost 10 cm needed for Co-57 source.

Consequently, for higher energy gamma sources such as brachytherapy room, the developed clay and ordinary concrete need to be doped with a higher Z element to ensure radiation safety for adjacent area.

5. References

[1] Stukenbroeker G L, Bonilla C F, Peterson Ruben W 1970 The use of lead as a shielding material, *Nucl. Eng. Des.* **13** 3-145.

[2] Hyun S J, Kim K J, Jahng T A, Kim H J 2016 Efficiency of lead aprons in blocking radiation - how protective are they? *Heliyon* **2**(5).

[3] McCaffrey J P, Mainegra-Hing E, Shen H 2009 Optimizing non-Pb radiation shielding materials using bilayers, *Med. Phys.* **36**.

[4] Yue K, Luo W, Dong X, Wang C, Wu G, Jiang M, Zha Y 2009 A new lead-free radiation shielding material for radiotherapy *Radiat. Prot. Dosim.* **133** p 256-260.

[5] Soylu H M, Lambrecht F Y 2015 Gamma radiation shielding efficiency of a new lead-free composite material *J. Radioanal. Nucl. Chem.* **305** p 529.

[6] Singh A K, Singh R K, Sharma B, Tyagi A K, Characterization and biocompatibility studies of lead free X-ray shielding polymer composite for healthcare application 2017 *Radiat. Phys. Chem.* **138** p 9-15.

[7] Chang L, Zhang Y, Liu Y, Fang J, Luan W, Yang X, Zhang W 2015 Preparation and characterization of tungsten/epoxy composites for g-rays radiation shielding *Nucl. Instrum. Methods Phys. Res. Sect. B Beam Interact. Mater. Atoms* **88**(93) p 356-357, http://doi.org/10.1016/j.nimb.2015.04.062.

[8] Tabbakh F, Babaee V, Naghsh-Nezhad Z 2015 Carbohydrate based materials for gamma radiation shielding *J. Phys.: Conf. Ser.* 611(01).

[9] Olukotun S F, Gbenu S T, Ibitoye F I , Oladejo O F, Shittu H O, Fasasi M K , Balogun F A 2018 Investigation of gamma radiation shielding capability of two clay materials, Nucl. Eng. Tech. **50** p 957-962.

[10] Tajudin S M, Sabri A H A , Abdul Aziz M Z, Olukotun S F, Ojo B M, Fasasi M K 2019 Feasibility of clay-shielding material for low-energy photons (Gamma/X) *Nucl. Eng. Tech.* **51** p 1633-1637.

[11] Wani A, Ara A, Usmani J, Lead 2015 toxicity: a review *Interdiscip. Toxicol.* **8** p 55-64.

[12] Billen P, Leccisi E, Dastidar S, Li S, Lobaton L, Spatari S, Baxter J B 2019 Comparative evaluation of lead emissions and toxicity potential in the life cycle of lead halide perovskite photovoltaics *Energy* **166** p 1089-1096. http://doi.org/10.1016/j.energy.2018.10.141.

[13] Flora G, Gupta D, Tiwari A 2012 Toxicity of lead: a review with recent updates *Interdiscip. Toxicol.* **5**(2) p 47-58.

[14] Akkurt H C 2011 Radiation attenuation of boron doped clay for 662, 1173 and 1332 keV gamma rays, Iran *J. Radiat. Res.* **9** p 37-40.

[15] Kaura T, Sharmab J, Singha T 2019 Review on scope of metallic alloys in gamma rays shield designing *Prog. Nucl. Energy* **113** p 95-113.

[16] Agar O, Sayyed M I, Akman F, Tekin H O, Kaçal M R 2018 An extensive investigation on gamma ray shielding features of Pd/Ag-based alloys *Nucl. Eng. Technol.* **51**(3) p 853-859. http://doi.org/10.1016/j.net.2018.12.014.

[17] Sayyed M I, Kaky K M, Akar E S, Akbaba U, Taki M M, Agar O 2019 Gamma radiation shielding investigations for selected germanate glasses *J. Non-Cryst.Solids* **512** p 33-40. November 2018, http://doi.org/10.1016/j.jnoncrysol.2019.02.014.

[18] Bashter I I 1997 Calculation of radiation attenuation coefficients for shielding concretes, *Ann. Nucl. Energy* **24**(17) p 1389-1401. http://doi.org/10.1016/S0306-4549(97)00003-0.

[19] Kurudirek M, Chutithanapanon N, Laopaiboon R, Yenchai C, Bootjomchai C 2018 Effect of Bi_2O^3 on gamma ray shielding and structural properties of borosilicate glasses recycled from

high pressure sodium lamp glass *J. Alloy. Comp.* **745** p 355-364. http://doi.org/10.1016/j.jallcom.2018.02.158.

[20] Ersundu A E, Büyükyıldız M, Ersundu M Ç, Akar E S, Kurudirek M 2018 The heavy metal oxide glasses within the WO3-MoO3-TeO2 system to investigate the shielding properties of radiation applications, *Prog. Nucl. Energy* **104** p 280-287. http://doi.org/10.1016/j.pnucene.2017.10.008.

[21] Hirayama H, Namito Y, Bielajew A F, Wilderman S J, Nelson W R 2010 EGS5 code system. SLAC Rep. SLAC-R-730 and KEK Rep. 2005, **8**,.

[22] Tajudin S M, Hanim A, Sabri A 2019 SIMU-RAD programme: A learning tool for radiation (photons and charged particles) interaction. *The J. of Polish Soc.of Med.Phys.* **25**(3), 189–192.

[23] Gerward L, Guilbert N, Jensen K B, Levring H 2004 WinXComda Program for calculating X-ray attenuation coefficients *Radiat. Phys. Chem.* **71** p 653-654.

[24] Japan Radioisotope Association 2011 *Radioiso. Pocket Data Book*, ed 11, (Tokyo: Maruzen) p 22-100.

[25] https://icru.org/.

[26] Tajudin S M , Namito Y, Sanami T, Hirayama H 2014 Quasi-monoenergetic 200 keV photon field using a radioactive source with backscatter layout, *Jpn. J. Appl. Phys.* **53** p 116401.

[27] https://physics.nist.gov/cgi-bin/Star/compos.pl.

[28] Tajudin S M, Tabbakh F 2019 Biological polymeric shielding design for an X-ray laboratory using Monte Carlo codes *Radiol. Phys. and Tech.* **12**(3) p 299–304.

IOP Conf. Series: Materials Science and Engineering **785** (2020) 012008 doi:10.1088/1757-899X/785/1/012008

The influence of clay addition in fly ash concrete mixture for nuclear shielding

Rogemah Ramli[1a], Faridah Mohamad Idris[2], Juniza Md Saad[1], M F K M Yapandi[3], S. Akmal Biyamin[3]

[1]Institute of Sustainable Energy, Universiti Tenaga Nasional, Jalan IKRAM-UNITEN, 43000 Kajang, Selangor, Malaysia.
[2]Malaysia Nuclear Agency, 43000 Kajang, Selangor, Malaysia.
[3]TNB Research Sdn Bhd, 43000 Kajang, Selangor, Malaysia.

[a]rogemah@uniten.edu.my

Abstract. The disposal of ashes from power plants has caused health issues and reduce the air quality of the surrounding areas. The ashes have a potential to be used in construction of concrete mixture as partial replacement for cement. This research studies the potential fly ash and clay mixture (FAC) as a nuclear shielding material in concrete mixture. The raw materials and the testing were all done in Agensi Nuklear Malaysia. All concrete mixtures had a constant water/cement ratio of 0.266. The hardened concrete mixtures were tested for compressive strength. Scanning electron microscopy (SEM), energy dispersive X-ray (EDX) and small angle neutron microscopy (SANS) are used to characterize the properties of the mixtures. The results suggested that mixture of FAC concrete with 10% of clay addition showing good potential for nuclear shielding compared to other ratios.

1. Introduction

In concrete studies, most of the paper focused on the composite mixture to improve the mechanical properties of the concrete [1]. However, in the nuclear industry, other parameters such as shielding properties and radiation absorption need to be taken into consideration. Any material can be used for radiation shielding if it has enough thickness to absorb the incident radiation to a safe level [2]. But for concrete, the variation in moisture contents add uncertainty to the calculations of attenuation coefficient.

The fundamental shielding prerequisites for radiation protection rely upon a few components. Radiation ray shielding is used to reduce the exposure to radiation. Diverse kinds of radiation are discharged with various levels of vitality behind them. Nowadays, the standards used for the selection of concrete radiation shields aggregates don't take into consideration the effect of the potential neutron activation on the size of the resulted radioactive wastes in these shields [3]. Most of the requirements of these standards concentrate on mechanical and shielding effectiveness of the resulting concrete.

The contribution of ordinary Portland cement (OPC) production worldwide to greenhouse gas emissions is estimated to be approximately 1.35 billion tons annually or approximately 7% of the total greenhouse gas emissions to the earth's atmosphere [4]. one of the efforts to produce more environmentally friendly concrete is to partially replace the amount of OPC in concrete with by-product materials such as fly ash. Fly ash is one of the numerous substances that can cause air, water, and soil pollution. The disposal of this waste material is a matter of great concern as it has become a major environmental problem [5]. Disposal of high amount of fly-ash from thermal power plants absorbs huge amount of water, energy and land area by ash ponds [6]. It is an airborne material and the utilization of fly ash in construction materials shows some example of success. Also, research on the use of the waste product which is fly ash can reduce the environmental problems. Our previous studies also reported that fly ash has potential as shielding properties in concrete mixture [7].

Clay materials are used for constructions due it its durability compared to cement especially under tropical condition [8]. Bentonite is an absorbent aluminum phyllosilicate, essentially impure clay consisting mostly of montmorillonite. The utilization of organofunctionalized bentonite as pozzolan is a good option of the point of view environmental, financial and technical, the bentonite is a clay mineral highly abundant in various parts of the world [9]. Most natural bentonites are found in nature to exist in the sodium and/or the calcium form. The performance of a calcium bentonite as a viscosity builder can often be enhanced by its conversion to the sodium form [10]. Kaolin are also has potential properties, are white raw materials, their essential constituent being fine grained white clay, which are amenable for beneficiation that make them ideal for an assortment of industrial applications [11]. Kaolin is basically consisting of the hydrated aluminosilicate mineral kaolinite with minor amounts of quartz, feldspar, mica, chlorite and other clay minerals [12]. It is recognized from different clay by its delicate quality, whiteness, and simplicity of scattering in water [13].

On this basis, the goal of this study is to investigate the influence of the addition of clays in fly-ash concrete mixture in further enhancing the nuclear shielding properties, gamma radiation and neutron radiation.

2. Methodology

2.1. *Materials*

Both fly ashes and bottom ashes were obtained from TNB Janamanjung Sdn. Bhd., Stesen Janakuasa Sultan Azlan Shah Power Plant in Perak. The raw ashes were grinded to about 2 mm size. Two different type of clay addition is used in this studies; bentonite and kaolin clay, with ratio of 1:1 for each combination (e.g., 10 % of clay consisted of 5% of bentonite and 5% of kaolin). Different percentage of clay combination were added in fly ashes concrete mixtures, but the percentage of fly ash remains constant for 20%. For the other materials which are barite, colemanite, sand, plasticizer, and water were kept constant. Bentonite clay and Kaolin clay was obtained from R&M Chemicals Sdn. Bhd., and cement used is collected from Lafarge Sdn. Bhd. Colemanite is used for fine aggregate that have a good neutron absorber capability, and is obtained from Malaysia Nuclear Agency, while Barite is used as coarse aggregate with the range size of 10-20 mm has a good gamma absorption for nuclear shielding [14]. Both aggregates were obtained from Malaysia Nuclear Agency. The summary of composition of concrete mixtures are as illustrated in Figure 1. Three concrete samples (15 cm x 15 cm x 15 cm) were prepared for each A to D batches to maintain the consistency and reliability of the results. The concrete samples were put in room temperature for 28 days during the curing process.

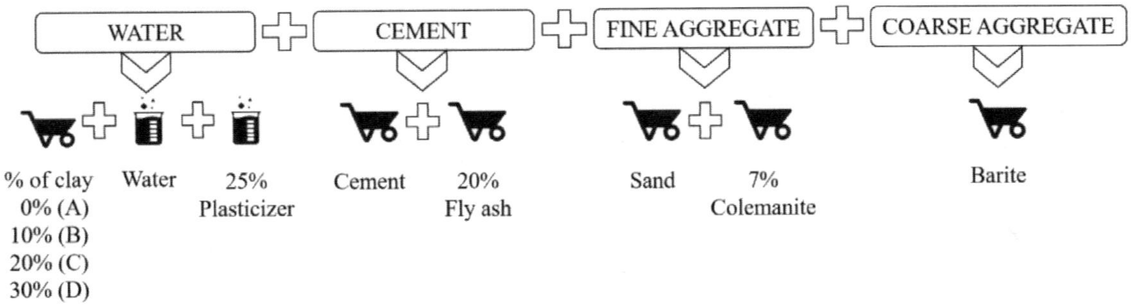

Figure 1. Sample preparation of fly ash and clay (FAC) concrete mixtures,

2.1.1. Characterisation. The concrete samples were characterized using small-angle neutron scattering (SANS) to investigate the microstructural radiation damage for nuclear applications [15]. Barium-1333 is used as the gamma source for gamma absorption test and Geiger-Mueller is used to measure and recorded the count rate of gamma rays. The surface morphology of the concrete samples were examined using scanning electron microscopy and energy dispersive x-ray analysis (SEM-EDX). In addition, compression test is also conducted to investigate the strength and hardening properties of the concrete samples.

3. Results and Discussion

3.1. Effect of clay addition on radiation

Table 1 and Figure 2 shows the percentage of neutron absorbed and average equivalent neutron radiation on the concrete samples. The results showed that the highest amount of absorption of neutron radiation is from sample that consist of 20% fly ash and 20% of clay (Sample C), while the sample that absorbed the least amount of neutron is from sample which consist of 20% of fly ash and 30% of clay (Sample D). Sample C also shows lowest average equivalent neutron radiation detected with only 143.74 µSv/hr, showing good absorbent for neutron compared to other samples. This explains that the higher the percentage of neutron radiation being absorbed, the lower the average equivalent neutron detected. The results also concluded that the 20% of clay is the optimum amount of clay addition to be added in the fly-ash concrete mixture to achieve highest neutron absorption.

Table 1. Amount of neutron absorbed over different percentage of clay addition

Concrete sample	Neutron absorbed (%)
0% Clay (Control) - (A)	96.88
10% Clay - (B)	95.80
20% Clay - (C)	98.02
30% Clay - (D)	96.92

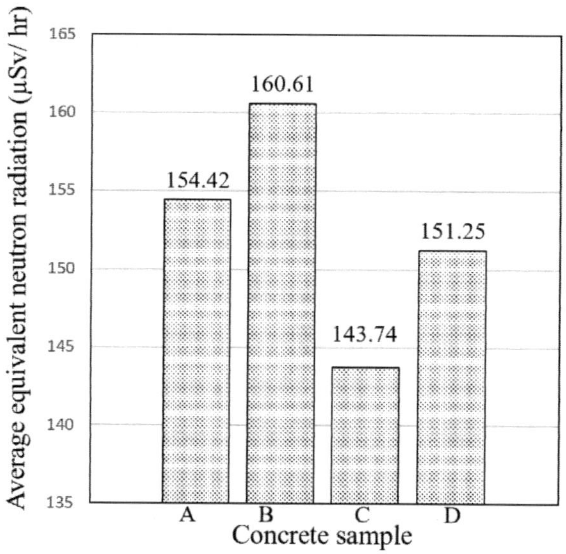

Figure 2. Average equivalent neutron radiation detected vs concrete sample.

Table 2 and Figure 3 shows the amount of gamma absorbed and gamma count rate of detector. Concrete sample B which consisted of 20% of fly-ash and 10% of clay addition shows highest amount of gamma adsorbed with 72.1%. The gamma count rate detected strengthen the finding, in which concrete sample B shows the minimum amount of count rate detected with only 1085 counts/min. Therefore, this explains that the more the amount of gamma radiation absorbed in the concrete mixture, the lower amount of gamma radiation pass through the concrete. However more addition of clay does not really improve the gamma radiation absorption.

Table 2. Amount of gamma absorbed over different percentage of clay addition

Concrete sample	Gamma absorbed (%)
0% Clay (Control) - (A)	70.69
10% Clay - (B)	72.1
20% Clay - (C)	71.80
30% Clay - (D)	70.61

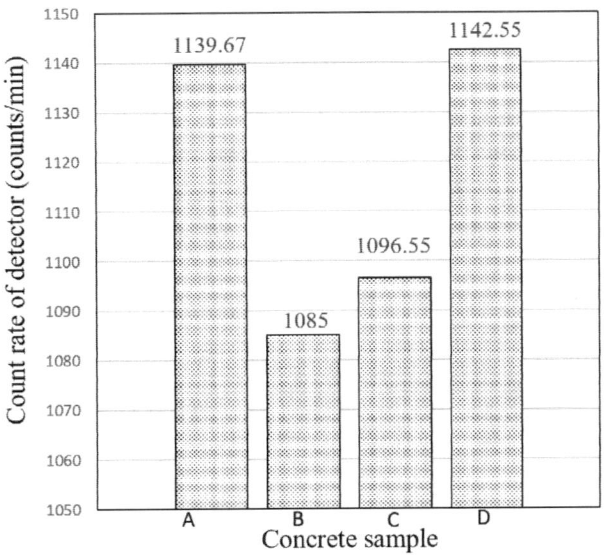

Figure 3. Amount of gamma absorbed over different sample ratios.

3.2. Effect of clay addition on surface morphology

Figure 4 shows the surface morphology of fly ash concrete mixture with different amount of clay addition. It can be seen that there are changes in surface morphology with the addition of clay (B, C and D), showing the blending formation between clay and fly ash concrete mixtures.

In addition, Table 3 presented the elemental composition of concrete samples with and without the addition of clay. From the table, there are addition of elemental component, ClK for 10%, 20% and 30% of clay addition. For 30% of clay addition, more elemental were found; MgK, SK and BaL. It is suggested that the addition of elemental component were from the clay (bentonite and kaolin).

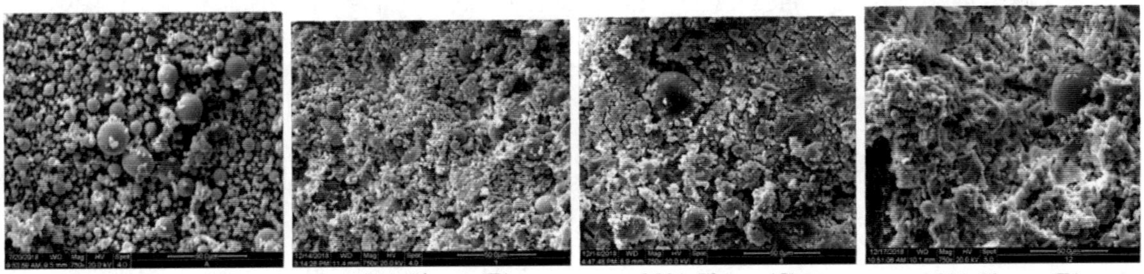

0% Clay – (A) 10% Clay – (B) 20% Clay – (C) 30% Clay – (D)

Figure 4. SEM results (x750 magnification) of fly ash and clay mixtures concrete

Table 3. Elemental composition of concrete samples in wt%.

Elemental	**0% Clay**	**10% Clay**	**20% Clay**	**30% Clay**
C K	19.96	nd	nd	12.68
O K	29.30	40.67	38.59	34.52
AlK	18.48	4.83	3.96	5.34
SiK	25.84	11.99	24.26	10.44
K K	2.32	nd	1.14	nd
CaK	0.66	34.67	24.54	13.63
FeK	3.45	4.79	2.32	1.19
ClK	nd	3.04	5.15	2.10
MgK	nd	nd	nd	1.71
S K	nd	nd	nd	5.67
BaL	nd	nd	nd	12.73

3.3. Effect of clay addition on the hardening properties
Table 4 presents the compressive strength results all four concrete samples. Based on the results, it is shown that concrete sample B with 10% clay addition mixture can stand more forces and strength compared to the control sample (without clay addition). However, no compressive strength was recorded for concrete sample C and D, i.e. these samples failed the test. It is suggested that with more addition of clay might cause the concrete structure to expand and caused the increment of pores between the concrete while increasing the void ratio and decreasing the strength of the sample.

Table 4. Compressive strength results for different sample.

Sample	Maximum compressive force (kN)	Compressive strength (N/mm^2)
A	340	15
B	540	24
C	-	-
D	-	-

4. Conclusion
This study has demonstrated that varying the ratios of clay in the fly ash concrete mixture has different reaction towards neutron and gamma radiation absorption. 20% of clay addition is most suitable for neutron shielding materials while 10% of clay addition is most suitable for gamma shielding materials. However, 10 % clay addition have high concrete strength, therefore it is recommended that 10% of clay addition in the flyash concrete mixture suit the most in nuclear shielding application. It is also suggested

that the properties of each bentonite and kaolin clay might affect the neutron and gamma absorption, hence the variation of the results presented.

5. References

[1] Suat Akbulut A H S 2015 A research on the radiation shielding effects of clay, silica fume and cement samples Rad. Phys. and Chemi. **117** p 88- 92.

[2] Singh Mann G S B K S M G S M H 2016 Experimental investigation of clay fly ash bricks Nucl. Eng. and Tech. **48** p 1230- 1236.

[3] Sukhpal Singh A K D S K S T G S M 2008 Barium–borate–flyash glasses: As radiation shielding materials, Beam Interactions with Materials & Atoms **266** p 140-146.

[4] Hardjito 2004D S E W D M J S A B R On the development of fly ash-based, ACI Mater. J. p 467-471.

[5] Ahmaruzzaman M 2010 A review on the utilization of fly ash, Progress In Energy and Combustion Science **36**(3) p 327-363.

[6] Basu M M P S 2009 Potential fly-ash utilization in agriculture: A global review, Progress in Natural Science **19**(10) p 1173-186.

[7] Ramli R, Mohamad Idris F, Yapandi M F K M, Saad J Akmal Biyamin S 2019 Investigation on different compositions of flyash concrete on gamma and neutron radiation, IOP Conf. Series: Mater. Sci. and Eng. **555** p 1-8.

[8] Olukotun S G F I O O H S M F S F 2018 Investigation of gamma radiation shielding capability of two clay, Nuclear Engineering and Technology **50** p 957-962.

[9] Lima-Guerra I M R R S D J 2014 Use of Bentonite and Organobentonite as Alternatives of Partial Substitution of Cement in Concrete Manufacturing, International Journal of Concrete Structures and Materials **8**(1) p 15- 26.

[10] Menking P M B J H R 1992 Process for producing bentonite clays exhibiting enhanced solution viscosity properties **8**(2).

[11] Prasad K H M S 1996 Kaolin: processing, properties and applications Appl. Clay Sci. **6**(2) p 87-119.

[12] Balcke G U N A K S H F D K I V P A F H F 2002 Adsorption of Humic Substances onto Kaolin Clay Related to Their Structural Features, Soil- Science Society of America Journal , vol. **66** p 1805-1812.

[13] Akbar J B A M A S A 2013 Evaluating the effect of bentonite on strength and durability of high performance concrete Adv. Struct. and Geotech. Eng. **2**.

[14] Koichi M K, Okuno H Y 2017 Development of novel neutron shielding concrete Nucl. Tech. **168** p 545-552.

[15] Coppola R, Klimenkov M, Lindau R, Möslang A, Rieth M, Vallic M 2018 Radiation damage studies in fusion reactor steels by means of small-angle neutron scattering (SANS) **551**, pp. 407-412.

Acknowledgement
The work reported in this paper is partially taken from an Undergraduate Final Year Project report submitted for Bachelor of Mechanical Engineering at Universiti Tenaga Nasional. The authors wish to thank to all who provided assistance in this research.

Study of neutron moderation properties of liquid moderators in room and cryogenics temperature

Danial Zikri Azmi[1], Mohd-Syukri Yahya[1, a)], Nasri A. Hamid[1], Faridah Mohamad Idris[2], Julie Adrianny Murshidi[2], Azraf Azman[2], Abdul Aziz Mohamed[3]

[1]College of Engineering, Universiti Tenaga Nasional (UNITEN), Selangor, Malaysia
[2]Malaysia Nuclear Agency (Nuklear Malaysia), Selangor, Malaysia.
[3]Malaysia Nuclear Society (MNS), Selangor, Malaysia

[a]syukri@uniten.edu.my

Abstract. – Neutron moderation is crucial to sustain the chain reactions in a nuclear fission reactor. To understand further basic characteristics of neutron moderation, this research investigated fundamental moderating properties of liquid moderators typically used in modern power reactors, namely water and salt solution. Specifically, we employed a small angle neutron scattering (SANS) technique at beam-port number 4 of Malaysia's PUSPATI TRIGA Reactor to study moderating characteristics of the liquid moderators at both room and cryogenic temperatures. To achieve cryogenic temperatures, a leakage-proof liquid sample holder was placed in a cryostat cooled by liquid nitrogen. Moderated and unmoderated neutron spectra were collected using Microspec-6 for analysis, which clearly shows that peak neutron counts dropped at cryogenic from room temperatures.

1. The setup

The aim of this research was to investigate the basic moderating characteristics of typical liquid moderators at room and cryogenic temperatures. The stated objective admittedly sounds rather elementary and quite academic, the authors believe that the experiences picked up during the experiment will be useful for future small angle neutron scattering (SANS) projects at the PUSPATI TRIGA reactor. This is because a low-flux research reactor such as TRIGA was normally not used for neutron moderation research. As such, this experiment could be a basic but relevant start.

A simple continuous flow (SCF) cryostat as depicted in Figure 1 was used to achieve the cryogenic temperature [1]. Main components of the SCF cryostat are stainless-steel vessel suitable for liquid nitrogen circulation, moderator holder and cover. Its cooling-ability was limited by cooling performance of the compressed liquid nitrogen. Data collection of the experiment was limited by the reactor beam time, as well as our experimental setup, namely the N-probe and Microspec-6 analyser. The N-Probe consists of two separate detectors to cover the thermal and fast spectra of neutron energy: a NE213-type liquid scintillator covers 800 keV to 20 MeV, while a 3He proportional counter covers thermal to 800 keV energy range. Both detectors' data were merged automatically to yield the neutron spectrum [1].

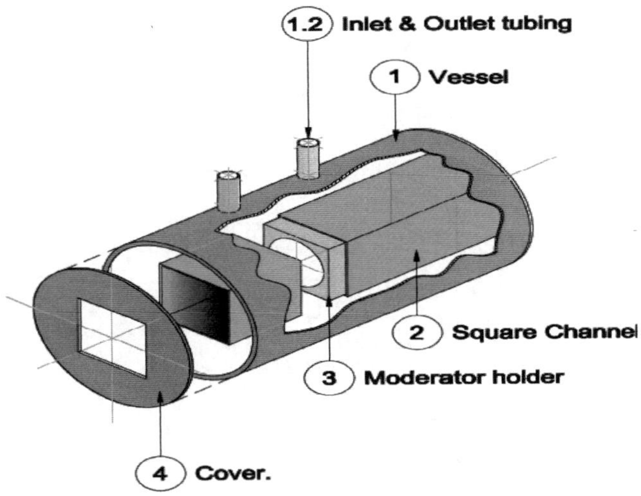

Figure 1. .Schematic of the SCF cryostat [1].

2. The experiment

The experiment was carried out at beam port #4 of the PUSPATI TRIGA reactor. Initial background neutron measurements were first collected. The experiment was later continued using the arrangement shown in Figure 2. The moderator samples were placed in front of the beam port.

Figure 2. .The experimental set up.

3. Results and discussion

Neutron spectra collected by Microspec-6 analyzer was tabulated as depicted in Figure 3. It was apparent that all plots exhibit similar characteristics: low count at fast region with a noticeable peak at the thermal region. Nonetheless, upon detailed investigation, it was obvious that spectrum at thermal range differed slightly for the two moderating samples (i.e., water and salt solutions). Summarized peak neutron counts are highlighted in Table 1. Peak neutron counts dropped at cryogenic from room temperatures for both samples as expected.

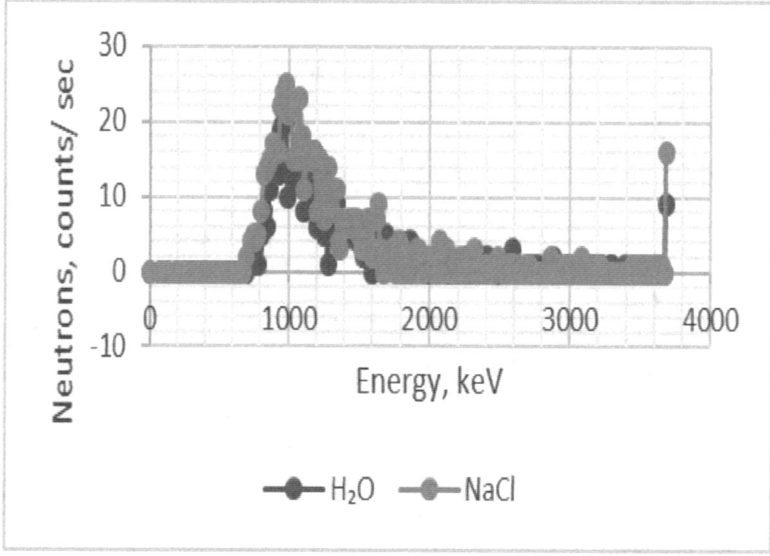

Figure 3. Neutron spectra collected by Microspec-6 analyzer.

Table 1. Post-moderation neutron counts by water and salt solution at room and cryogenic temperatures after 20 minutes of neutron irradiation.

Material	Condition	Peak count/second
No Moderator	Room temperature	35
	Cyrogenic temperature	34
H_2O	Room temperature	22
	Cyrogenic temperature	20
NaCl	Room temperature	25
	Cyrogenic temperature	24

4. Conclusion

Peak neutron counts dropped at cryogenic from room temperatures since moderator samples slowed down fast neutrons as expected.

5. References

[1] Zakaria Dris *et al.* 2016 Cryostat system for investigation on new neutron moderator materials at Reactor TRIGA PUSPATI *AIP Conf. Proc.* **1704**(1), id.030015, Jan doi: 10.1063/1.4940084.

[2] Yavar R, Khalafi H, Kasesaz Y, Sarmani S, Yahaya R, Wood A K, and Khoo K S 2012 Verification of MCNP simulation of neutron flux parameters at TRIGA MK II reactor of Malaysia. *Appl. Radiat. Isot.* **70**(10) p 2488–93.

[3] Yavar R, Sarmani S, Wood A K, Fadzil S M, Masood Z, and Khoo K S 2011 Neutron flux parameters for k0-NAA method at the Malaysian Nuclear Agency research reactor after core reconfiguration, *Radiat. Meas.* **46**(2) p 219–223.

[4] Parajon M H, Abad E and Bermejo F J 2014 A review of the cold neutron moderator materials : neutronic performance and radiation effects *Phys. Procedia* **60** p 74–82.

[5] Parma V Cryostat design

[6] Mocko M, Daemen L L, Hartl M, Huegle T and Muhrer G 2010 E i n n m m.

[7] Korobkina E, Medlin G, Wehring B, Hawari A I, Huffman P R, Young A R, Beaumont B and Palmquist G 2014 Ultracold neutron source at the PULSTAR reactor: Engineering design and cryogenic testing *Nucl. Instr. Meth. Phys. Res. Sect. A Accel. Spectrometers, Detect. Assoc. Equip.* **767** p169–175.

[8] Kimura M 1992Neutron production, moderation, and characterization of sources *Lns.Pnpi.Spb.Ru.*

[9] Kasesaz Y, Khala H, Rahmani F, Ezati A, and Keyvani M 2014 A feasibility study of the Tehran research reactor as a neutron source for BNCT **90** p 132–137.

[10] 2000 Unique portable neutron spectrometer p 31278.

Acknowledgments
The experiment was completed partly using an IAEA financial grant via a coordinated research project (CRP) titled "Advanced Moderators for Intense Cold Neutron Beams in Materials Research".

IOP Conf. Series: Materials Science and Engineering **785** (2020) 012010 doi:10.1088/1757-899X/785/1/012010

ESCORT: Energy sweep compact rapid cycling hadron therapy

Leo Kwee Wah[1,a)], Ken Takayama[2], Toshikazu Adachi[2], Tadamichi Kawakubo[2] and Tanuja Dixit[3]

[1]Nuclear Agency, Bangi 43000 Kajang. Selangor. Malaysia.
[2]High Energy Accelerator Research Organization (KEK), Tsukuba, Ibaraki, Japan.
[3]Society for Applied Microwave Electronics Engineering & Research (SAMEER), Mumbai, India.

Type the corresponding author's e-mail address here

Abstract. The Energy Sweep Compact Rapid Cycling Hadron driver for future cancer therapies based on the induction synchrotron concept is described. This fast cycling synchrotron that allows the energy sweep beam scanning. Assuming a 1.5 T bending magnet, the ring can deliver heavy ions of 430 MeV/u at 10 Hz. A beam fraction is dropped from the barrier bucket at the desired timing and the increasing negative momentum deviation of this beam fraction becomes enough large for the fraction to fall in the electrostatic septum extraction gap, which is placed at the large D(s) region. The programmed energy sweeping extraction makes spot scanning beam irradiation on a cancer area in depth possible.

1. Introduction

3D spot scanning of hadron beams to cancer tissues of human organs is of most concern in this society [1]. We will focus on spot scanning by the energy sweeping extraction from a fast cycling induction synchrotron [2]. A hadron bunch captured in the barrier bucket is continuously accelerated by the induction flat voltage and a fraction of the beam bunch is spilled out from the stable barrier bucket by non-adiabatically changing timing of the acceleration voltage controlling trigger signal in the desired time period. Equilibrium orbits of spilled out particles move inward depending on the dispersion function D(s) and those particles enter into the electrostatic septum gap region to be further deflected inward, and then propagate through the extraction region downstream consisting of extraction device such as a Lamberson magnet to put on the extraction beam line. Start of the extraction and a number of spilled out particles are simply determined by controlling of the gate signal. Thus, we can obtain a driver beam for the cancer therapy with the function of 3D spot scanning, the energy of which changes continuously in the same acceleration cycle. Details of this scheme has been proposed as compact driver in Reference [3]. Its essence will be described here.

IOP Conf. Series: Materials Science and Engineering **785** (2020) 012010 doi:10.1088/1757-899X/785/1/012010

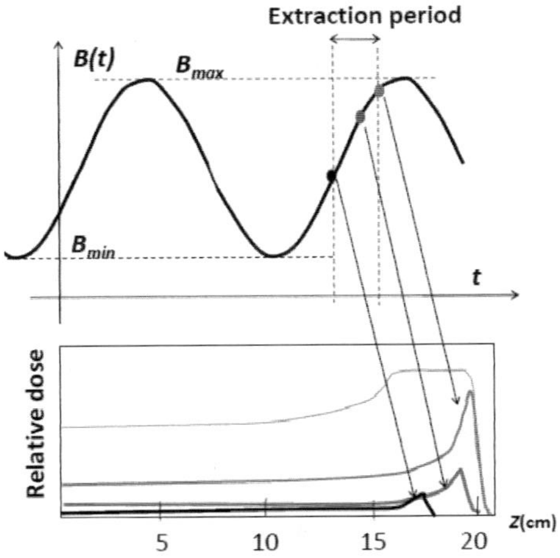

Figure 1. Energy sweep extraction in the same acceleration cycle and integrated dose along the path, where *B(t)* is the magnetic flux density of the guiding magnet.

2. Beam Parameters

The ring has the two-fold symmetry as shown in Fig. 2. Its super lattice has a mirror symmetry, which consists of two sorts of F-B-D-B-F cell in Fig. 3. The lattice profile is similar to that of the CERN design PIMMS [4] Two long dispersion-free straight sections are occupied by the injection system and induction acceleration cells for acceleration and confinement. One long straight section with a large flat dispersion function is occupied by the extraction system for energy sweeping scanning. The other is partially occupied by the kicker magnets for the fast extraction, which is used in a combination with the septum magnet placed downstream with a betatron phase advance of 90 degrees.

Properties of the lattice design and machine parameters [3] as shown in Fig. 2 and Table 1must comprise of:

- Dispersion-free region for induction acceleration devices and injection device.
- Localized large flat dispersion region for the extraction device with the length of 4 m.
- Local betatron phase advance of $\pi/2$ for the fast extraction.

IOP Conf. Series: Materials Science and Engineering **785** (2020) 012010 doi:10.1088/1757-899X/785/1/012010

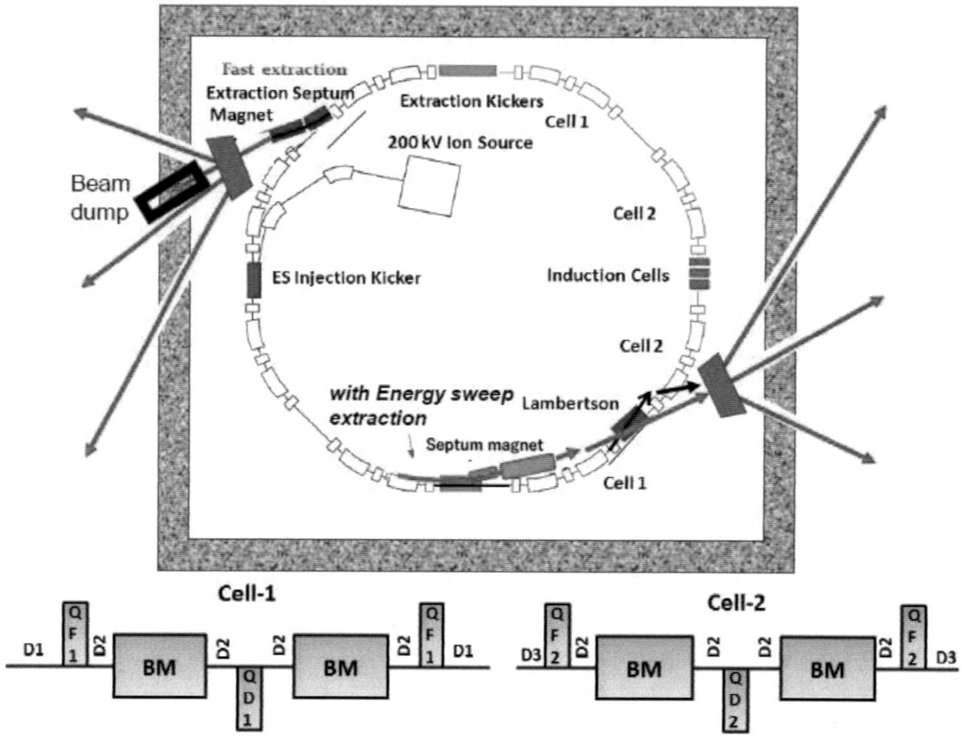

Figure 2. Outline of the driver ring and cell structure.

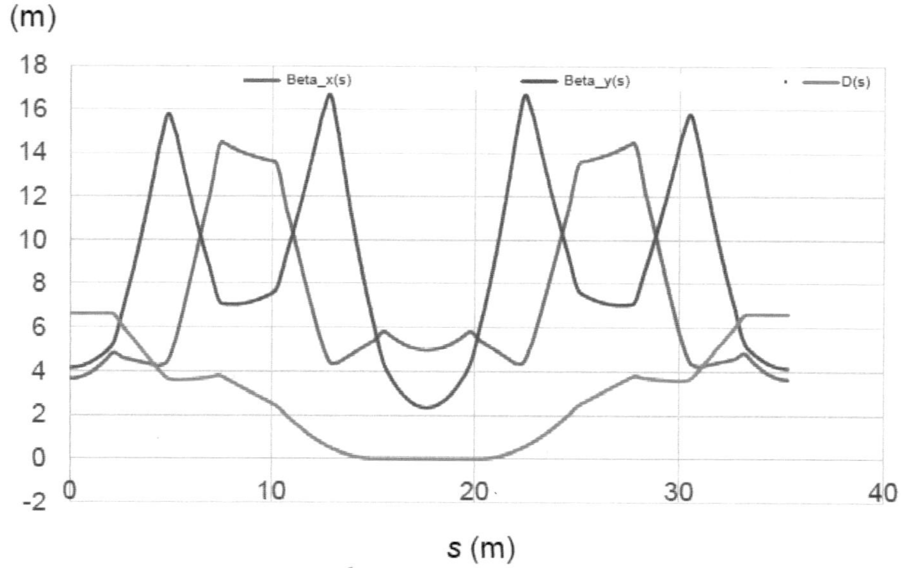

Figure 3. Super-symmetry lattice functions.

IOP Conf. Series: Materials Science and Engineering **785** (2020) 012010 doi:10.1088/1757-899X/785/1/012010

Table 1. Machine and beam parameters.

Machine	Beam parameters
Energy	430 MeV/nucleon for $A/Q = 2$ ion
C_0	52.8 m
Ion species	Gaseous/Metal ions
Ion source	Laser ablation IS, ECRIS
Injector	200 kV (electrostatic)
Ring	fast cycling (10 Hz)
	$B_{max} = 1.5$ Tesla
	$\rho = 4.42928$ m
	FODOF cell with edge focus of B
	Mirror symmetry
	$v_x/v_y = 1.88024/1.69714$
	2m long dispersion-free region
	4m long flat large dispersion region
	α_p=0.0396646
	γ_T=5.0211
Acceleration	Induction cells driven by SPS
	employing SiC-MOSFET
	$V_{acc} = \rho C_0 dB/dt$ (max 7 kV)
Vacuum	10-8 Pascal

3. Simulation for extraction

3.1. Energy sweeping extraction

In this driver ring, a hadron bunch is trapped in the barrier bucket and accelerated with the induction voltage pulse as shown in Fig. 4. Particles in a normal acceleration, where an entire bunch is accelerated to the end of acceleration cycle, behave as shown in Fig. 5. In the energy sweep extraction mode, the flat edge of V_{acc} pulse is moved to the rising edge of positive V_{bb} pulse just in 1 turn at the extraction timing. Insufficient magnitude of the acceleration voltage allows an artificial and continuous leak of a fraction of macro-particles beyond the assumed extraction timing.

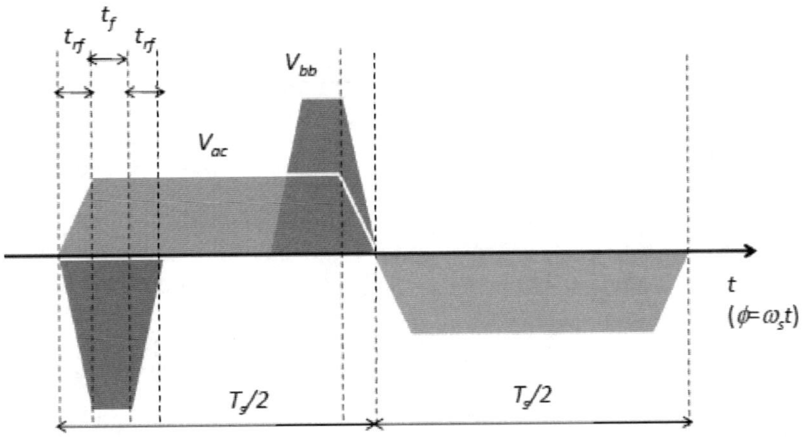

Figure 4. V_{ac} and V_{bb} profile in time before extraction, where T_s and ω_s are the revolution time period and angular revolution frequency of the synchronous particle t_{rf} is the rising and falling time period of both voltage.

IOP Conf. Series: Materials Science and Engineering **785** (2020) 012010 doi:10.1088/1757-899X/785/1/012010

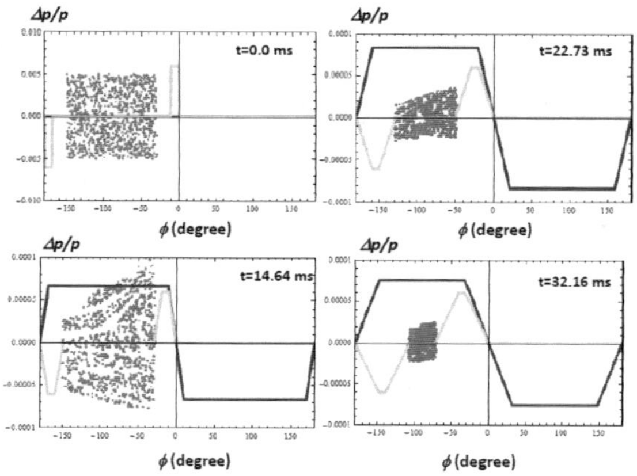

Figure 5. Phase plots of tracked macro-particles in the phase space with V_{ac} *(Blue)* and V_{bb} *(Green)*, where the voltage heights are shown in a relative unit.

3.2. Spill drop from the barrier bucket

For this simulation, the gate signals for the V_{ac} voltage pulses are changed beyond the starting time of extraction so that the flat edge of V_{ac} becomes to be equal to the rising bottom of positive V_{bb}. Particles entering into the positive barrier region, where the V_{ac} voltage profile has a negative slope are affected by an insufficient acceleration voltage. As the result, a lager negative momentum deviation is generated. These particles leave the barrier bucket region or trapped region to move downward further. Eventually, they arrive the boundary region of $\Delta p/p$= -10^{-2}, beyond which a particle entering into the electrostatic septum region is kicked inward in the horizontal direction by the electrostatic fields. Typical examples of the phase plot with the spill drop from the barrier bucket are shown in Fig. 6. One finds that the small fraction of macro-particles continuously drifts down in the momentum space. At this stage, the spill is uncontrolled because any parameters are not optimized.

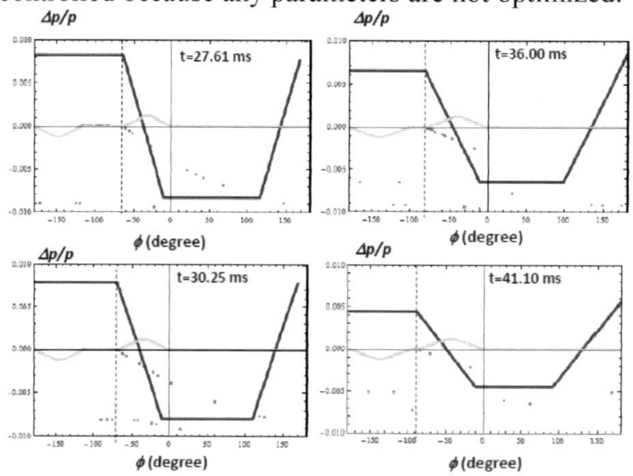

Figure 6. Phase plots of macro-particles leaving the barrier bucket region, where the bunch core is invisible, because its momentum spread is quite small. The position of broken lines, from which particles leak must be noted.

4. Spill Control

The spill size can be controlled, where the turning-off time of V_{ac} is changed in a programmed manner. If some of particles being trapped in the barrier bucket are not given a required energy matching to the guiding magnet pattern, they will leave the trapping region or the barrier bucket and drifting downward in the longitudinal phase space. This is a key feature of spill control. Acceleration is uniquely determined by V_{ac}. Its downhill profile in time is steep and its starting phase of falling-down in V_{ac}, which is indicated by ϕ_{ext} in Fig. 7, is maneuvered by gate-control for the switching power supply. This timing is always adjusted in the programmed manner or by means of feed-back from the spill monitoring system. The procedure for spill control is shown below.

- ϕ_{ext} is instantaneously moved to near the left edge of
- the right barrier voltage pulse (ϕ_{bb}) at the starting time of spill extraction.
- Then, ϕ_{ext} is changed to satisfy the expected spill profile in the programmed manner or watching an actually extracted spill profile.

As shown in Fig. 8, The integrated spill profile as a function of time for three extraction cases with different timing of V_{ac} but constant ϕ_{ext} and 1,000 macro particles. It is clear that the spill profile depends on the extraction parameter δ. This parameter can be controlled to meet a requirement on the spill profile. Such a typical example is given in Reference 3, where the constant spill is maintained up to the end of acceleration. Actually the programmed spill profile is realized by a program loaded on the field programmed gate array (FPGA). Real time feedback control may be also possible by watching an actual spill in the acceleration cycle.

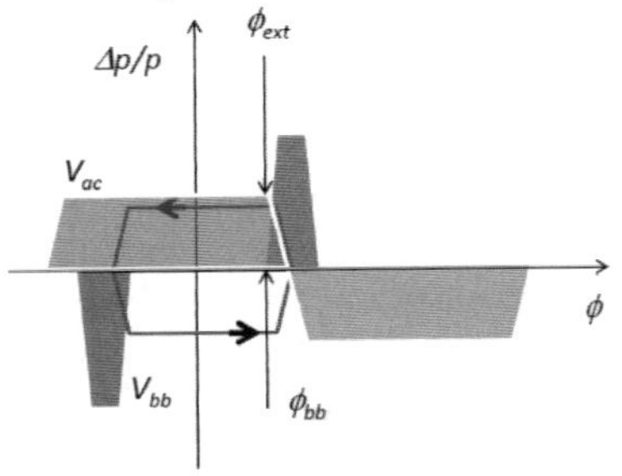

Figure 7. Phase space with the V_{ac} profile adjusted for extraction

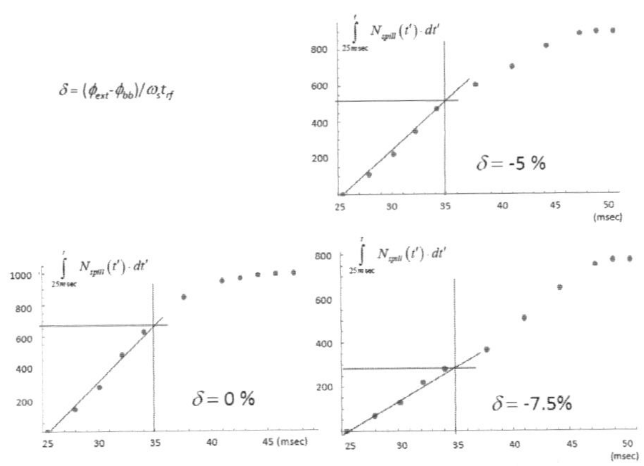

Figure 8. Integrated spill for different δ

5. Crucial devices for energy sweep extraction

From the beam dynamics point of view, it is the most important issue how the extracted beam with different energy in the acceleration cycle is guided to the beam handling region in the transverse direction for the 2D spot scanning along the extraction beam line. If all guiding devices including the extraction devices are ramped in the same way as the main guiding magnets in the ring, the ideal orbit should be common at any extraction timing and the extracted beams can enjoy the same lattice function of the extraction beam line through the entire acceleration cycle. Extraction devices desired for the present purpose of the electrostatic septum and Lambertson magnet, which seem to be also practical from the engineering point of view, are considered here.

5.1 Electrostatic septum

The maximum voltage generated by a DC power supply such as the Cockloft-Walton is usually added across the ES gap [3]. It seems to be difficult to introduce the additional pulse voltage generator in the ES high voltage circuit, because the additional voltage pulse has to have the well controlled time transient profile varying in a wide range of more than 50 kV as to correspond to the extraction from a middle stage in the acceleration cycle.

For this purpose, a novel charging method to employ a combination of high power solid-state switches and discharging switch has been studied [3]. Its equivalent circuit as shown Fig. 9. Its performance in a nominal operation is described as below.

- The DC power supply V_1 always charges the electrostatic septum through the charging resistor R_1 to a voltage of order of -50 kV before extraction.
- Switch 1 transfers charge the load from the other DC power supply V_2 so that the required voltage profile *V(t)* proportional to $\beta\gamma^2$ is satisfied beyond the extraction starting time. Time constant of the voltage build-up is determined by the circuit constants such as the resistance of the charging resistor R_2, floating inductance L, parameters of the coaxial cable connecting between the V_2 and the load, and the capacitance of the electrostatic septum. Switch 1 is discretely and repeatedly tuned on and off in a well programmed way.
- After completing the extraction, the discharging switch S_2 is turned out until the voltage of the load reaches lower than V_1.
- Just after the acceleration cycle, the above process is repeated.

In order to confirm the feasibility of this programmed charging process, the SPICE simulation has been carried out, especially focusing on the charging process from V_1 to V_2. The SI-Thyristor is assumed as a solid-state switching element for S_1, which has a rising/falling time of 200 nsec. Finite but relatively fast switching characteristics give no notable effects on the charging performance of current concern. Simulation results are shown in Fig. 11, where R_1, L, and C are assumed to be 9.58 MΩ, 1μH, and 1200 pF, respectively. It is noted that the capacitance of 1200 pF is very close to that of the existing KEK 12 GeV PS electrostatic septum with the similar size as the present demand. S_1 is turned on and off 8 times during the assumed extraction time period, as shown in Fig. 10. It is emphasized that the discrepancy from the ideal voltage profile is less than 1%. Further improvement seems to be possible by introducing more frequent switching performance.

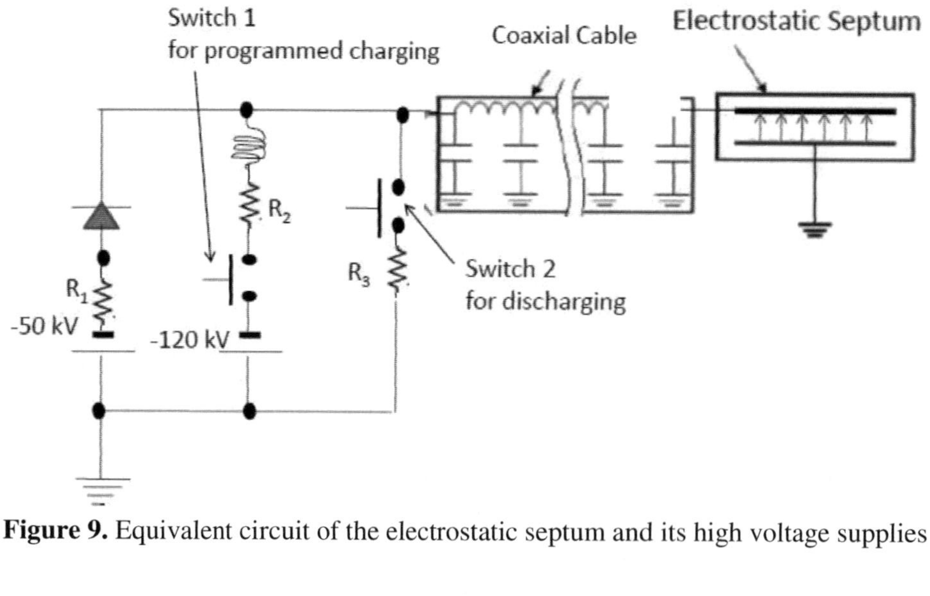

Figure 9. Equivalent circuit of the electrostatic septum and its high voltage supplies

The switching performance is simulated by using a simulation code, SPICE.

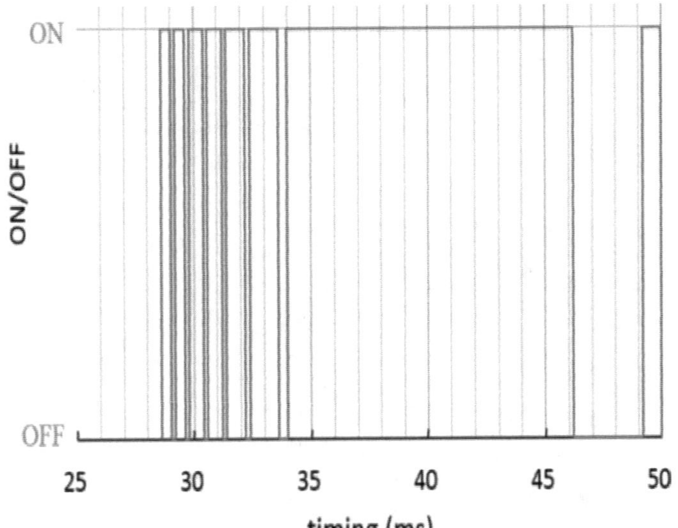

Figure 10. Time table of Switch On/Off

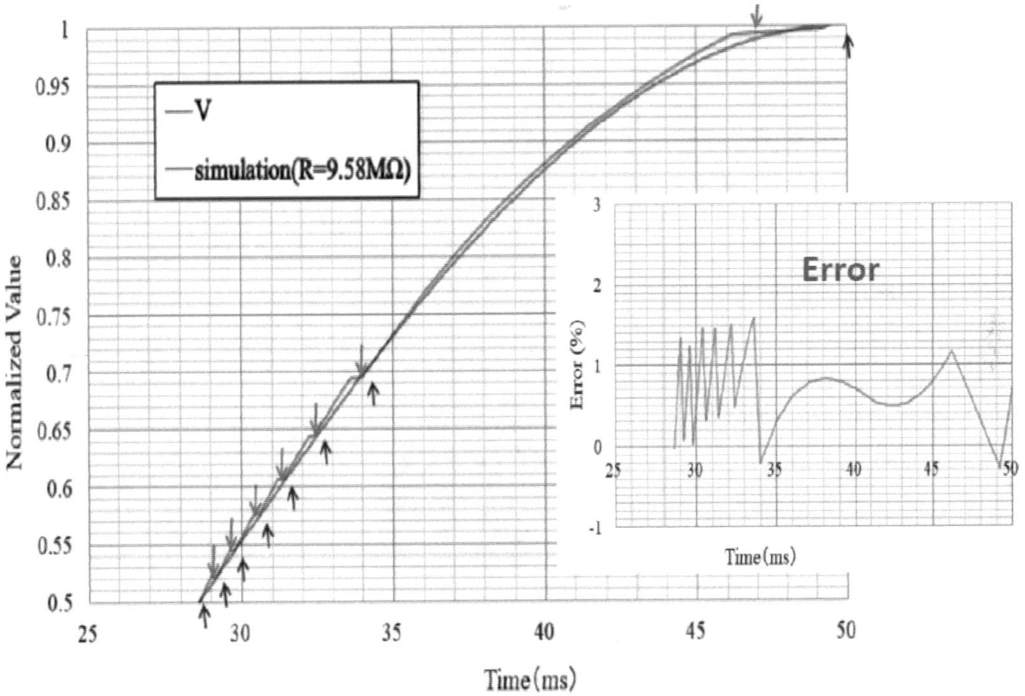

Figure 11. Ideal and simulated voltage patterns with the discrepancy between both

5.2 Lambertson magnet

A usual septum magnet is excited by the transient current, which is generated in the circuit architecture consisting of a condenser bank, resistor, diode, and thyratron as a switch, and transformer (magnet itself). The field profile is uniquely determined by the circuit parameters and switching time of the thyratron [5]. The peak field is used for extraction. Its time constant of change in the magnetic fields is different from 10 Hz or does not meet the current demand. It is unknown whether or not additional circuit components or parameters can be chosen to meet the demand on the ideal extraction field profile.

Here a Lamberston magnet as seen in Fig. 12 is considered as an extraction magnet device instead of the septum magnet. It is emphasized that the Lambertson magnet is excited in a similar way as that of the main magnets of the ring. This is a crucial point why this magnet is used in the present proposal. The Lamberston magnet extracts the beams in the vertical direction in a limited space. However, the angle and position of the beam line at the edge of the magnet are still interfere to the adjacent quardrupole magnet (QF1) and bending magnet as seen as in Figs.13 and 14. In such a case, the beam aperture for extraction must be put in the magnet return yoke. Perturbations of this modification on the circulating beam are limited and their compensations are possible.

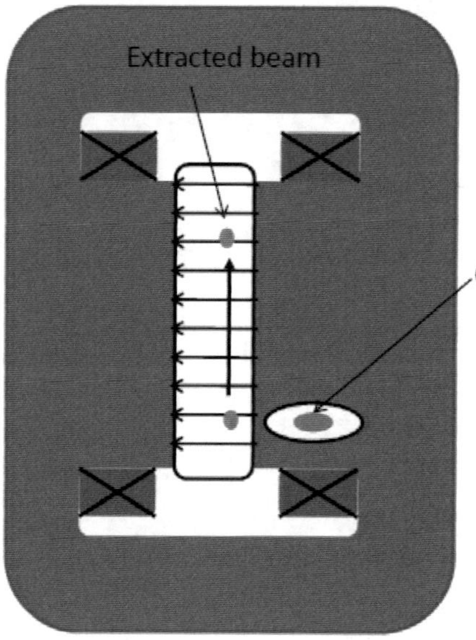

./

Figure 12. Cross-section of Lambertson magnet.

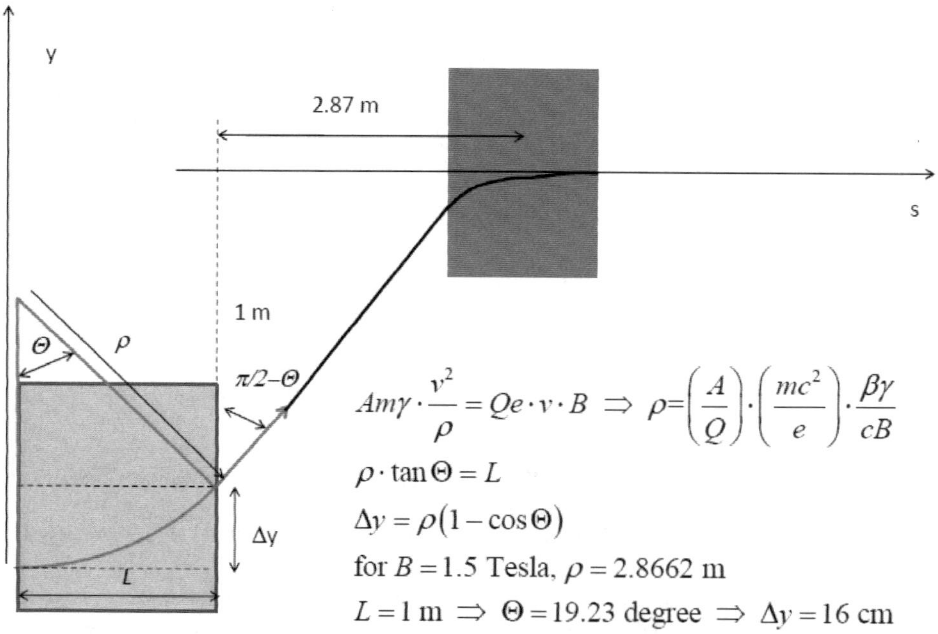

Figure 13. Extraction beam orbit and downstream.

IOP Conf. Series: Materials Science and Engineering **785** (2020) 012010 doi:10.1088/1757-899X/785/1/012010

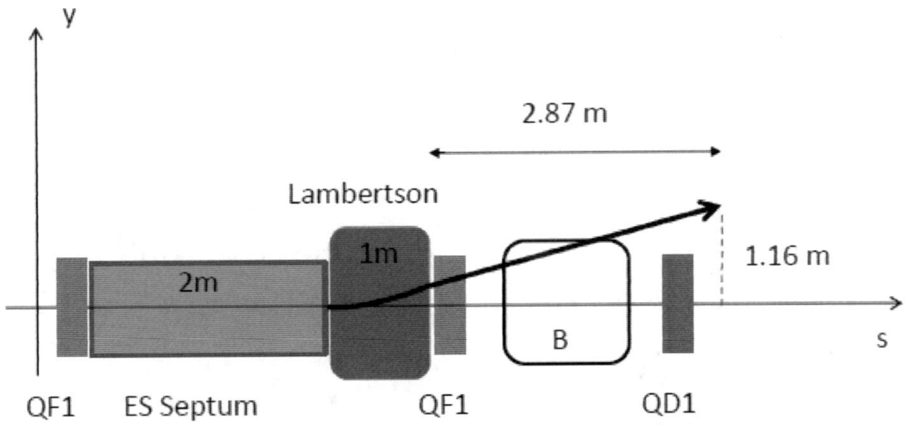

Figure 14. Extraction orbit on the vertical plane

6. Conclusion

The hadron machine so-called ESCORT with continuous energy scanning for cancer therapies has been designed and its performance has been confirmed by the macro-particle simulations. We believe that this is realized utilizing the modern technology of variable voltage power supply employing a solid-state switch. However, there are still several issues related with beam transverse emittance that must be studied.

7. References

[1] Hiramato K 2009 *Workshop on Hadron Beam Therapy of Cancer* (Erice, Italy)

[2] Takayama K, Yoshimoto T, Barata M, Leo K W, Liu X, Iwashita T, Harada S, Adachi T, Arai T, Arakawa D, Asao H, Kadokura E, Kawakubo T, Nakanishi H, Okada Y, Okamura K, Okazaki K, Takagi A, Takano S and Wake M 2014 *Phys. Rev. ST-AB* **17** p 010101.

[3] Leo KW, Adachi T, Kwakubo T, Monma T, Dixit T, and Takayama K 2015 A compact energy hadron driver for cancer therapies with continuous energy sweep scanning *Phys. Rev. ST-AB.*

[4] Edwards D A and Edwards H T 2008 Particle Colliders for High Energy Physics *Rev. of Accel. Sci. and Tech.* (World Scientific) **1** p 99–120

[5] Proto-ion medical machine study (PIMMS), CERN/PS 2000-007 (DR) (2000). https://cdsweb.cern.ch/record/449577/files/ps-2000-007.pdf

[6] Dougall N J, Bruggink S, Ebmeier K P 2004 *Am. J. Geriatr Psychiatry* **12** p554-70.

Radiological and heavy metals analysis in soil and water at the former tin mining site in the urban area (Danau Kota Lake, Kuala Lumpur)

N H Abdul Hamid[1,2], N Mohd Hidzir[1,2,a)], N A Mohd Radzali[1,2] and A K Mokhtar[1,2]

[1] Nuclear Technology Research Centre, Faculty of Science and Technology, Universiti Kebangsaan Malaysia.
[2] Department of Applied Physics, Faculty of Science and Technology, Universiti Kebangsaan Malaysia.

[a]syahidah@ukm.edu.my

Abstract. Former mining sites are known to be concentrated with enriched naturally occurring radioactive materials (NORM) and heavy metals, as a result of past mining activities. A number of the sites are reclaimed as an open area, which in turn raised concerns regarding public safety. Thus, it is crucial to evaluate the safety of the public engaging in activities around the former tin mining areas. A study is carried out to determine the concentrations of NORM and heavy metals in soil and water collected from Danau Kota Lake, Kuala Lumpur, known as a former mining site. Both NORM and heavy metals were analysed using gamma spectrometry and inductively coupled plasma mass spectrometry (ICP-MS), respectively. The average concentration of ^{226}Ra and ^{232}Th in water collected was 0.7 ± 0.2 Bq/L and 0.1 ± 0.03 Bq/L, respectively. ^{40}K, however, was not detected in the water. The average activity concentration of ^{226}Ra, ^{232}Th, and ^{40}K in soil was 60.8 ± 2.3 Bq/kg, 42.9 ± 1.4 Bq/kg, and 338.4 ± 11.7 Bq/kg, respectively. From the values obtained, it is observed that the activity concentrations for both soil and water are still within the range limit reported by WHO and UNSCEAR. In regards to heavy metals analysis, manganese concentration was the highest in both water and soil with a value of 1.18 μg/L and 147.7 mg/kg, respectively. To evaluate the radiological hazard, the radium equivalent activity index (R_{eq}), gamma index (I_γ), absorbed dose rate (D), effective dose rate (D_{eff}) and external hazard index (H_{ex}) are calculated and valued at 148 Bq/kg, 1.1 Bq/kg, 69 nGy/h, 0.1 mSv/year and 0.4, respectively. All radiological hazard indices are found to be below the safety limits. Therefore, the lake area is considered safe for public activities except for I_γ.

1. Introduction

Abandoned mining sites in Malaysia are actively reused or redeveloped into various land uses such as recreational parks, housing areas, and agricultural farms. The initiative of reusing those mining sites helps in restoring the land into a more economically usable state in this country. In spite of that, the soil disturbed throughout past mining activities contain enriched concentrations of naturally occurring radioactive materials (NORM) and heavy metals in the soil. The NORM and heavy metals in the soil arise through processes such as extraction of metals and soil smelting. The environmental degradation that occurs may pose detrimental impacts on the public's health. According to Zalina *et al.*, tin mine sites contains a higher NORM concentration than other areas [1]. Additionally, people are at the risk of being exposed to hazardous gamma rays produced by NORM in soil, such as ^{238}U and its decaying products, ^{226}Ra, ^{232}Th and ^{40}K.

A lot of research regarding the impact of NORM and heavy metals on human health was conducted in the past, indicating how pressing the problems are when it comes to contamination of NORM and heavy metals towards the environment [2-5]. International bodies such as the World Health Organization (WHO), Environmental Protection Agency (EPA) and United Nations (UN) have developed many guidelines for environmental monitoring and assessment for NORM and heavy metals contamination that can threaten human health. The effects of heavy metal exposure such as lead, cadmium, mercury, and arsenic cause damages to the kidneys and bones and poses the risk of cancer [6]. The natural radionuclide and heavy metal pollution will then accumulate in aquatic life and subsequent plants to humans if taken as a source of nutrition [7].

Recreational lakes are more susceptible to the effect of urbanisation and anthropogenic activities due to their geographic areas, in which they are commonly located at the center of commercial areas [8]. In the case of recreational parks developed from mining lands, the human health risk towards hazards in soil and water via direct and indirect exposure are not be taken lightly. The pathway of toxic chemicals exposure changes depending on the type of actions performed and the degree of contact. Recreational activities can be categorised into two types, whole-body contact sports and non-body contact sports. Example of whole-body contact sports are swimming, surfing, and canoeing, whereas, non-contact sports are like fishing, walking, birdwatching, and picnicking (incidental contact). Generally, the exposure of skin and whole-body contact activities increases with the amount of probable water ingested. Most outcomes on matter regarding health problems arise due to activity exposure from swimming activities and the ingestion of water. Hence, WHO drinking water quality has been used as guidelines in screening-level risk assessment and monitoring recreational exposure [9].

It is evident how vital NORM and heavy metals studies are in ensuring the safety of the lake area for public uses. Therefore, the present work intends to determine the content of NORM and heavy metals in water and soil at Danau Kota Lake, Kuala Lumpur, which is a newly re-established recreational area from former tin mines that operated during the 1980s. Besides that, the radiological hazard assessment was conducted. To evaluate the safety of the lake area, the results obtained were further compared to the range limits reported by several credible guidelines.

2. Experimental

2.1. Materials and method

Water and soil samples were collected from six different stations represented as T1, T2, T3, T4, T5, and T6 as depicted in figure 1. The average concentration of NORM and heavy metals from each station represents the total concentrations of the lake area. The results obtained for concentration of NORM and heavy metals in both water and soil sample were compared with several guidelines such as National Lake Water Quality Criteria and Standards Malaysia (NLWQS), National Water Quality Standards Malaysia (NWQS), guidelines for Drinking Water Quality-World Health Organisation (WHO 2017), Alberta Soil Remediation guidelines, guidelines for safe recreational water environments (WHO 2003), Nuclear Energy Agency (NEA-OCED) Reports 1979, National Council on Radiation Protection and Measurements (NCRP) Reports 1999, Annex B- Exposures from natural radiation sources (UNSCEAR 2000) and Putrajaya Lake Water Quality Standards (PLWQS 2013) to evaluate the safety of the lake area.

2.2. Pre-treatment of samples

For soil samples, 2 kg of soil was taken from each station. The soil sample was taken in a diameter of 30 cm according to IAEA 2004 [10]. The physical parameters of water include pH, conductivity, dissolved oxygen, and temperature were determined using a pH meter (WITEG W 100), a conductivity meter (YSI SCT Model 33), and an oxygen meter (HANNA instruments HI 91942), respectively.

In NORM analysis, the water sample was filtered using a plasma membrane. 2000 mL of water was concentrated by heating until its volume decreases to 200 mL. For soil samples, it was weighed and dried in an oven before sieving it through 800 μm siever to obtain a uniform sample size. The water and soil samples were kept in polyethylene bottles (7 cm diameter × 9 cm height) for 30 days until a secular equilibrium state is achieved before it is further analysed by a Highly Pure Germanium (HPGe) detector [11].

For heavy metals analysis, 100 mL of water sample was filtered using filter paper, whereas, soil samples were digested by adding 6 mL of hydrochloric acid and 4 mL nitric acid and was further digested in Microwave Digestive System for two hours. The clear solution formed was kept in a bottle before being analysed by Inductively Coupled Plasma-Mass Spectrometry (ICP-MS) [12].

Figure 1. Geographical map of (a) Peninsular Malaysia, (b) Selangor state, and (c) location of the sampling site.

2.3. Characterisation

The concentration of NORM in water and soil samples was measured using a High Purity Germanium (HPGe) detector (Canberra Model GC1018) with the efficiency of 30% at Nuclear Science Laboratory UKM. The concentration of NORM in the water sample can be calculated using Equation 1:

$$A_s(\frac{Bq}{L})=\frac{N}{\mathcal{E} \times P\gamma \times V}, \quad (1)$$

where N is the corrected net peak area background (cps), $\mathcal{E}$ is the efficiency of the detector for a γ-ray interest (cps/Bq), $P\gamma$ is the branching ratio of elements (%), and V is the volume of the water sample (L). The concentration of NORM in the sample was determined from net count of the sample using multichannel analyser (Software Genie 2000) by referring to the peak energy for ^{226}Ra (609 keV and 1764 keV) for soil and water, respectively, ^{232}Th (2614 keV), and ^{40}K (1460 keV) [13]. The NORM in soil sample can be calculated using the comparative method where the counts obtained by the sample are compared with the standard sample from IAEA. The concentration of the NORM in the sample can be calculated using Equation 2:

$$W_s(\frac{Bq}{kg})=\frac{M_P}{M_s}\times\frac{A_s}{A_p}\times W_p\,, \quad (2)$$

where W_S is the concentration of the radionuclide in the soil sample, M_P is mass of the sample, M_S is mass of the standard, A_P is the activity of the sample, A_S is the activity of the standard, and W_P is the concentration of radionuclide in the standard sample. Heavy metals in water and soil samples were determined using the ICP-MS model (ELAN® DRC-E, Perkin Elmer Elan 9000) at Environmental Science Laboratory UKM. Five heavy metals were analysed which were Manganese (Mn), Arsenic (As), Copper (Cu), Nickel (Ni), and Lead (Pb).

2.4. Radiological hazard analysis

In the present study, six quantities were recognized as the parameters to assess the radiological safety of the lake site. First, the radium equivalent activity index (Ra_{eq}) was introduced to compare the specific activity of a sample with different concentrations of ^{232}Th, ^{226}Ra, and ^{40}K. Ra_{eq} is defined as the total activity of the three radionuclides based on the estimation that 370 Bq/kg of ^{226}Ra, 259/kg of ^{232}Th, and 4810 Bq/kg of ^{40}K produce the same dose rates. The upper limit for Ra_{eq} is 370 Bq/kg. Ra_{eq} was determined using formula 3. Gamma Ray Index (I_γ) estimates that the value of gamma ray is correlated with the value of the natural radionuclide in soil-375 and can be calculated using formula 4. External Hazard Index (H_{ex}) is widely used to show external exposure. H_{ex} can be calculated using formula 5. The safe value for I_γ and H_{ex} is ≤ 1 [14-16].

Using Equation 6, Absorb dose (D) was measured using the conversion factor published by UNSCEAR 2000 with a height of 1 meter above the ground [17]. Next, Annual Effective dose (D_{eff}) was determined to consider the outdoor occupancy factor and coefficient of absorbed dose and can be calculated using Equation 7:

$$Ra_{eq}(\frac{Bq}{kg}) = A_{Ra} + 1.43A_{Th} + 0.077A_K, \quad (3)$$

$$I_\gamma = \frac{ARa}{150} + \frac{A_{Th}}{100} + \frac{A_K}{1500} \leq 1, \quad (4)$$

$$H_{ex} = \frac{ARa}{370} + \frac{A_{Th}}{259} + \frac{A_K}{4810} \leq 1, \quad (5)$$

$$D(\frac{nGy}{h}) = 0.462A_{Ra} + 0.621A_{Th} + 0.0417A_K, \quad (6)$$

$$D_{eff}(\frac{mSv}{t}) = D(\frac{nGy}{h}) \times 8760(\frac{h}{t}) \times 0.2 \times 0.7(\frac{Sv}{Gy}) \times 10^{-6} \quad (7)$$

3. Results and discussion

3.1. Physical parameters of water

Physical parameters studied were pH values, water conductivity, temperature, and dissolved oxygen (DO). Results indicate that the lake water is alkaline with a pH of 8.8 ± 1.4. The average value for water conductivity, dissolved oxygen and the temperature of the lake water during the sampling is 38 ± 1, 4.13 ± 13 mg/L, and 32 ± 1 °C, respectively. The pH does not give a direct impact on recreational users, as there are no swimming activities conducted there.

3.2. NORM and heavy metals concentration

The comparison between the concentration of NORM and heavy metals in soil and water samples in Danau Kota's lake and other lakes in Malaysia are shown in Table 1. The concentration of NORM in water shows that the concentration of ^{226}Ra (0.77 ± 0.22 Bg/L) was the highest, followed by ^{232}Th (0.1 ± 0.03 Bq/L). Meanwhile, ^{40}K was below detection reading where the value was almost similar to the background reading. Guidelines from Putrajaya Lake Water Quality Standards (PLWQS) and National Lake Quality Water Standard (NLQWS) dictate that the concentration of radium in the lake water should be less than 0.1 Bq/L [18, 19]. The concentration of ^{226}Ra is shown to be higher than the guidelines. Referring to the World Health Organization (WHO) guidelines, the concentration values of the samples are lower than the permitted value [20]. The value of NORM in water, excluding ^{40}K, from this study was observed to be slightly higher than the study conducted at KUIS lake, which is also a former mining site [12]. Although fishing competitions are frequently conducted, no body contact activities such as swimming are conducted here. Thus, the dermal contact contributes very little exposure. For soil samples, the results show that all of the concentrations of NORM are below the average value stated by UNSCEAR 2000 except for ^{40}K. ^{40}K shows the highest concentration followed by ^{226}Ra and ^{232}Th. The high value of ^{40}K might be due to the natural composition of ^{40}K in the earth's crust and the use of NPK fertilizer in grass planting during the redevelopment of this recreational park. Alsaffar et al. (2016) explained that the increase of concentration of NORM in soil with the use of fertilizer enhance the soil properties for plant growth [21].

Five heavy metals that show significant concentration in both water and soil samples are shown in Table 1. Those five heavy metals are manganese (Mn), Arsenic (As), Nickel (Ni), Copper (Cu) and Lead (Pb). The concentration of heavy metals in both water and soil samples show similar trends where the concentrations are found in the order of Mn > As > Cu > Ni > Pb. The highest concentration is Mn, and the lowest is Pb. The concentration of Mn is still below the guidelines values from NLQWS and no health value guidelines made by WHO for Mn for drinking water as the concentration of Mn is not of health concerning levels. The high concentration of Mn presence in the water and soil samples was probably due to the natural abundance of existing Mn in the environment [22]. As, Ni, Cu, and Pb in water were found below the safe guidelines value provided by NLQWS and WHO.

All the heavy metals in soil samples except for As, do not exceed the values stated by Alberta Soil Remediation guidelines. The calculated heavy metal concentrations in this study also show a higher concentration of Mn and Ni when compared to that of Bestari Jaya, also a former mining area. The concentration for As obtained in this study and that of Bestari Jaya are almost similar and both concentrations exceed the Alberta Guidelines value. The concentration of As in this study is within the contamination level, thus poses a health risk to the lake users. Past mining activities can contribute to the high concentration of As in the collected soil samples. A study by Trajce Stafilov et al proved that mining activities increase the concentration of As in the soil [23].

Referring to National Lake Water Quality Criteria Standards (NLWQS) Malaysia 2015, taking physical parameters, NORM and heavy metals concentrations in the water into account, Danau Kota Lake can be classified into category D. This is due to its good preservation of aquatic life and effective integrated lake management strategies conducted. The guidelines from Interim National Quality Water Standard (INQWS) Malaysia states that lake water is classified into class III, where fishing activities are allowed. Extensive treatment is also needed if the lake is to be used as a source of water supply [18].

Table 1. Concentration of NORM and heavy metals in water and soil samples.

Type of samples	Samples	Concentration of NORM (Bq/L^* or Bq/kg^{**})			Concentration of Heavy metals ($\mu g/L^*$ or mg/kg^{**})				
		^{226}Ra	^{232}Th	^{40}K	Mn	As	Ni	Cu	Pb
Water	This study	0.77 ± 0.22	0.1 ± 0.03	bdl	1.18	0.71	0.04	0.14	0.01
	KUIS lake [12]	0.19 ± 0.10	0.04 ± 0.02	0.62± 0.37	bdl	0.58	-	0.49	-
	NLWQS [24]	<1	-	-	100	400	50	20	50
	WHO [20]	1	1	-	-	10	70	2000	10
Soil	This study	60.8 ± 2.3	42.9 ± 1.4	338.4 ± 11.7	147.7	66.3	9.9	13.5	0.2
	Bestari Jaya [25]	-	-	-	85	79	6.7	125	105
	Alberta [26]	300	300	1700	-	17	45	63	140
	UNSCEAR 2000 [17]	67	82	310	-	-	-	-	-

* unit for water samples
** unit for soil sample
bdl below detection limit

3.3 Radiological hazard analysis

Radiological hazard analysis for the soil was further evaluated to study possible influence of NORM towards human upon direct contact with the soil while engaging in recreational activities. Table 2 presents that all radiological hazards analysis show safe values except for I_{γ}. The recorded Ra_{eq} was 225.5 Bq/kg, which is less than the recommended maximum levels by NEA-OCED (370 Bq/kg) [27]. Similar trend was observed for the H_{ex} with values less than the unity [28]. In addition, the D_{eff} value recorded in this study is lower than the range proposed by the UNSCEAR 2000 (1 mSv/y).

Value of D obtained in this study is lower than the Malaysian average (93 nGy/h) and slightly higher than the world average (59 nGy/h) as reported by UNSCEAR 2000 [17]. In comparison with a study conducted by Muneer Aziz et al. at eastern region peninsular Malaysia (172 ± 90 nGy/h), the value of D obtained in this study is lower [29]. However, I_{γ} was measured to be ≥ 1, with an average of 1.1 Bq/kg that exceeded the unity value as recommended by the European Commission which indicates that the soil is not safe to be used as building materials for construction [16]. All radiological hazard analysis in this study is seen to be lower than the results from KUIS lake study.

Table 2. Radiological Hazard Analysis.

Location	Ra_{eq} (Bq/kg)	I_{γ}	H_{ex}	D (nGy/h)	D_{eff} (mSv/y)
This study	148	1.1	0.4	69	0.08
KUIS Lake **[12]**	225	1.6	0.6	104.1	0.1
Recommended value	370 NEA-OCED [27]	≤1 EU [16]	≤1 NCRP [28]	93 UNSCEAR [17]	1-10

4. Conclusion

Results from this study indicate that the lake site is safe for public use. Most of the concentration of NORM and heavy metals in water and soil samples are below the average values stated in the health guidelines. Despite the high concentration of ^{226}Ra in water samples and As in soil samples, the site is still considered safe for public use due to the absence of direct contact activities. The soil sample is not suitable to be used as building materials based on the I_{γ}, which exceeded the safe value. For future studies, sediment inside the lake can be a good potential in determining the concentrations of NORM and heavy metals. This, in return, will help analyse the safety of the lake as most of the toxicants found in the water will deposit into the sediment. Moreover, studies on the aquatic life samples will provide useful information regarding the transfer factor of the NORM and heavy metals.

5. Reference

[1] Rahmat Z *et al.* 2010 *Sci. Malaysiana* **39** p 599-605.
[2] Zhang X-J *et al.* 2019 *J. of Chem. Eng. of Japan* **52** p 362-368.
[3] Hashim M *et al.* 2018 *Indonesian J. of Geo.* **50**.
[4] Paraguassú L *et al.* 2019 *Env. Earth Sci.s* **78** p 1-10.
[5] Dike C G *et al.* 2019 *Human and Eco. Risk Assess. An Int. J.* p 1-11.
[6] Rahman Z and Singh V P 2019 *Env. Monit. Assess.* **191** p 419.
[7] Schmus W R V 1995 *Global Earth Geophys.: A Handbook of Physical Constants* (USA: American Geophysical Union) p 283-291.
[8] Fazli B *et a.l* 2018 *Lakes & Reservoirs: Research & Management* **23** p 24-33.
[9] WHO 2003 *Guidelines for safe recreational water environments* **1** ed WHO (Geneva: WHO) p 218.
[10] IAEA 2004 *Soil sampling for environmental contaminants* ed IAEA (Vienna: IAEA) p 75.
[11] Yasir M S *et al.* 2008 *Malaysia J. Anal. Sci.* **12** p 172-178.
[12] Radzali N A M, Hidzir N M, and Rahman I A 2018 *Malaysia J. Anal. Sci.* **22** p 828-838.
[13] Hassan N N and Siong K K Year *Conf. Proc. on Meas. of natural radioactivity and assessment of radiation hazard indices in soil samples at Pengerang, Kota Tinggi, Johor* **1584** ed A A Mohamed *et al.* (Kuala Lumpur: AIP Publishing) p 190-5.
[14] Diab H M *et al.* *J. Nucl. Radiat. Phys.* **3** p 53-62.
[15] Guptaa M *et al.* 2010 *Indian J. Pure Appl. Phys.* **48** p 482-5.
[16] Commission E 1999 *Radiological Protection Principles Concerning the Natural Radioactivity of Building Materials - Radiation Protection* **112** ed N Safety *et al.* (Geneva: E Commission) p 16.
[17] UNSCEAR 2000 *Annex B-Exposures from natural radiation sources* **1** ed UN (New York: U N Publication) p 654.
[18] DOE 2000 *National Water Quality Standards for Malaysia* **1** ed WEPA (Malaysia: WEPA) p 6.
[19] Baginda A R A 2013 *Putrajaya Lake & Wetland: Water Quality Status*.
[20] WHO 2017 *Guidelines for drinking water quality* **4** ed WHO (Geneva: WHO) p 499.
[21] Alsaffar M S *et al.* 2016 *Pollution* **2** p 10.
[22] Standing D *et al.* 2006 *Modern Soil Microbiology* **2** ed J D V Elsas *et al.* (Boca Raton: CRC Press) p 646.
[23] Stafilov T, Aliu M, and Sajn R 2010 *Int. J. Environ. Re.s Public Health* **7** p 4050-61.
[24] Suratman S, Sharip Z, and Majizat A 2015 *National lake water quality criteria and standards* **1** ed NAHRIM *et al.* (Malaysia: NAHRIM *et al.*) p 25.
[25] Ashraf M A, Maah M J, and Yusoff I B 2010 *Int. J. Basic Appl. Sci.* **10** p 7-27.
[26] Alberta 2016 *Alberta Tier 1 Soil and Groundwater Remediation Guidelines* **1** ed Alberta (Edmonton: Alberta) p 197.
[27] NEA and OECD 1979 *Exposure to radiation from the natural radioactivity in the building materials* **1** ed C O *et al.* (Paris: OECD) p 34.
[28] NCRP 1999 *Recommended screening limits for contaminated surface soil and review of factors relevant to site-specific studies* **129** ed NCRP (Bethesda: NCRPM) p 353.
[29] Muneer Aziz Saleh *et al.* 2019 *Radiochimica Acta* **107** p 503-516.

Acknowledgement

The authors wish to thanks Faculty of Science and Technology, UKM especially Nuclear Science Programme, GUP-2017-085 and FRGS/1/2019/STG02/UKM/02/6 for the funds. The authors also acknowledge the facilities of the School of Environmental Science, UKM for assistance with ICP-MS analysis.

IOP Conf. Series: Materials Science and Engineering **785** (2020) 012012 doi:10.1088/1757-899X/785/1/012012

Determination of factors affecting indoor doses from Malaysia's ceramic tiles containing natural radionuclides

Shittu Abdullahi[1,2] and Aznan Fazli Ismail[2,3,a)]

[1]Department of Physics, Faculty of Science, Gombe State University, P.M.B. 127 Gombe, Nigeria.
[2]Nuclear Technology Research Centre, Faculty of Science and Technology, Universiti Kebangsaan Malaysia 43600 Bangi, Selangor, Malaysia.
[3]Centre for Frontier Sciences, Faculty of Science and Technology, Universiti Kebangsaan Malaysia 43600 Bangi, Selangor, Malaysia.

[a]*aznan@ukm.edu.my & shittub2k@gmail.com*

Abstract. The level of natural radioactivity and radiological risks attributed to forty (40) different ceramic tiles using gamma-ray spectroscopy employing high-purity germanium detector were studied. The average activity concentrations of ^{226}Ra, ^{232}Th and ^{40}K range from 37.5 ± 0.3 to 215.9 ± 5.8 Bq kg^{-1}, 42.2 ± 0.1 to 181.8 ± 3.8 Bq kg^{-1} and 349.5 ± 25 to 1589.2 ± 21.1 Bq kg^{-1}, respectively. The radium equivalent activity and activity concentration index were calculated to estimate the potential radiological hazards to the dweller. Furthermore, the factors affecting the total effective dose equivalent (TEDE) such as ventilation rate, room size and dweller position were investigated from the measured activity concentrations using RESRAD-BUILD computer code. The simulation results of TEDE from the variations of ventilation rate in a room range from 0.26 ± 0.01 to 0.61 ± 0.01 mSv y^{-1}, on the other hand, the percentage variation of TEDE due to dweller position and room size are 35% and 33%, respectively. The calculated radiological risks parameters were all below the recommended maximum limit. Therefore, the radiological impact attributed from the ceramic tiles under study is negligible.

1. Introduction

Natural background radiations are responsible for over 80% of the radiation exposure received by members of the public [1]. The critical ones' present in the environment are Uranium (^{238}U) and Thorium (^{232}Th) decay progeny and isotope of Potassium (^{40}K). These radionuclides are naturally found in all earth materials including soil, rock and building materials. However, the radiation exposure from natural radionuclides are unavoidable since they can be almost found everywhere. Building materials such as tiles can significantly increase exposure due to natural radionuclides. Moreover, the level of contributions of the exposure from building materials to the general public depends on the selection of raw materials, quantity of each material in the building, geological origin, manufacturing process and added opacifier during industrial processes. The emphasis here is ceramic tiles which are mostly used for decorative purposes in both indoor and outdoor environment. Ceramic tiles are produced from earth materials by pressing into shape and firing at high temperature [2]. The surface of the tile may be glazed or unglazed depending on the desired needs [3]. The important properties of ceramic tiles are less water absorption level (< 0.5 to 10%), high chemical and mechanical properties [2–4]. Therefore, the aim of this study is to determine the factors affecting the indoor doses attributed to Malaysia's ceramic tiles containing natural radionuclides.

2. Material and Method

The materials under study are ceramic tiles sourced from various locations in Malaysia. The number of the samples considered in the study are forty (40). The samples were cleaned and dried in an oven at 105 °C for 48 hours until constant weight is achieved. The choice of the temperature and time were in accordance with International Atomic Energy Agency (IAEA) [1,5,6]. The samples were grinded, sieved using 500 μm mesh, weight, packed and sealed in an air tight merinelli beaker. The samples were then kept in a laboratory for 30 days to achieve secular equilibrium; the steps mentioned are important in the present study, since the activity concentration of ^{226}Ra and ^{232}Th were determined from their daughter nuclides.

The prepared samples were studied using High-Purity Germanium (HPGe) detector located at Nuclear Science Program, Universiti Kebangsaan Malaysia. Each sample was prepared in three replicates and counted for 12 hours. Prior to counting of the samples, energy calibration of the HPGe detector was performed using a standard mixture of ^{22}Na, ^{60}Co and ^{137}Cs radionuclides. Furthermore, the background measurements were also performed by routine counting of empty merinelli beaker for 12 hours and the background contributions were deducted during spectrum analysis of the samples. The activity concentrations of ^{226}Ra and ^{232}Th were determined through gamma-ray energy peaks of 1765 keV (^{214}Bi) and 2615 keV (^{208}TI), respectively. While, the activity concentration of ^{40}K was determined directly from 1460 keV gamma-ray energy peak.

The present study adopted a relative method for determination of ^{226}Ra, ^{232}Th and ^{40}K natural radionuclides. The certified IAEA-375 soil of known activity concentrations was prepared in two replicates counted several times using the preset counting time of 12 hours. The activity concentrations of IAEA-375 soil are 20, 20.5 and 424 Bq kg^{-1} for ^{226}Ra, ^{232}Th and ^{40}K, respectively. The obtained gamma-ray energy peaks of the IAEA-375 soil and its activity concentrations (C_{std}) in Bq kg^{-1} were compared with the obtained gamma-ray energy peaks of the samples with a view to determine the individual activity concentrations of the sample (C_s) in Bq kg^{-1} as shown in Eq. (1) [1][7][8].

$$C_s = \left(\frac{M_{std} \times A_s \times C_{std}}{M_s \times A_{std}}\right) \tag{1}$$

where M_{std} and M_s are Mass of the standard (IAEA-375 soil) and sample in kg, respectively. A_{std} and A_s are activity of the standard and sample in count per second (cps), respectively.

3. Result and discussion

The measured activity concentrations of ^{226}Ra, ^{232}Th and ^{40}K are presented in Fig. 1(a and b) with values ranging from 37.5 ± 0.3 to 215.9 ± 5.8 Bq kg^{-1}, 42.2 ± 0.1 to 181.8 ± 3.8 Bq kg^{-1} and 349.5 ± 25 to 1589.2 ± 21.1 Bq kg^{-1}, respectively. It can be observed that the variations of the natural radionuclides are not uniform across the studied tiles. The observed non-uniform behavior may be associated with different raw materials selection and manufacturing processes. Furthermore, most of the measured activity concentrations in the studied tiles are above the world average values of ^{226}Ra, ^{232}Th and ^{40}K in building materials as reported by United Nations Scientific Committee on the Effects of Atomic Radiation (UNSCEAR) of 50, 50 and 500 Bq kg^{-1} for ^{226}Ra, ^{232}Th and ^{40}K, respectively [9]. On the contrary, the measured activity concentrations are comparable with reported Malaysia's soil activity concentrations of 67, 82 and 310 Bq kg^{-1} for ^{226}Ra, ^{232}Th and ^{40}K , respectively [10].

3.1 Radiological Hazard Parameters

The potential radiological hazard associated with ceramic tiles were studied through radium equivalent activity (Ra_{eq}) and activity concentration index (ACI) as shown in Eq. (2) and Eq. (3), respectively [4][7].

$$Ra_{eq}(Bq\ kg^{-1}) = \left(\frac{C_{Ra}}{370} + \frac{C_{Th}}{259} + \frac{C_K}{4810}\right) \times 370 \quad (2)$$

$$ACI = \frac{C_{Ra}}{300} + \frac{C_{Th}}{200} + \frac{C_K}{3000} \quad (3)$$

where C_{Ra}, C_{Th} and C_K are the activity concentrations of ^{226}Ra, ^{232}Th and ^{40}K in Bq kg^{-1}, respectively. The two radiological hazard parameters are the main parameters used to determine the external radiological hazard due to gamma radiation from building materials. The contributions of building materials to external radiological hazard is considered negligible if the Ra_{eq} in the materials is less than 370 Bq kg^{-1} [7]. On the other hand, the ACI of building materials should be less than the exemption level of 2 for the materials used in limited quantities such as tiles and board, while the action level is 6. However, the building materials used in bulk quantities such as cement, bricks and sand have the same exemption and action level of 1[4,11]. The variations of Ra_{eq} and ACI from the ceramic tiles are presented in Fig. 1(c) and Fig. 1(d), respectively with values ranging from 146 to 565 Bq kg^{-1} and 0.5 to 2.0 for Ra_{eq} and ACI, respectively. The highest and the least value of Ra_{eq} and ACI are found in tile 5 and tile 2, respectively. And, over 90 % of both Ra_{eq} and ACI reported herein are below the recommended maximum limits. Therefore, the external radiological hazards due to gamma radiation from the studied ceramic tiles are negligible.

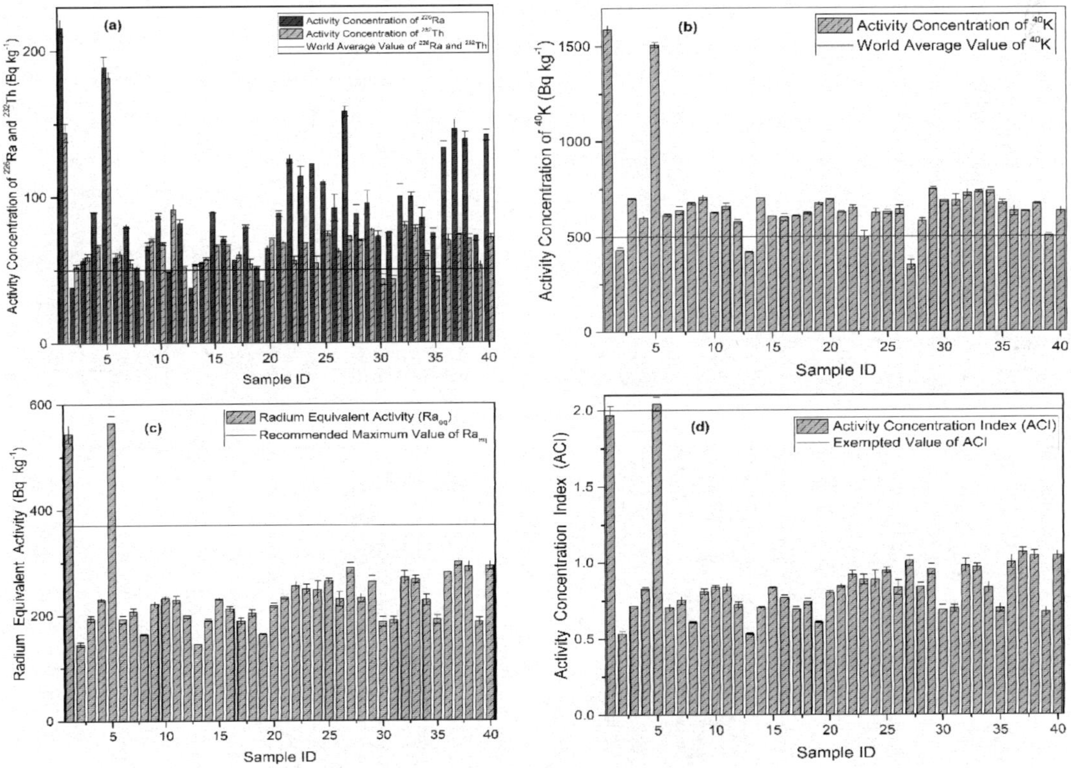

Fig. 1 (a and b) Variations of activity concentrations of ^{226}Ra, ^{232}Th and ^{40}K in various ceramic tiles; (c) variations of Radium equivalent activity from various ceramic tiles; (d) variations of activity concentration index from various ceramic tiles.

3.2 Evaluation of Factors Influencing Total Effective Dose Equivalent using RESRAD-BUILD Computer Code

RESRAD-BUILD Computer code is a model designed to determine the indoor doses from Residual Radioactivity in Buildings (RESRAD-BUILD) [12]. The RESRAD code is developed by Argonne National Laboratory United States (US). The design of the code was sponsored by the department of energy and federal agencies in the US [12]. The code is flexible and permit user to calculate indoor doses over the desired integrated time.

The measured activity concentrations were projected to RESRAD-BUILD computer code from initial year to 70 years to evaluate the effects of room size, dweller position and air exchange rate (ventilation rate) on the total effective dose equivalent (TEDE). TEDE refers to the combined indoor external and internal doses related with the measured activity concentrations of ^{226}Ra, ^{232}Th and ^{40}K in the ceramic tiles as discussed in the previous section. The code determined the indoor doses by assuming a uniform distribution of the radiation doses in the buildings. However, the code may over-estimate the indoor doses from Thoron (^{220}Rn) due to its known variability nature, the variability nature of ^{220}Rn is associated with its short half-life of 56 s, which would make it difficult to attain a uniform distribution in the buildings.

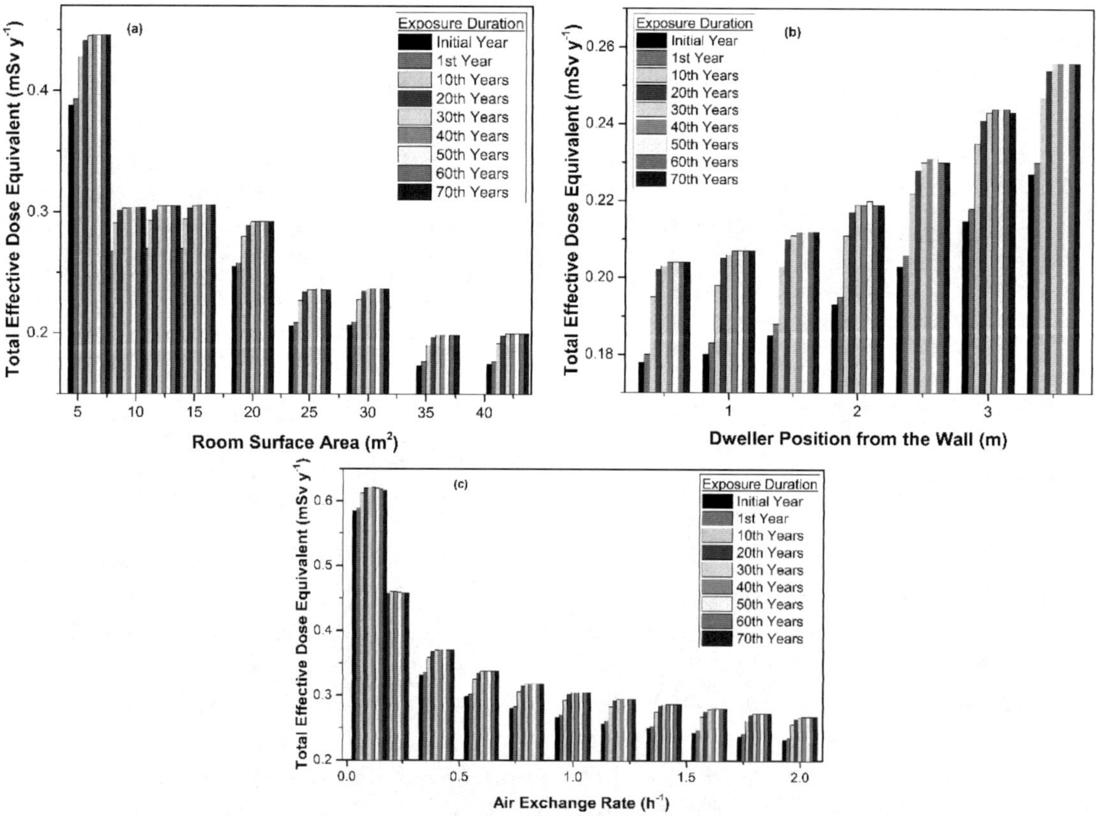

Fig. 2 (a) Effects of room surface area on total effective dose equivalent over time; (b) effects of dweller position relative to the wall on total effective dose equivalent over time; (c) effects of air exchange rate on total effective dose equivalent over time.

The variations of room size and air exchange rate considered in this study are 6 to 42 m^2 and 0.1 to 2.0 h^{-1}, respectively. The role of room size, dweller positions and air exchange rate on the variations of

TEDE over time are presented in Fig. 2(a), Fig. 2(b) and Fig. 2(c), respectively with values ranging from 0.2 to 0.4 mSv y^{-1}, 0.2 to 0.3 mSv y^{-1} and 0.3 to 0.6 mSv y^{-1} for TEDE due to room size, dweller position and air exchange rate, respectively. The variations of TEDE with room size is presented in Fig. 2(a), it can be observed that the TEDE decrease as the room size increase from 6 m^2 up to 36 m^2 however from 36 m^2 upwards the TEDE seem to be relatively stable. In addition, the variations of TEDE over time indicate nearly identical behavior. The impact of dweller position from the wall on TEDE is presented in Fig. 2(b) with TEDE increase as the dweller is moving away from one extreme end of the room toward to the other end. The initial position of the dweller was assumed to be the location of the air exchange systems of the room. Therefore, the air exchange systems play a significant role in the variations of TEDE as show in Fig. 2(c). It can be observed that the TEDE decrease with increasing air exchange rate. Even though, the highest TEDE recorded from the studied factors was below the recommended maximum limit of 1 mSv y^{-1} as reported by European Commission (EC) [13], there are still significant variations observed from initial year to 70 years as shown in Fig. 2(a), Fig. 2(b) and Fig. 2(c), respectively. The behavior of the TEDE with the studied factors were also in agreement with reported previous studies [14][15].

4. Conclusion

The activity concentrations of ^{226}Ra, ^{232}Th and ^{40}K in the studied tiles materials were mostly above the world average values of 50, 50 and 500 Bq kg^{-1}, respectively. On the contrary, the calculated radium equivalent activity and activity concentration index were generally below the recommended maximum values except for tile 1 and 5. The factors influencing the indoor doses were also investigated using RESRAD-BUILD computer code. It was observed that increase in room size and ventilation decrease the indoor doses receive by the dweller. Similarly, the closer the dweller position is to the wall the higher the indoor doses. Therefore, the radiological hazards associated with ceramic tiles under study is negligible.

5. References

[1] Abdullahi S, Ismail AF, Samat S. 2019 Determination of indoor doses and excess lifetime cancer risks caused by building materials containing natural radionuclides in Malaysia *Nucl. Eng. Technol.* **51**(1) p 325–36.

[2] Righi S, Guerra R, Jeyapandian M, Verità S, Albertazzi A 2009 Natural radioactivity in Italian ceramic tiles. *Radioprotection* **44**(5) p 413–9. Available from: https://www.radioprotection.org/articles/radiopro/pdf/2009/05/radiopro44078.pdf [cited 2017 Jul 20].

[3] Ismail A F, Abdullahi S, Samat S, Yasir M S 2018 Radiological dose assessment of natural radioactivity in Malaysia's tiles using Resrad-Build computer code *Sains Malaysiana* **47**(5) p 1017–23. Available from: http://www.ukm.my/jsm/english_journals/vol47num5_2018/contentsVol47num5_2018.htm

[4] Abdullahi S, Ismail A F, Mohd S F, Samat S 2019 Assessment of the long-term possible radiological risk from the use of ceramic tiles in Malaysia. Nucl Sci Tech **30**(2). Available from: https://doi.org/10.1007/s41365-019-0558-6 [cited 2019 Jan 19].

[5] IAEA 2000 Reference materials IAEA-375: Radionuclides and trace elements in soil. [Internet]. Vienna;. Available from: https://nucleus.iaea.org/rpst/Documents/rs_iaea-375.pdf

[6] IAEA 1989 Measurement of Radionuclides in Food and the Environment (Vienna: IAEA) Available from: http://www-pub.iaea.org/MTCD/Publications/PDF/trs295_web.pdf [cited 2017 Oct 10].

[7] Abdullahi S, Ismail A F, Samat S, Yasir M S 2018 Assessment of natural radioactivity and associated radiological risks from tiles used in Kajang, Malaysia *AIP Conf. Proc.* Available from: http://aip.scitation.org/doi/abs/10.1063/1.5027916

[8] Abdullahi S, Ismail A F, Samat S 2019 Assessment of naturally occurring radionuclides in

Malaysia'S building materials *Radiat. Prot. Dosimetry.*
[9] UNSCEAR 1993 Sources and effects of ionizing radiation (Exposure from natural sources of radiation) (New York) Available from: http://www.unscear.org/docs/publications/1993/UNSCEAR_1993_Annex-A.pdf [cited 2017 Sep 29].
[10] UNSCEAR 2000 Sources and effects of ionizing radiation (Exposures from natural radiation sources) (New York) Available from: https://www.unscear.org/docs/publications/2000/UNSCEAR_2000_Annex-B.pdf
[11] Abdullahi S, Ismail A F, Samat S 2019 Radiological characterization of building materials used in Malaysia and assessment of external and internal doses *Nucl. Sci. Tech.* **30**(3) p15.
[12] Yu C, LePoire D J, Cheng J J, Gnanapragasam E, Kamboj S, Arnish J, *et al.* 2003 *User's Manual RESRAD-BUILD ver. 3.* Available from: http://www.doe.gov/bridge
[13] EC 1999 Radiation Protection 112: Radiological protection principles concerning the natural radioactivity of building materials (Finland) Available from: https://ec.europa.eu/energy/sites/ener/files/documents/112.pdf [cited 2017 Jul 20].
[14] Stranden E 1979 The influence of variations in the ventilation rate in rooms upon the respiratory dose from inhalation of radon daughters *Phys Med Biol.* **24**(5) p 913–20.
[15] Risica S, Bolzan C, Nuccetelli C 2001Radioactivity in building materials: room model analysis and experimental methods. *Sci. Total Environ.* **272**(1–3) p 119–26. Available from: http://www.ncbi.nlm.nih.gov/pubmed/11379899

Acknowledgements
This work was supported by the Universiti Kebangsaan Malaysia under grant number DPK-2019-005. The authors would like to acknowledge all lab technician of Nuclear Science Program, UKM for their technical support throughout the works. Shittu Abdullahi also wishes to appreciate and acknowledges Gombe State University, Gombe, Nigeria for providing the Ph.D fellowship.

IOP Conf. Series: Materials Science and Engineering **785** (2020) 012013 doi:10.1088/1757-899X/785/1/012013

Peat soil stabilization using electrokinetic stabilization (EKS) treatment at Parit Lapis Kadir, Batu Pahat, Johor, Malaysia

Abdul Wahab[1,2,a)], Zaidi Embong[1,2], Abbas Ali Naseem[3], Saiful Azhar Bin Ahmad Tajudin[4], Qamar Uz Zaman[5],

[1]Research Centre for Soft Soil, Universiti Tun Hussein Onn Malaysia.
[2]Faculty of Applied Sciences and Technology, Universiti Tun Hussein Onn Malaysia.
[3]Department of Earth Science, Quaid-E-Azam University, Islamabad, Pakistan.
[4]Faculty of Civil and Environmental Engineering, Universiti Tun Hussein Onn Malaysia.
[5]Department of Geology, Northwest University Xian PR China.

[a]wahab.dir555@gmail.com

Abstract. Peat is found all over the world, which covered about 4.5% of the total world land and considered one of the most challenging soil, it creates many problems in constructions projects due to its characteristics like low shear strength, high water content and high compressibility. Hence, buildings and roads construction on peat has become more complex and differential settlements, slope failures, global instabilities, and long-term excessive settlements have occurred. Developing countries such as Malaysia, also facing such type of problems in building and road constructions. To overcome this problem, several matters related to soft soil strength, its essential to be resolved in proper way. This can be done by performing soil treatment to enhance its strength and physical properties. The application of soil improvement by using the electrokinetic stabilization (EKS) method has the potential to overcome problems such as low shear strength, high moisture content and high compressibility of the low permeable soil. The principles of EKS method involve applying a low potential gradient through electrodes which are inserted in the low permeable soil that cannot readily be drained. The voltage gradient of 110 V and 150 V was applied to the peat soil with the help of aluminium electrodes for the operational period of 3 and 6 hours. The peat sample was collected from Parit Lapis Kadir, Batu Pahat, Johor state, Malaysia. Some soil parameters like shear strength (τ) moisture content (WN), liquid limit (WL), maximum dry density (MDD), optimum moisture content (OMC) and dynamic properties were examined for pre and post EK. The result shows that shear strength was improved from 10.7 to 58 and 64 kPa, moisture content was reduced from 476.84 to 336.22 and 306.38%, the liquid limit was improved from 152.33 to 198.10 and 213.96%, maximum dry density was increased from 7.97×10^{-7} to 7.29×10^{-7} and 8.39×10^{-7} kg/m3 and optimum moisture content was improved from 27.56 to 114.01 and 125.25%. The shear wave velocity (SWV) was increased from 75.5 m/s to 109.5 m/s. The experimental results suggest the potential of developing EKS technique to stabilize shear strength of soft peat is effectively and efficiently.

1. Introduction

Peat has been considered as an organic remnant that suffers decomposition process throughout times under overburden pressure. The composition of peat normally consists of organic matters and it's exceeded sometimes up to 75% especially when the woods grows in marshes places in conditions where deficiencies of oxygen exist [1]. Peat consists of high moisture content about 250% to 985.40% and high compressibility [2], with the low shear strength of 5 to 20 kPa, which often results in creates difficulties when construction work is undertaken on peat deposit [3]. In comparison with other soft soils, peat is often viewed as problematic soil, mainly because of their inferior poor construction [4]. Peat is considered as one of the most challenging soil for the construction projects due to its characteristics in the result instability problems such as excessive settlement, local sinking, and long-term settlement occurred, if some moderate load is subjected to it or when buildings or roads are constructed on it [5].

In recent years, major settling or tilting of buildings, bridges, and road embankments have been occurred due to low shear strength and high compressibility of soil. In this regard, it is necessary to determine the soil improvement alternatives, technically and economically, from the ultimate state design in accordance with geotechnical categories. To stabilize and remediate the existing peat soil prior to the commencement of any construction activities the distortion under stress, reduce compressibility, control shrinking and swelling, control permeability and reduce water pressure [6].

Peatland is found in all parts of the world which covered about 4.5% of the whole world land [5]. Malaysia is considered 9th country in the world with highest amount total area of peat [7]. In entire country peat covers about 8% or 3 million hectares of the total land. Out of this land, 6300 hectares of the peatland is found in Batu Pahat, Pontian, and Muar districts in West Johor state as tabulated in Table 1. Sarawak state is the largest peatland area in the country, consisting about 13% of the peatland or 1.66 million hectares of the total peatland in Malaysia [8].

Table 1. The peatland distribution in Malaysia [8]

State	Peatland (ha)
Sarawak	1,697,847
Selangor	164,708
Pahang	164,113
Johor	143,974
Sabah	116,965
Terengganu	84,693
Perak	69,597
Kelantan	9,146
Negeri Sembilan	6,245
Federal Territory	381
Total Peat Land	**2,457,669**

Many construction projects and coastal high-rise buildings whose basics are frequently supported by peat, are facing issues when not given physical existence. Likewise, in Malaysia, the construction industry is facing many challenges related to peat stabilization and major economic activities; and social developments focusing on the coastal area. One of the major problems for the construction industry is the settlement of peat foundation for the development of highways and buildings during or after construction when the soil is not properly treated. To ensure the solution of these problems or at least minimizing their overall negative impact on the industry, electrokinetic stabilization methods should be applied for the peat stabilization [9].

Some previous researchers [10] have applied EKS treatment for the improvement of low permeable soil. The clay samples were collected from East Java, Indonesia. Calcium chloride was used as an electrolyte solution while aluminum was used as an electrode with the electric gradient of 13 V for the operational period of 12 hours to 3 days, intervalley. The authors concluded that the shear strength was improved up to 56% while the liquid limit was increased. The authors Moayedi in 2014 performed a laboratory experiment to improve the shear strength of peat soil by using EK method. The peat samples were collected from Selangor, Malaysia. Graphite was used as an electrode with the electric gradient of

60 V for the operational period of 7 days. The results show the strength of peat soil was improved up to 29 kPa [11]. Keykha in 2014 conducted electrokinetic stabilization laboratory-based method, for the determination of improving the strength and reduce the moisture content of the soft soil. Two graphite laminate were used as electrode and calcium carbonate used as electrolyte solution with the electric gradient of 60 V, for the period of 4 days, which shows the shear strength was improved from 6 kPa to 60 kPa [12].

Amnart Rittirong conducted the laboratory-based electrokinetic technique for the assessment of electrokinetic strengthening of soft clays. The clay soil was collected from Ankara, Turkey. Circular electrokinetic experimental setup was arranged with multiple electrodes were used, while C_aCl_2 was used as a stabilizing agent (electrolyte solution) with a voltage gradient of 10 V for the operational period of 10 days, the authors concluded that shear strength was increased up to 36, 50, 58 and 85 kPa respectively in different phases [13].

2. Method and material

There were twenty different peat locations were investigated at Parit Botak area, Batu Pahat, Johor, Malaysia, which was based on low shear strength. The peat sample was collected from Parit Botak area. The shear strength was measured with the help of (field vane shear) at the depth of 20 cm, 30 cm and 40 cm, respectively, according to EK cell size as shown in Figure 2. The in-situ measurement of water content was not possible so, accordingly, the sample was carefully transferred to the laboratory. Where some standardized laboratory tests such as shear strength, moisture content, liquid limit and bender element test exist for observing the physical, chemical and dynamic properties of peat.

Figure 1. In-situ shear strength measurement.

3. Electrokinetic stabilization (EK) treatment

The EK experiments were carried out at the Research Center for Soft Soil (RECESS), UTHM. The Electrokinetic experiment was conducted in a laboratory environment at an ambient temperature of 27 °C. The EK cell designed for this research was made up of the transparent acrylic plate with rectangle shape open at the top with 40 cm of depth and 42 cm of width while the thickness of the acrylic plate was 1.5 cm. Further it was divided into 3 major slots which as an anode, cathode and untreated peat compartment. The two outside slots were reserved for the anode and cathode while the middle section was reserved for the untreated peat sample. The anode and cathode sections were 10 cm while contaminated soil compartment was 17 cm as shown in Figure 2.

The electrodes were inserted vertically into the untreated peat. The two aluminum passive electrodes were placed at the extreme edges of the peat specimen. The post-EK treatment was divided into two phases, where in phase I, the voltage gradient of 150 V was applied for the operational period of 6 hours as shown in Figure 4, while in phase II, the voltage gradient of 110 V was applied with load of 50 kg for the operational period of 3 hours as shown in Figure 3.

IOP Conf. Series: Materials Science and Engineering **785** (2020) 012013 doi:10.1088/1757-899X/785/1/012013

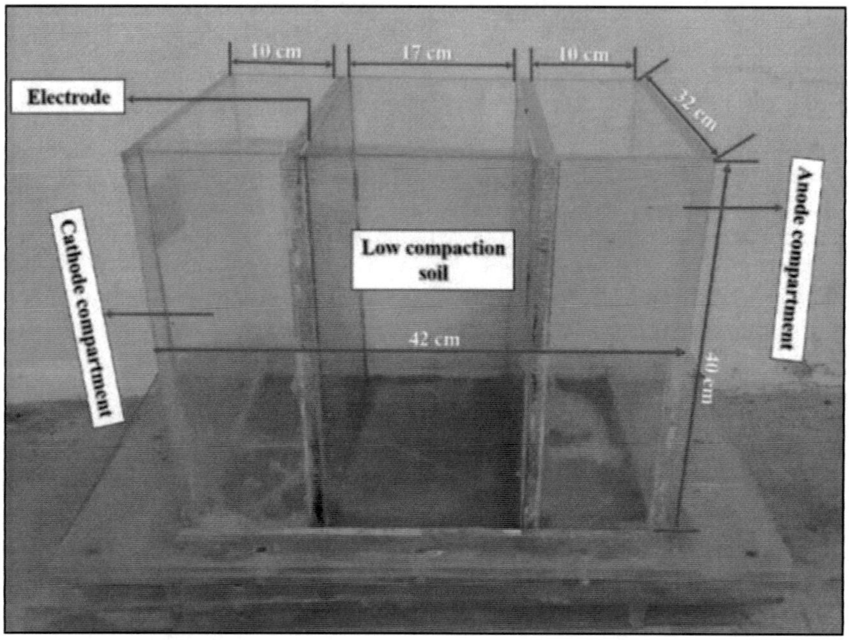

Figure 2. EK stabilization compartments and its dimensions.

Figure 3. Load test inside the EK compartment for post-EK.

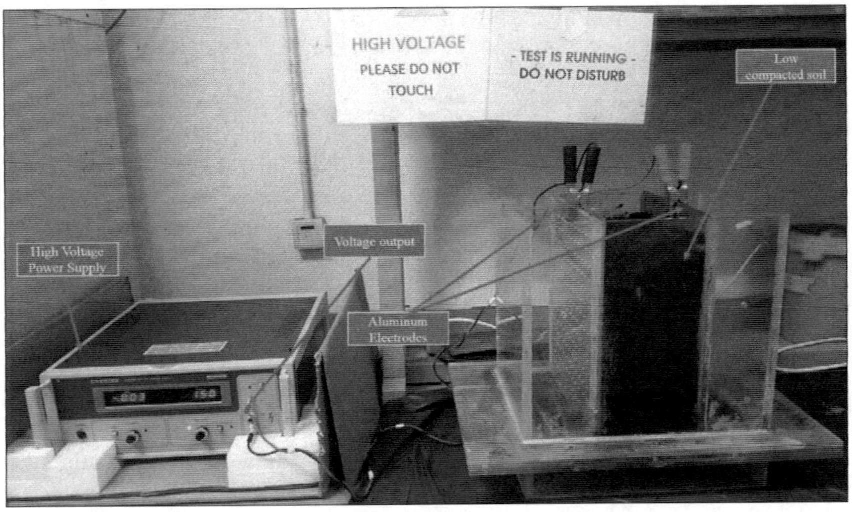

Figure 4. A complete set of EKS treatment (with voltage gradient of 150 V)

4. Result and discussion

4.1. Physical and dynamic properties of untreated peat

The physical and dynamic properties tests were conducted for pre EK (untreated peat), to examine the physical and dynamic properties of peat. The physical properties tests include shear strength test, moisture content test, liquid limit test and compaction tests while dynamic properties test shows the shear wave velocity in peat sample as tabulated in Table 2.

Table 2. The physical and dynamic properties for untreated peat.

Properties	**Parit Lapis Kadir**	**Value and Reference**	
Shear strength (kPa)	10.7	3-17	[1]
Organic Content (%)	97.286	96.64	[14]
Moisture Content (%)	476.849	200-700	[14]
Liquid limit (%)	152.333	173.75	[15]
Maximum Dry Density *(kg/m³)*	7.79×10^{-7}	$4.90 - 8.12\times 10^{-7}$	[14], [15]
Shear wave velocity *(m/s)*	75.5	26.02 - 96.89	[16]

4.2. Physical and dynamic properties of treated peat (Phase I)

The physical and dynamic properties of peat were observed after applied the load of 150 V for the operational period of 6 hours. The soil parameters i.e. shear strength, moisture content, liquid limit, maximum dry density, optimum moisture constant and shear wave velocity were observed improved as compared to pre-EK as tabulated in Table 3.

Table 3. The physical and dynamic properties for pre and post EK (phase I)

Parameters	Parit Lapis Kadir	
Shear strength (τ) (kPa)	**Pre EK**	**Post EK (Phase I)**
	10.7	58
Moisture content (W_N), (%)	476.849	370.942
Liquid Limit (W_L), (%)	152.333	198.106
Maximum dry density ρ_d *(kg/cm³)*	7.79×10^{-7}	7.29×10^{-7}
Optimum moisture content (%)	27.569	114.011
Shear wave velocity *(m/s)*	75.5	88.5

4.3. Physical and dynamic properties of treated peat (Phase II)
The physical and dynamic properties of peat were observed after applied the voltage gradient of 110 V with a load of 50 kg for the operational period of 3 hours. The load was applied before applied the current. After removing the load, the current of 110 V was applied for 3 hours. The properties i.e. shear strength, moisture content, liquid limit, maximum dry density and shear wave velocity were observed improved as compared to phase I as shown in Table 4.

4.3.1. Shear strength (kPa) of treated peat. The load of 50 kg was applied before providing the voltage gradient. After removing the load form EK cell, the peat was observed compacted. The electric gradient of 110 V was applied through Aluminium electrode for the operational period of 3 hours. The shear strength was measured with the help of (field vane shear apparatus) inside EK cell in various positions. The results show that shear strength was improved from 10.7 kPa to 64 kPa as shown in Figure 5.

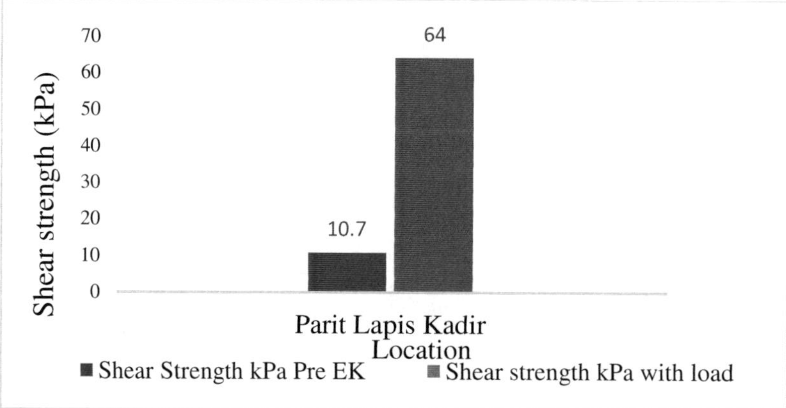

Figure 5. Shear strength (kPa) for pre and post EK.

4.3.2. Moisture content of treated peat. The moisture content was reduced with an applied load of 50 kg along with the voltage gradient of 110 V for the operational period of 3 hours. The peat was observed compacted, moisture content and compressibility were reduced due to a load of 50 kg, the contaminations was moved from cathode to anode. The initial moisture content of untreated peat soil was observed 476.849% while applying the voltage gradient of 110 V for the operational period of 3 hours continuously, the moisture content was reduced up to 306.384%. Figure 6 shows moisture content for post EK (treated soil).

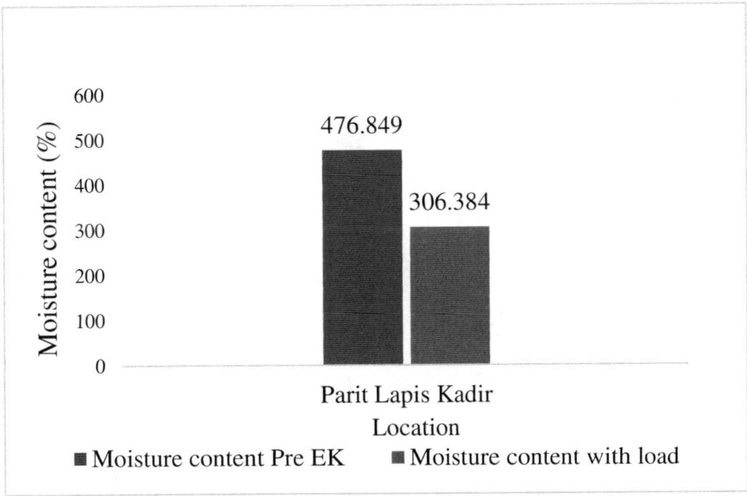

Figure 6. Moisture content for Pre and Post EK.

4.3.3. Liquid Limit (LL) of treated peat. The liquid limit was increased with an applied load of 50 kg along with a voltage gradient of 110 V for the operational period of 3 hours respectively. Here, the initial liquid limit of untreated peat was observed 152.333%, while after treatment the liquid limit was observed improved up to 213.965%. Figure 7 shows that the liquid limit was increased for post EK.

4.3.4. Compaction behaviour of treated peat. The maximum dry density and optimum moisture content were observed increased with an applied load of 50 kg along with a voltage gradient of 110 V for the operational period of 3 hours. The result shows that MDD was increased up to 8.39 kg/m^3 while OMC was increased to 125.253%.

4.3.5. Shear wave velocity for treated peat. The bender element test was performed for treated peat samples after applied the load of 50 kg with the voltage gradient of 110 V for the operational period of 3 hours. As observed in pre-EK, that shear wave velocity for Parit Lapis Kadir untreated peat was 75.5 m/s while it was increased up to 109.5 m/s as shown in Figure 7.

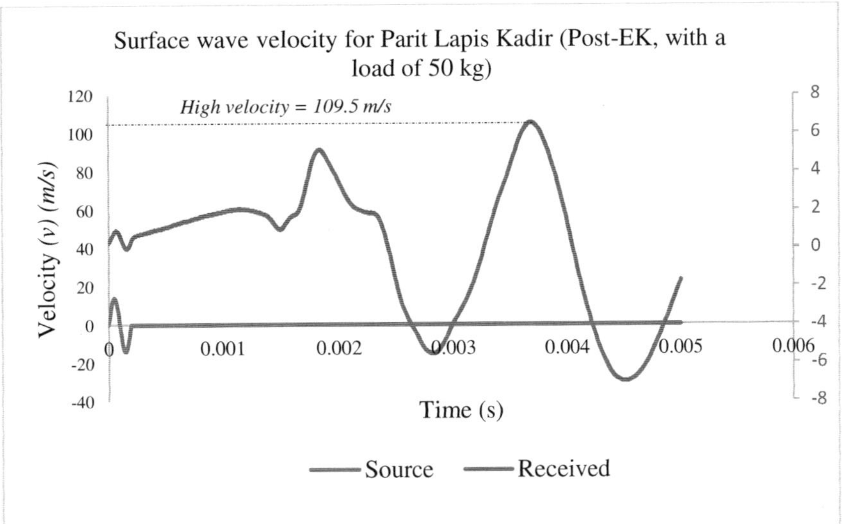

Figure 7. Surface wave velocity for Parit Lapis Kadir (Post-EK, Phase II).

Table 4. Physical and dynamic properties for treated peat (Post EK, Phase II).

Parameters	Parit Lapis Kadir	
Shear strength (kPa)	**Pre EK**	**Post EK (Phase I)**
	10.7	64
Moisture content (%)	476.849	306.384
Liquid Limit (%)	152.333	213.965
Maximum dry density ρ_d *(kg/cm³)*	7.79× 10^{-7}	8.39× 10^{-7}
Optimum moisture content (%)	27.569	125.253
Shear wave velocity *(m/s)*	75.5	109.5

4.4. Compression between pre and post-EK

The pre and post result shows, that physical properties of peat were more observed in post-EK phase II, with the voltage gradient of 110 V with an applied load of 50 kg for the operational period of 3 hours.

(i) The shear strength was improved from 10.7 to 58 kPa in phase I, with applied voltage of 150 V for the operational period of 6 hours, while up to 64 kPa in phase II, with applied voltage 110 V with applied load of 50 kg for the operational period of 3 hours as shown in Figure 8.

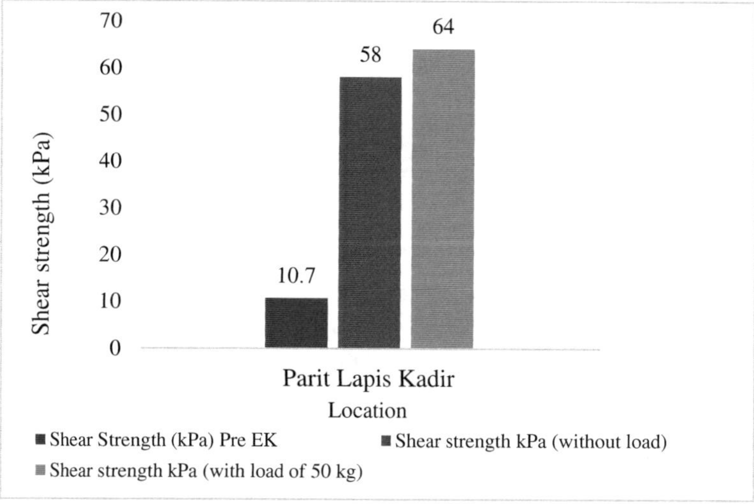

Figure 8. Shear strength of peat between pre and post-EK.

(ii) Similarly, moisture content also improved in post-EK phase II, with voltage gradient of 110 V with load of 50 kg for the operational period of 3 hours, the moisture content was reduced from 476.849 to 306.384%, while it was reduced to 370.942% with applied voltage gradient of 150 V for the operational; period of 6 hours as shown in Figure 9.

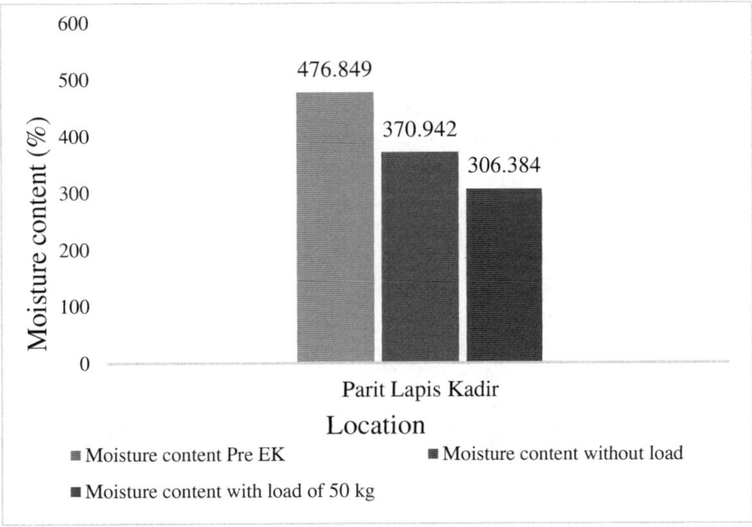

Figure 9. The moisture content of peat for pre and post-EK

(iii) The liquid limit was observed increased in phase II, with a voltage gradient of 110 V with a load of 50 kg for the operational period of 3 hours, LL was improved from 152.333 to 213.965 %, while it was improved up to 198.106 % with applied voltage gradient of 150 V for the operational; period of 6 hours as shown in Figure 9.

(iv) Maximum dry density *(kg/cm^3)* was observed 7.79× 10^{-7} for pre EK, while it was decreased to 7.29× 10^{-7} *(kg/m^3)* in phase I, while increased up to 8.39× 10^{-7} *(kg/m^3)* in phase II. The optimum moisture content was recorded 27.569 % for re-EK, while for post-EK phase I, it was observed 114.011% and 125.253% in phase II.

(v) The shear wave velocity was observed 75.5 m/s for untreated peat soil (Pre-EK), where it was increased up to 88.5 m/s in phase I post-EK, while in phase II, it was increased up to 109.5 m/s.

5. Conclusions and recommendations

The result shows improvement when the voltage gradient of 110 V was applied for an operational period of 3 hours with the help of the aluminium electrode. As tabulated in Table 5, It was observed that shear strength was improved when the electric gradient was applied to the peat soil. The moisture content was decreased because the contaminations was moved outside. The liquid limit was found increased due to high particle compactness. The shear wave velocity was found increased which is directly reflect the strength. The outcome of this study potential may contribute toward the selective properties of particular soil in order to support a specific infrastructure and construction.

Table 5. The physical properties for treated peat

Parameters	Parit Lapis Kadir		
SS (kPa)	**Pre EK**	**Post EK**	
		Phase I	**Phase II**
	10.7	58 (54% ↗)	64 (59% ↗)
MC (%)	478.849	370.94 (77% ↗)	306.384 (64% ↗)
LL (%)	152.333	198.106 (13% ↗	213.965 (14% ↗)
MDD ρ_d *(kg/cm³)*	7.79× 10^{-7}	7.29× 10^{-7} (-1% ↗)	8.39× 10^{-7} (1% ↗)
OMC (%)	27.569	114.011 (41% ↗)	125.253 (45% ↗)
SWV *(m/s)*	75.5	88.5 (11.72% ↗)	109.5 (14% ↗)

(i) For all types of soft soil, by applying the voltage gradient EKS, that may increase the shear strength easily and can be beneficial to improve their physical properties.
(ii) If the high voltage gradient (>150 V) is to be applied on soft soil, it may easily enhance the physical, chemical as well as dynamic properties of soil.
(iii) In context, if the high load (>50 kg) it can result to achieve maximum stabilization at very less cost.

6. References

[1] Moayedi H, Nazir R 2018 Malaysian experiences of peat stabilization, state of the art. Geotech. and Geolog. Eng. **36**(1) p 1-11.
[2] Zainorabidin A, Mohamad H M 2017 Engineering properties of integrated tropical peat soil in Malaysia. *Electr. J. of Geotech. Eng.* **22** p 457-466.
[3] Razali S N M, Bakar I, Zainorabidin A 2013 Behaviour of peat soil in instrumented physical model studies. *Procedia Eng.* **53** p 145-155.
[4] Mesri G, Ajlouni M 2007 Engineering properties of fibrous peat. *J. of Geotech. and Geoenviron. Eng.* **133**(7) p 850-866.
[5] Abdel-Salam A E 2017 Stabilization of peat soil using locally admixture *HBRC J.*
[6] Mosavat N, Oh E, Chai G 2012 A review of electrokinetic treatment technique for improving the engineering characteristics of low permeableproblematic soils *Int. J. of GEOMATE,* **2**(2) p 266-272.
[7] Mohamad H M 2015 Post cyclic behaviour of Malaysian peat soil (Universiti Tun Hussein Onn Malaysia).
[8] Melling L 2016 Peatland in Malaysia *Trop. Peatland Ecosys* (Springer) p 59-73.
[9] Hua L J, Mohd S, Tajudin S A A, Mohamad S N A, Bakar I, Masirin M I M, Mahmood A A-W 2016 Construction of infrastructure on peat: case studies and lessons learned *MATEC Web of Conf.*
[10] Thuy T T T, Putra D P E, Budianta W, Hazarika H 2013 Improvement of expansive soil by electro-kinetic method *J. of Appl. Geol.* **5**(1) p 50 – 59.
[11] Moayedi H, Kassim K A, Kazemian S, Raftari M, Mokhberi M 2014 Improvement of peat using Portland cement and electrokinetic injection technique *Arabian J. for Sci. and Eng.* **39**(10) p 6851-6862.
[12] Keykha H A, Huat B, Asad A 2014 Electrokinetic stabilization of soft soil using carbonate-producing bacteria *Geotech.and Geological. Eng.* **32**(4) p 739-747.
[13] Aski, T, Turer D 2016 Effect of electrode configuration on electrokinetic stabilization of soft clays. *Quart. J. of Eng, Geol. and Hydrogeol.y,* 49(4), 322-326.
[14] Zainorabidin A, Zolkefle S N A 2014 Dynamic behaviour of western Johore peat, Malaysia.
[15] Deboucha S, Hashim R, Alwi A 2008 Engineering properties of stabilized tropical peat soils

Electr. J. of Geotech. Eng. **13** p 1-9.
[16] Boulanger R W, Arulnathan R, Harder Jr L F Torres R A Driller MW 1998 Dynamic Properties of Sherman Island Peat *J. of Geotech. and Geoenviron. Eng.* **124**(1) p 12-20.

Acknowledgement
The authors would like to express his gratitude to Research Centre for Soft Soil (RECESS), UTHM that provides all the facilities to ensure the success of this research and thanks also goes to those who assist directly and indirectly.

IOP Conf. Series: Materials Science and Engineering **785** (2020) 012014 doi:10.1088/1757-899X/785/1/012014

Removal of Thorium from industrial waste via electrosorption technique

Eli Syafiqah Aziman[1] and Aznan Fazli Ismail[1, 2,a)]

[1]Nuclear Science Program, Faculty of Science and Technology, Universiti Kebangsaan Malaysia, 43600 UKM Bangi, Selangor, Malaysia.
[2]Center of Frontier Science, Universiti Kebangsaan Malaysia, 43600 UKM Bangi, Selangor, Malaysia.

[a]aznan@ukm.edu.my

Abstract. Activated carbon electrode has been tested as a potential sorbent for removal of Th ions from the actual wastewater sampel by electrosorption technique. The objective of this research is to investigate the performance of activated carbon electrode for remediation of real thorium sample. Adsorption of thorium ion by activated carbon electrodes was investigated with contact time up to 180 minutes by using positive potential at 0.2V (vs. Ag/AgCl), 0.4V (vs. Ag/AgCl) and 0.6v (vs. Ag/AgCl). Result shows that the condition has the highest percentage of thorium ion removal is at 0.6V (vs. Ag/AgCl) with highest specific capacity at 0.5 mg Th/g Carbon. Based on this research, the removal of thorium through electrosorption technique has the tendency to remove thorium ion and parameter such as contact time of electrosorption and voltage applied should be studied in more depth, where the removal of thorium is expected to increase.

1. Introduction

Wastewater is a by-product in the form of liquid produced by human activities be it by domestic, industrial, commercial or agricultural activities. Wastewater may contain physical, chemical or biological pollutants which exceed the limit that has been set by the authorities. If this wastewater is not handled properly, they may cause environmental contamination. There has been reports made regarding contamination of water supply with high concentrations of thorium in Malaysia. Bukit Merah, Malaysia 1992, was an incident where water supply were contaminated with a high concentration of thorium which leads to congenital disabilities and leukemia among nearby residents [1, 2].

Electrosorption is one of the promising techniques that enables the removal of thorium from wastewater aside from other conventional technique such as adsorption, ion-exchange and membrane filtration [3]. This process uses an electric current as a force that helps drive the metal ion to be adsorbed at the negative carbon electrode. Electrosorption is a non-faradaic process, the nature of electrostatic force where lead to electrical double layer capacitance (EDLC) causes the movement of ion to be adsorb to the surface of carbon electrode [4,5]. Electrosorptive capacity on activated carbon strongly depends on the physical properties such as surface area, pore size distribution, surface functional groups, surface polarity and surface wettability [6,7]. The objective of this paper is to investigate the performance of fabricated activated carbon electrode for remediation of thorium real sample through electrosorption technique.

2. Material and method

A carbon slurry was prepared by mixing activated carbon powder (DARCO 100; Sigma Aldrich, St Louis, MO, USA) with polyvinylidenefluoride (PVDF, M.W. ¼ 534,000, Sigma Aldrich) in an N,Ndimethylacetamide (DMAc, 99%; Alfa Aesar, Ward Hill, MA,USA) solution at 21°C. A stainless steel gauze (40 mesh woven with 0.25 mm diameter wire, Type 304; Alfa Aesar) was coated with the carbon slurry and dried in an oven at 80°C for 6 hours to form an activated carbon electrode. Solution contain thorium in this research were obtained from real sample of leach solution of radioactive waste. Concentration of thorium in the leached solution was 1 mg/L with the presence of other REE earth elements such as U, Ce, Nd, La, Sm and others at various concentration.

Batch-mode adsorption experiment was carried out using a three-electrode electrochemical cell. Three electrodes used in this experiment were functioning fabricated activated carbon electrode with platinum wire (A-002233; Bio-Logic) and Ag/AgCl (RE-1S; Bio-Logic, Grenoble, France) as counter electrode and reference electrode respectively. The electro-chemical cell was connected to a potentiostat (VersaSTAT 3; Princeton Applied Research). The activated carbon electrode was kept under positive bias potential at 0.2V, 0.4V and 0.6V (vs Ag/AgCl). Thorium removal efficiency ($Th_{removal}$) and the total thorium adsorbed, Th_q (mg/g) at time, t was determined as shown in Eq. 1 and Eq. 2 respectively [8].

$$Th_{removal} = \frac{C_0 - C_t}{C_0} \times 100\% \quad (1)$$

$$Th_q = \frac{(C_0 - C_t)}{W} \times V \quad (2)$$

where,

c_0 = initial thorium concentration in aqueous waste (mg/L)
c_t = Thorium concentration after electrosorption at time, t (mg/L)
V = volume of aqueos waste
w = mass of activated carbon electrode (g) at time, t

3. Result and discussion

3.1. Characterisation of activated carbon electrode

The surface area and porosity of the activated carbon electrode was derived from the adsorption branch of the isotherm as shown in Fig. 1, based on the Brunauere Emmette Teller (BET) method using physisorption analyser (ASAP 2020 from Micromeritic). The BET surface area for activated carbon electrode is shown in Table 1. The surface morphology by Scanning Electron Microscope (SEM) analysis of the carbon electrode is shown in Fig. 2 shows that the stainless steel gauze was excellently covered by activated carbon electrode. However, the surface area of activated carbon electrode is reduced. This is due to the use of PVDF polymer as an agent to bind the activated carbon particles effectively.

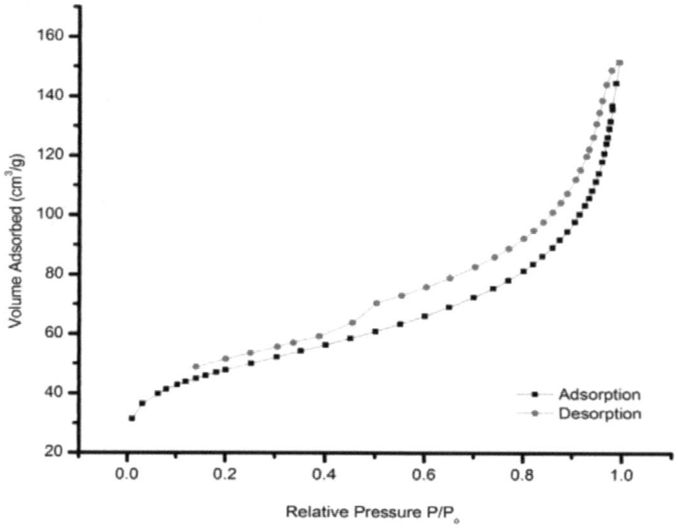

Figure 1. Nitrogen adsorption–desorption isotherms of activated carbon electrode.

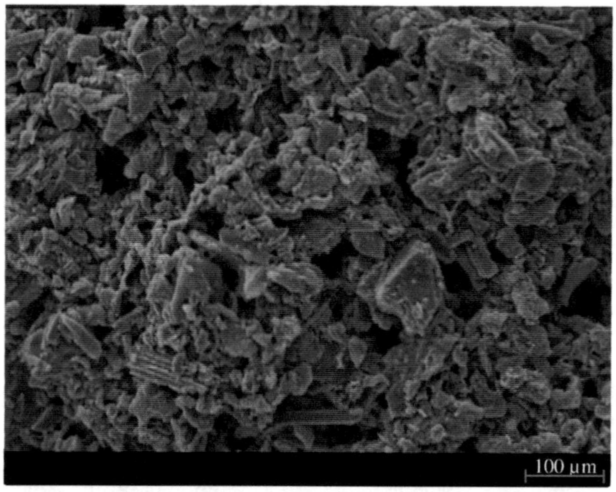

Figure 2. Surface morphology of activated carbon at 100µm.

Table 1. Pore structure of activated carbon powder and the fabricated activated carbon electrode.

Characteristic	AC powder	Activated carbon electrode
Surface area	1,024 m2/g	158.07 m²/g
Total pore volume	0.8 cm3/g	0.22 cm³/g
Pore size	3.1 nm	5.79 nm

3.2. Batch-mode electrosorption experiments

Effect of potential voltage on electrosorption of thorium from aqueous waste is shown in Fig. 3. The optimum voltage applied in this research was 0.6V vs Ag/AgCl. The highest specific capacity obtained was 0.5 mg/g with removal percentage at 22% like shown in Fig. 4.

Aside from activated carbon electrode acting as an adsorbent, organic waste adsorbent such as banana peel nano sorbent (BPN) were also used to remove thorium ion from wastewater generated by mining activity [8]. Even though removal of thorium ion by BPN adsorption technique is higher compare activated carbon electrode in this research however BPN requires a longer time of contact specifically 24 hours. The initial concentration used in activated carbon electrode is lower, specifically at 1 mg/L while the initial concentration of thorium used in BPN adsorbent is at 18.8 mg/L. BPN recorded 80% removal of thorium was due to the higher probability of thorium ion collision with BPN adsorbent compared activated carbon electrode. Aside from that, the initial concentration was different, the impurities in the sample is also a crucial factor. Actual sample in this research contains other elements besides thorium ion, therefore ion competition to be adsorb by activated carbon may be higher compared to thorium ion when comparing to the BPN experiment. Thus, activated carbon electrode has its own advantages depending on the parameter chosen in this research. Table 2 shows comparison of experimental parameter between activated carbon with previous research.

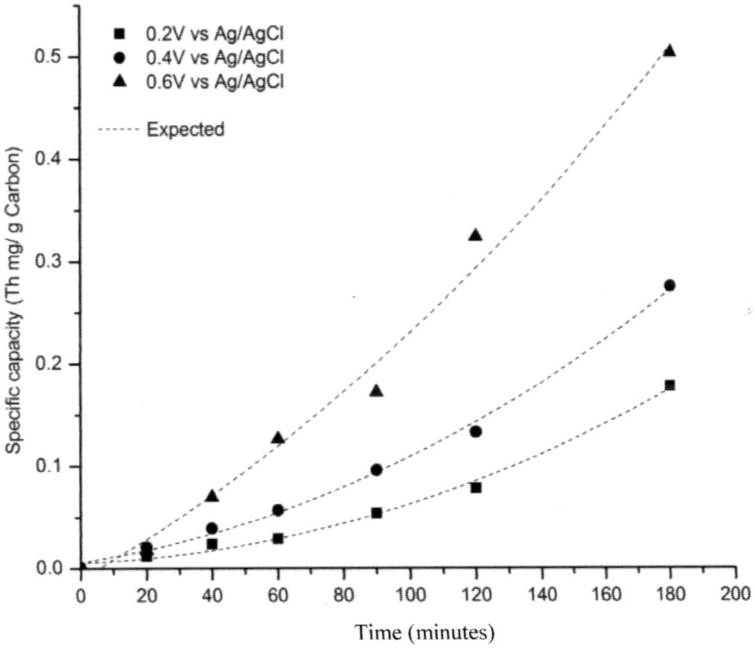

Figure 3. Effect of potential voltage of Thorium.

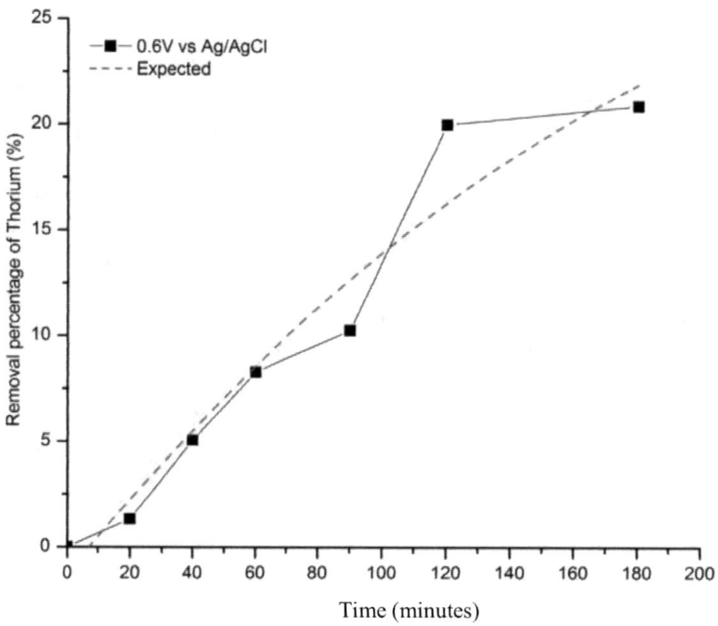

Figure 4. Removal percentage of thorium ion from sulphate aqueous waste.

Table 2. Comparison experimental parameter by previous research.

Adsorbent	Initial concentration (mg/L)	Specific capacity (mg Th/g adsorbent)	Time of contact (minutes)	Impurities in sample	References
Magnetic ion-imprinted chitosan resin	20.0	39.0	Not Stated	-	[9]
BPN	18.8	10.1	1440	Th,U	[8]
Activated carbon electrode	1.0	0.5	180	Th,U,REE	**This study**

4. Conclusion

Activated carbon electrode has a high tendency in the removal of thorium from actual waste water samples through electrosorption technique. The best condition recorded in this research is where the voltage used was 0.6V (vs. Ag/AgCl) and the highest specific capacity was 0.5 mg Th/ g carbon. It is should be noted that the specific capacity value was not at the maximum yet. Various parameter should be studied still. Also, theoretically, an increase in the contact time of electrosorption and voltage applied should increase the removal of thorium.

5. References

[1] Keith-Roach M, Grundfelt B, Ferrari A, Kousa A 2015 Past experience of environmental, health and safety issues in REE mining and processing industries and an evaluation of related EU and international standards and regulations *Final report of the EuRare project; Kemakta Konsult AB: Stockholm, Sweden* (Geological Survey of Finland).

[2] Poh L K 2015 *In Search of Environmental Justice in Malaysia: The Cases of Broga and Bukit Merah* (thesis dissert.).

[3] Aziman E S, Mohd Salehuddin A H J, Ismail A F 2019 Remediation of Thorium (IV) from wastewater: current status and way forward, separation and purification reviews, doi: 10.1080/15422119.2019.1639519.

[4] Guan L, Yu L and Chen G Z 2016 Capacitive and non-capacitive faradaic charge storage. *Electrochimica Acta*. **206** p 464–478.

[5] Lee J H, Choi J H 2013 Ion-selective composite carbon electrode coated with TiO_2 nanoparticles for the application of electrosorption process *Desalination and Water Treatment.* **51**(1-3), p 503–510.

[6] Peng L, Chen Y, Dong H, Zeng Q, Song H, Chai L, Gu J D 2015. Removal of trace As (V) from water with the Titanium Dioxide/ACF composite electrode *Water, Air and Soil Pollution.* **226**(7).

[7] Zou L 2011 Developing nano-structured carbon electrodes for capacitive brackish water desalination *Expanding Issues in Desalination* ed Robert Y N p 301–318.

[8] Oyewo O A, Onyango M S, Wolkersdorfer C 2016 Application of banana peels nanosorbent for the removal of radioactive minerals from real mine water *J. of Env. Radioact.* **164** p 369–376.

[9] Guolin H, Zhongsheng C, LeLe W, TingTing L, Jeffrey S 2016 Removal of Thorium (IV) from aqueous solution using magnetic ion-imprinted chitosan resin. *J Radioanal. Nucl. Chem.* doi: 10.1007/s10967-016-4993-0.

Acknowledgements
This work was supported by the Universiti Kebangsaan Malaysia under grant number GUP-2018-113. The authors would like to acknowledge all lab technician of Nuclear Science Program, UKM for their technical support throughout the works.

IOP Conf. Series: Materials Science and Engineering **785** (2020) 012015 doi:10.1088/1757-899X/785/1/012015

Solvent extraction of Thorium from Malaysian xenotime using Tributyl Phosphate (TBP)

Norhazirah Azhar[1,a)], Khaironie Mohamed Takip[1], Roshasnorlyza Hazan[1], Wilfred Selyvester Paulus[1], Nur Aqilah Sapiee[1] and Jacqueline Kones[1]

[1]Industrial Technology Division, Malaysian Nuclear Agency, Bangi, 43000 Kajang, Selangor, Malaysia

[a]norhazirah@nuclearmalaysia.gov.my

Abstract. Growing interest in environmental concern and vast industrial application makes Thorium popular and widely studies nowadays. In this study, solvent extraction process was done to determine the optimized parameters for the extraction of thorium. Parameters studied are the concentrations (M) of nitric acid (HNO_3), percent (%) of Tri-Butyl Phosphate (TBP) in diluent and the volume ratio of acid to organic phase (V_A:V_O). The mixed rare earth (RE) oxalate powder was dissolved in HNO_3 and Thorium elements was extracted using TBP diluted in kerosene. Energy dispersive x-ray fluorescence (ED-XRF) instrument was used to analyze the initial and final concentration of Thorium left in the aqueous solution. From the ED-XRF quantitative values, calculations were made to determine the extraction efficiency of Thorium. In the end of experiment, it is found that the optimized parameters are at 7M HNO_3, 20% TBP diluted in kerosene and at V_A:V_O ratio of 2:3.

1. Introduction

Xenotime, bastnaesite and monazite are among minerals that contains rare elements and radioactive elements such as Thorium and Uranium. Due to the significant amount of rare earth in these minerals, it makes them an important source of RE. In Malaysia, xenotime and bastnaesite mineral was obtained from tin mining industrial by product and higher in heavy RE content in comparison to monazite [1]. Xenotime have high content of Yttrium aside of heavy RE elements from Lanthanide group which makes it termed YPO_4. Economic value of Yttrium, Gadolinium, Dysprosium, Erbium and other REE in the mineral is rather high in demand due to their unique catalytic, chemical, electrical, magnetic and optical properties [16]. These heavy RE contents in xenotime can be separated further after recovery of thorium from the mineral ass a spin-off.

As an alternative in replacing Uranium as fuel for nuclear power plant, Thorium element become popular and widely studies nowadays. Various technique was done to recover Thorium, Uranium and REE from various sources such as ion-exchange, selective precipitation, chromatography, electrophoresis and liquid-liquid extraction methods [2]. Liquid-liquid extraction is the technology initially developed to recover Uranium from ore-leach liquor [6]. The method used was then adapted for recovery of other REE including Thorium. Studies in separation of Thorium from various sources was made possible via some technique including solvent extraction, direct precipitation and ion exchange [14]. The purification of thorium was made by repeated transfers between organic and aqueous medium.

Compared to direct precipitation and ion exchange, solvent extraction technique is more eminent due to its speed, large throughput and non-complicated procedure [12].

According to Gupta *et. al.* [5], popular alkylphosphorus extractant used for separation of uranium and thorium includes TBP, Di-2-Ethylhexylphosphoric acid (DEHPA) and Tri-N-Octylphosphine Oxide (TOPO). Tributyl Phosphate (TBP) is a type or organophosphorus extractant which is prominent in nuclear industries in the recovery of Uranium and Thorium from rare earth minerals [8]. Since TBP is less flammable and more stable makes it preferable during the World War II to separate Uranium [15]. Various studies are done to improve the extraction ability of each extractant. Previous studies done via TBP solvent to extract various types of elements such as Molybdenum [4], and Cerium(IV) [3] from nitric acid media. TBP have low vapor pressure which makes it safer for handling and storing with reduced flame risk [13]. Researchers studied different parameters and media to improve the extraction efficiency of TBP.

Famous use of TBP in extraction were done mainly for the recovery of Uranium since it is highly selective to uranium [13]. In this study, the solvent extraction via TBP in kerosene was chosen to recover Thorium instead. Separation of Uranium from Thorium via extraction is somewhat difficult to achieved because of the limited variance oxidation states which thorium poses [9]. However, for this study, Malaysian xenotime undergone cracking, leaching, selective precipitation and decomposition proses prior in obtaining mixed RE and Thorium oxides powder which no longer contains Uranium. Figure 1 shows the composition of Thorium and other RE oxides composition elements after alkaline fusion. This mixed oxides will further undergo extraction process for recovery and purification of thorium.

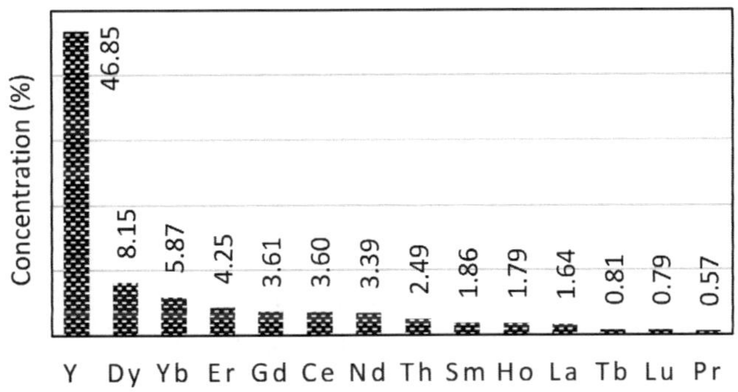

Figure 1. Composition in percent of mixed rare earth and thorium oxides after alkaline fusion.

The paper will be focusing on optimized extraction parameters which include molarity of HNO_3, percentage of TBP in the diluent and the ratio of acid to organic V_A:V_O. Single stage extraction will be done and analyze under ED-XRF to determine the concentration (ppm) before and after extraction process.

2. Experimental

The extraction process was done by mixing aqueous solution with extractant and allows extraction to take place. RE mixed Thorium Oxalate was processed in prior through alkaline fusion technique [7]. The xenotime mineral fused with sodium hydroxide in a furnace, washed with distilled water to remove loose phosphate content and further leached with concentrated at with heating. Final form of oxalate was precipitated from the acidic solution using oxalic acid. The oxalate undergo decomposition into oxide form and analyzed under ED-XRF for the determination of thorium content. The analysis was done using Thermo Scientific ED-XRF model ARL Quant ED-XRF Analyzer.

Initial steps for the solvent extraction procedure are the preparation of aqueous and organic phase. Aqueous solutions were made by dissolving oxalate powder with HNO_3. Then, extractants were prepared by mixing TBP in kerosene as the diluents. Extraction process was the done by mixing both

IOP Conf. Series: Materials Science and Engineering **785** (2020) 012015 doi:10.1088/1757-899X/785/1/012015

liquids with 30 minutes of vigorous stirring using magnetic stirrer in a beaker at room temperature (25 ± 2°C). The two-layer solution was transferred into a separatory funnel and waited until appeared clearer and transparent before flush out into two different sample bottle. The flow chart in Figure 2 shows the overall process of solvent extraction. This experiment has three different batch of solvent extraction process done.

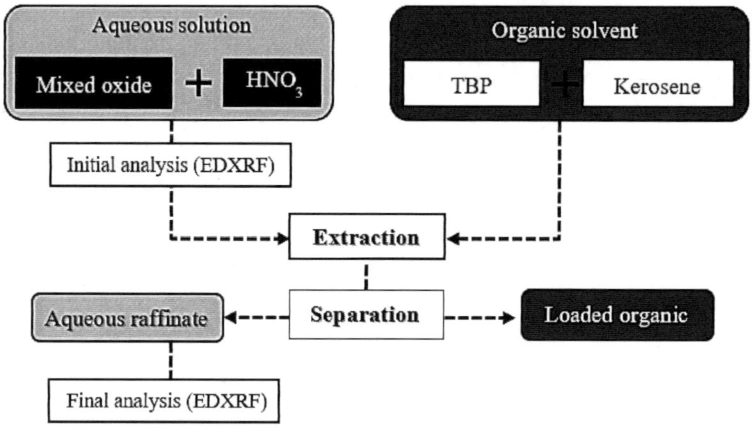

Figure 2. The overall process of solvent extraction.

The analysis to determine Thorium concentration was done via ED-XRF for initially prepared aqueous solution and the aqueous solution after undergone solvent extraction. The extraction efficiency (E) was determined from Equation (1):

$$E\,(\%) = \frac{C_{Th,\ initial} - C_{Th,\ final}}{C_{Th,\ initial}} \times 100\% \qquad (1)$$

where,

$C_{Th,\ initial}$ = Concentration of thorium in initial aqueous solution before extraction and,
$C_{Th,\ final}$ = Concentration of thorium in final aqueous solution after extraction.

The first batch extraction's parameter study was done to determine the effect of different HNO_3 concentration toward the efficiency of solvent extraction process. Five different molarity used were 1, 3, 5, 7 and 9M HNO_3. The concentration of TBP in kerosene was set at 20% whilst the ratio of V_A:V_O was kept at 1:1.

Second batch extraction was done to determine optimum percentage of TBP in kerosene during extractant preparation. Six different organics prepared which are at 5%, 10% 15%, 20%, 25% and 30% TBP in kerosene. Molarity of HNO_3 selected based on optimum extraction efficiency value from the first batch extraction. Fixed variables in this process is the amount of oxalates (0.5 gram) and V_A:V_O of 1:1.

Third batch extraction was done to determine the optimum V_A:V_O ratio. The variation ratio was made at six different ratios of 1:3, 1:2, 2:3, 1:1, 3:2 and 2:1. The optimum percentage of TBP in the diluent and optimum molarity of HNO_3 from previous experiment was used while keeping the mixed oxalate at 0.5g.

3. Results and discussion

3.1. Optimization of HNO_3 concentrations towards extraction

Tabulated data in Table 1 for first parameters studied showed that a single stage extraction give efficiency of Thorium at 28.23 % when 1 M HNO_3 used. After increased the concentration to 3 M and 5 M, the efficiencies become 72.39 and 82.12 % respectively. Studies on 7 M HNO_3 shows significant extraction efficiency with 83.13 % then dropped to 74.75 % for 9 M HNO_3. The $C_{Th, initial}$ and $C_{Th, final}$ were shown in Table 1.

Table 1. ED-XRF results of aqueous solution before and after extraction of Thorium using different concentration of HNO_3.

Molarity of HNO_3	Concentration (ppm)		E (%)
	$C_{Th, initial}$	$C_{Th, final}$	
1M	56.50	40.55	28.23
3M	52.70	14.55	72.39
5M	45.30	8.10	82.12
7M	40.90	6.90	83.13
9M	44.15	11.15	74.75

Figure 3 shows the plotted E (%) against concentration of HNO_3 for molarity study. The graph indicates that efficiency improved when molarity of HNO_3 increases. The increment was highest at 7 M HNO_3. After increased to 9 M, extraction efficiency shows a significant drop. The increasing efficiency of thorium (1 M to 7 M HNO_3) might resulted a higher distribution coefficient between aqueous and organic phase [11]. However, at higher concentration of acid, the extraction mechanism might change to solvation instead of cationic exchange due to formation of neutral complex with nitrate anions [10].

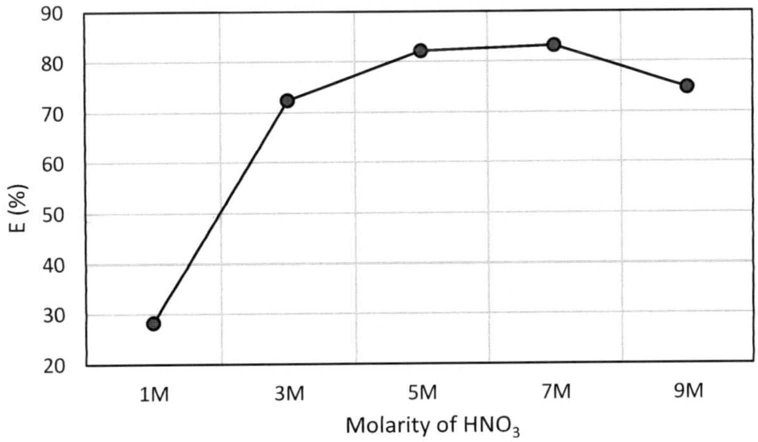

Figure 3. Effect of HNO_3 concentration upon extraction efficiency of Thorium.

3.2. Optimization of TBP concentration in kerosene towards extraction

Second extraction parameter study was done using different percentage of TBP in kerosene. From Table 2, the value of extraction efficiency is 33.22% when 5% of TBP in kerosene was used. The efficiency become 73.73% when 10% TBP added and further increased to 91.78 for 15% TBP. The extraction efficiency lingered around 98.78, 97.62 and 94.84% when 20, 25 and 30% TBP in kerosene was introduced respectively.

Table 2. ED-XRF results of aqueous solution before and after recovery of thorium using different concentration of TBP in kerosene.

Percent of TBP in kerosene	Concentration (ppm)		E (%)
	$C_{Th, initial}$	$C_{Th, final}$	
5%	45.15	30.15	33.22
10%	42.45	11.15	73.73
15%	48.05	3.95	91.78
20%	40.90	0.50	98.78
25%	39.85	0.95	97.62
30%	42.60	2.20	94.84

Figure 4 shows the graph of extraction efficiency against percentage of TBP in the extractant. The graph showed an increment when more TBP is used for extraction of Thorium. Start from 5% until 20% TBP. However, the efficiency starts to decreased slightly when more TBP used. This might due to the increasing viscosity of extractant which affect the diffusion rate of Thorium between the two phases [11] Thus, the significant efficiency of Thorium extraction was achieved at 20% TBP.

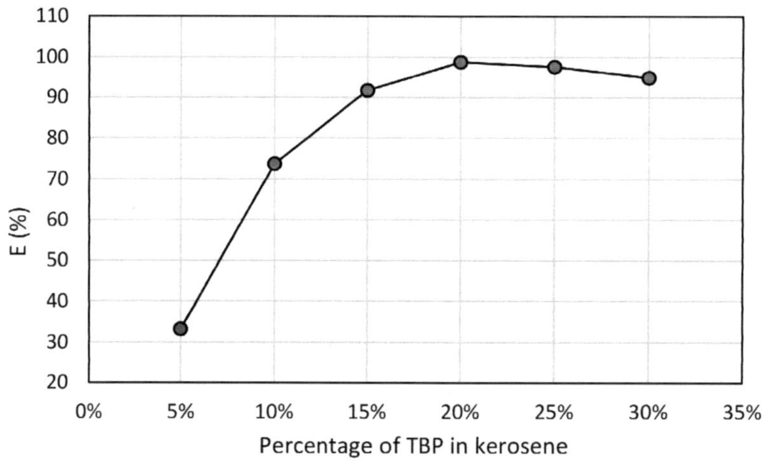

Figure 4. Effect of TBP concentration (%) in kerosene towards extraction efficiency of thorium.

3.3. Optimization of aqueous solution to extractant ratio towards extraction

The third parameter study is about the volumetric ratio of aqueous solution containing the Thorium (V_A) with the organic extractant (V_O). Six different ratio (V_A:V_O) was used, 1:3, 1:2, 2:3, 1:1, 3:2 and 2:1 From data in Table 3, the extraction efficiency relatively high for V_A:V_O of 1:3, 1:2 and 2:3 which is 93.21, 91.18 and 94.75 % respectively. When more aqueous added such as 1:1, 3:2 and 2:1, the efficiency dropped significantly from 83.13 to 72.96 until 58.70 %.

Table 3. ED-XRF results of Thorium in aqueous solution before and after extraction at different volumetric ratio of aqueous solution to extractant (V_A:V_O).

Ratio of acid to organic (V_A:V_O)	Concentration (ppm)		E (%)
	$C_{Th, initial}$	$C_{Th, final}$	
1:3	44.20	3.00	93.21
1:2	44.80	3.95	91.18
2:3	38.10	2.00	94.75
1:1	40.90	6.90	83.13
3:2	43.45	11.75	72.96
2:1	39.95	16.50	58.70

From Figure 5, the graph shows linear pattern for when 1:3, 1:2 and 2:3 ratio used. After that, the extraction efficiency dropped linearly when 1:1, 3:2 and 2:1 ratio was used. It can be concluded from the graph, when volume of aqueous solution is more than volume of extractant (2:3 until 2:1), the ability of TBP to extract thorium become less efficient.

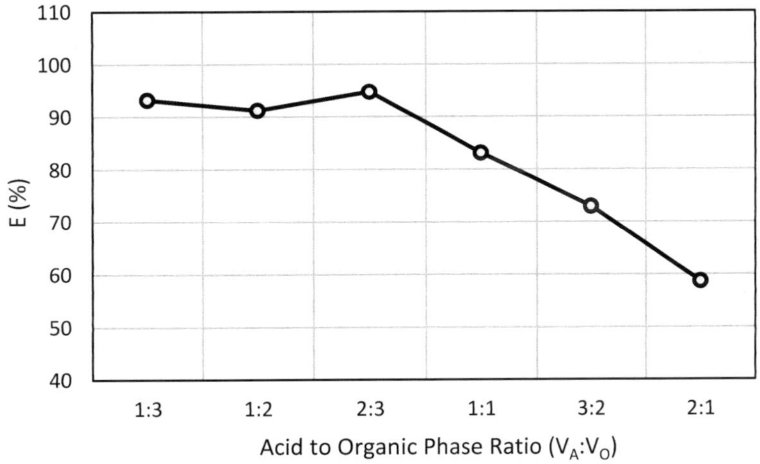

Figure 5. Effect of different volume of acid to organic phase (V_A:V_O) towards thorium extraction efficiency.

4. Conclusion

The optimization of solvent extraction technique focusing on Thorium from crude oxalate products of xenotime via alkaline fusion was achieved. From the experiments, the optimum acid concentration was at 7 M HNO_3 with extraction efficiency of 83.13 %. In the other hand, optimum 20 % of TBP content in the kerosene resulted in best efficiency with 98.78 % thorium recovered. Lastly, the optimum volume ratio between aqueous solution and extractant used is at 2:3 (V_A:V_O) with recovery percent of 94.75 %. Further study can be done by using different parameters and also by doing scrubbing to removed several elements prior to stripping process in order to increase the Thorium stripping efficiency. The extracted Thorium Oxalate from loaded organic will be further studied through precipitation technique in order to achieve Thorium of higher purity.

References

[1] Che Nor Aniza C Z B, Wadeeah M A, Amran A M and Mohd Izzat F M R 2016 Production of rare earth elements from malaysian monazite by selective precipitation. *Malaysian J. of Analytical Sci.* **21**(1) p 44-50.

[2] Cheira M F, Orabi A S, Hassanin M A & Hassan S M 2018 *J. of Chem. Data Collection.* **13-14** p 84-103.

[3] Deqian L 2017 A Review on Yttrium Solvent Extraction Chemistry and Separation Process. *J. of Rare Earth* **35**(2) p 107-119.

[4] Goletskii N D, Mashirov L G, Zilberman B Y, Fedorov S Y, Ryabkov D V, Makarychev-Mikhailov M N, Puzikov E A, Blazheva I V 2010 Extraction of Molybdenum from supersaturated solutions in nitric acid with Tributyl Phosphate solutions. *J. of Radiochem.* **52** (2) p155-61.

[5] Gupta B, Malik P, and Deep A 2002 Extraction of uranium, thorium and lanthanides using Cyanex-923: Their separations and recovery from monazite. *J. of Radioanal. and Nuc. Chem.* **251** (3) p 451-456.

[6] Janubia C B S A and Carlos A M 2010 Thorium and uranium extraction from rare earth elements in monazite sulphuric acid liquor through solvent extraction. *J. of Minerals Eng.* **23** p 298-503

[7] Meor Yusoff M S, Kaironie M T, Nursaidatul K, Ahmad Khairulikram Z and Nur Aqilah S 2015 An alternative alkaline fusion process for the production of heavy rare earth, thorium, uranium and phosphate from Malaysian xenotime *Proc. of The 13th Int. Conf. On Env., Ecosys. and Dev.* 15, 23-25 April 2015 (Kuala Lumpur)p 163-137.

[8] Michaud S, Miguirditchian M, Deblonde G, Dinh B, Heres X and Andreoletti G 2012 Modelling of thorium extraction by TBP. *Procedia Chem.* **7** p 251-257.

[9] Nasab M E 2014 Solvent extraction separation of Uranium(VI) And Thorium(IV) with neutral organophosphorus and Amine ligands. *J. of Fuel* **116** p 595-600.

[10] Nasab M E, Sam A, Milani S A 2011 Determination of optimum process conditions for the separation of thorium and rare earth elements by selvent extraction. *J. of Hydrometal.* **106** p 141-147

[11] Pandey N K, Murali R, Augustine E, Ganesh S and Joshi J B 2017 Kinetics of interphase transfer of zirconium between nitric acid and Tributyl Phosphate solutions. *J. Radioanal Nucl Chem.* **314** p 1991. https://doi.org/10.1007/s10967-017-5625-z

[12] Patil C B, Mohapatra P K and Machanda V K 2008 Transport of thorium from nitric acid solution by non-dispersive solvent extraction using hollow fiber contactor. *J. of Desalination* **232** p 272-280.

[13] Stas J, Dahdouh A and Shlewit H 2005 Extraction of Uranium (VI) from nitric acid and nitrate solutions by Tributylphosphate/kerosene. *J. of Periodica Polytechnica Ser. Chem. Eng.* **49**(1) p 3-18.

[14] Wadeeah M A, Che Nor Aniza C Z B, Amran Ab M and Sukiman S 2017 Solvent extraction of thorium from rare earth elements in monazite thorium concentrate. *Malaysian J. Anal. Sci.* **21**(6) p 1250-56

[15] Wright A and Paviet-Hartmann P 2010 Review of physical and chemical properties of Tributyl Phosphate/diluent/nitric acid systems. *J. of Separation Sci. And Tech.* **45** p 1753-62

[16] Xie F, Zhang T A, Dreisinger D, Doyle F 2014 A critical review on selvent extraction of rare earth from aqueous solutions. *J. of Mineral Eng.***56** p 10-28.

IOP Conf. Series: Materials Science and Engineering **785** (2020) 012016 doi:10.1088/1757-899X/785/1/012016

Evaluation of glass waste formed containing NORM

Nur Syuhada Izzati Ruzali[1], Syazwani Mohd Fadzil[1,a)], Rohyiza Baan[2]

[1]Nuclear Science Programme, Universiti Kebangsaan Malaysia, 43600 UKM, Bangi Selangor, Malaysia.
[2]Malaysian Nuclear Agency, Bangi, 43000 Kajang, Selangor, Malaysia.

[a]syazwanimf@ukm.edu.my

Abstract. In oil and gas production, crude oil is an important ingredient in the manufacture industry. The sludge from the oil is semi-solid waste and classified as a low-level waste that consists of a mixture of oil, sediment and, water containing radioactive element (Naturally Occurring Radioactive Materials (NORM); and high hydrocarbon concentration. It is categorized and considered as hazardous waste that required careful handling and treatment before disposal. Vitrification process was carried out in this study to immobilize the sludge containing radioactive waste. The objectives in this study is to optimize waste loading and to evaluate the glass waste formed after vitrification process. It is found that 20 wt% of oil sludge at 1200°C is the best optimize waste loading in the experiment because the glass waste formed are completely homogeneous phase.

1. Introduction

Naturally Occurring Radioactive Materials (NORM) can be found as abundant element in the earth. Uranium, thorium and potassium are radioactive elements that have long-lived and produced decay products such as radium and radon [5]. Normally, NORM can also found in a few industries including oil and gas and called as technologically enhanced naturally occurring radioactive material (TENORM) because it is produced by the human activities in the environment [5].

In oil and gas production, crude oil is an important ingredient in the manufacture industry. It is extracted from the bottom of the sea and transferred to the processing plant. It is refined and separated by boiling point to produce a large number of consumer products such as petrol and kerosene into asphalt and chemical reagents that used to make plastics and pharmaceuticals. This industry produces a variety of materials and consumed about 88 million barrels each day to the world [1]. However, it had serious concerns about the production of large amounts of oil sludge as a consequences of crude oil exploration, production and processing activities [2].

The sludge is semi-solid waste and classified as a low-level waste that consists of a mixture of oil, sediment and, water containing radioactive element; and high hydrocarbon concentration [3]. It is categorized and considered as hazardous waste that required careful handling and treatment before disposal because it can cause the most serious environmental problem and harmful to human health [1]. Due to high demand in the production of oil around the world, the annual output of sludge is estimated at around 1 billion tons per year [6]. Management and disposal of this hazardous waste remain challenging problem worldwide even as variety of treatments had been developed. Therefore, vitrification process was carried out in this study to immobilize the sludge containing radioactive waste. Vitrification is one of the processes that turns radioactive into the solid waste form which remain stable for many thousand years [4]. Besides

IOP Conf. Series: Materials Science and Engineering **785** (2020) 012016 doi:10.1088/1757-899X/785/1/012016

that, it also can reduce the volume of hazardous materials such as NORM. Thus, the objectives in this study are to optimize waste loading and to evaluate the glass waste formed after vitrification process.

2. Materials and methods

Oil sludge and glass pieces (host) were supplied from Malaysia Nuclear Agency (MNA). The sludge was thick, semi-solid form, dark brown lumps and in the wet condition. So, the sludge must first be dried in a conventional oven at 80°C for a few days. The aim of this drying process is to reduce the moisture content of waste until it less than 10 mf wt% (refers to the moisture free weight percentage). After that, the oil sludge and the glass host were homogenized by crushing to fine powders using mortar and pestle, then sieved through a 500 μm sieve. Then, the various composition of oil sludge and glass host were created and sealed in the container or plastic before vitrification process.

Table 1 shows the various composition contains 20-30 wt% of oil sludge and 70-80 wt% of glass host that were mixed together in the 50 ml alumina crucible and melted at 1100 °C -1200°C for an hour. Then, the glass waste form after vitrification process were observed and evaluated.

Table 1. Various composition of oil sludge (S) and host (H).

Composition, wt% / Temperature, °C	1100	1200
20 S + 80 H	S1	S2
25 S + 75 H	S3	S4
30 S + 70 H		S5

The glass waste form was evaluated using a few analysis such as observation through optical microscope, XRD and Field Emission Scanning Electron Microscopy (FESEM) (model Merlin, Zeiss). The optical microscope is used to investigate the formation of crystal, while XRD is used to identify the crystal phase, if formed. The FESEM is used to observe the morphology surface of sample and homogeneity. All instruments were analyzed at Centre for Research and Instrumentation Management (CRIM).

3. Result and discussion

3.1. Observation after melting process

Figure 1 showed the observation of various composition of samples after thermal process with different temperature. Fig. 1 (a) and (b) showed the samples of S1 and S2 that contained 20wt% and 25wt% of sludge and 75wt% and 80wt% of host respectively at 1100°C. Both samples were milky in white color and had rough and bumpy surfaces. The host was also not incompletely dissolved during the thermal process and had black spot obviously in the samples. This shows that oil sludge is not incompletely mixed with the host at this temperature.

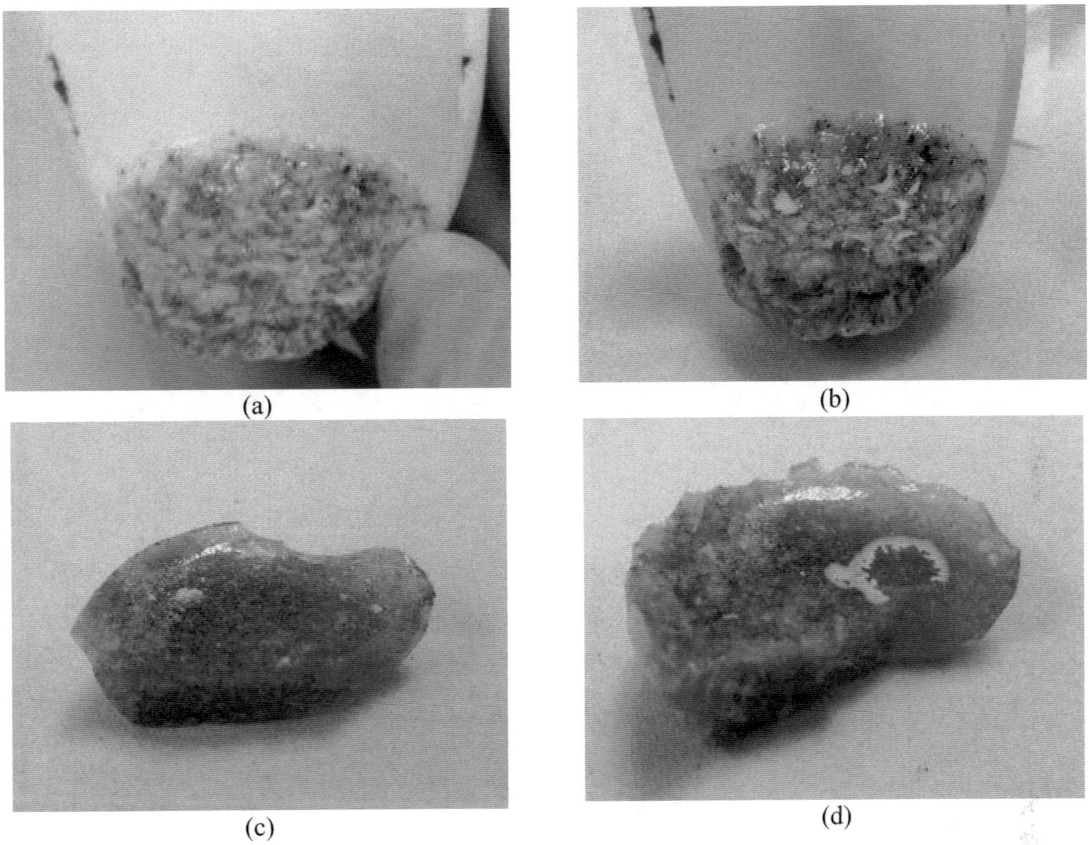

Figure 1. Observation of sample after melting process.

Figure 1(c) and (d) showed the sample S3 and S4 that were melted at 1200°C with 20 and 25 wt% of oil sludge mixed with 75 and 80 wt% of host respectively. Fig. 1(c) had bubbled and transparent green color compared to Fig. 1(d) which is more cloudy and in green color. Both samples had smooth and shiny surface than the previous samples. This indicates that oil sludge and host were completely dissolved during the melting process in Fig. 1(c), but somehow contained undissolved particle at higher sludge content (Fig. 1(d)). The observation in Fig. 1 indicated that the glass waste formed that contained 20wt% of oil sludge at 1200C was the best optimize waste loading in this study because the glass waste formed were homogenous phase. According to Singh and Chaturvedi study [7], the vitrification process of 15% waste, 35% Na2CO3 and clay started at 850°C and increased the temperature when increased the waste content.

3.2. Evaluation test

Figure 2 showed observation from optical microscope (OM). It revealed the presence of tiny crystals and bubbles in the sample. The morphology of the crystals is not clear due to very low concentration and small.

IOP Conf. Series: Materials Science and Engineering **785** (2020) 012016 doi:10.1088/1757-899X/785/1/012016

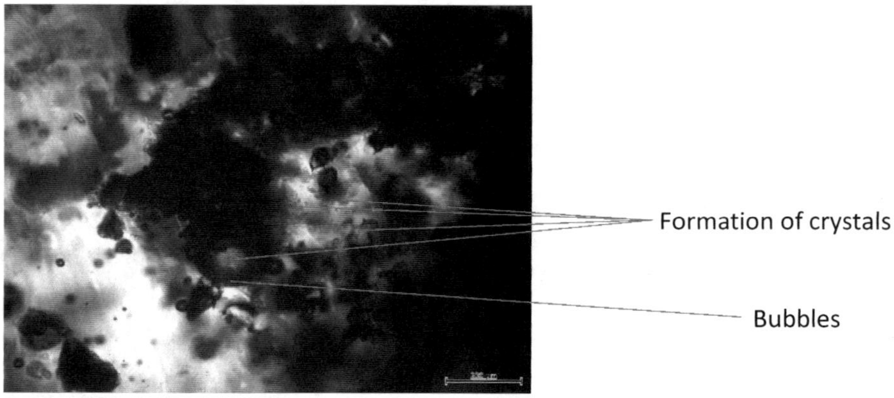

Figure 2. Observation of crystalline phase by Optical Microscope (OM).

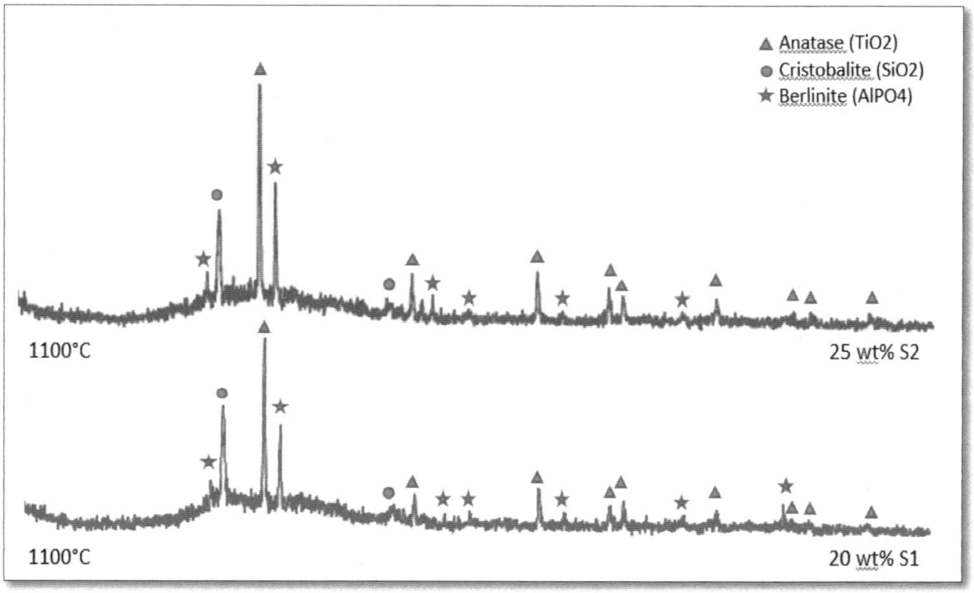

Figure 3. Crystalline phase formation of glass waste formed at 1100°C.

The different behavior of waste form with different concentration of oil sludge is attributed to specific glass/sludge interaction. At 1100°C, S1 presents with waste amount of 20wt%, as shown in Fig.3, indicates a high crystalline phase of cristobalite and berlinite at the early stage of crystallization. Berlinite is formed by covalent network of oxygen-bridged alternating phosphoric acid and alumina that shares many chemical and physical properties of silica dissolve in the sample. XRD spectrum for S2 in Fig. 3 shows the intensity of cristobalite and berlinite peaks are increase due to the increasing the concentration of waste.

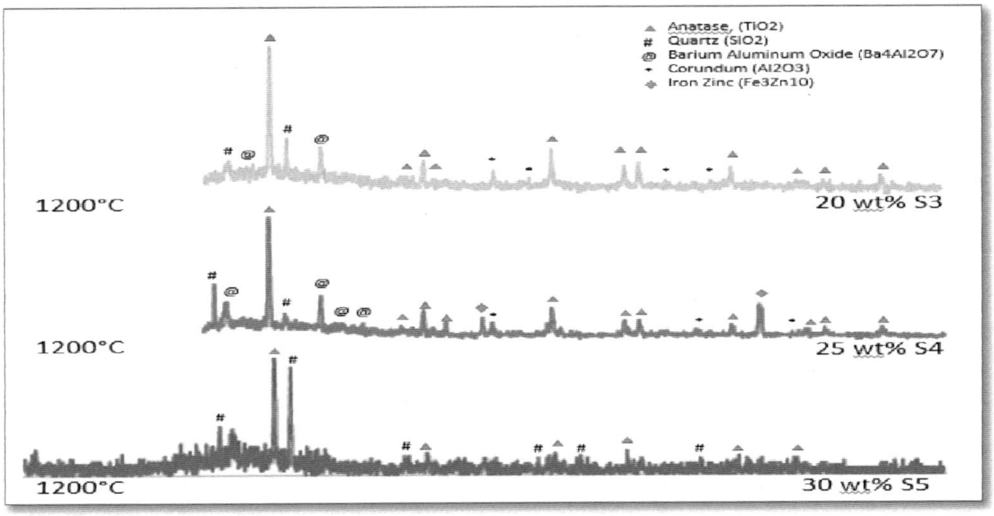

Figure 4. Crystalline phase formation of glass waste formed at 1200°C.

XRD patterns in Fig. 4 that composed of 20wt%, 25wt% and 30wt% of oil sludge respectively at 1200°C. The main crystal phase in S3 and S4 are quartz (SiO_2) and tetrabarium dialuminum oxide (Ba_4 Al_2 O_7). Formation of corundum (Al_2 O_3) typically containing traces of Fe, Ti, Vn and Cr in S3 and S4 due to the high content of Al in the oil sludge. Iron zinc (Fe_3 Zn_{10}) is appears slowly in the S4. This can be attributed to the crystallization of the distinctive of iron and zinc-rich phase in the host and oil sludge. However, S5 indicates only early stage of crystallization of quartz is formed because the sample was incompletely homogenous between host and oil sludge and this might contributed by high concentration of oil sludge showed in Fig. 4. Cristobalite and quartz are present at all melting temperature, although the peaks intensity is lower with higher melting temperature due to its partial dissolution elements.

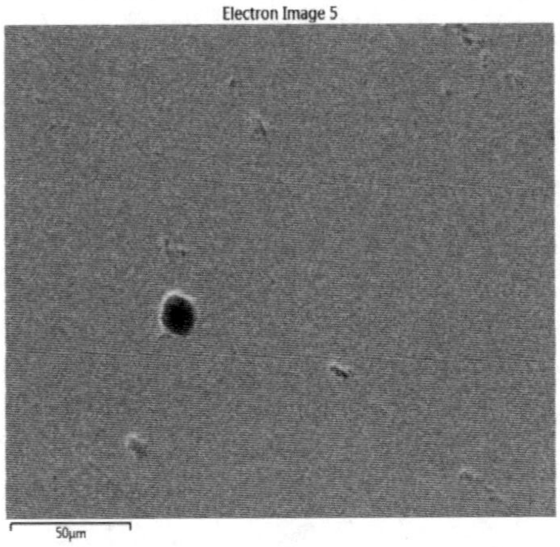

(a)

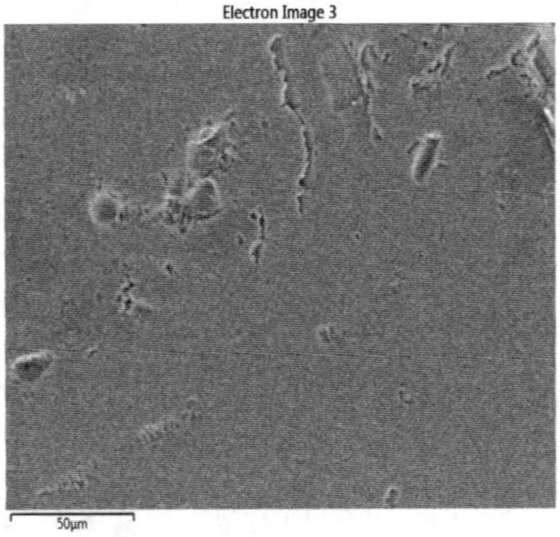

(b)

Figure 5. FESEM image of different composition at 1100°C (a) S1 (20 wt% sludge and 80 wt% host), (b) S2 (25wt% sludge and 75 wt% host.

IOP Conf. Series: Materials Science and Engineering **785** (2020) 012016 doi:10.1088/1757-899X/785/1/012016

Figure 5(a) and 5(b) showed FESEM image of different composition of waste which are magnified by 500x. S1 and S2 as shown in Fig. 5(a) and Fig. 5(b) respectively, indicates it had air bubbles and some corrosion formed in the sample might due to sample polishing before analysis. EDX analysis in Table 2 showed both sample had consistent homogeneous distribution throughout the sample. The crystal phases that were detected through the XRD and OM might be too small to be seen by FESEM.

Table 2. Element concentration through EDX analysis.

Element	Weight Percentage, wt%	σ
O	43.7	0.6
Si	38.0	0.5
Na	7.3	0.2
Ca	6.4	0.3
Al	1.9	0.2
P	1.7	0.2
K	0.6	0.1
U	0.4	0.5
Cs	0.1	0.4

4. Conclusion

As a conclusion, 20 wt% of oil sludge at 1200°C is the best optimize waste loading in the experiment because the glass waste formed are completely homogeneous phase. SiO_2, $Ba_4 Al_2 O_7$ and a small formation of $Al_2 O_3$ are the main crystalline phase that occurred in the both sample as shown in XRD analysis. The elements are distributed uniformly through EDX analysis and indicates these sample are almost amorphous because the size of crystal that can be seen in the FESEM are too tiny.

5. References

[1] Badrul I 2015 Petroleum sludge, its treatment and disposal: a review *Int. J. of Chem. and Sci.* **13**(4): p 1584-1602.

[2] Bipin K, Raj M B 2013 Petroleum oily sludge and the prospects of microwave for its remediation. *Int. J. of Eng. Res. and Tech.* **2**(11) p 359 – 370.

[3] Gopang I A, Mahar A S, Akhtar K S, Omer M, Azeem M S 2016 Charaterization of the sludge deposits in crude oil storage tanks *J. of Fac. of Eng.and Tech.* **23**(1) p 57-64.

[4] Harrison M T 2014 Vitrification of high level waste in the *UK Procedia Mater. Sci.* **7** p 10-15.

[5] Maria H P G, Andressa A D A, Yana B B, Clovis A H, Jose M D O G 2005 Radioactivity concentration in liquid and solid phases of scale and sludge generated in the petroleum industry *J.of Env. Radioact.* **81** p 47-54.

[6] Pazoki M, Hasanidarabadi B 2017 Management of toxic and hazardous contents of oil sludge in Siri Island Glob. J. of Env., Sci, and Manag., **3**(1) p 33-42.

[7] Singh I B, Chatruvedi K 2015 Immobilization of toxic metals of industrial waste via low temperature vitrification Indian J. of Chem. Tech. **22** p 162-166.

Radiation safety based on the sky shine effect in reactor

Nur Nadia Nasir[1], Siti Nor Kalillah Mahyuddin[2] and Siti Amira Othman[1,a)*]

[1]Department of Physics and Chemistry, Faculty of Applied Sciences and Technology, Universiti Tun Hussein Onn Malaysia, Pagoh Educational Hub, KM 1, Jalan Panchor, 84600, Panchor, Johor.
[2]Siti Healthcare Sdn Bhd, Klinik Siti Banting, No. 50, Jalan Bunga Pekan, Banting, 42700 Banting, Selangor.

[a]sitiamira@uthm.edu.my

Abstract. In the reactor operation, neutrons and gamma rays are the most dominant radiation. As protection, lead and concrete shields are built around the reactor. However, the radiation can penetrate the water shielding inside the reactor pool. This incident leads to the occurrence of sky shine where a physical phenomenon of nuclear radiation sources was transmitted panoramic that extends to the environment. The effect of this phenomenon is caused by the fallout radiation into the surrounding area which causes the radiation dose to increase. High doses of exposure cause a person to have stochastic effects or deterministic effects. Therefore, this study was conducted to measure the radiation dose from sky shine effect that scattered around the reactor at different distances and different height above the reactor platform. In this paper, the analysis of the radiation dose of sky shine effect was measured using the experimental method.

1. Introduction

Sky shine is a phenomenon of radiation that is only transmitted panoramic (widespread) into the environment from nuclear sources regardless of the type of radiation source and it is difficult to see because the radiation is invisible [1]. Defines sky shine as a phenomenon of radiation that penetrates the containment building. Sky shine is scattered far from its source and after experiencing multiple scattering in the atmosphere, it falling into the surrounding area. When the nuclear reactor operates it produced the radiation field and can penetrate the roofs and walls of containment buildings that are not covered with shields or the shields used is very weak to absorb the radiation. In easier words, when the reactor is operating, the source of radiation generated from the reactor core will be released through the roof and reactor walls. Then, the scattered radiation drops to the surrounding area are known as the sky shine effect. Therefore, radiation sources need to be protected regardless of the intensity of radiation emitted from the source. According to [2], there is no specified dose limit for the dose of sky shine effect but the benchmark for the sky shine effect is when radiation sources are left open, losing encasement, leaks from the reactor and the shield is not thick enough.

Nuclear reactors undergo nuclear fission process when they are operating. This phenomenon occurs when the effect of the chain reaction in the fission process during the reactor core operates and emits neutron, gamma, alpha and beta rays [3]. This incident is likely to cause the resulting light to be reflected out of the roof and reactor walls. The production of the rays will cause the surrounding dose to increase and change the classification area of the work. In order to reduce the sky shine effect scattered around, the thick shielding and shielding with the ability to absorbing excess radiation should be used.

Exposure to the high doses will cause a worker likely to receive a stochastic or deterministic effect [4]. To reduce the exposure dose, employees should follow the basic principles of radiation protection, exposure dose control and work area classification. Classification of work areas is important for the safeguard measures and the appropriate safety provisions for each area.

2. Methodology

In order to carry out experimental in this study, the detection device used for monitoring gamma dose and neutron dose is a telescopic survey meter as shown in Figure 1(a). The telescopic survey meter provides readings directly. Additionally, other side materials are measuring tape, PVC conduit pipe as in Figure 1(b) and adhesive tape.

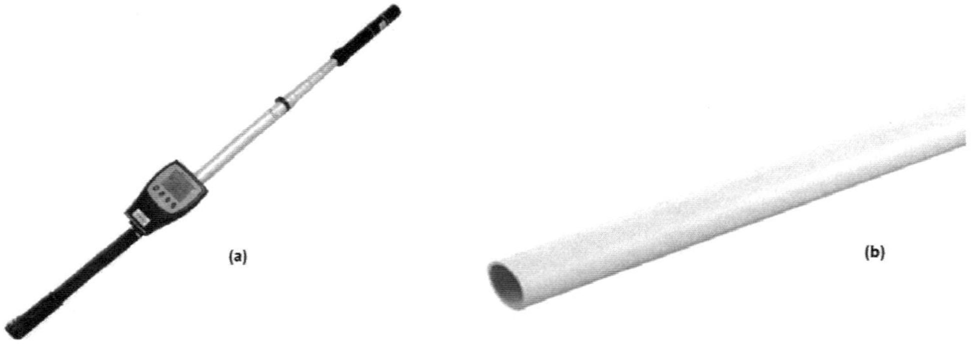

Figure 1. (a) Telescopic survey meter and, (b) PVC conduit pipe.

3. Results and Discussion

3.1 Measurement of radiation dose

In this study, dose measurements were taken at 750 kW. The distance is divided into ten points for the measurement. Gamma and neutron radiation dose measurements were detected using telescopic survey meters. Before taking the radiation dose of the sky shine in the reactor, the background dose was taken on the outside of the reactor. Based on Figure 2, the red circle in the diagram is the location of the study during the conduct of the monitoring in the reactor. Besides that, Figure 3 shown ten points that have been chosen for the measurement.

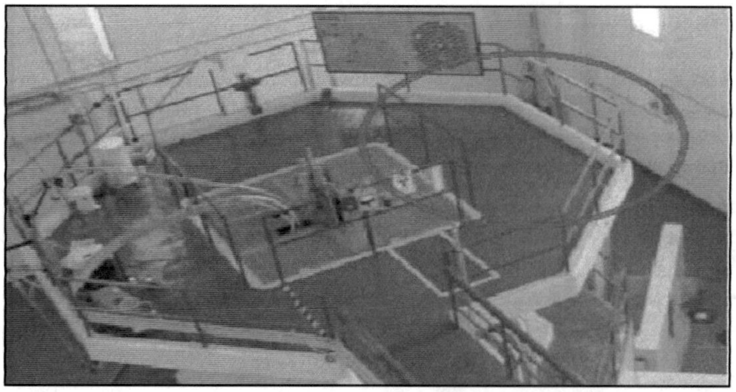

Figure 2. Red circle is location of the study area.

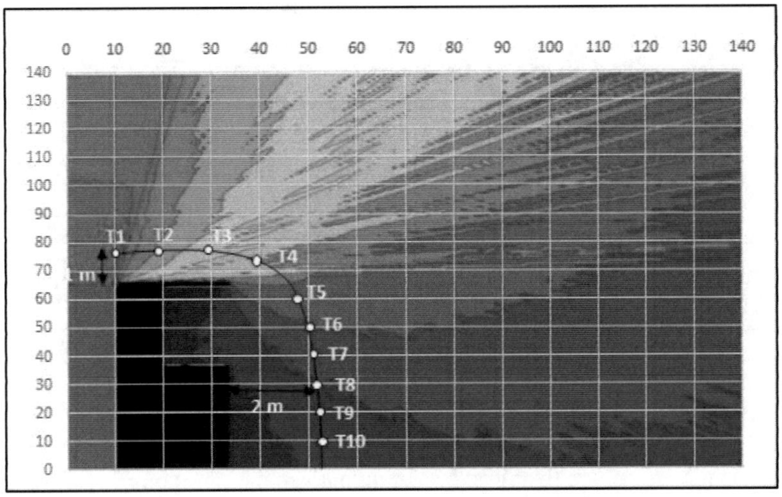

Figure 3. Point for radiation dose measurement.

Neutron radiation is the most dominant radiation generated by nuclear reactors. This radiation has a very strong penetration. Strong and thick shielding is needed to absorb neutron radiation. In this reactor, Barytes concrete has been used as a shield [5]. The main shield of the reactor is the vertical built water pool, thick concrete shields and radius-like pool water. The water depth above the core and the concrete thickness was designed to provide more radiation protection to workers and when the reactors operating at 1 MW under the normal conditions. In this study, the shields built in the reactor were able to absorb the neutron light generated from the reactor because the neutron ray dose and the effects of the sky shine could not be seen and detected. The use of water as a moderator and coolant in the reactor is not only effective in slowing down the neutrons and cooling down but also as a shield to the generated radiation inside the reactor core so it will not endanger the reactor operators on the pond during repair and others activity.

Based on Figure 4, the data taken when the reactor operates at 750 kW, result obtains that the dose of radiation decreases at a height of 1 m and at different distances. The radiation charge obtained at the surface of the reactor platform is approximately 1226 μSv/h and decreases as the distance increases up to 9 m is 16 μSv/h.

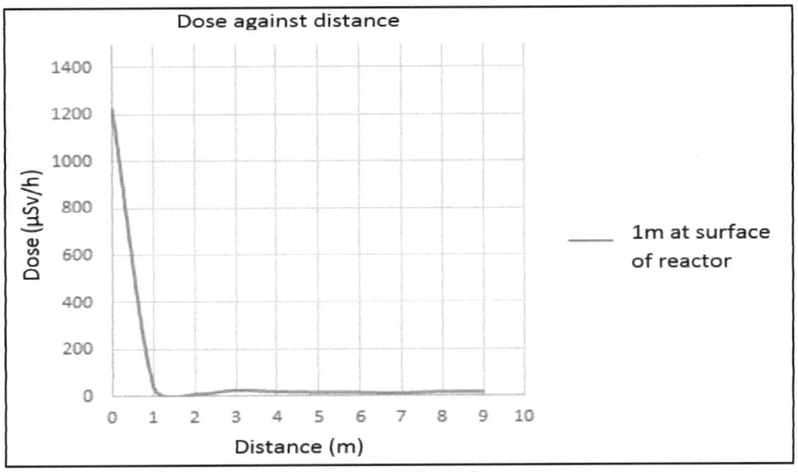

Figure 4. Point for radiation dose measurement.

4. Conclusion

As a result of this study, the determination of the radiation dose of sky shine effect at different distances found that when the distance increases, the radiation dose of the sky shine decreases. For the determination of the radiation dose of sky shine effect at different altitudes, it is found that when height increases, the radiation dose of the sky shine decreases. This proves that the intensity of radiation decreases when the distance or elevation increases. According to [5], the maximum distance of the sky shine effect depends on the activity factor of the radiation source, the type of radiation source used, the height of the shield wall and the loss of a shield.

5. References

[1] Cossairt J D, Couch J G, Elwyn A J and Freeman W S 1985 Radiation measurements in a labyrinth penetration at a high-energy proton accelerator Health Phys. **48**(2).
[2] Prabhakar Rao G and Sarkar P K 1998 Estimation and analysis of neutron skyshine at particle accelerators (India: Bhabha Atomic Research Centre).
[3] Mwambinga S A and Yoo S J 2013 Public dose assessment modeling from skyshine by proton accelerator. *Trans. of the Korean Nucl. Soc. Autumn Meeting Gyeongju, Korea* p 24-25.
[4] Sharanjeet Kaur B and Ariffin A E 2000 Dos dedahan radiasi dan penglihatan warna pekerja di MINT *Buletin Kesihatan Masyarakat Isu Khas 2000* p 129- 134.
[5] Halmshaw R 1995 *Industrial Radiography: Theory and Practice* (United Kingdom: Chapman & Hall) 2nd ed.

Acknowledgement

The authors would like to thank the Agensi Nuklear Malaysia for facilities provided and gratefully acknowledged Universiti Tun Hussein Onn Malaysia for the financial support from a research grant (H074 and H417) that make the research possible.

Water quality monitoring of Reactor TRIGA PUSPATI

M. R. Hasniyati[1,a)], H. S. Na'im[1], J. T. Oon[2], M. H. M. Zin[1], H. Zaredah[1] and L. Tonny[1]

[1]Reactor Technology Centre, Malaysian Nuclear Agency, Bangi, 43000, Kajang, Selangor, Malaysia.
[2]School of Chemical and Energy Engineering (SCEE) Universiti Teknologi Malaysia 81310 Johor Bahru, Johor Malaysia.

[a]hasniyati@nuclearmalaysia.gov.my

Abstract. At present, Reactor TRIGA PUSPATI (RTP) is the only nuclear research reactor in operation in Malaysia since 1982. Each year, the reactor is operated on average for about 600 hours mainly for the advanced neutron and gamma radiation studies as well as for isotope production, sample activation and student training. About one-third of the RTP's core volume is occupied by water. The water-cooling and purification systems maintain low water conductivity, remove impurities, maintained the optical clarity of the water and dissipating the reactor heat. They consist of a water surface skimmer, pump, filter, demineralizer, heat exchanger unit, associated piping and valves as well as other miscellaneous instrumentations. Thus, the aim of this study is to investigate water quality and maintains the water chemistry to the required specifications at all times in order to ensure longevity and performance of the reactor system.

1. Introduction

Cooling water chemistry and its effects on fuel elements cladding and reactor components materials performance are important factors for safe and reliable operation of research reactor. This is because at elevated temperature, water is an aggressive medium and may causes corrosion problems in fuel elements cladding and reactor structural materials. For this reason, reliability and safety for research reactor are achieved by using proper fuel cladding and structural materials. Also, cooling water regimes must be adjusted and proved to prevent dangerous corrosion problems. Thus, the aim of this study is to investigate water quality and maintains the water chemistry to the required specifications at all times in order to ensure longevity and performance of the reactor system.

Water is used as a primary coolant in most of research reactor and as a medium in the secondary circuit as well as in associated auxiliary systems. At elevated temperatures water is an aggressive substance when it is in contact with structural materials, which means that the reliability of many systems in research reactor depends strongly on cooling water control. Water coolant chemistry and corrosion problems are issues of special importance in the safe and reliable operation of nuclear power plants. So that water regimes for research reactor must be developed and proved to be satisfactory. Normally, reliability and safety are achieved by using proper fuel cladding and structural materials and by taking special measures to prevent dangerous corrosion, erosion and other processes.

High purity water coolant should ensure adequate corrosion resistance of all the materials used such as stainless and carbon steels, zirconium, copper and nickel alloys. Table 1 shows the sources of impurities that could enter into the reactor and the measures that could be taken to prevent or decrease the accumulation of impurities.

Table 1. Source of impurities that could enter into the reactor and the measures that could be taken to prevent or decrease the accumulation of impurities.

Source of impurities	Chemical composition of impurities	Measures to prevent or decrease impurity accumulation
Initial water for circuit	Soluble salts (Na, K, Mg, Ca, Cl etc)	Efficient purification of water
Make-up water	Soluble salts (Na, K, Mg, Ca, Cl)	Efficient purification of water
Cladding materials	Corrosion products	Suppression of corrosion by water conditioning
Structural Materials	Corrosion products: Fe, Co, Cr, Ni, Mu, Cu, etc	Suppression of corrosion by water conditioning

Research reactors have a variety of applications, such as training, neutron activation analysis and radioisotope production. In the majority, most of the research reactors use water as a core cooling fluid, moderator and biological shielding. RTP is a pool type reactor, where the reactor core sits at the bottom of a 7-meter high aluminium tank and this is surrounded by a biological shield made of high density concrete.

2. Methodology

Reactor's water (350ml) was sampled at the reactor tank before and after operation at three different places; surface of the pool reactor, 3 meter from the surface of the pool reactor and at the bottom of the pool reactor using vertical water sampling bottle as shown in Figure 1.

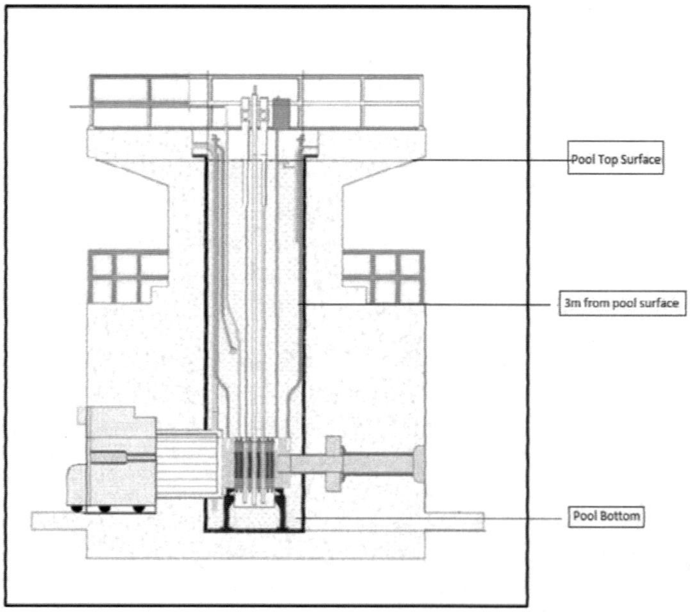

Figure 1. Reactor's water sampling location.

Reactor's water was then being inserted into a transparent 350ml acrylic sample bottles and sent to reactor physics laboratory for further analysis. The sample bottles were tightly sealed to prevent spillage. These steps have been replicates at 3 times and the temperature was recorded. In average, the temperature recorded for all the sampling is around 35°C. A gamma ray spectroscopy consists of hyper pure Germanium (HpGe) detector with 8192 channels computerized multi-channel analyzer were used to qualitatively determine the radionuclides and quantitatively calculate the specific activities. All samples were position at a distance of 5cm from the detector surface. A 5cm thick lead shield was used to reduce the effects of surrounding background radiation. The counting process was conducted for 90 minutes while the detector dead time kept below that 20%. The specific activity for each radionuclide was calculated using equation [2] as shown below:

$$As = \frac{(Cs - Cb)}{t\ \varepsilon\gamma\ P\gamma\ Ms} \tag{1}$$

where;

As = specific activity ($Bqkg^{-1}$)
Cs = net count intensity from samples (cps)
Cb = count rate intensity of background (cps)
t = counting period
$\varepsilon\gamma$ = efficiency of the detector for particular gamma energy
$P\gamma$ = branching ration
Ms = mass of the reactor water samples

3. Results and Discussion

The concentration level of various impurity elements and radionuclides production rate in reactor water is important to maintain the safe operational of the reactor. The determination of activated nuclides using gamma spectrometry has found several radionuclides including of ^{24}Na, ^{27}Mg, ^{41}Ar, ^{56}Mn and ^{60}Co at three different places are shown in Table 2 below. The highest specific activity was found to be in Ar-41 at pool surface, 3 meter form surface and bottom of the pool reactor. It is known that Ar-41 is produced by activation of stable Ar-40 nuclide [1]. Certain amount of activated argon radionuclides decays while in the water and the rest might escape from the reactor pool water. The increase of water temperature will reduce the solubility of argon in water which consequently accelerates the increase of argon concentration in reactor hall [3].

On the other hand, Na-24 nuclide can be produced through the reaction of salt impurities in reactor water or an activation of structural aluminum through Al-27 and Na-24 reaction [4]. Radionuclides of Mg-27, Mn-56 and Co-60 were produced from activation of metals or alloys that has been used in Systems, Structures and Components (SSCS) in reactor pool tank.

Table 2. Specific activities of radionuclides at three different places.

Nuclides	Energy (keV)	Half-life (m/h/y)	Abundance	Specific Activity ($Bqkg^{-1}$)		
				Pool Surface	3m from surface	Bottom pool
Na-24	1368.60	14.96h	100.00	2186.48±0.001	2036.56±0.001	2109.87±0.001
Mg-27	843.76	9.46m	71.40	563.18±0.001	523.90±0.001	536.31±0.001
Ar-41	1293.64	1.83h	99.16	40985.12±0.001	51236.78±0.001	43741.42±0.001
Mn-56	846.76	2.58h	98.87	2133.78±0.001	2366.90±0.001	2190.41±0.001
Co-60	1173.24	5.27y	99.90	226.54±0.001	189.82±0.001	197.33±0.001

IOP Conf. Series: Materials Science and Engineering **785** (2020) 012018 doi:10.1088/1757-899X/785/1/012018

4. Conclusion

The concentration level of various impurity elements and radionuclides production rate in reactor water is important to maintain the safe operational of the reactor. The determination of activated nuclides using gamma spectrometry has found several radionuclides including of Na-24, Mg-27, Ar-41, Mn-56 and Co-60.

5. References

[1] Ajijul Hoq M, Malek Soner M A, Rahman A, Salam M A and Islam S M A 2016 Estimation of Na-24 activity concentration and release rate from the TRIGA Mark-II research reactor J. of Environ. Radioact. **153** p 68-72.

[2] Jabbar T, Akhter P, Khan K, Jabbar A and Saleem K 2009 Radiological impact of composite food served at PINSTECH *Food and Chem. Toxicol.* **47** p 1205-1208.

[3] Hamzah N S, Jalal Bayara A M, Rabir M., Abdul Khalil M H, Mustafa M K A, Yusrib N S, Abd Halim N AF, Afrizal NH, Wan Ibrahim W AL 2018 Coolant Impurities and Radionuclides Production Rate Monitoring at Reactor TRIGA PUSPATI. . Research and Development Seminar 2010, Malaysian Nuclear Agency.

[4] Zalina L, Muhamat O, Mohd Zaidi I, Mohd Abd Wahab Y, Julia A B K and Nik Marzukee N I 2010 Characterization of spent ion exchange resins from PUSPATI TRIGA Reactor Nuklear Malaysia *Res. and Dev. Sem 2010* (Malaysian Nuclear Agency).

Acknowledgement

The authors are pleased to acknowledge all the research officers and assistant engineers at Reactor Technology Centre, Technical Support Division, Malaysian Nuclear Agency for their supports in this project.

IOP Conf. Series: Materials Science and Engineering **785** (2020) 012019 doi:10.1088/1757-899X/785/1/012019

Analysis of seismic events in Malaysia using International Monitoring System (IMS) data of the Comprehensive Nuclear-Test-Ban Treaty Organization (CTBTO)

M Z Zolkaffly[1,a)] and F I A Rashid[1]

[1]Planning and International Relations Division, Malaysian Nuclear Agency, Bangi, 43000 Kajang, Selangor, Malaysia.

[a]zulfakar@nuclearmalaysia.gov.my

Abstract. The International Monitoring System (IMS) of the Comprehensive Nuclear-Test-Ban Treaty Organization (CTBTO) is designed to be non-discriminatory and does not single out any country or region for enhanced monitoring. The seismic network, when completed, will consist of 50 primary and 120 auxiliary seismic monitoring stations in 76 countries around the world to detect and locate underground nuclear explosions. The data from the IMS stations can also be used for scientific and civil purposes. In this study, we present the results of data analysis of selected seismic events in Malaysia from 2015 to 2018 using IMS seismic stations data. Important seismicity parameters including the epicentral, magnitude distribution and seismic phases were presented in this study. We also perform comparison of the seismic events recorded by the IMS with similar seismic events detected by others seismic network at local, regional and international level. This study demonstrates that other than for nuclear explosion monitoring, the IMS data can also be used for civil and scientific applications and can make significant contribution in seismological research.

1. Introduction

Article IV of the Comprehensive Nuclear-Test-Ban Treaty (CTBT) provides that the establishment of a verification regime shall be capable of meeting the requirements of the Treaty when it enters into force [1]. The IMS developed for this verification regime uses four modern technology, namely seismic, hydro-acoustics, infrasound and radionuclide technologies to monitor the globe for any sign of a nuclear explosion. As shown in Figure 1, the IMS consists of 321 monitoring stations and 16 radionuclide laboratories built worldwide [2]. The CTBT's International Data Centre (IDC) in Vienna began the routine automatic and interactive processing of IMS data since June 1999 [3]. From the IDC, data and analysis are forwarded to the member states, in near real time for the automated data. For the seismic events, the detected and located events are also systematically included in the Reviewed Event Bulletin (REB). Apart of its primary mandate to monitor for any sign of nuclear test explosion, the seismicity catalog of the IDC also constitutes to a wide range of global seismological research and analyses.

IOP Conf. Series: Materials Science and Engineering **785** (2020) 012019 doi:10.1088/1757-899X/785/1/012019

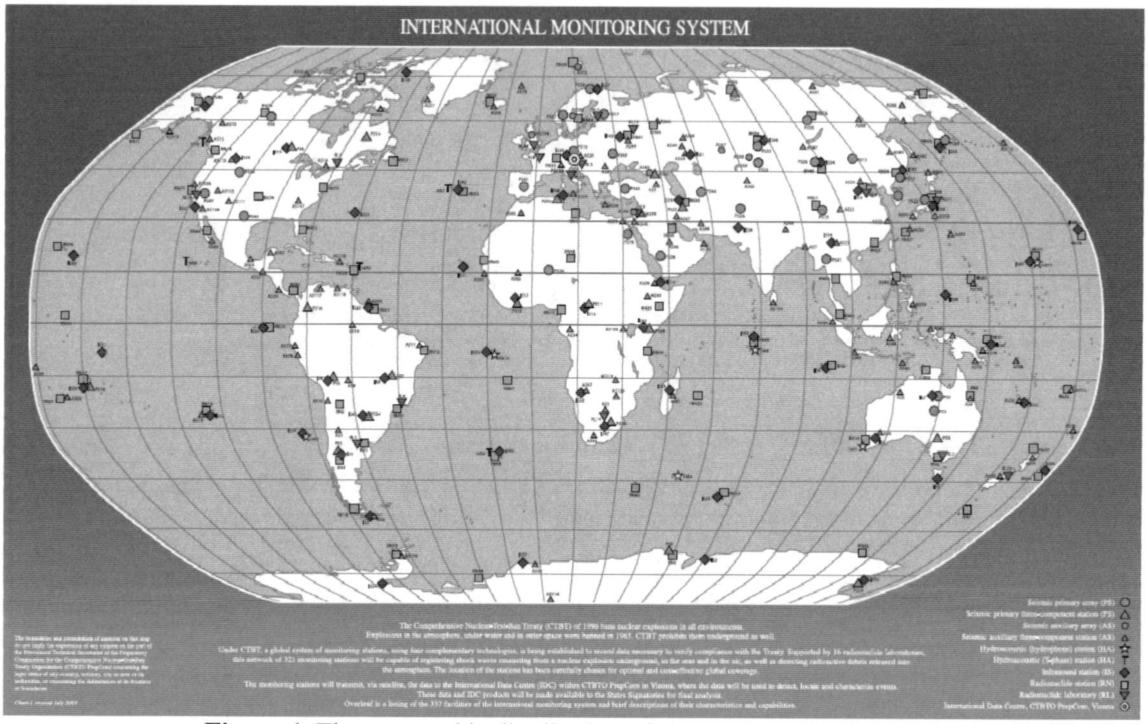

Figure 1. The geographic distribution of the CTBT IMS Network.

Malaysia is situated on the southern edge of the Eurasian and considered to have low seismicity profile. It is closed to the most two seismically active plate boundaries, the inter plate boundary between the indo-Australian and Eurasian plates on the west and the inter-plate boundary between the Eurasian and Philippines plates on the east [4]. The tectonic framework for the whole of Malaysia covers between longitudes 90 E to 140 E and latitudes of 12 S to 20 N [5]. Figure 2 shows the major tectonic in Malaysia [6].

1.1. Seismic activity in Malaysia

Peninsular Malaysia is classified as a seismically stable as no earthquake has originated from the area [7]. However, it is still considered vulnerable as it lies close to Sumatran fault and Sumatran Subduction zone. Large earthquakes that originated from these two active areas have created considerably ground motion over western part of Peninsular Malaysia.

East Malaysia (Sabah and Sarawak) on the other hand, is classified as moderately active in seismicity [7]. Figure 3 shows the epicenter of felt earthquakes ever recorded in East Malaysia. Sabah is prone to earthquake activities compared to other parts of Malaysia, where it has suffered several earthquakes of moderate magnitude. Some of these earthquakes have caused structural damage to buildings and other infrastructures and injuries to humans [8]. In 1976, one of the worst earthquakes occurred in Sabah when a 5.8 magnitude on Richter scale earthquake shocked Lahad Datu. Later in 1991, an earthquake of magnitude 4.5 on Richter scale shook Ranau, resulting in structural damage to a school. Eastern Sabah is also exposed to tremors caused by earthquakes originated in the southern Philippines and the Straits of Macassar, Sulu Sea and Celebes Sea.

Similarly, Sarawak also has experienced several moderate-magnitude earthquakes of local origin [7]. Beside of having earthquakes of local origin, Sarawak was also affected by long-distance earthquake that originated from Southern Philippine and the Straits of Macassar, Sulu Sea and Celebes Sea.

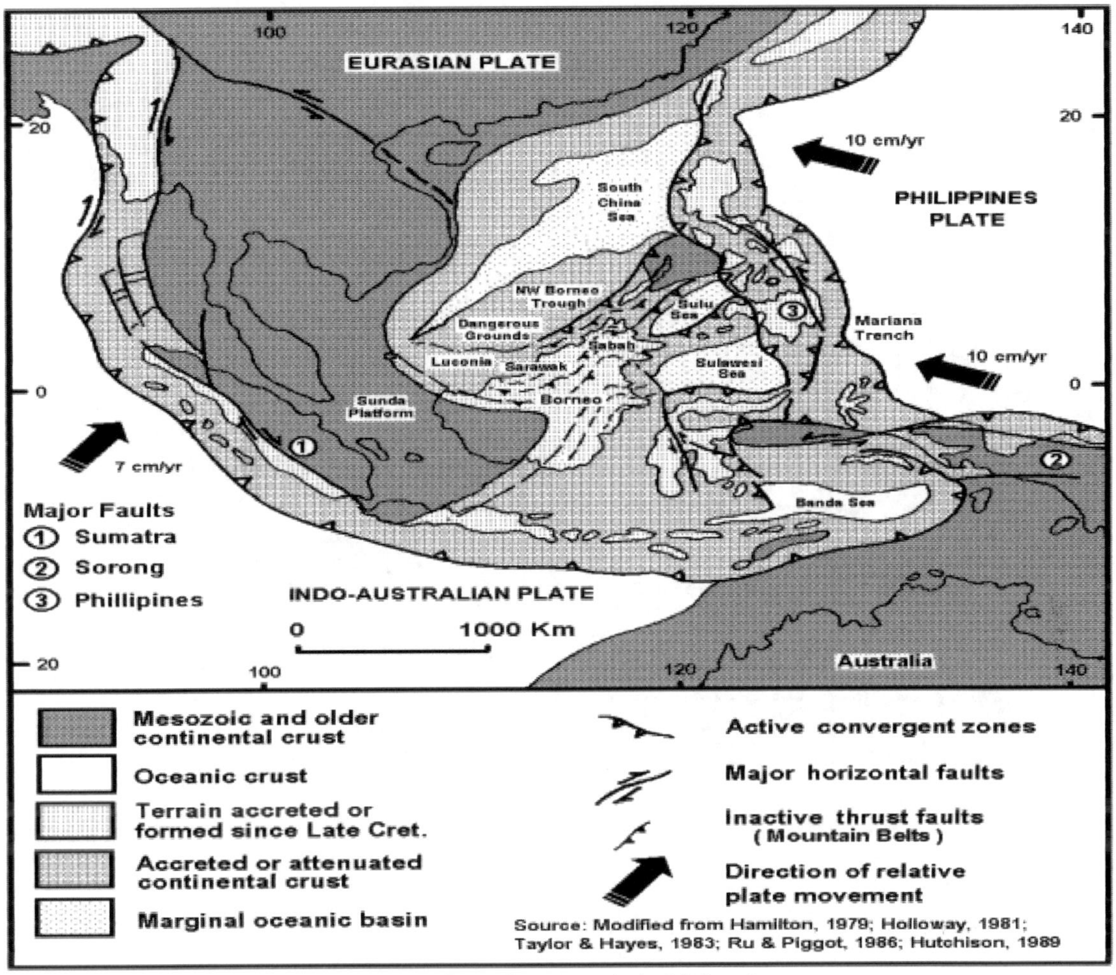

Figure 2. Major tectonic in Malaysia

To locate seismic epicenter and its magnitude, the seismologists require seismic data from the seismic stations. The quantity and quality of the data will determine the accuracy of the epicenter's location. The quantity of data means the number of time arrival of seismic phase used in the seismic determination, while the quality means the distribution of the station with respect to the epicenter. The Malaysian Meteorological Department (MMD) serves as national seismology information centre for the country. MMD provides information, advice and consultation related to earthquake to the Government, authorities, general public, as well as related users such as architects and planners for well-being, safety and sustainable development of the country.

IOP Conf. Series: Materials Science and Engineering **785** (2020) 012019 doi:10.1088/1757-899X/785/1/012019

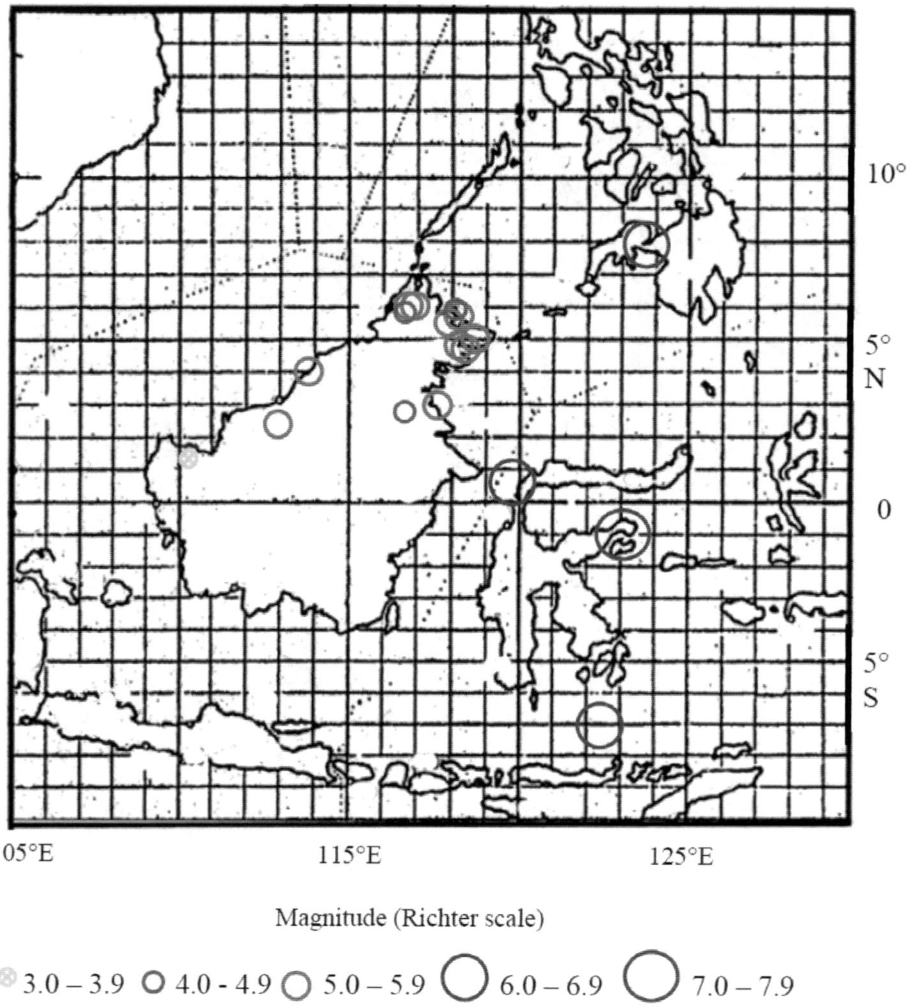

Figure 3. Epicentre of felt earthquakes in East Malaysia.

2. Methods

The CTBTO is operating the IDC in Vienna, Austria. Monitoring data from IMS stations are transmitted to the IDC for storing, processing and analysis. As a result, various products are produced at the IDC, including event bulletins and listings. In this study, the listing of seismic events occurred in Malaysia for the time period of January 1, 2015 to December 31, 2018 were obtained from the IDC Reviewed Event Bulletin (REB). Event bulletins produced by other seismic networks, namely the U.S. Geological Survey (USGS) and Incorporated Research Institutions for Seismology (IRIS), were also obtained for the similar time period. Such events were screened to ensure that only the detected events fall within Malaysia's region are considered in this study. This is done by confirming the reported coordinates of each of recorded events are identical to the latitude and longitude of Malaysia's region. For event listed in CTBTO REB, we then retrieved the waveform data for the listing events and performed analysis using Geotool software to identify important seismicity parameters including location, depth and seismic phases. For other events listed by other seismic networks, we also perform the same step by compiling the important seismicity parameters. Lastly, we compared the seismicity parameters of events recorded by REB and other networks. The methodology used in this study is shown in Figure 4.

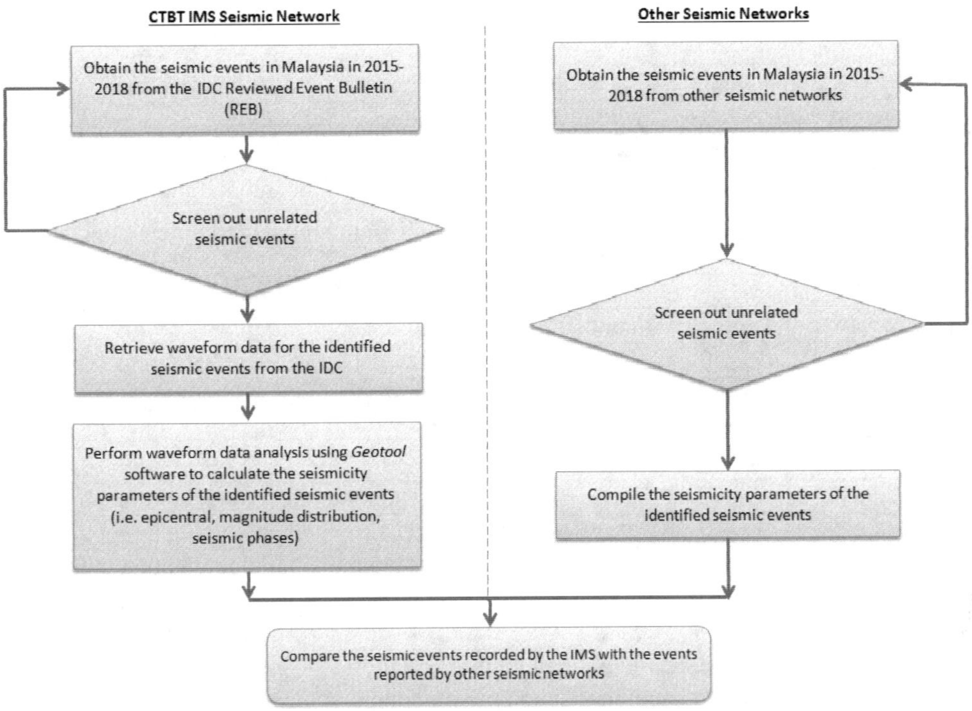

Figure 4. Methodology of the study.

3. Results and discussion

As reported in the REB of seismic events in Malaysia from the period of 2015 to 2018, the CTBT IMS has recorded 16 seismic events originated in Malaysia, where all of them were in East Malaysia (Sabah and Sarawak). Out of these 16 events, only 2 were located in Sarawak, each in 2015 (Event ID 11997735) and 2018 (Event ID 15916404). The rest of the events were located in Sabah. From our further analysis on the identified waveforms utilizing the Geotool software, all of these events are classified as natural earthquakes with moderate body wave magnitude (mb) ranging from the lowest 3.4 mb to maximum magnitude of 5.2 mb. Table 1 shows the list of earthquakes recorded by the CTBT IMS network from 2015 to 2018, including its important seismicity parameters obtained from our waveform analysis.

Table 1. 2015-2018 earthquake events in East Malaysia recorded by the CTBT IMS network.

CTBT IMS Network										
Year 2018	**Event ID**	**Date**	**Time**	**Latitude**	**Longitude**	**Magnitude**				**Depth**
1	15530461	08/03/2018	13:06:10	5.8856	116.5306		mb 4.4	mbtmp 4.4	Ms 4.3	0.0
2	15848813	27/05/2018	7:53:42	5.2794	119.0089		mb 3.8	mbtmp 3.8	Ms 3.2	0.0
3	15916404	16/06/2018	7:47:58	1.0666	111.3856		mb 3.8	mbtmp 3.8	Ms 2.8	0.0
Year 2017	**Event ID**	**Date**	**Time**	**Latitude**	**Longitude**	**Magnitude**				**Depth**
1	14228945	26/03/2017	9:30:49	4.7966	118.7287	ML 4.1	mb 3.7	mbtmp 4.0	Ms 3.4	52.8
2	14888716	24/09/2017	0:25:23	5.9464	118.3581		mb 3.5	mbtmp 3.5	Ms 2.7	0.0
Year 2016	**Event ID**	**Date**	**Time**	**Latitude**	**Longitude**	**Magnitude**				**Depth**
1	13497338	26/08/2016	1:39:32	6.4172	117.3407		mb 3.6	mbtmp 3.6	Ms 2.8	0.0
Year 2015	**Event ID**	**Date**	**Time**	**Latitude**	**Longitude**	**Magnitude**				**Depth**
1	11783042	19/03/2015	21:56:00	5.4795	118.3777	ML 3.7	mb 3.8	mbtmp 3.8	Ms 2.9	0.0
2	11997735	14/05/2015	10:25:07	1.8141	110.5797		mb 3.6	mbtmp 3.6		0.0
3	12047012	04/06/2015	23:15:42	6.039	116.6377	ML 4.4	mb 5.2	mbtmp 5.2	Ms 5.5	0.0
4	12048513	05/06/2015	13:12:14	6.6386	117.191	ML 4.5	mb 3.7	mbtmp 3.9	Ms 3.1	0.0
5	12048801	05/06/2015	15:13:32	6.0524	116.6861	ML 4.7	mb 3.9	mbtmp 3.9		0.0
6	12050340	06/06/2015	5:45:13	6.2528	116.8675	ML 4.1	mb 4.1	mbtmp 4.2	Ms 3.4	0.0
7	12068968	12/06/2015	18:25:34	6.4085	117.3888	ML 4.1	mb 3.4	mbtmp 3.6		0.0
8	12068970	12/06/2015	18:29:24	6.0878	116.8661		mb 4.4	mbtmp 4.7	Ms 4.3	75.0
9	12098600	23/06/2015	9:32:34	5.9739	116.5596		mb 3.7	mbtmp 3.9		49.1
10	12199515	26/07/2015	10:04.9	6.2731	117.0466		mb 3.9	mbtmp 3.9	Ms 3.4	0.0

Figure 5 (a) and (b) show the snapshot of process of locating the origin location (epicenter) of 2 seismic events recorded by the CTBT IMS network. These epicenters of selected earthquakes are one of the seismic event parameters obtained from the Geotool-analysis performed in this study. Snapshot of seismic waveform analysis using Geotool software for both events are also shown in Figure 6. Figure 7 shows the epicenter of all 16 events recorded in East Malaysia by the CTBT IMS network for the period of this study. This map was tabulated based on the seismicity parameters determined from our waveform analysis.

The year of 2015 has recorded the highest number of earthquake occurrence in East Malaysia throughout the study period. The earthquake event recorded by the IMS network on June 4, 2015 23:15:43 UTC (Event ID 12047012) is consistent with the Ranau's moderate earthquake on June 5, 2015 7:15 am local time as reported by the MMD. The MMD reported that the epicenter of this earthquake was located at 16 km northwest from Ranau with the depth of 54 km beneath the earth. Figure 8 shows the news clip reported the earthquake that shook Kota Kinabalu and other parts of Sabah's west coast on June 5, 2015 after an earlier magnitude 5.9 earthquake hit near Ranau district [9].

In the next step, in order to compare those earthquakes recorded by the CTBT IMS network with other global networks, we compiled all the seismic events by the USGS and IRIS networks recorded in Malaysia for the same study period. Both the earthquake catalogs of USGS and IRIS reported in total 12 earthquakes were taking place in East Malaysia, slightly lower than the number of earthquakes recorded by the CTBT IMS network. Furthermore, similar with the CTBT IMS, there was no seismic event recorded took place in Peninsula Malaysia by both the USGS and IRIS networks. Table 2 and 3 show the list of earthquake events including its seismic parameters recorded and reported by the USGS and IRIS.

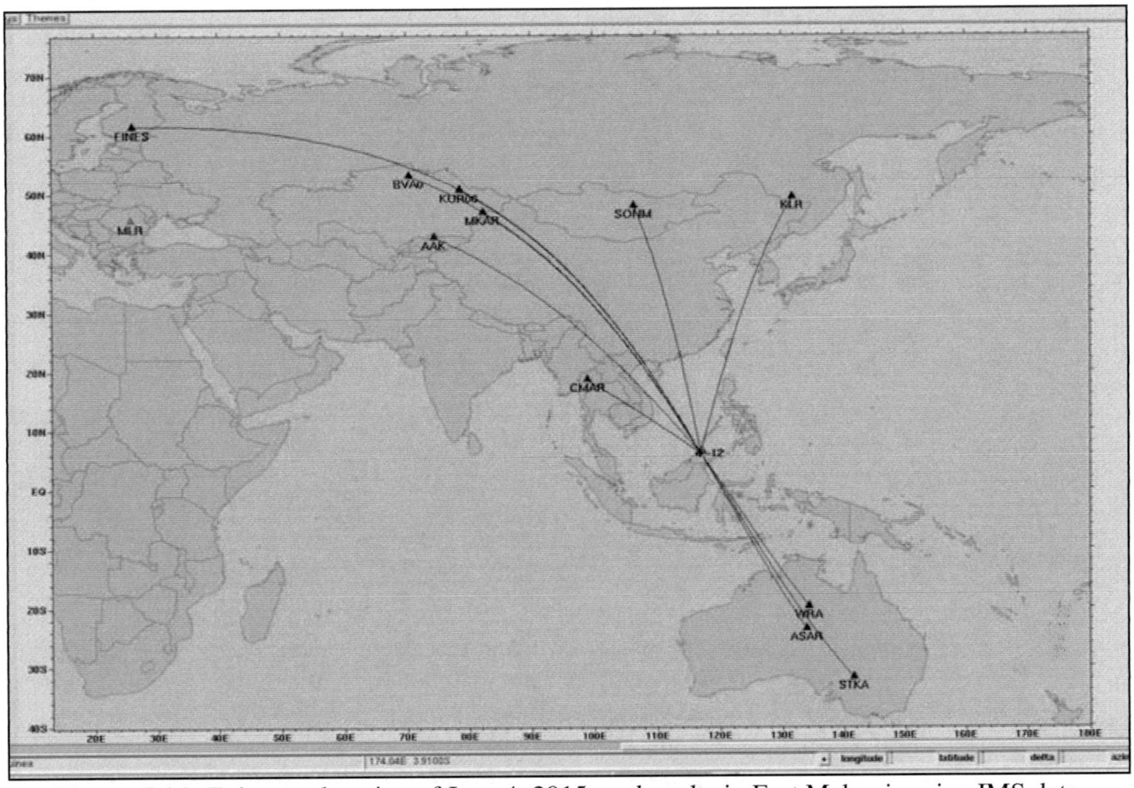

Figure 5 (a). Epicenter location of June 4, 2015 earthquake in East Malaysia using IMS data.

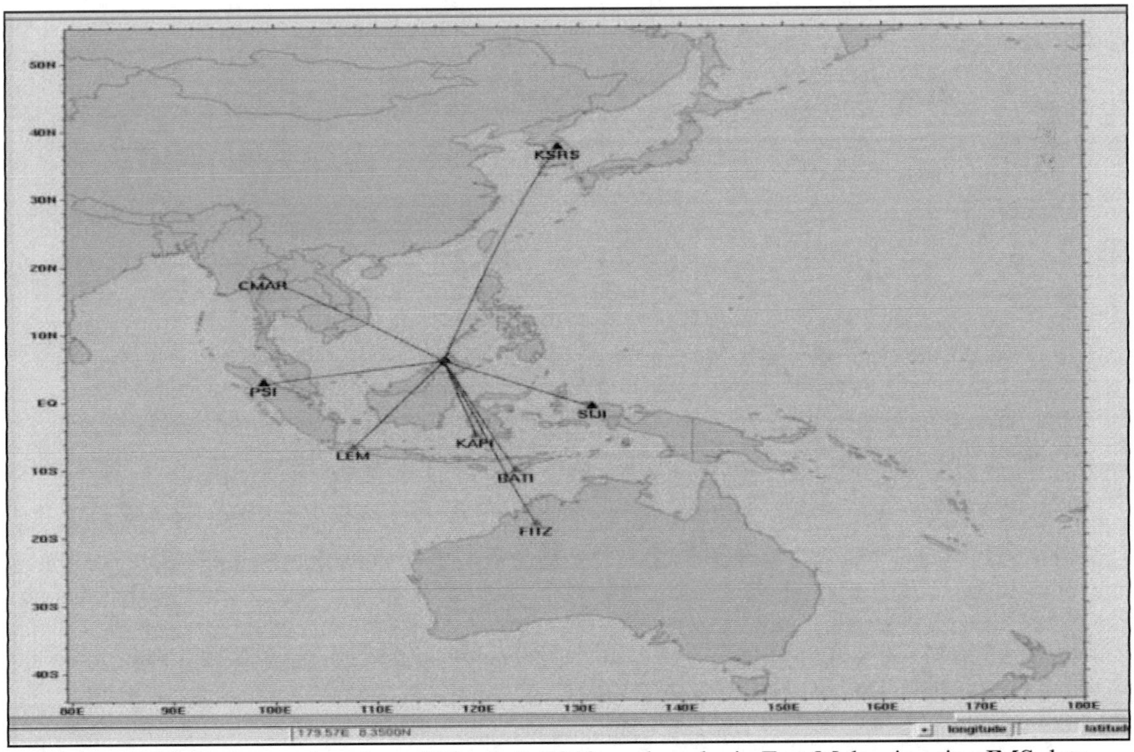

Figure 5 (b). Epicenter location of March 8, 2018 earthquake in East Malaysia using IMS data.

Table 2. 2015-2018 earthquake events in East Malaysia recorded by the USGS network.

U.S. Geological Survey (USGS)						
Year 2018	**Date**	**Time**	**Latitude**	**Longitude**	**Magnitude**	**Depth**
1	08/03/2018	13:06:13	6.0674	116.6109	mb 5.2	10
2	08/03/2018	14:53:26	6.106	116.554	mb 3.2	5
Year 2017	**Date**	**Time**	**Latitude**	**Longitude**	**Magnitude**	**Depth**
1	26/03/2017	9:30:48	4.9334	118.7791	mb 4.6	33.96
Year 2016	**Date**	**Time**	**Latitude**	**Longitude**	**Magnitude**	**Depth**
1	04/03/2016	0:43:36	4.9182	118.4359	mb 4.1	34.96
Year 2015	**Date**	**Time**	**Latitude**	**Longitude**	**Magnitude**	**Depth**
1	19/03/2015	21:56:07	5.5346	118.6135	mb 4.1	50.77
2	04/06/2015	23:15:44	5.9867	116.5409	mww 6	10
3	05/06/2015	15:13:36	6.1402	116.7228	mb 4.4	18.23
4	06/06/2015	5:45:15	6.1416	116.6689	mb 4.6	10
5	12/06/2015	18:25:37	6.1504	116.692	mb 4.4	15.01
6	12/06/2015	18:29:16	6.2053	116.6814	mb 5.3	7.25
7	23/06/2015	9:32:31	6.1277	116.5537	mb 4.5	15.32
8	26/07/2015	16:10:12	6.2782	116.8568	mb 4.6	14.96

Table 3. 2015-2018 Earthquake events in East Malaysia recorded by the IRIS network.

Incorporated Research Institutions for Seismology (IRIS)						
Year 2018	**Date**	**Time**	**Latitude**	**Longitude**	**Magnitude**	**Depth**
1	08/03/2018	13:06:13	6.04	116.61	mb 5.2	10
Year 2017	**Date**	**Time**	**Latitude**	**Longitude**	**Magnitude**	**Depth**
1	26/03/2017	9:30:48	4.93	118.78	mb 4.6	33.96
Year 2016	**Date**	**Time**	**Latitude**	**Longitude**	**Magnitude**	**Depth**
1	04/03/2016	0:43:35	4.92	118.44	mb 4.1	34.96
Year 2015	**Date**	**Time**	**Latitude**	**Longitude**	**Magnitude**	**Depth**
1	19/03/2015	21:56:04	5.6182	118.6962	mb 4.1	35
2	04/06/2015	23:15:45	6.0439	116.6651	Mw 6.0	18.1
3	05/06/2015	13:12:16	6.2967	116.6288	mb 3.5	10
4	05/06/2015	15:13:35	6.2153	116.8726	mb 4.4	10
5	06/06/2015	5:45:15	6.188	116.7836	mb 4.6	10
6	12/06/2015	18:25:39	6.039	116.5783	mb 4.4	35
7	12/06/2015	18:29:18	6.0674	116.621	Mw 5.2	28.3
8	23/06/2015	9:32:33	6.0491	116.5472	mb 4.5	35
9	26/07/2015	16:10:11	6.0664	116.8082	mb 4.6	10

Again, by taking Ranau's June 5, 2015 earthquake as an example, the parameters of seismic event detected by USGS and IRIS networks in East Malaysia on this date are also identical to the one recorded by the CTBT IMS network, as initially described above. Table 4 shows the deviation of epicenters of all earthquake events (2015 – 2018) in Sabah recorded by the USGS and IRIS networks relative to the epicenter located by the CTBT IMS detection. This deviation could be due to different number of stations used by global seismic network in locating such earthquakes in Sabah. This also owe to the fact that the more stations reporting the more accurate the epicenter location [10]. Other factor that affected the accuracy of epicenter location includes the phases used to locate the event, station distribution, and largest and secondary azimuthal gap.

Figure 9 shows a map that plots 3 three different locations of epicenter of Ranau's June 5, 2015 earthquake calculated by the USGS and IRIS respective data as relative to the epicenter calculated by the CTBT IMS Network.

Table 4. The deviation of epicenters of all earthquake events (2015 – 2018) in East Malaysia recorded by the USGS and IRIS networks as relative to the epicenter defined by the CTBT IMS detection.

IMS	Year	2015										2016	2017		2018		
	No.	1	2	3	4	5	6	7	8	9	10	1	1	2	1	2	3
USGS		√		√		√	√	√	√	√	√		√		√		
IRIS				√	√	√	√	√	√	√	√		√		√		
Deviation of epicenter relative to IMS (USGS)	Straight Length (KM)	26.98		11.87		12.06	24.85	81.7	24.48	19.21	20.94		16.41		23.47		
	Heading (degrees)	75.48		243.4		19.34	241.3	250.9	304.3	358.4	273.5		19.97		14.58		
Deviation of epicenter relative to IMS (IRIS)	Straight Length (KM)			3.24	71.63	28.08	11.45	98.02	27.12	9.8	34.22		16.12		19.84		
	Heading (degrees)			70.64	240	45.39	233.2	246.1	267.3	352.1	229.7		20.5		26.34		

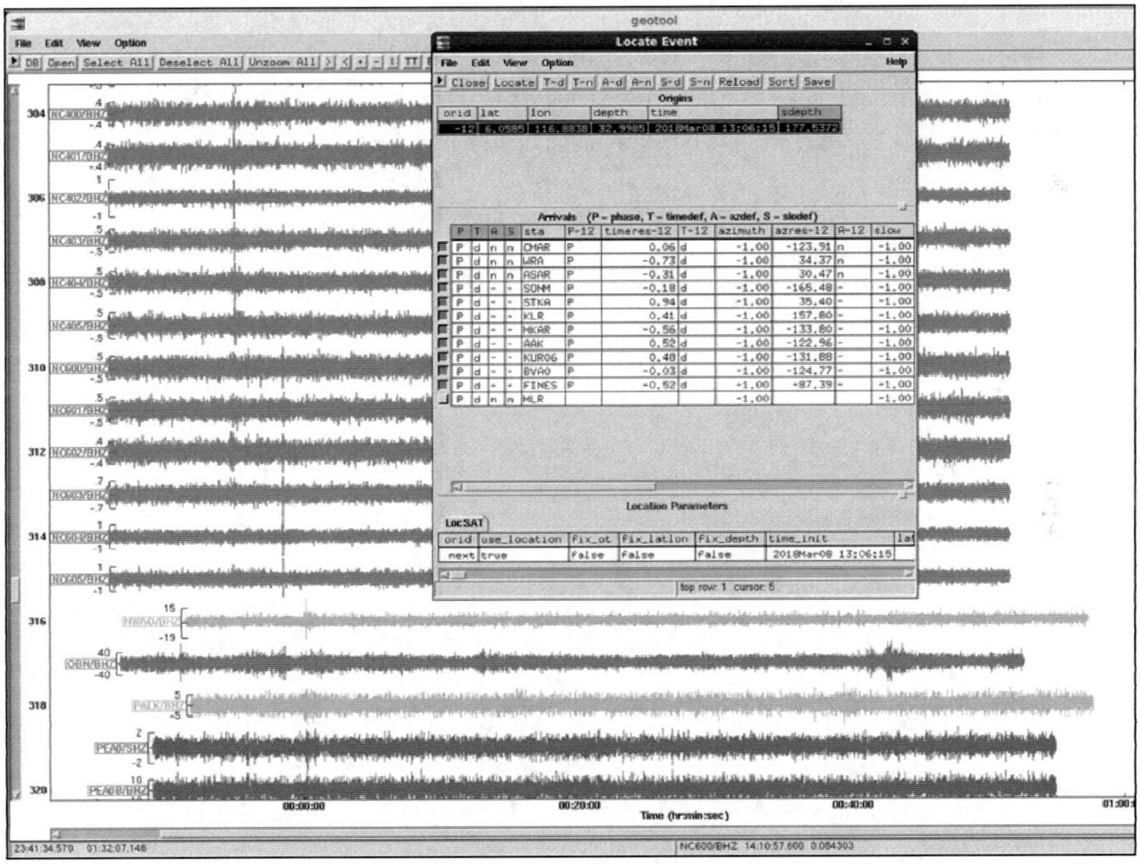

Figure 6. Snapshot of seismic waveform analysis using Geotool software.

IOP Conf. Series: Materials Science and Engineering **785** (2020) 012019 doi:10.1088/1757-899X/785/1/012019

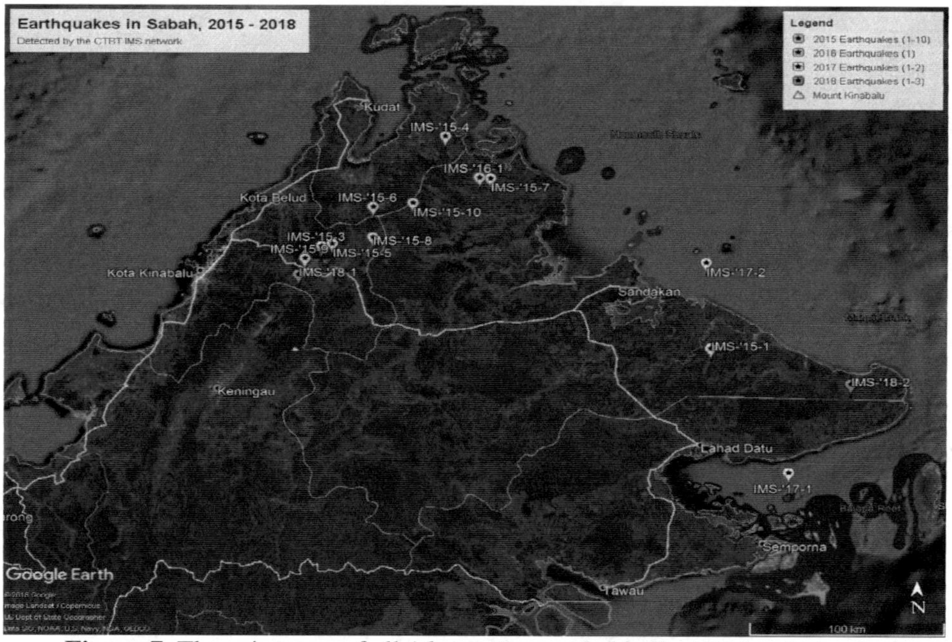

Figure 7. The epicenters of all 16 events recorded in East Malaysia by the CTBT IMS from 2015 – 2018.

KOTA KINABALU: Tremors shook Kota Kinabalu and other parts of Sabah's west coast on Friday after a magnitude 5.9 earthquake hit near Ranau district.

Initial reports from the Malaysian Meteorological Services Department said the quake struck 16km northwest of Ranau at 7.15am.

The US Geological Survey put the Richter scale at 6. No tsunami warning was issued.

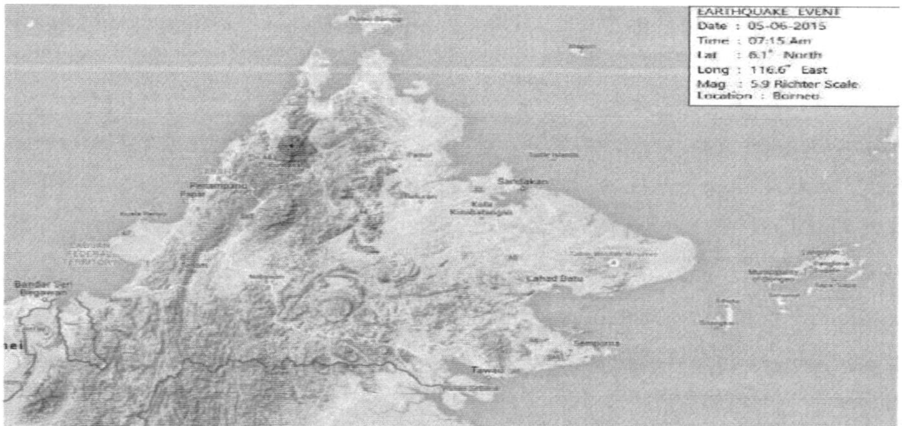

Courtesy of Malaysian Meteorological Department

It said that the strong quake struck at a depth of 10km, with its epicentre located 19km from the town of Ranau, and 54km from Kota Kinabalu.

Apart from the state capital, the tremors were felt in the northern Kudat and Kota Marudu districts, and as far away Beaufort in the south.

The tremors shook buildings and rattled windows prompting people to run out from their houses, shops, and even at the Kota Kinabalu International Airport terminal.

Figure 8. News clip reported the earthquake that shook Kota Kinabalu and other parts of Sabah's west coast on June 5, 2015.

IOP Conf. Series: Materials Science and Engineering **785** (2020) 012019 doi:10.1088/1757-899X/785/1/012019

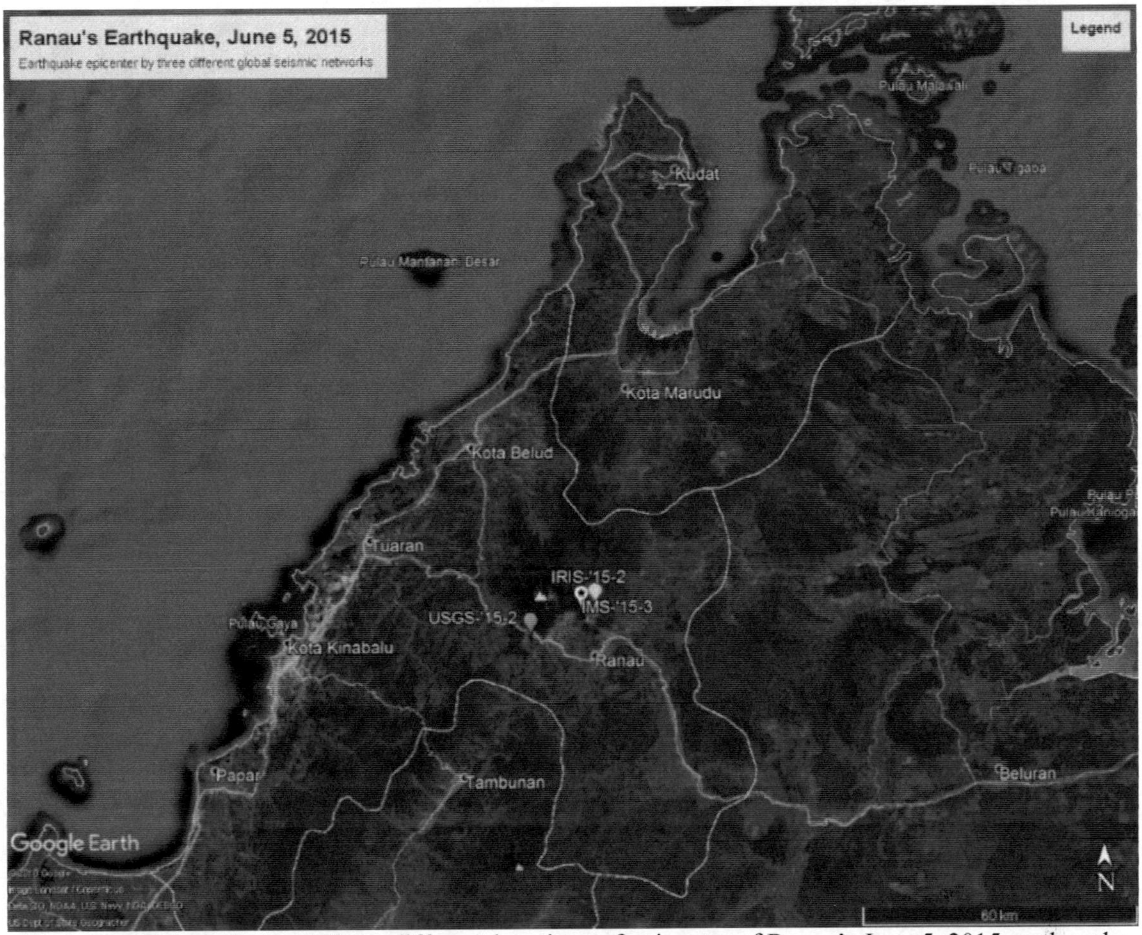

Figure 9. A Map plotting 3 three different locations of epicenter of Ranau's June 5, 2015 earthquake calculated by the USGS and IRIS as relative to the epicenter calculated by the CTBT IMS network.

4. Conclusion

During the period of 2015-2018, the IMS network had detected 16 earthquakes in East Malaysia. From our comparison between IMS, USGS and IRIS, we found that there was slight deviation at detection in terms of epicenter location. This deviation is due to several factors including the number of stations used for locating seismic event.

This study also demonstrates that other than for nuclear explosion monitoring, the IMS data can also be used for civil and scientific applications and can make significant contribution in seismological research.

5. References

[1] Zolkaffly M Z and Rashid F I A 2019 The Comprehensive Nuclear-Test-Ban Treaty (CTBT): seismic monitoring *IOP Conf. Ser.: Mater. Sci. Eng.* **555** p 012010.

[2] Map of global monitoring network, url: https://www.ctbto.org/fileadmin/content/reference/outreach/ims_mapfront.pdf

[3] Sherif A and Daya S 2016 Study of seismicity in the NW Himalaya and adjoining regions using IMS network *J. of Seismol.* **21** p *10.1007/s10950-016-9603-7.*

[4] Aminaton M, Tan S, Kasim F and Mohd Yunus N Z 2013 Seismic impact in Peninsular Malaysia p *10.13140/2.1.3094.9129.*

IOP Conf. Series: Materials Science and Engineering **785** (2020) 012019 doi:10.1088/1757-899X/785/1/012019

[5] Academic Science of Malaysia 2009 Seismic and tsunami hazards and risks study in Malaysia p *978-983-9445.*
[6] Tongkul F 1999 Regional geological correlation of Paleogene sedimentary rocks between Sabah and Sarawak *Geol. Soc. Malays Bull.* ***43*** p 31–39.
[7] Earthquake monitoring in Malaysia, url: http://www.myiem.org.my/download/downloadlink.aspx?fn=984_Earthquake%20Monitoring%20In%20Malaysia.pdf&id=984
[8] Leyu C H 1980 Earthquake swarm in eastern Sabah, June-October,1976
[9] Strong earthquake strikes Sabah, url: https://www.thestar.com.my/news/nation/2015/06/05/sabah-quake
[10] Control and reduction of armaments: detection of and inspection for underground nuclear explosions. Replies from seismologists to subcommittee questionnaire, url: https://books.google.com.my/books?id=WhHAMDRq1cEC&pg=PR1#v=onepage&q&f=false

IOP Conf. Series: Materials Science and Engineering **785** (2020) 012020 doi:10.1088/1757-899X/785/1/012020

Adsorption of Radioactive Element by Clay: A Review

Muhamad Yusmadi Md Yusof[1], Mohd Idzat Idris[1,a)], Faizal Mohamed [1] and Marlida Md Nor[1]

[1] Faculty of Science and Technology, Universiti Kebangsaan Malaysia, 43600 UKM Bangi, Selangor, Malaysia

[a]idzat@ukm.edu.my

Abstract. Clay is one of the most important industrial minerals that have been used to improve the quality of product and economize the production cost. Clay and their minerals, both in its natural and modified forms, have the ability to absorb various radioactive materials from aqueous solution, such as Uranium, Thorium and Caesium as discussed extensively in this review. This article presented an overview of properties and classifications of clay, current research literature on using clay minerals as an absorber, and a descriptive analysis of their adsorption behaviour. Three type of clay are the focused in this review namely Bentonite, Kaolin and Zeolites due to their excellent qualification in absorbing radioactive materials such as Uranium, Thorium and Caesium.

1. Introduction

Radioactive elements occur naturally in the earth's rocks, soils and water in varying concentrations. Many industrial operations, tend to accumulate naturally occurring radioactive materials (NORM) at concentration above normal in by-product waste streams. Massive amounts of NORM wastes produced annually by manufacturing industries around the world should receive the attention of international and national environmental protection agencies and regulatory bodies. Since the mid-1980s, both federal and state regulatory agencies have become increasingly concerned about the presence of NORM [1].

The long-lived radionuclides in radioactive waste are considered to be hazardous pollutants, and their migration along with groundwater flow is strongly affected by the adsorption of geological materials. The existence of radioactive materials in aquatic environment has been known to cause various health problems to humans and animals. They can be absorbed and accumulated in the human body and cause serious health effects such as cancer, organ damage, nervous system damage, and in extreme cases, death [2]. Additionally, improper disposal of NORM may contaminate soil and water and lead to higher indoor radon levels in nearby buildings [3].

Therefore, radioactive waste treatment is gaining worldwide attention due to the harmful effect of radiation associated with these wastes to human health and environment [4]. Several techniques such as ion exchange, solvent extraction, adsorption and electrochemical treatment have been used to remove these hazardous pollutants [5]. Among these, adsorption has been proven to be a successful method for removal of radioactive materials from wastewater. This technique is highly effective, inexpensive, availability and easy to operate compared to other techniques [6-7].

Clay is a naturally occurring earth's subsurface material composed primarily of fine-grained minerals, which has high plasticity index when moist and become hard when dried or fired [8]. Clay and clay minerals play an important role in the environment and can be used as an effective adsorbent for the removal of radioactive materials in water solution. The use of clays as adsorbent has significant advantages in terms of low production cost, availability, non-toxic nature, high specific surface area and excellent adsorption properties [9]. In addition, clays also have great potential for ion exchange (cations and anions) mechanism, thus, many researchers and scientists worldwide has focused on the use of natural or modified clay materials as an adsorbent for water treatment [10].

The main goal of this review is to provide detailed information on natural or modified forms of clay minerals and their excellent adsorption capacities for various radioactive materials such as Thorium, Uranium and Caesium from aqueous solution. This review presents the characteristic and classification of various types of clay minerals, current research literature on adsorption studies using Bentonite, Kaolin and Zeolites as an absorber, and a descriptive analysis of their adsorption behaviour using X-ray Diffraction (XRD).

2. Properties and classification of clays

Clay has a variety of physical properties such as plasticity, hardness, fineness of grain, shrinkage under fire and under air-drying, colour after firing and catalytic properties [11]. Clays and clay minerals have small particle size and complex porous structure with high specific surface area, which allows strong physical and chemical interactions with dissolved species. These interactions are due to the electrostatic repulsion, crystallinity and adsorption or specific cation exchange reactions [12].

Grim (1962) proposed a classification of clay minerals to outline the nomenclature and differences between various clay minerals. According to Grim, the major groups of clay minerals are kaolinite, montmorillonite, and illite. Almost, all clays composed of one or more members of these three groups [13]. Figure 1 shows Grim's classification and lattice structure of clay minerals [13].

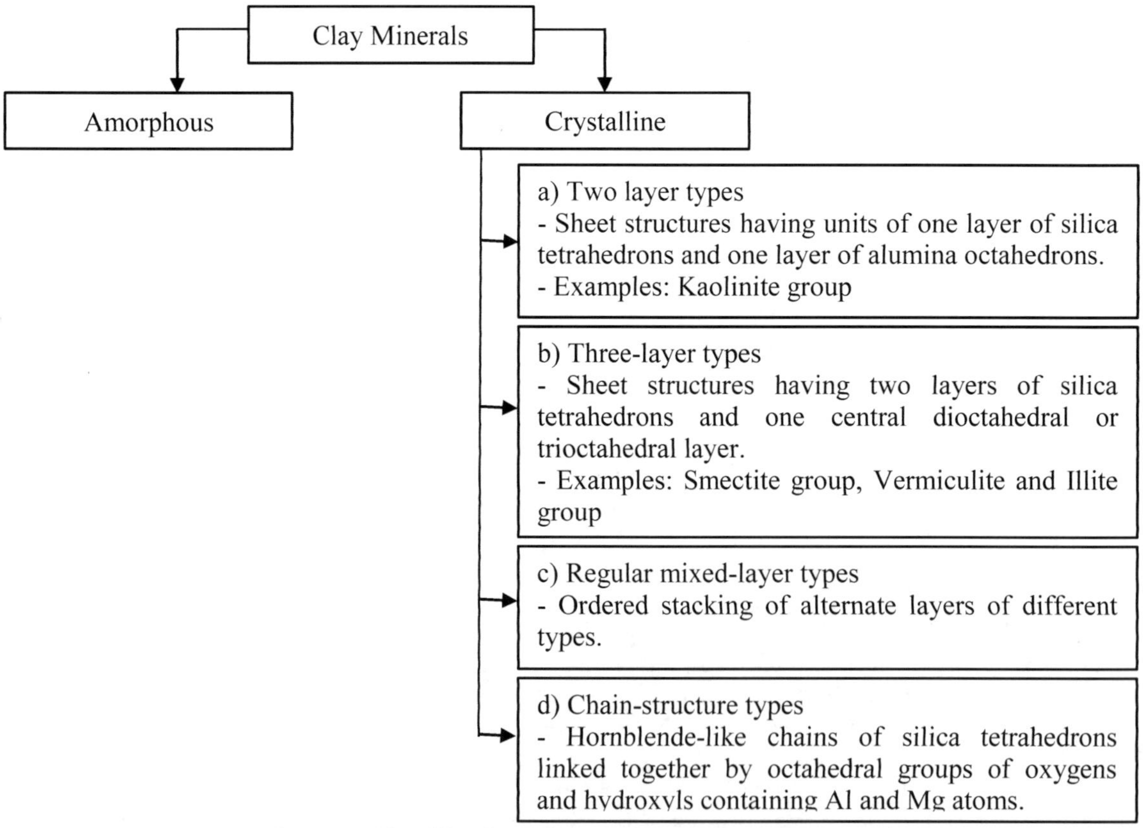

Figure 1. Classification of clay minerals according to Grim.

Clays are hydrous aluminosilicates which composed of mixtures of fine-grained clay minerals, crystals of other minerals and metal oxides [14]. On the basis of such qualities, clays are variously divided into classes or groups such as kaolinite, mica (illite), vermiculite, smectites (montmorillonite), chlorite, and pyrophyllite (talc) as shown in Figure 2 [15].

Figure 2. Clay minerals; (a) kaolinite, (b) mica (illite), (c) vermiculite, (d) smectite (montmorillonite), (e) chlorite, and (f) pyrophyllite (talc).

2.1. Structure of clays

The structure of some most important clay minerals is briefly summarised as follows:

- *Kaolinite* structure possesses great advantages in many processes, due to its high chemical stability, cation exchange capacity, and low expansion coefficient [16]. Kaolinite layers composed of one tetrahedral sheet linked to an octahedral sheet. Therefore, they classified as 1:1 type layer silicates. The structure of kaolinite is a tetrahedral silica sheet, alternating with an octahedral alumina sheet (Figure 3a).
- *Montmorillonites (Smectite)* have a 2:1 layer structure consist of an octahedral alumina sheet sandwiched between two opposing tetrahedral silica sheets [17]. (Figure 3b).
- *Illites* are also 2:1 type minerals, but the interlayers are bonded together with potassium ion (K+) to satisfy the charge and lock the structure [18]. (Figure 3c).

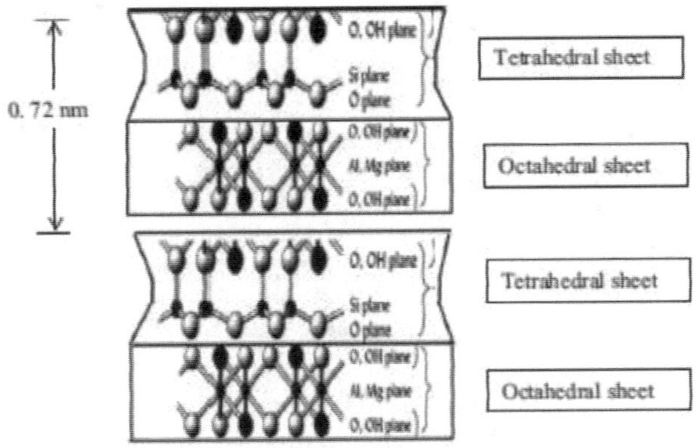

(a) Koalinite (1:1)

IOP Conf. Series: Materials Science and Engineering **785** (2020) 012020 doi:10.1088/1757-899X/785/1/012020

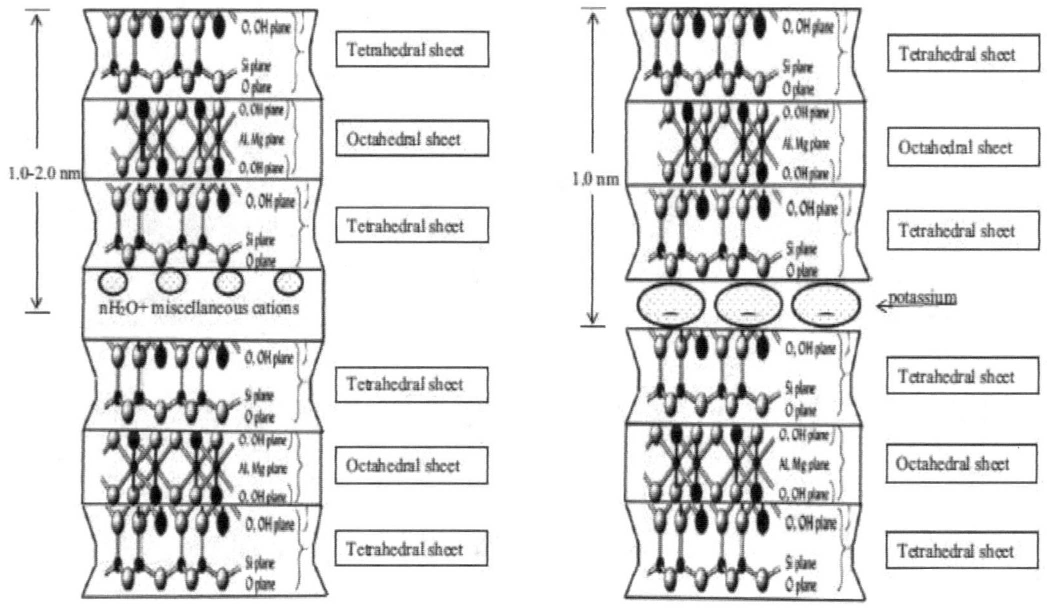

(b) Montmorillonites (Smectite) (c)Illite

Figure 3. Diagrammatic representation of crystal structures (a) Kaolinite (b) Montmorillonite (smectite) (c) Illite.

3. Cases study (Research literature)

The use of natural and modified clay as a potential adsorbent in the removal of metal ions from aqueous solution have been extensively studied for many years [19-21]. There are three types of clay that has been focused in this review which are Bentonite, Kaolin and Zeolites, as adsorbents for uptake of Uranium, Thorium and Caesium from aqueous solution.

3.1. Bentonite

Bentonite has attracted great interest in nuclear waste management due to its high swelling property, good self-sealing ability and cation exchange capacity [22]. Bentonite is also used as buffer or backfill material according to its function and location in the nuclear waste disposal field [23].

Khalili *et al.* studied the sorption behavior of purified Jordanian bentonite towards UO_2^{2+} and Th^{4+} metal ions in aqueous solutions using both batch and column techniques [17]. The main objective of this study was to investigate the sorption capability of purified Jordanian bentonite at different pH (i.e. 1.0, 2.0, and 3.0), at different temperatures (i.e. 25°C, 35°C, and 45°C) and at different contact times (i.e.). The purification of raw bentonite was done by the removal of quartz. This include mixing the raw bentonite with water and then centrifuge it at 750 rpm, the process being repeated until white purified bentonite is obtained. XRD technique is used for the characterization of purified bentonite.

The characteristic XRD peaks for the minerals in raw bentonite are shown in Table 1. Raw Jordanian bentonite consists of montmorillonite (M) as major clay mineral, whereas quartz (Q) and calcite are present as impurities, as well as anorthite is present as minor constituent. The XRD patterns for raw bentonite and purified bentonite (stirring with distilled water) are shown in Figure 4.

Table 1. The major XRD peaks for raw bentonite.

Mineral	2θ	D (Å)
Montmorillonite (bentonite) (M)	19.89	4.457
	35.36	2.536
	40.30	2.236
	61.84	1.499
Quartz (Q)	20.88	4.251
	26.67	3.340
	50.18	1.817
Anorthite	13.71	6.452
	21.98	4.041
	28.43	3.137
Calcite	29.84	2.991

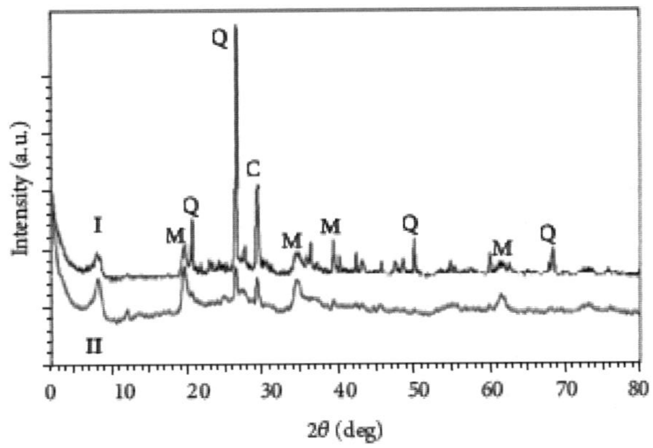

Figure 4. XRD patterns for (I) raw bentonite and (II) purified bentonite, M: montmorillonite, Q: Quartz, and C: calcite.

From their experiments, the Jordanian bentonite has shown high metal ion uptake capacity toward uranium(VI) than thorium(IV). The influence of different pH on metals uptake also showed that the metal ion uptake by bentonite increased with pH and reached a maximum at pH = 3 for Th^{4+} and UO_2^{2+} as shown in Table 2. The maximum absorption capacity of bentonite was high for Th^{4+} and the extent of metal ions uptake followed the order: $Th^{4+} > UO_2^{2+}$ at pH = 3 and 25°C. The equilibrium for each metal ion on the surface of bentonite occurs at 18 hours to achieve maximum uptake level.

Table 2. Uranium(VI); and Thorium(IV) percentage uptake by bentonite at pH 1.0, 2.0 and 3.0 at 25°C and concentration 50 ppm.

pH	Percentage Uptake By Bentonite	
	Uranium(VI)	Thorium(IV)
1	40%	57%
2	55%	75%
3	81%	78%

3.2. Kaolin

A number of studies on kaolin and kaolinite used to adsorb metal ions from aqueous solution have been reported [24-27]. The most common mineral in the kaolin group of minerals is kaolinite. Typically, kaolinite has a low cation exchange capacity (CEC), which is 0.03–0.15 meq/g [28][29]. Although kaolinite has low CEC, their adsorption capacity can be improved if chemically modified [30-31].

A studied conducted by Wang et al. (2009) proved that the modified kaolin has greatly enhances the adsorption of uranium (VI) [17]. They modified kaolin by calcination process and followed by acid-activation. The modified kaolin samples were dispersed in uranium(VI) solution of a given concentration and pH value for 1 hour at 25 °C, then centrifuged at 4000 rpm for 10 minutes. The adsorption capacity of modified kaolin for uranium(VI) increased with the contact time and reached adsorption equilibrium within 60 minutes (Table 3). Compared with natural/non-modified kaolin, the adsorption was strongly increased after modification. The adsorption also increased with increasing pH as shown in Table 4. Calcined and acid-activated kaolin is a favorable adsorbent for uranium(VI) removal from aqueous solutions.

Table 3. Influence of contact time on the adsorption of uranium (VI) [UO_2^{2+}] =20 mg/L, Adsorbent mass=50 mg, pH=5.0, T=25 °C).

Time (min)	Absorption capacity (mg/g)
10	2.70
20	3.00
30	3.25
40	3.30
50	3.35
60	3.40

Table 4. Influence of pH on the adsorption of uranium (VI) [UO_2^{2+}] =20 mg/L, Adsorbent mass=50 mg, T=25 °C).

pH	Removal efficiency
2	4%
3	6%
4	32%
5	53%
6	65%
7	70%

3.3. Zeolite

Zeolite is a natural and synthetic hydrated aluminosilicates with a three-dimensional framework. Along with the isomorphous replacement of Si by Al, negative charges are generated, and zeolites have the capacity to adsorb metal cations including heavy metal cations [32-33].

Johan *et al.* conducted a Cs^+ adsorption experiment using two natural zeolites, namely mordenite and clinoptilolite [21]. Similar experiments were also performed using synthetic zeolite A, synthetic zeolite X, and artificial zeolite Na-P1 as references. Synthetic zeolite A and zeolite X were purchased from WAKO pure chemicals industry, Japan. Meanwhile, Na-P1 was an artificial zeolite made from coal fly ash by mixing 10.0 g of coal fly ash with 80 mL of 2M NaOH, followed by heating at 100 ºC for 24 hours.

The purpose of this study was to investigate the potential of natural zeolites on the adsorption of radioactive Cs^+. The results show that the CEC of the natural zeolites was much lower compared to the synthetic and artificial zeolites as shown in Table 5. The higher CEC of the synthetic zeolites (zeolite A and zeolite X) is due to the lower Si/Al ratios (around 1.0 to 1.2) and higher crystallinity, as shown in the XRD patterns of Figure 5. On the other hand, the lower CEC of the natural zeolites is due to their higher Si/Al ratios and impurities contained.

Table 5. CEC values of the zeolite samples.

Zeolite samples	CEC (cmol $kg^{-}1$)
Mordenite	266
Clinoptilolite	211
Na-P1	325
Zeolite A	615
Zeolite X	556

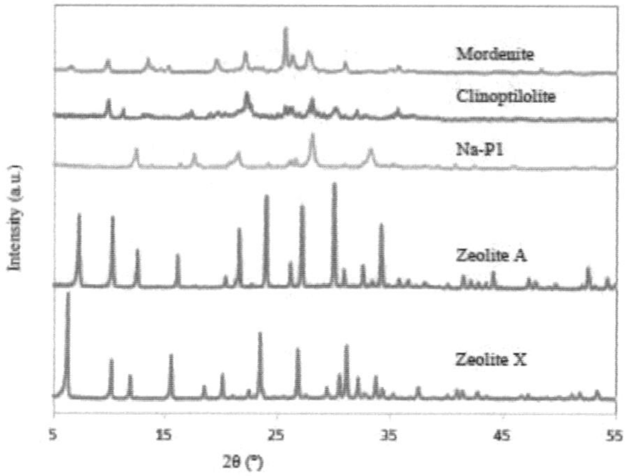

Figure 5. Powder XRD patterns of the zeolite samples.

Although the CEC of the natural zeolites was much lower compared to the synthetic and artificial zeolites, the adsorption capacity of Cs^+ was much greater than the synthetic and artificial zeolites as illustrated in Figure 3. The order of the adsorption capacity of Cs^+ was mordenite > clinoptilolite > Na-P1 >> zeolite A > zeolite X. It indicates that natural zeolites are promising materials in the removal of radioactive Cs^+.

4. Discussions

Adsorption is one of the most important processes of metal uptake with several properties that take place at the mineral-solute interface. The performance of clay materials and the adsorption process in the reported studies have been verified by examining the effect of various factors such as pH, temperature, contact time and initial concentration. From the experiment conducted by Khalili et al. [19], the influence of different pH on metals uptake indicated that the metal ion uptake by bentonite increased with pH and reached a maximum at pH = 3 for Th^{4+} and UO^{2+}_2 . Meanwhile, experiment conducted by Wang et al. [20], showed that the adsorption of uranium(VI) by modified kaolin also increased with increasing pH at pH = 7. In terms of contact time, the equilibrium for each metal ion on the surface of bentonite occurs at 18 hours to achieve maximum uptake level. Meanwhile, the adsorption capacity of modified kaolin for uranium(VI) increased with contact time and reached adsorption equilibrium within 60 minutes. In most of the reported works, adsorption experiments that were conducted by using the batch technique have confirmed their applicability and selectivity of the maximum adsorption capacities towards targeted pollutants [19]

Many clay materials have shown their excellent potential for the removal of metal pollutants without any modifications, but in many studies, their removal capacities can be improve through certain chemical modifications as discussed in this review. Wang et al. [20] modified kaolin by calcination process and followed by acid-activation. The result showed that compared with natural kaolin, the adsorption was strongly increased after modification. Therefore, the adsorption properties also depend on the extent of chemical treatment, activation, and modification of the adsorbent.

5. Conclusion

Through the analysis of 30 papers from different clay prospect studies, many clay materials have shown their excellent potential in removing metal ions from aqueous solutions. Clay, inexpensive, locally available, and effective adsorbent material, will surely offer many promising benefits in the future. In addition, high specific surface area, non-toxic nature, excellent adsorption capacity and large potential for ion exchange give tremendous importance to clay minerals and can be demonstrated in future research. Bentonite, Kaolin and Zeolites have shown an excellent performance in absorbing radioactive materials such as Uranium, Thorium and Caesium as strongly discussed in this review.

6. References

[1] Smith K P 1992 An overview of naturally occurring radioactive materials (NORM) in the petroleum industry Argonne National Laboratory, (United States). *Env. Assessment and Information Sci. Div.* p 5.
[2] Gunatilake S K 2015 Methods of removing heavy metals from industrial wastewater *Methods* **1**(1).
[3] Limbanyen K, Soponkanabhorn T, Chanyotha S, Chankow N and Krobbuaban P 2007 Application of the Systematic Technique for the Characterization of Naturally Occurring Radioactive Materials (NORM) at Bongkot Offshore field and Songkhla Petroleum Development Support Base *SPE Asia Pacific Health, Safety, and Sec. Env. Conf. and Exhib.* (Society of Petroleum Engineers).
[4] International Atomic Energy Agency 2012 Management of low and intermediate level radioactive wastes with regard to their chemical toxicity IAEA-TECDOC-1325; (Vienna, Austria: IAEA).
[5] Rao R A K, Ikram S and Uddin M K 2014 Removal of Cd (II) from aqueous solution by exploring the biosorption characteristics of gaozaban (Onosma bracteatum) *J. Env. Chem. Eng.* **2** p 1155–1164.
[6] Gupta V K, Kumar R, Nayak A, Saleh T A and Barakat M A 2013 Adsorptive removal of dyes from aqueous solution onto carbon nanotubes: a review *Adv. Colloid Interface Sci.* **193** p 24–34.
[7] Bhattacharya A K, Mandal S N and Das S K 2006 Adsorption of Zn(II) from aqueous solution by using different adsorbents, *Chem. Eng. J.* **123** p 43–51.
[8] Mann H S, Brar G S, Mann K S and Mudahar G S 2016 Experimental investigation of clay fly ash bricks for gamma-ray shielding *Nuclear Engineering and Technology* **48**(5) p 1230-1236.
[9] Crini G, and Badot P M 2010 *Sorption process and pollution, conventional and non-conventional sorbents for pollutant removal from wastewaters* (Franche-Comté: Presses Univ.)
[10] Srinivasan R Advances in application of natural clay and its composites in removal of biological, organic, and inorganic contaminants from drinking water 2011 *Adv. Mater. Sci. Eng.* **2011** p 1–17.
[11] Odoma A N, Obaje N G, Omada J I, Idakwo S O and Erbacher J 2013 PALEOCLIMATE reconstruction during Mamu formation (cretaceous) based on clay mineral distributions *J. Appl. Geol. Geophys.* **1** p 40–46.
[12] Murray H H 2007 *Applied Clay Mineralogy: Occurrences, Processing and Application of Kaolins, Bentonites, Palygorskite-Sepiolite, and Common Clays,* (Elsevier Press).
[13] Grim 1962 *Applied Clay Mineralogy* (McGraw Hill Press).
[14] Mockovciakova A, and Orolinova Z 2009 Adsorption properties of modified bentonite clay *Chem. Technol.* **1** p 47–50.
[15] Shichi T and Takagi K 2000 Clay minerals as photochemical reaction fields *J. Photochem. Photobiol. C: Photochem. Rev.* **1** p 113–130.
[16] Miranda-Trevino, J C, and Coles C A. 2003 Kaolinite properties, structure and influence of metal retention on pH *Applied Clay Science* **23**(1-4) p 133-139.
[17] Brigatti M F, Galan E, Theng B K G 2013 Structure and Minerelogy of Clay Minerals *Developments in Clay Science* **5** p 21-68.
[18] Bear F E. 1964 Chemistry of the Soil *Soil Science* **98**(1) p 70.
[19] Khalili F I, Salameh N A H and Shaybe M M. 2012 Sorption of uranium (VI) and thorium (IV)

by Jordanian bentonite *Journal of Chemistry* **2013**.

[20] Wang G, Wang X, Chai X, Liu J and Deng N 2010 Adsorption of uranium (VI) from aqueous solution on calcined and acid-activated kaolin *Applied Clay Science* **47**(3-4) p 448-451.

[21] Johan E, Yamada T, Munthali M W, Kabwadza-Corner P, Aono H and Matsue N 2015 Natural zeolites as potential materials for decontamination of radioactive cesium *Procedia Environmental Sciences* **28** p 52-56.

[22] Sun D, Cui H and Sun W 2009 Swelling of compacted sand bentonite mixtures *Applied Clay Science* **43**(3-4) p 485–492.

[23] Mollins L H, Stewart D I and Cousens T W 1996 Predicting the properties of bentonite-sand mixtures *Clay Minerals* **31**(2) p 43–252.

[24] Adebowale K O, Unuabonah I E and Olu-Owolabi B I 2006 The effect of some operating variables on the adsorption of lead and cadmium ions on kaolinite clay *J. Hazard. Mater. B* **134** p 130–139.

[25] Bouchenafa-Saib N, Khouli K and Mohammedi O 2007 Preparation and characterization of pillared montmorillonite application in adsorption of cadmium *Desalination* **217 p** 282–290.

[26] Turan P, Dogan M and Alkan M 2007 Uptake of trivalent chromium ions from aqueous solutions using kaolinite *J. Hazard. Mater.* **148** p 56–63.

[27] Bhattacharyya K G and Gupta S S 2008b Adsorption of a few heavy metals on natural and modified kaolinite and montmorillonite: a review *Adv. Colloid Interface Sci.* **140** p 114–131.

[28] Grim R E 1968 *Clay Mineralogy* 2, ed (New York: McGraw-Hill).

[29] Jiang M Q, Wang Q P, Jin X Y and Chen Z L 2009 Removal of Pb(II) from aqueous solution using modified and unmodified kaolinite clay *J. Hazard. Mater.* **170** p 332–339.

[30] Adebowale K O, Unuabonah I E and Olu-Owolabi B I 2005 Adsorption of some heavy metal ions on sulfate- and phosphate-modified kaolin *Appl. Clay Sci.* **29** p 145–148.

[31] Caballero I, Colina F G and Costa J 2007 Synthesis of X-type zeolite from dealuminated kaolin by reaction with sulfuric acid at high temperature *Ind. Eng. Chem. Res.* **46** p 1029–1038.

[32] Kabwadza-Corner P, Munthali M W, Johan E, and Matsue N 2014 Comparative study of copper adsorptivity and selectivity toward zeolites *American Journal of Analytical Chemistry* **5** p 395-405.

[33] Erdem E, Karapnar N and Donat R 2004 The removal of heavy metal cations by natural zeolite *Journal of Colloid and Interface Science* **280** p 309-314.

Acknowledgements
The work was financially supported by GGPM-2017-042. The author express gratitude to the staffs of the Universiti Kebangsaan Malaysia (UKM) for their assistance throughout the research.

IOP Conf. Series: Materials Science and Engineering **785** (2020) 012021 doi:10.1088/1757-899X/785/1/012021

Mobile robot for radiation mapping in indoor environment

Nur Aira Abd Rahman[1,2,a)], Khairul Salleh Mohamed Sahari[1], Muhammad Fairuz Abdul Jalal[1], Anwar Abdul Rahman[2], Mohd Izwan Abd Adziz[2], and Mohd Zaid Hassan[2]

[1] Universiti Tenaga Nasional (UNITEN), 43000 Kajang, Selangor, Malaysia
[2] Malaysian Nuclear Agency, 43000, Bangi, Selangor, Malaysia.

[a]nur_aira@nm.gov.my

Abstract. Radiation mapping is the process of measuring radiation intensity level at distributed sampling points in a predefined region of interest (ROI). This procedure is crucial in any radiological related emergency to locate contamination hotspot(s) or in routine inspections in any radiation facilities. When workers knowingly enter a high radiation zone to perform radiation mapping, they are exposed to the risk of radiation exposure. Mobile robot is a potential solution that could eliminate the health and safety risk and subsequently improve the measurement accuracy. This paper reports implementation of mobile robot for radiation mapping by using a commercial mobile robot platform, Turtlebot2 in conjunction with Robot Operating System (ROS). Mobile robot is capable to generate physical map of the targeted environment by using Simultaneous Localization and Mapping (SLAM). Simultaneously, Geiger Muller detector is utilized to perform radiation measurement. The radiation data is directly plotted on the physical map hence the radiation intensity distribution throughout the ROI could be assessed. Functionality of the radiation mapping robot is evaluated in real-world experiments and the results will be presented in the paper.

1. Introduction

Radiological emergency requires immediate and planned response to avoid significant damage on health, psychological, and financial consequences. The primary objective of emergency response are to regain control of the situation and mitigate the consequences in order to avoid or minimize the risk of deterministic and stochastic effects of radiation exposure to human being and the environment [1]. During any radiological related emergency, the process of radiation mapping is critical to measure radiation intensity distribution throughout the accident's parameter. Radiation map will help to establish safety boundary, identify contaminated 'hotspots', localize the origin of radioactive source, as well as to estimate the dose rate distribution throughout the perimeter [2][3].

The role of robot is proved to be significant after Fukushima Daiichi nuclear accidents in 2011. Both inside and outside of accident site (buildings housing the nuclear reactors) are not accessible to human due to potential risk of high radiation exposure. Among the robots that were deployed are iRobot Packbot [4], Quince [5], and Talon [6]. These robots were utilized to perform damage assessment, radiological surveys, as well as some decontamination and clean-up tasks. However, robot missions after Fukushima Daiichi accidents revealed several shortcomings on the available robot technology targeted for emergency response. One of the most prominent issue is dependency on teleoperation which requires good communication signal and highly skilled operators to control and maneuver the robots [5][7]. Due to this, some of the robots failed to be recovered and were lost during the mission.

Researches on autonomous mobile robot navigation are typically broken down into three important elements; mapping, localization, and path planning. In terms of mapping, Simultaneous Localization and Mapping (SLAM) has enables robot to create the map of unknown environments [8–10]. To localize the position of robot in the map, a number of methods existed [11][12]. Finally, path planning will tell the robot where to go and how to avoid obstacles along the way [13][14]. Given solutions to these three elements, a lot of current researches focus on utilizing mobile robot in wide range of applications such as in home and healthcare services, industrial, agriculture, and even in space exploration.

The aims of this research are to develop robotics platform for radiation mapping by integration of mobile robot and radiation detector. Next, we design an algorithm for coverage radiation mapping in indoor environment and validate both robot and algorithm in experiments and real-world application. Section 2 presents the robot setup and algorithm. Section 3 contains results and discussion from experimental evaluation. Finally, Section 4 will provide the conclusion and our plan for future works.

2. Research methodology

The research methodology to develop and implement autonomous mobile robot for radiation mapping is done in three stages. Initially, the platform is established by integrating mobile robot with radiation detector module. Next, we enable autonomous navigation for the mobile robot by utilizing ROS Navigation Stack. Lastly, we create the procedure of autonomous radiation mapping by mobile robot by using SLAM and in-house coverage radiation mapping algorithm. Each stage will be described in detail in the following sub-sections.

2.1. Integration of mobile robot and radiation detector module

Turtlebot2 (Turtlebot) is chosen as the mobile robot platform in this research. It is a commercial mobile robot platform that is widely used for research purposes due to its versatility. This robot consists of 2 differential drive wheels and equipped with navigation sensors such as on-board RGBD camera, wheels encoder, gyro sensors, wheel drop sensors, bump sensors and cliff sensors. Turtlebot is used in conjunction with Robot Operating System (ROS), an open source framework licensed under BSD. Simulation of robot navigation codes is done with ROS Gazebo whereas real time data visualization is done with ROS RViz.

Geiger Muller (GM) detector LND7121 is utilized as the radiation detector. The detector electronics module has been developed in our previous work [15]. This module consists of GM, high voltage supply of 500 Volts for detector, and signal processing unit that converts analog output of detector to digital signal (TTL pulse). A counter timer is implemented in microcontroller to count pulse generated by GM. The output of this module is gross count of radiation event represented in count per second (CPS) unit.

GM electronics module is connected to Turtlebot host computer via USB connection as shown in Figure 1. The module is installed on one of Turtlebot shelves. ROS package *rosserial_arduino* is utilized to implement ROS topics publisher. Radiation detector output data is published into Turtlebot ROS environments as */RadiationData* topic at frequency of 1 Hertz.

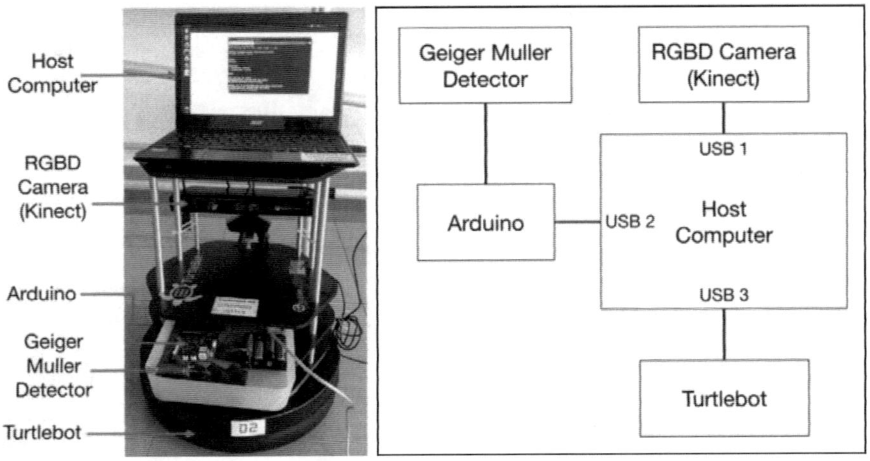

Figure 1. Physical setup and block diagram of Turtlebot, GM, and host computer.

2.2. ROS navigation stack for autonomous navigation in physical environment

ROS Navigation Stack is a set of ROS nodes and algorithms that computes how to move robot from its original position (current pose) to desired destination (goal pose). The navigation requires physical map of the environment typically referred as static map. Static map enables robot to localize its current pose in the map and execute navigation commands to the goal pose. In addition, the static map is also useful to visualize the result of radiation mapping with respect to environment geometry.

Navigation planning and execution involves path planner and costmap. Planner generates path from current pose to goal pose while referring to costmap. On the other hand, costmap provides map with static and dynamic obstacles information along the path. *Move_base* is the primary component of ROS Navigation Stack. When *move_base* receives a new goal pose, global planner computes a complete path from current pose to goal pose based on static map. As *move_base* execute the path, local costmap re-calculate the path as needed based on real time sensors data. Visualization of each component in RViz as Turtlebot moves from current pose to goal pose is shown in Figure 2.

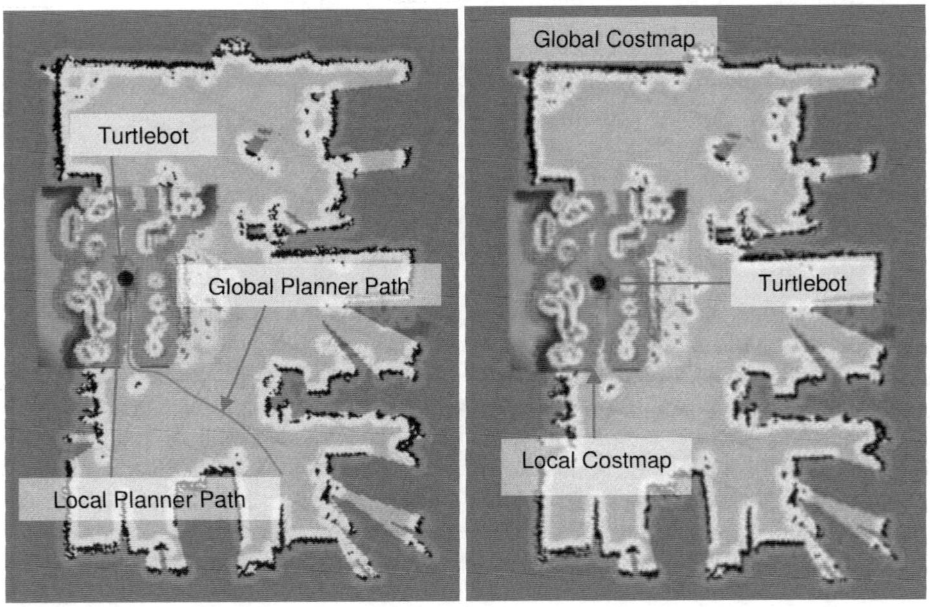

Figure 2. RViz Visualization of global planner, local planner, global costmap and local costmap.

2.3. Procedures of Autonomous Radiation Mapping
The complete procedures of autonomous radiation mapping involve three sequential steps as shown in Figure 3. As shown in Figure 3, the steps are mapping of physical environment, autonomous radiation mapping by mobile robot, and data processing and analysis.

Figure 3. Procedures of radiation mapping.

2.3.1. Mapping of Physical Environment
In order to build the physical map, a popular method called SLAM has been adopted [8]. SLAM enables map creation from visual and odometry data as robot moves around the region of interest. ROS *gmapping* package is an implementation of a SLAM algorithm to create 2-dimensional occupancy grid maps. The map uses pixel values to classify unoccupied spaces, occupied spaces, and unexplored spaces. Figure 4 shows map created by Turtlebot with *gMapping* package. Pixel resolution in Figure 4 (right) is 5 cm^2 whereas the map resolution is 512 pixels width and 480 pixels wide. Each pixel in the map could be translated into respective X-Y coordinates to define current pose and goal pose of the robot.

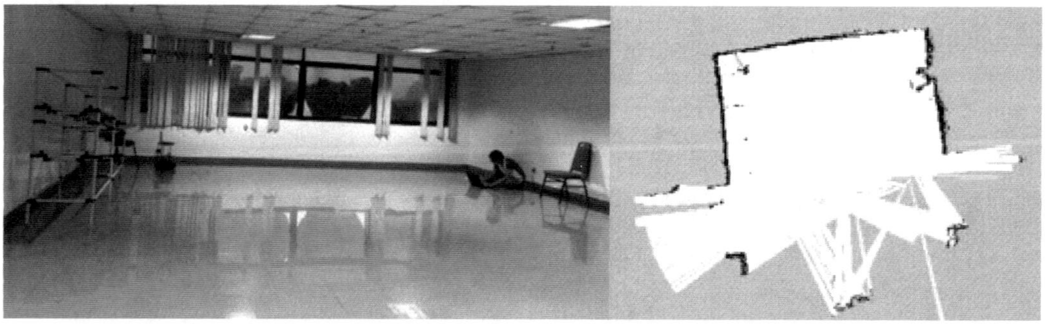

Figure 4. Real world environment (left) and the resulting Occupancy Grid Map created by Turtlebot with *gMapping* package (right).

2.3.2. Autonomous radiation mapping by mobile robot
Radiation mapping algorithm enable autonomous operation of the radiation mapping. Flowchart of the algorithm is shown in Figure 5. This algorithm requires occupancy map as the input. Occupancy map will be segmented into grids. To ensure coverage of the resulting radiation map, all unoccupied cells will be mapped by mobile robot. The sampling points are centroid of every unoccupied cell. Next, path that connect all the sampling points is generated for mobile robot navigation. During the scanning process, detector count data is accumulated for 3 seconds and recorded along with the X and Y coordinates of the sampling points. If any of the sampling points become unreachable during the mapping process, the action server goal will be pre-empted (cancelled) and robot will proceed to the next sampling point.

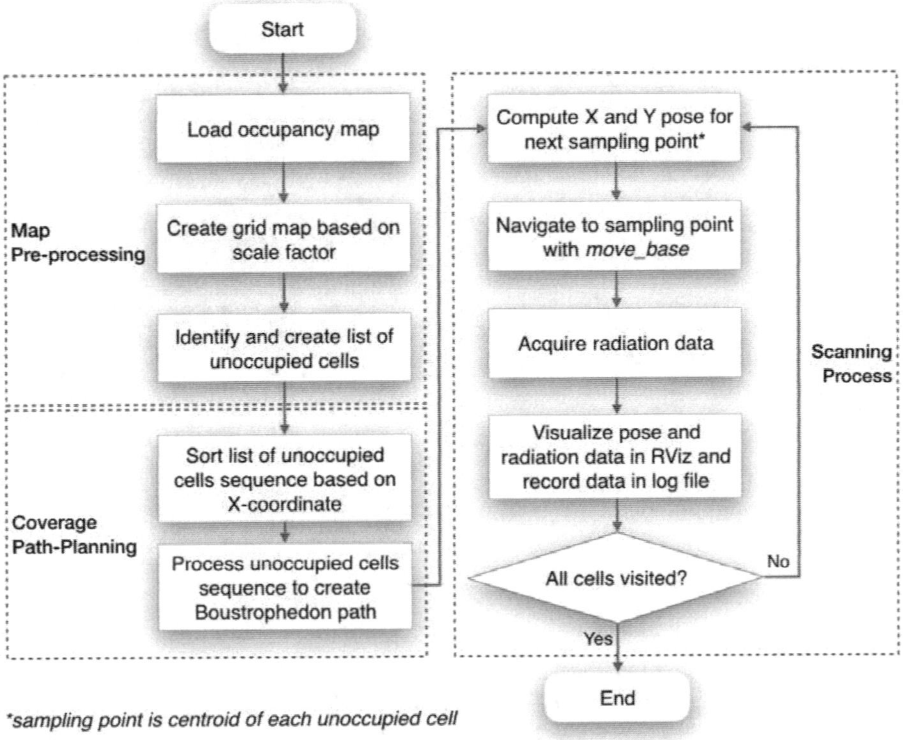

Figure 5. Flowchart of coverage radiation mapping algorithm.

2.3.3. Data processing and analysis

To generate the final radiation map, count data is plotted against the X-Y coordinates. The background counts are subtracted from the counts. At the moment, a simple python program is written to interpolate the discrete data and generate a continuous map. Location of the hotspot(s) are highlighted by map colour that represent the radiation intensity.

3. Result and discussion

3.1. Verification of the system

A series of experiments were conducted to verify the radiation map produce by the autonomous robot. These experiments were performed in a controlled environment where sealed radioactive sources are placed in known locations i.e. coordinates in the map. Then, robot is deployed into the environment to run radiation mapping autonomously. Resulting radiation map is analysed in terms of accuracy of the hotspot location with respect to actual position of the radioactive sources.

As shown in Figure 6, all three coordinates of the radioactive sources have been identified and visually highlighted by colour variations that represents the radiation intensity at the respective sampling points. X-pose and Y-pose in the map axis represent the cell X-Y coordinates respectively where the origin (0,0) is the start position in SLAM *gmapping* process (refer section 2.3.1). The results proved that combination between robot and coverage radiation mapping algorithm has manage to generate radiation map of the target location.

IOP Conf. Series: Materials Science and Engineering **785** (2020) 012021 doi:10.1088/1757-899X/785/1/012021

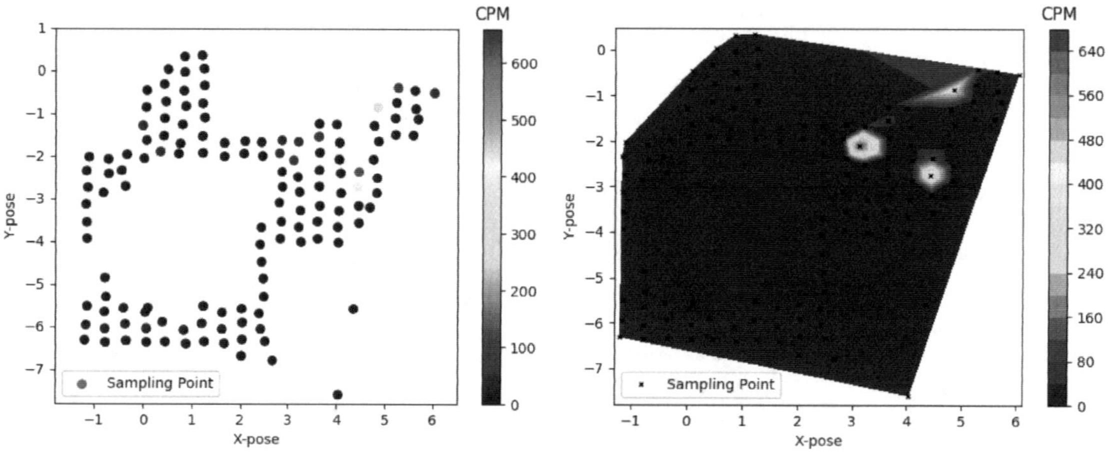

Figure 6. Resulting radiation map with discrete data (left) and interpolated data (right).

3.2. Radiation mapping at operational facilities

Finally, we investigate the potential application of autonomous radiation mapping for optimizing the safety of radiation worker. Radiation mapping were performed at two operational radiation related facilities. The first location is Alpha Spectrometry Counting Laboratory (Figure 7) that performs analysis of environmental samples such as soil, ground water, as well as flora and fauna. These samples are only associated with background or low level of radioactivity. On the other hand, the second location is Mobile Hotcell (Figure 8) that involves higher radioactivity sources stored in Long Term Source Storage (LTSS). Both selected locations are indoor with relatively smooth flooring surface.

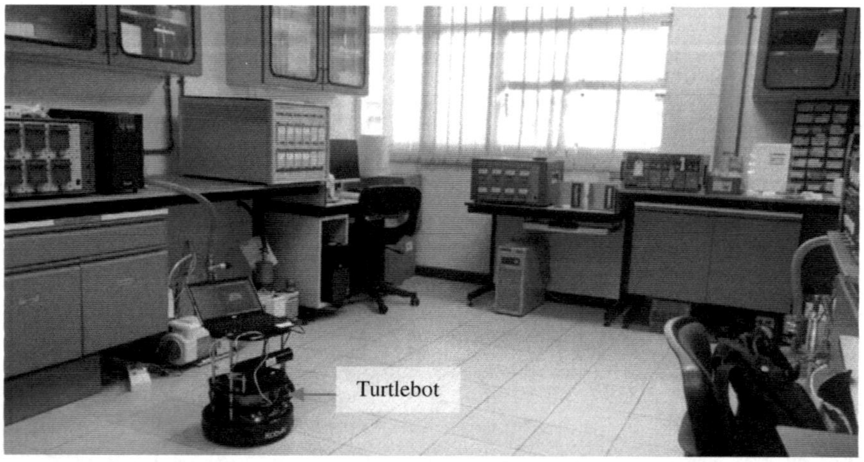

Figure 7. Radiation mapping in Alpha Spectrometry Counting Laboratory (Location 1).

Figure 8. Radiation mapping in Mobile Hotcell (Location 2).

The results are shown in Figure 9 and 10. Both maps indicated elevated radiation level at expected areas where radioactive sources are stored in the respective locations. Based on Geiger Muller detector manufacturer datasheet, conversion factor of 0.010417 could be used to convert from CPM to microsievert per hour (μSv/hr). Highest dose rate at Location 1 is recorded at 1.5 μSv/hr whereas highest dose rate for Location 2 is 8.5 μSv/hr.

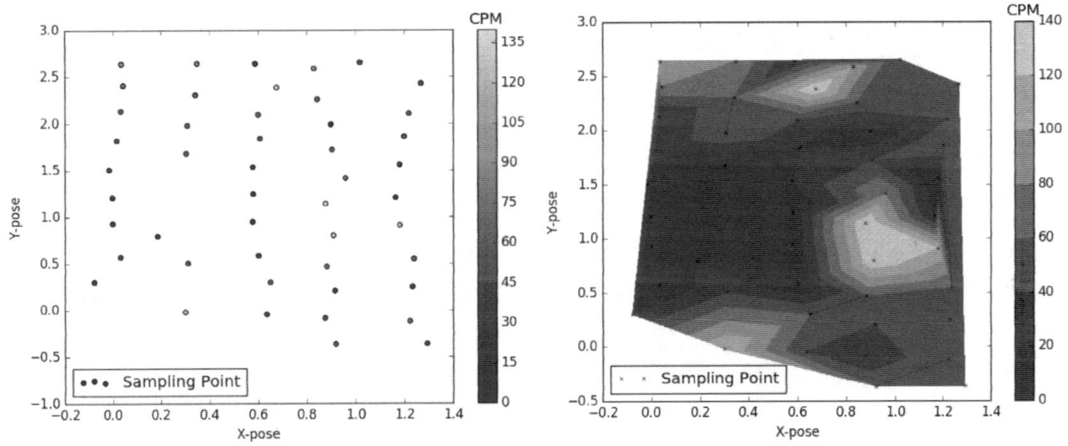

Figure 9. Radiation map for Alpha Spectrometry Counting Laboratory.

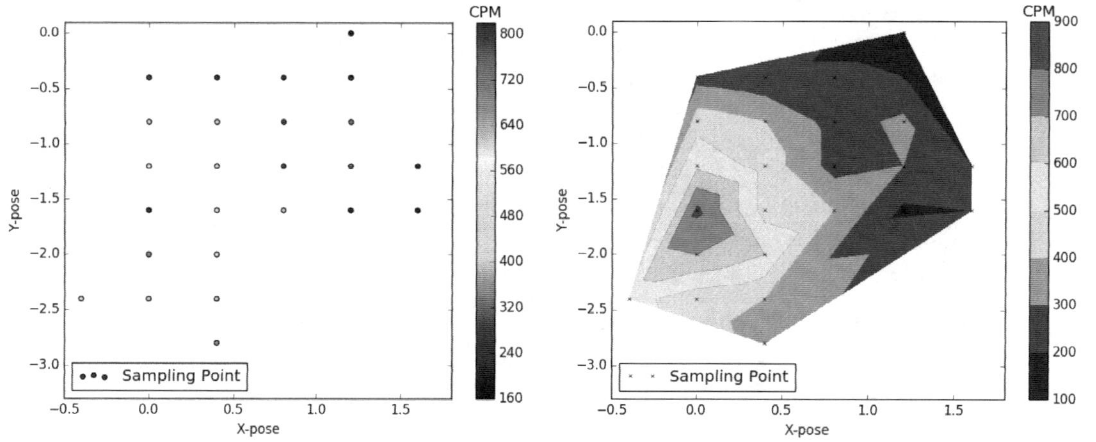

Figure 10. Radiation map for Mobile Hotcell.

In terms of optimizing the safety of radiation worker, this map could be utilized to coordinate safe routes and duration during any activity related to the location. Thus, radiation exposure could be efficiently reduced and minimized in accordance to As Low As Reasonable Achievable (ALARA) practise.

4. Conclusion and recommendation

This paper concludes that a mobile robot for autonomous operation of radiation mapping has been successfully implemented by integration of Turtlebot, Geiger Muller detector, and coverage radiation mapping algorithm. This robot is targeted to enable coverage radiation mapping in indoor environment by using SLAM to generate the physical map of the location. Experimental evaluation proved that the radiation map produced by the robot has successfully highlighted the hotspots i.e. coordinate(s) of radioactive source in the physical map. Finally, applicability of the robot is investigated at two operational radiation related facilities. The resulting map could help radiation worker to coordinate their work and reduce the radiation exposure when working in the respective radiation facility. Thus, the robot is useful in terms of radiation safety for worker in two areas; automation of the process of mapping and planning the route or position during activity in radiation related facilities.

5. References

[1] IAEA 2015 *Preparedness and Response for a nuclear or radiological emergency, no. GSR Pa.* (Vienna, Austria: IAEA General Safety Requirements).

[2] Takahashi T, Nemoto Y, Nakano H, and Tsukiyama T 2014 Development of a new simulation software system to evaluate radiation doses and facilitate decontamination tasks in reactor buildings *Prog. Nucl. Sci. Technol.* **4** p 543–547.

[3] Minamoto G, Takeuchi E, and Tadokoro S 2014 Estimation of ground surface radiation sources from dose map measured by moving dosimeter and 3D map *EEE/RSJ Int. Conf. on Intelligent Robots and Sys.* p 1889–1895.

[4] Duckworth D, Shrewsbury B, and Murphy R 2013 Run the robot backward *IEEE Int. Symp. Safety, Secur. Rescue Robot.*

[5] K. Nagatani *et al.* 2011 Redesign of rescue mobile robot Quince *IEEE Int. Symp. Safety, Secur. Rescue Robot* p 13–18.

[6] Ohno K, Kawatsuma S, Okada T, Takeuchi E, Higashi K, and Tadokoro S 2011 Robotic control vehicle for measuring radiation in Fukushima Daiichi Nuclear Power Plant *9th IEEE Int. Symp. Safety, Secur. Rescue Robot* p 38–43.

[7] Kawatsuma S, Fukushima M and Okada T 2012 Emergency response by robots to Fukushima-Daiichi accident: summary and lessons learned *Ind. Robot An Int. J.* **39**(5) p 428–435.

[8] Cadena C. *et al.* 2016 Past, Present, and future of simultaneous localization and mapping: towards the robust-perception age *IEEE Trans. Robot.* **32**(6) p 1309–1332.

[9] Mur-Artal R, Tardos J D and Tardós J D 2017 Orb-slam2: An open-source slam system for monocular, stereo, and rgb-d cameras *IEEE Trans. Robot.* **33**(5) p 1255–1262.

[10] Grisetti G, Stachniss C and Burgard W 2014 Improved techniques for grid mapping with Rao-Blackwellized particle filters p 1–12.

[11] Silva Almeida J, Bezerra Marinho L, Mendes Souza J W, Assis E A and Reboucas Filho P P 2018 Localization System for autonomous mobile robots using machine learning methods and omnidirectional sonar *IEEE Lat. Am. Trans.* **16**(2) p 368–374.

[12] Pedrosa E, Pereira A, and Lau N 2017 Efficient localization based on scan matching with a continuous likelihood field *IEEE Int. Conf. on Auton. Robot Sys. and Competitions, ICARSC 2017* p 61–66.

[13] Bogaerts B, Sels S, Vanlanduit S, and Penne R A 2018 Gradient-based inspection path optimization approach *IEEE Robot. Autom. Lett.* **3**(3) p 2646–2653.

[14] Thoa T, Copot C, Trung D andDe Keyser R 2016 Heuristic approaches in robot path planning : A survey *Rob. Auton. Syst.* **86** p 13–28.

[15] Abd Rahman N . *et al.* 2016 Arduino based radiation survey meter Arduino based radiation survey meter *AIP Conf. Proc.* ***1704*** p 030012.

Acknowledgements
The work is supported by Ministry of Education Malaysia through FRGS research grant FRGS/1/2019/TK04/UNITEN/01/2. The authors would like to acknowledge Malaysian Nuclear Agency for for the use of their laboratory, their radioactive sources, and their support in conducting the experiment. In addition, the authors would also like to acknowledge the CAMARO Research Group of Universiti Tenaga Nasional for the facilities and equipment.

IOP Conf. Series: Materials Science and Engineering **785** (2020) 012022 doi:10.1088/1757-899X/785/1/012022

Application of fuzzy logic for power change rate constraint in core power control at Reaktor TRIGA PUSPATI

Mohd Sabri Minhat[1,2], Nurul Adilla Mohd Subha[1, a)], Fazilah Hassan[1], Anita Ahmad[1]

[1]School of Electrical Engineering, Faculty of Engineering, Universiti Teknologi Malaysia, 81310 Skudai, Malaysia.
[2]Reactor Instrumentation and Control Engineering, Reactor Technology Centre, Malaysian Nuclear Agency, 43000 Kajang, Malaysia.

[a]nuruladilla@utm.my

Abstract. The 1MWth Reaktor TRIGA PUSPATI (RTP) in Malaysia Nuclear Agency has been in operation more than 37 years. The existing core power control uses a conventional controller known as Feedback Control Algorithm (FCA). It is technically challenging to keep the core power output stable and operate within the acceptable error bands for the safety demand of the RTP. At present, the power tracking performance of the system could be considered unsatisfactory where constant gains of power change rate constraint and control rod speed constraint are used. Hence, a study of a new power change rate constraint design to achieve safe control rod speed range is conducted to improve the current performance. In this paper, a new power change rate constraint (PCRC) method using fuzzy logic is proposed to control the core power. The Takagi-Sugeno (T-S) type Fuzzy model is chosen due to its capability to work well with linear controller and making the computational control algorithm efficient. The model for core power control consists of mathematical models of the reactor core, FCA, and control rods selection algorithm. The mathematical models of the reactor core are based on point kinetics model, thermal-hydraulic models and reactivity models. The performance of power tracking and actuation signal for control rod drive input are compared between the conventional PCRC (cPCRC) and Fuzzy PCRC using MATLAB. In conclusion, the proposed Fuzzy PCRC has satisfactory performance in core power tracking for controlling the nuclear reactor with high reliability and safety.

1. Introduction

The current Reaktor TRIGA PUSPATI (RTP) operating at the Malaysian Nuclear Agency is a TRIGA Mark II nuclear reactor type. It is widely used as non-power nuclear reactor with applications include production of radioisotopes, neutron radiography, basic research on the properties of materials and for education and training.

The core power control is important for safe operation of the reactor and to keep the nuclear reactor operating within its safety limit at any time and under any circumstances. The malfunction of reactor control system or any serious error made by reactor operator may lead to abnormal behaviour by the reactor, namely reactivity accidents [1]. If this error occurs, it may lead to loss of control of the chain reaction in the reactor core, which results in inability to shut down the reactor and loss of integrity of any barriers that prevent the release of radioactive fission products. Thus, the malfunction of control system has drawn attention to the researcher in developing and designing the core power control to constraint reactor safety parameter for safe and reliable operation of the nuclear reactor.

The most common safety parameter constraints such as reactivity limiting [2], reactor period limiting [3], power change rate limiting [4] and control rod speed limiting [5][6] are used to design robust core power control. Single control rod speed constraint is the most popular safety constraint which is much easier to design for implementation in a practical system. Hence, in this paper, both control rod speed and power change rate constraints will be considered to improve the tracking performance at RTP.

The RTP uses a conventional controller known as Feedback Control Algorithm (FCA) [7] which is capable to control core power up to 1MW based on conventional power change rate constraint (cPCRC). Currently, the power control system is employing a constant gain in cPCRC. However, this method yield unsatisfactory power tracking performance due to its frequent control rod movements which due to power oscillations and longer settling time to reach power demand. To solve this problem, a Takagi-Sugeno (T-S) type fuzzy approach has been established as in [8]. The fuzzy approach is chosen due to its capability to work well with a linear controller and making the computational control algorithm more efficient [9].

The power change rate is considered in designing the core power control in nuclear reactor for safe operation and within the safety limits. A few studies on power change rate constraints in core power control have been carried out. For example, in [10] and [11], a robust nonlinear Model Predictive Control (MPC) is designed by considering the optimization constraints such as xenon oscillation and maximum control rod speed. However, this method requires excessive computational effort in the on-line application.

In addition, although there are several methods dealing with the power change rate constraint which have been developed by using a reactor power profile such as the sigmoidal type trajectory in [12] and [13], these methods are not applicable for practical application since the user need to generate the appropriate load profile which significantly increases the sensitivity to input variations on the system. To cater this issue, the fuzzy-based power change rate constraint has been proposed based on tangential and sigmoid function [4]. However, the presented results require more computational time and still can be further optimized. Furthermore, in the aforementioned works [10-13], only the control rod speed (reactivity insertion) constraint is considered for the core power control.

In this paper, the power change rate constraint using fuzzy logic based on simplest linear membership function is proposed which lead to significant improvements such as applicability as a practical system, easy setting of limiting condition value for safe operation and less computational time consumption. Also, multiple constraint parameters combination of power change rate and control rod speed have been studied.

This paper is organized as follows. The modelling of RTP is presented in section 2 and the RTP core power control system is briefly described in section 3. The conventional power change rate constraint (cPCRC) and Fuzzy PCRC designs are presented in section 4. The experiment results and discussion on the implementation of cPCRC and Fuzzy PCRC to RTP model are given in section 5. Finally, conclusions are at the end of the paper.

2. Modelling of Reaktor TRIGA PUSPATI

2.1. Non-linear model

The point reactor core for a TRIGA Mark II reactor has been already modelled by [14] and is derived based on the combination of point kinetic, thermal-hydraulic and reactivity equations. The reactor core model can be simplified as:

IOP Conf. Series: Materials Science and Engineering **785** (2020) 012022 doi:10.1088/1757-899X/785/1/012022

$$\left.\begin{aligned}
\frac{d\psi}{dt} &= \frac{\rho-\beta}{\Lambda}\psi + \textstyle\sum_{i=1}^{6}\lambda_i\eta_i \\
\frac{d\eta_i}{dt} &= \frac{\beta_i}{\Lambda}\psi - \lambda_i\eta_i \quad i = 1, \dots, 6 \\
\frac{dT_m}{dt} &= \frac{N_0(1-f)}{M_m C_m} + \frac{T_f - T_m}{M_m C_m/K} - \frac{\Gamma}{M_m}(T_m - T_{in}) \\
\frac{dT_f}{dt} &= \frac{N_0 f}{M_f C_f} - \frac{T_f - T_m}{M_f C_f/K} \\
T_m &= wT_{out} + (1-w)T_{in} \\
\Gamma &= \frac{Kw(T_f - T_m)}{C_m(T_m - T_{in})} \\
\rho &= a_h \Delta h_{cr} + \alpha_m(T_m - T_m^0) + \alpha_f(T_f - T_f^0) \\
\delta\dot{\rho}_r &= W z_d
\end{aligned}\right\} \tag{1}$$

The actual core power, N, can be expressed as [15]:

$$N = \psi N_0 \tag{2}$$

All the parameters used in this paper are summarized in Table 1.

Table 1. Nomenclature for equations 1 and 2.

ψ	Relative neutron density	β	The total fraction of effective delayed neutron (*Δk/k*)
W	Reactivity worth of the control rod bank (m^{-1})	β_i	The *i*-th group of the delayed neutron (*Δk/k*)
N_0	Nominal core power (*W*)	α_f	Reactivity due to change in temperature fuel (*Δk/k/°C*)
N	Actual core power (*W*)	α_m	Reactivity due to change in temperature moderator (*Δk/k/°C*)
C_m	Moderator specific heat capacity ($J\ kg^{-1}K^{-1}$)	η_i	The *i*-th group of normalized precursor concentration (m^{-3})
C_f	Fuel specific heat capacity ($J\ kg^{-1}K^{-1}$)	T_f	Average temperature of fuel (*°C*)
M_m	Moderator total mass (*kg*)	T_f^0	Average temperature of fuel at initial equilibrium state (*°C*)
M_f	Fuel total mass (*kg*)	ρ	Total reactivity
Γ	Coolant mass flow rate (kgs^{-1})	$\delta\rho_r$	Reactivity produced by the movement of the control rod bank
K	Global heat transfer coefficient ($W\ {}^0C^{-1}$)	T_{out}	Average outlet temperature of coolant (*°C*)
λ_i	Decay constant of the *i*-th group of delay neutron precursor (s^{-1})	T_m	Average temperature of coolant (*°C*)
w	Weighting factor for computation of moderator temperature	T_m^0	Average temperature of coolant at initial equilibrium state (*°C*)
f	Fraction of power deposited in the fuel	T_{in}	Average inlet temperature of coolant (*°C*)
Λ	Mean neutron generation time (*s*)	z_d	Velocity of the control rod bank (ms^{-1})
a_h	Reactivity worth of the control rod (*Δk/k/m*)	Δh_{cr}	Control rod position differences from the critical position (*m*)

3. RTP core power control system

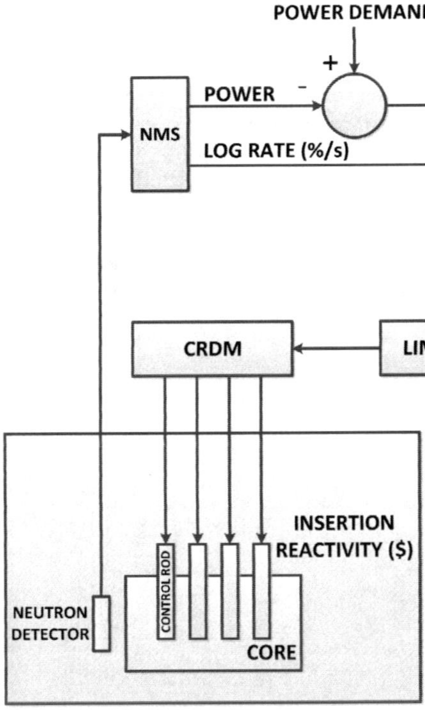

Figure 1. Block Diagram of Core Power Control System in RTP.

The closed-loop block diagram of the core power control system in RTP known as Feedback Control Algorithm (FCA) is illustrated in Figure 1. The input for this system is the Power Demand (PDM) and the output is the measured thermal neutron power from the core. The neutron detector feeds Neutron Measurement System (NMS) in which processed signals are used to provide measurements of the core power (*N*) and the rate of power change (Log Rate). The error between the PDM and the power output, and then constant gain rate (*G2*) represents the conventional power change rate constraint (cPCRC) are used as the inputs for the RTP controller. The output from the controller which is the control rod velocity is fed to the velocity limiter to constraint the reactivity insertion rate in the core. Only one control rod is allowed to move (inserted to or withdrawn from the core using Control Rod Drive Mechanism, CRDM) at any one time at the speed of 2 mm/s, equals to 0.4071 mm/cycle. To date, by using cPCRC, the core power control has 1% full power chattering error with longer settling time when the reactor power is increased in the case of a sudden change in power demand.

In order to improve the core power control performance, the new design of PCRC is proposed and the FCA parameters need to be taken into account during the design stage. The FCA parameters are given in Eq. 3 [16]

$$\left.\begin{aligned} u_{PI} &= (\mathrm{G3})E + (\mathrm{G4})\int_{t_0}^{t} E\,dt \\ E &= \left[(\mathrm{G1})\log\left(\frac{\mathrm{PDM}}{N}\right)\right]_{\pm 1} - (\mathrm{G2})\frac{1}{N}\frac{dN}{dt} \end{aligned}\right\} \quad (3)$$

where *G1*, *G2*, *G3*, and *G4* are the control gains used as constraints to limit the power change rate at below 12.5% Full Power (FP)/s limit, *E* is the error signal in terms of step numbers, t_0 is start time to calculate error signal, and t is final time (time interval) to calculate error signal. The output of the Proportional-Integral (PI) controller which determines the control rod velocity, u_{PI}, is made by using Eq. 3. The details of the RTP core power control system is illustrated in Figure 2 [17].

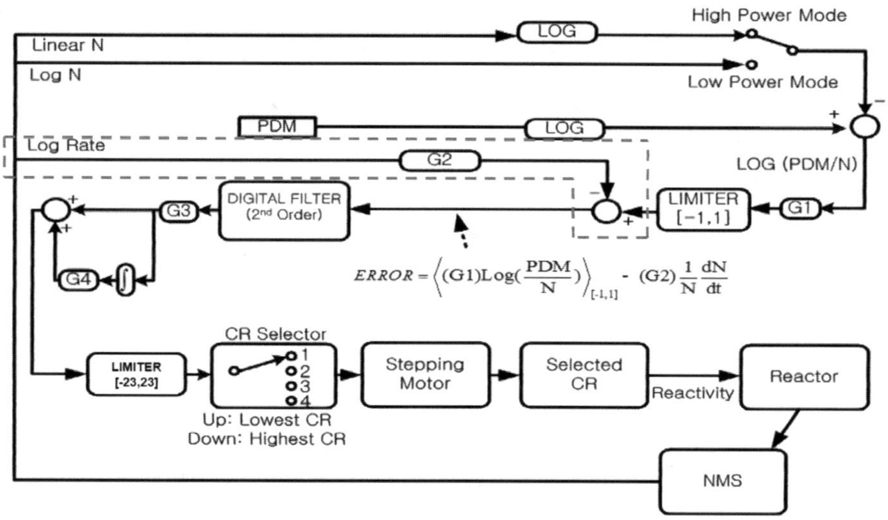

Figure 2. Details of RTP power core control system.

4. Conventional PCRC (cPCRC) and fuzzy PCRC

In this section, a new method of PCRC is proposed using fuzzy logic so that the power change rate can remain below the critical value in order to maintain the safe operation of the reactor. Due to the insertion and withdrawal motion of the control rods, the power change rate plays an important role in the core power control system. The characteristic exponential rise of the reactor power is defined in the following:

$$N(t) = N_0 e^{\frac{t}{\tau}} \tag{4}$$

where the reactor period τ is defined as the time taken for the reactor power to rise by a factor of *e* which is the base of the natural logarithm. From Eq. 4, it can be derived as:

$$ln\left(\frac{N}{N_0}\right) = \frac{t}{\tau} \tag{5}$$

Therefore, the power change rate signal over the time interval *t* can be defined as:

$$\xi_{rate} = \frac{d\, ln\, N}{dt} = \frac{1}{N}\frac{dN}{dt} = \frac{1}{\tau} \tag{6}$$

From Eq. 6, the power change rate signal should be limited to a certain level during the whole control process. Therefore, the cPCRC from Figure 2 is given in the following equation;

$$(\xi_{rate})_{constraint} = \left[G2\frac{1}{N}\frac{dN}{dt}\right]_{\pm 1} \tag{7}$$

To observe the limitation of cPCRC, the constant gain (*G2*) of Eq. 7 was set to 0.08 (12.5%/s) and then to 0.064 (15.63%/s). The visual comparison of the two different gains is illustrated in Figure 3.

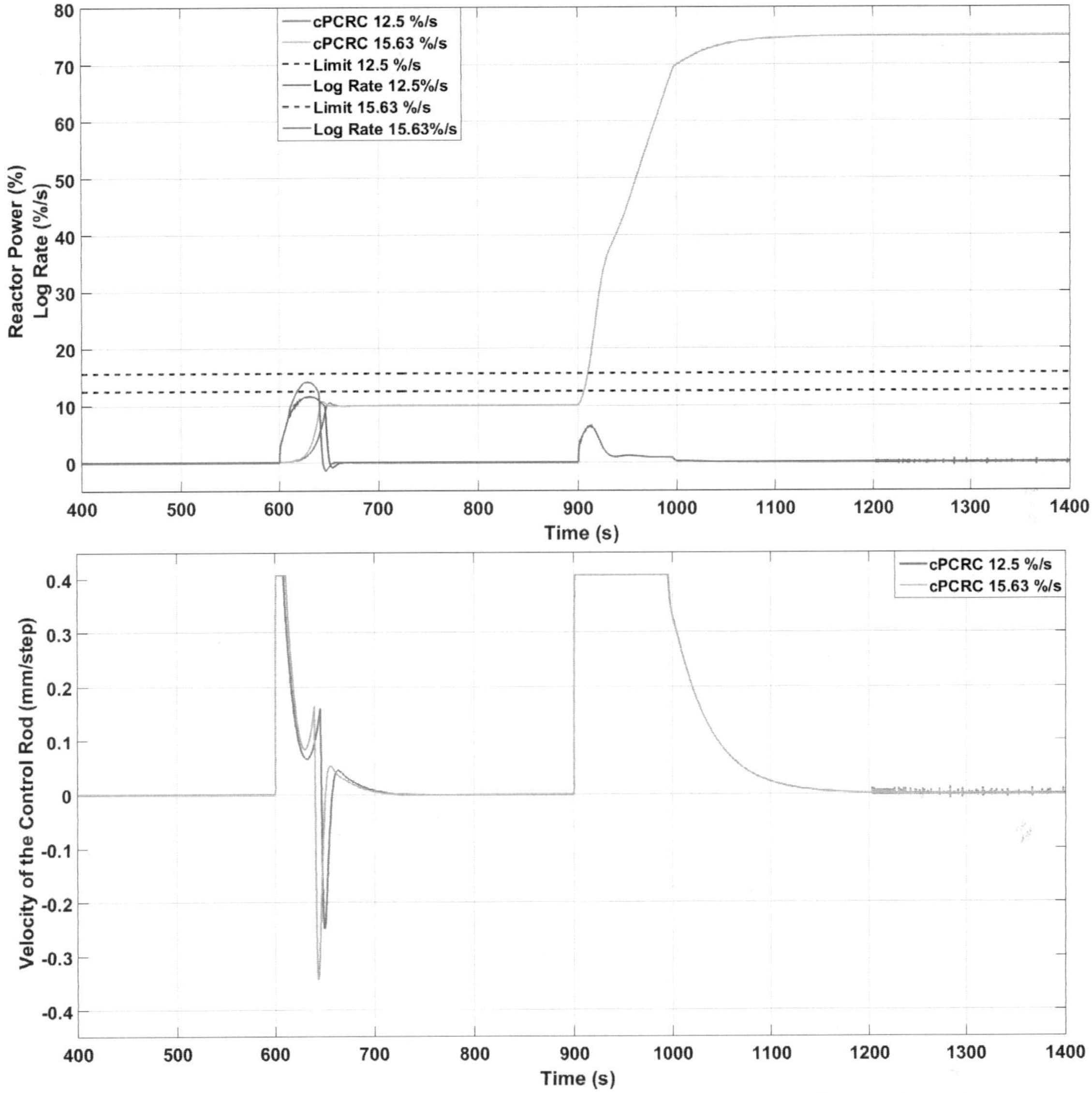

Figure 3. Comparison of different log rate values to limit the rate of power increment.

By referring to Figure 3, it is proven that increasing power change rate constraint gain in Eq. 7 has no noticeable effect on tracking control performance. However, there is a small overshoot at 10% FP.

In RTP, the cPCRC uses a single constant gain for various steps of power changes. Both the higher and lower rate of power change values give a small penalty to reduce control rod speed. In case of emergency, the maximum value of the power change rate constraint is set to 33.3% FP/s [18] using independent reactor protection system (RPS) so that the power increase can be terminated by shutting down the system to prevent the reactor power from increasing to dangerous level.

The Fuzzy PCRC uses different weighted gain constant for various step power change in which the design is using knowledge-based decision making by capitalizing operator experience. The control rod speed can be varied at different levels of penalty based on the rate of power change. The structure of

Fuzzy PCRC is illustrated in Figure 4. The output from the velocity limiter is fed to Fuzzy-Based Power Change Rate Limiter. The output from the second limiter is then applied to the control rod drive mechanism (CRDM). The advantage of this new structure is to eliminate the presence of unconsidered power change rate constraints after the controller implements the cPCRC. Furthermore, the power change rate can be adjusted to an appropriate level and thus in this work, it is set to be constrained within ±12.5% FP/s as predetermined safety limit by the Final Safety Analysis Report [18].

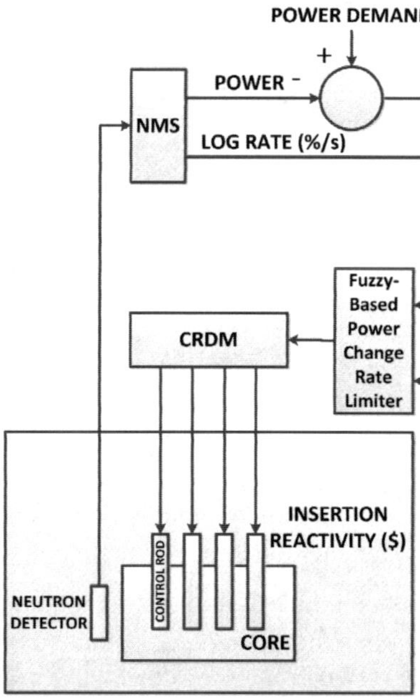

Figure 4. Block Diagram of Core Power Control using Fuzzy PCRC.

In fuzzy systems, the most commonly used parameterized membership functions are triangles (straight lines), trapezoids, bell curves, Gaussian, and sigmoidal function [19]. In this paper, three types of membership functions are studied; straight lines, bell curves and Gaussian.

The membership functions are formed using straight lines for Fuzzy PCRC as depicted in Figure 5. These straight lines of membership functions have the advantage of their simplicity.

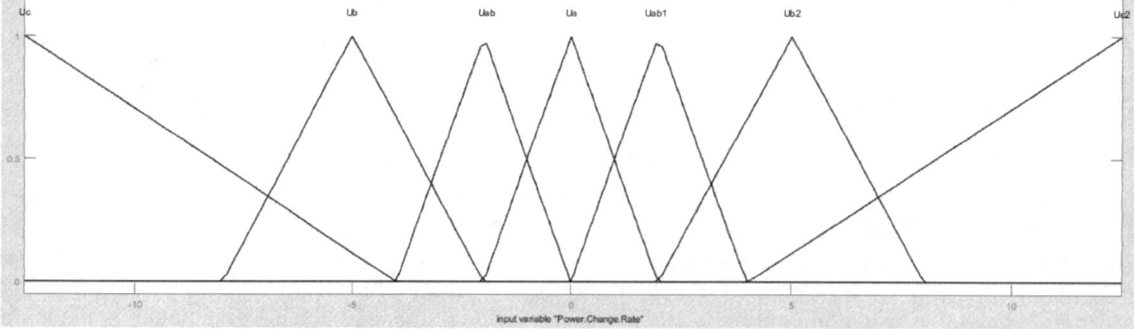

Figure 5. Straight line membership functions for Fuzzy PCRC.

The straight line membership functions for Fuzzy PCRC consists of four fuzzy sets where a power change rate ξ_{rate} is given as input variable and are introduced as follows [20].

$$\left.\begin{array}{l}
u_a = f(\xi_{rate}; a_a, 0, c_a) = \begin{cases} 0, & \xi_{rate} \le a_a \\ -\frac{\xi_{rate}-a_a}{a_a}, & a_a < \xi_{rate} < 0 \\ \xi_{rate}, & \xi_{rate} = 0 \\ \frac{c_a-\xi_{rate}}{c_a}, & 0 < \xi_{rate} < c_a \\ 0, & \xi_{rate} \ge c_a \end{cases} \\
u_{ab} = f(\xi_{rate}; a_{ab}, b_{ab}, 0) = \begin{cases} 0, & \xi_{rate} \le a_{ab} \\ \frac{\xi_{rate}-a_{ab}}{b_{ab}-a_{ab}}, & a_{ab} < \xi_{rate} < b_{ab} \\ \xi_{rate}, & \xi_{rate} = b_{ab} \\ \frac{\xi_{rate}}{b_{ab}}, & b_{ab} < \xi_{rate} < 0 \\ 0, & \xi_{rate} \ge 0 \end{cases} \\
u_b = f(\xi_{rate}; a_b, b_b, c_b) = \begin{cases} 0, & \xi_{rate} \le a_b \\ \frac{\xi_{rate}-a_b}{b_b-a_b}, & a_b < \xi_{rate} < b_b \\ \xi_{rate}, & \xi_{rate} = b_b \\ \frac{c_b-\xi_{rate}}{c_b-b_b}, & b_b < \xi_{rate} < c_b \\ 0, & \xi_{rate} \ge c_b \end{cases} \\
u_c = f(\xi_{rate}; a_c, b_c, c_c) = \begin{cases} 0, & \xi_{rate} \le a_c \\ \frac{\xi_{rate}-a_c}{b_c-a_c}, & a_c < \xi_{rate} < b_c \\ \xi_{rate}, & \xi_{rate} = b_c \\ \frac{c_c-\xi_{rate}}{c_c-b_c}, & b_c < \xi_{rate} < c_c \\ 0, & \xi_{rate} \ge c_c \end{cases}
\end{array}\right\} \quad (8)$$

where a_i, b_i, and c_i (for $i = a, ab, b, c$) are chosen as the lower limit, the center, and the upper limit respectively of the membership functions. Before applying the implication fuzzy operator, the rule weight need to assigned first. Each rules has weight, w_i is a number from 0 through 1.0, which is applied to each part of the antecedent (a single fuzzy degree of membership between 0 and 1.0) is satisfied for each rule. Generally, this w_i is 1.0 and thus has no effect on the implication process. However, in this study changing its weight value to something other than 1.0 can decrease the effect of one rule relative to the others. The variable w_i (for $i = 1, 2, 3, 4$) is adjustable weighting parameter and can be varied at different levels of penalty based on the rate of power change. The proper w_i has been assigned to each rule based on resulting input-output surface in Figure 7 on Fuzzy PCRC (straight line). The fuzzy rules for straight lines membership functions are expressed as:

$$\left.\begin{array}{c}
Rule\ 1. If\ (Power.Change.Rate\ is\ u_c)\ then\ (output1\ is\ V_s^3)\ (w_3 = 1.0) \\
Rule\ 2. If\ (Power.Change.Rate\ is\ u_b)\ then\ (output1\ is\ V_s^2)\ (w_2 = 0.15) \\
Rule\ 3. If\ (Power.Change.Rate\ is\ u_{ab})\ then\ (output1\ is\ V_s^4)\ (w_4 = 0.01) \\
Rule\ 4. If\ (Power.Change.Rate\ is\ u_a)\ then\ (output1\ is\ V_s^1)\ (w_1 = 0.001) \\
Rule\ 5. If\ (Power.Change.Rate\ is\ u_{ab1})\ then\ (output1\ is\ V_s^4)\ (w_4 = 0.01) \\
Rule\ 6. If\ (Power.Change.Rate\ is\ u_{b2})\ then\ (output1\ is\ V_s^2)\ (w_2 = 0.15) \\
Rule\ 7. If\ (Power.Change.Rate\ is\ u_{c2})\ then\ (output1\ is\ V_s^3)\ (w_3 = 1.0)
\end{array}\right\} \quad (9)$$

The complex membership functions used for the power change rate constraint at the nuclear reactor is already modeled by [4]:

IOP Conf. Series: Materials Science and Engineering **785** (2020) 012022 doi:10.1088/1757-899X/785/1/012022

$$\left.\begin{aligned} u_a &= -0.5\, tanh\big(0.8(|\xi_{rate}| - c_1)\big) + 0.5 \\ u_b &= \frac{1}{\left|1+\left[\frac{|\xi_{rate}|-c_2}{1.2}\right]^6\right|} + 0.5 \\ u_c &= 0.5\, tanh\big(0.8(|\xi_{rate}| - c_3)\big) + 0.5 \end{aligned}\right\} \quad (10)$$

where c_1, c_2, and c_3 are chosen as the center of the membership functions. The hyperbolic tangent [21] and sigmoidal function [22] can be defined in terms of the exponential function as

$$\left.\begin{aligned} tanh(|\xi_{rate}|) &= \frac{2}{1+e^{-2|\xi_{rate}|}} - 1 \\ sigmoid(|\xi_{rate}|) &= \frac{1}{1+e^{-a(|\xi_{rate}|-c)}} \end{aligned}\right\} \quad (11)$$

In [4], the proposed method using complex functions such as hyperbolic tangent and sigmoid function for power change rate constraint is useful in adjusting the constraint bound but is computationally expensive and does not add any more precision. Thus, in reactor core power control, a simple symmetric function is preferable in building the membership function. The two membership functions are built upon the second proposed Fuzzy PCRC: a simple Gaussian curve [20] and a generalized bell curve [23].

In Figure 6, the membership functions are formed based on a simple Gaussian curve and generalized bell curve. Both Gaussian and bell membership functions are popular methods for specifying the fuzzy sets. Both curves have the advantage of being smooth and nonzero at all points.

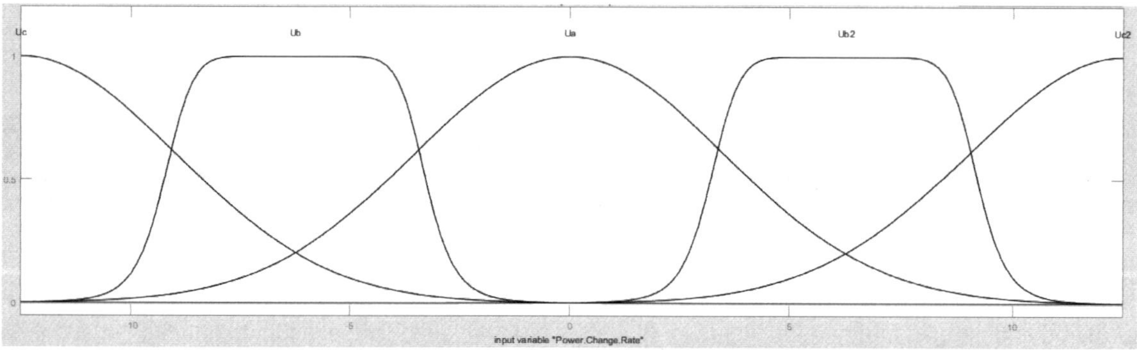

Figure 6. Gaussian and bell curves membership functions for Fuzzy PCRC.

The proposed membership functions in Figure 6 can be expressed as:

$$\left.\begin{aligned} u_a &= e^{-\frac{(|\xi_{rate}|)^2}{2\sigma^2}} \\ u_b &= \frac{1}{\left|1+\left[\frac{|\xi_{rate}|-c_b}{a}\right]^{2b}\right|} \\ u_c &= e^{-\frac{(|\xi_{rate}|-c_c)^2}{2\sigma^2}} \end{aligned}\right\} \quad (12)$$

where c_b, and c_c are positive constants and are chosen as the center of the membership functions.

In addition, the following scaled control inputs are considered:

$$V_s^i = w_i u_{step} \quad (13)$$

where w_i (for $i = 1, 2, 3$) is an adjustable weighting parameter for the constraint of the power change rate based on resulting input-output surface in Figure 7 on Fuzzy PCRC (Gaussian and bell curve). The fuzzy rule bases for Gaussian and bell curves membership functions are then expressed as

$$\left.\begin{aligned} &Rule\ 1. If\ (Power.Change.Rate\ is\ u_c)\ then\ (output1\ is\ V_s^3)\ (w_3 = 1.0) \\ &Rule\ 2. If\ (Power.Change.Rate\ is\ u_b)\ then\ (output1\ is\ V_s^2)\ (w_2 = 0.5) \\ &Rule\ 3. If\ (Power.Change.Rate\ is\ u_a)\ then\ (output1\ is\ V_s^1)\ (w_1 = 0.001) \\ &Rule\ 4. If\ (Power.Change.Rate\ is\ u_{b2})\ then\ (output1\ is\ V_s^2)\ (w_2 = 0.5) \\ &Rule\ 5. If\ (Power.Change.Rate\ is\ u_{c2})\ then\ (output1\ is\ V_s^3)\ (w_3 = 1.0) \end{aligned}\right\} \tag{14}$$

Then, using a center-average defuzzifier [8], a Fuzzy PCRC is designed separately for both straight lines and Gaussian and bell curves membership functions as follows:

$$\hat{u}_{step} = u_{step} - \left(\frac{u_a V_s^1 + u_b V_s^2 + u_c V_s^3 + u_{ab} V_s^4}{u_a + u_b + u_c + u_{ab}}\right) \Rightarrow \hat{u}_{step} = \left[1 - \left(\frac{u_a w_1 + u_b w_2 + u_c w_3 + u_{ab} w_4}{u_a + u_b + u_c + u_{ab}}\right)\right] u_{step} \tag{15}$$

By referring to Eq. 12, if the rate of power change is large, the membership function u_c becomes dominant i.e., have large value. In order to limit the rate, the scaled control input V_s in Eq. 13 is introduced by multiplying the weighting parameter, $w_{i=3}$ in Eq. 14 with the control rod speed is calculated by the controller, u_{step}. The larger value of V_s consequently changed and penalized the final control input $\hat{u}_{step}$ in Eq. 15 to become zero. Therefore, it can be noticed that the control of the rate of power change can be effectively accommodated in the design of the constrained control input. The Fuzzy PCRC has the advantage that adjust the constraint bound more easily compared to the cPCRC. The adjustment of the constraint bound in this work should also be investigated for three other types of PCRC control surfaces as shown in Figure 7 when considering the choice of rules and membership functions in the design stage. Then, only one input-output surface is to be selected to fit as a solution for the problem at cPCRC.

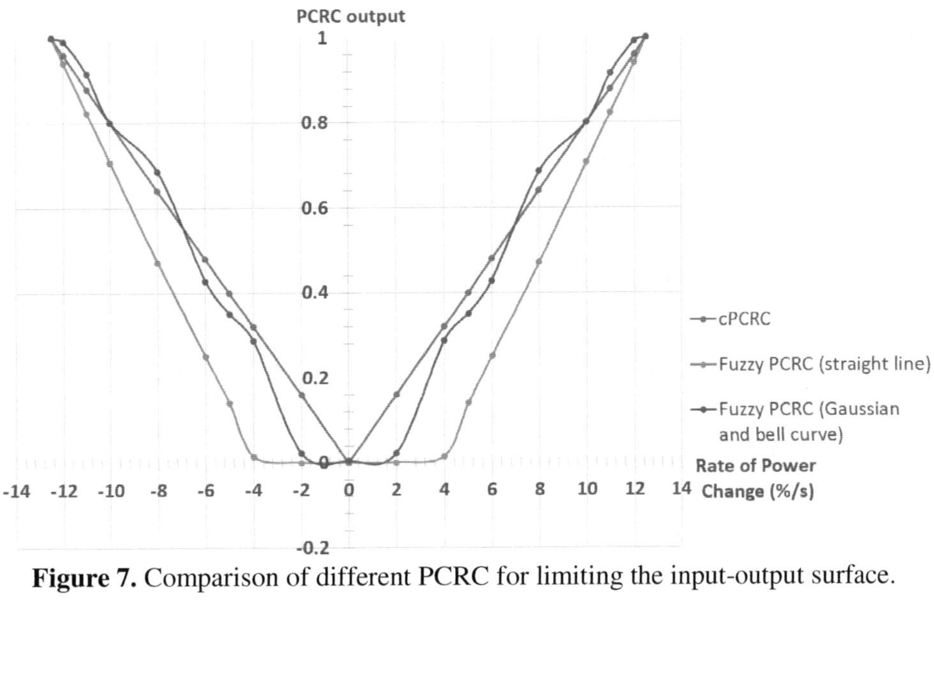

Figure 7. Comparison of different PCRC for limiting the input-output surface.

5. Experiment results and discussion

It is recommended to start a design procedure with the linear surface and then tune the gains. Figure 8 shows the overall results for the core power control performance with different PCRCs. By referring to Figures 9 to 11, the performance of all PCRC methods are observed at three different scenarios; by increasing the power to 10% FP, by reducing the power to a low power (50% FP to 10% FP) and by increasing the power to 75% FP respectively.

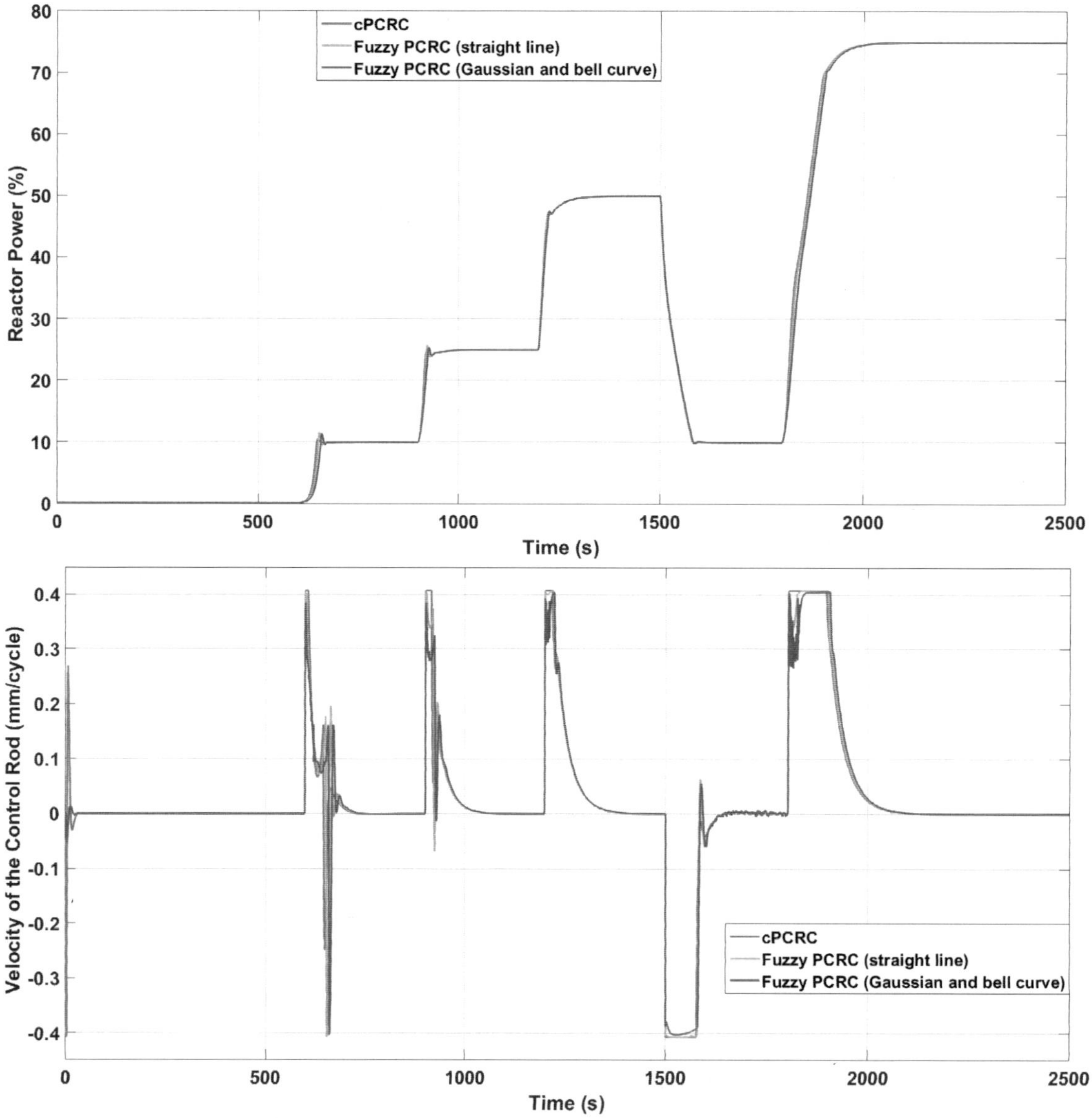

Figure 8. Comparison of different PCRC for limiting the rate of power increment at various steps of power change.

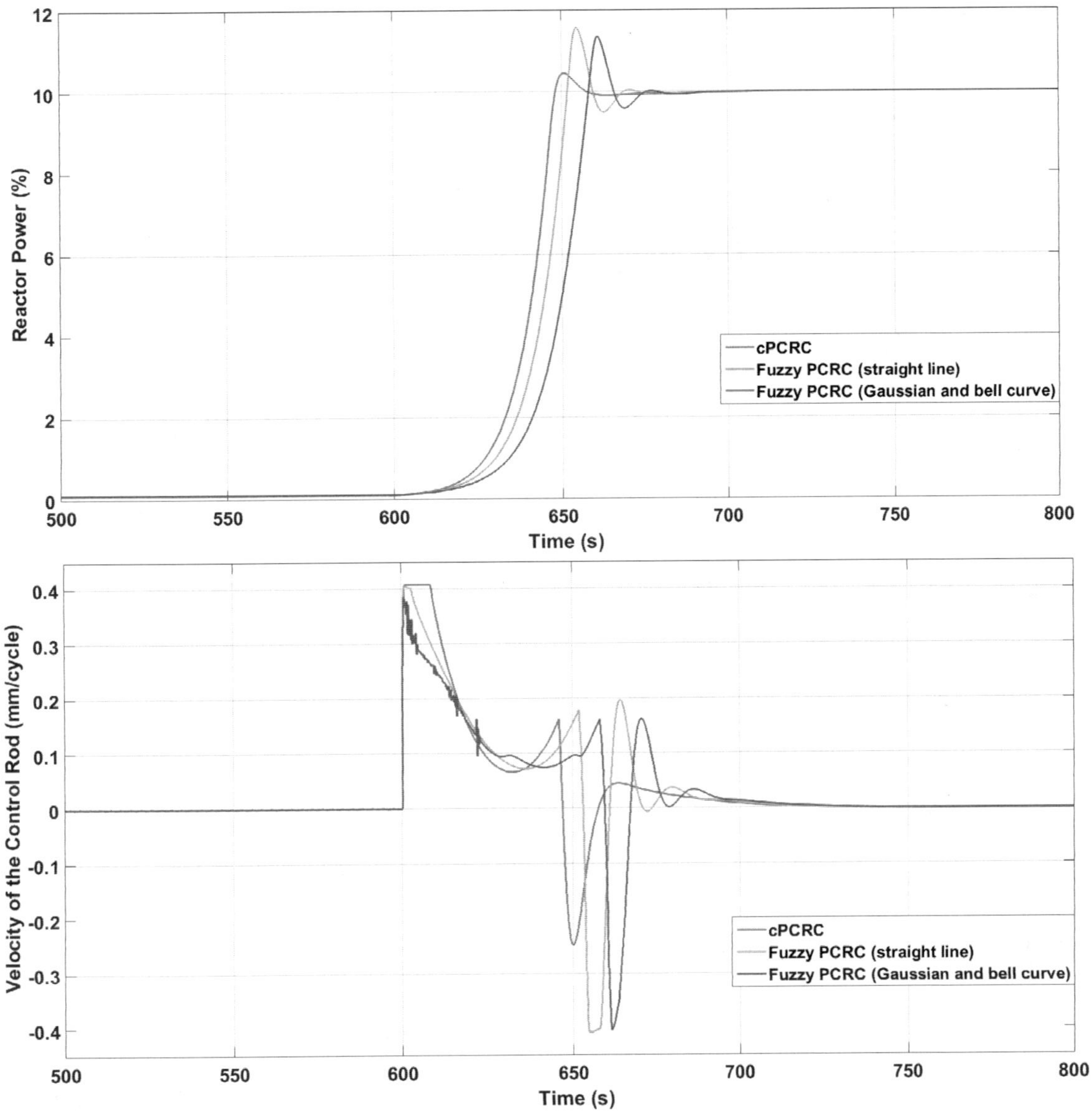

Figure 9. Comparison of different PCRC methods for limiting the rate of power increment (by increasing the power to 10% FP).

IOP Conf. Series: Materials Science and Engineering **785** (2020) 012022 doi:10.1088/1757-899X/785/1/012022

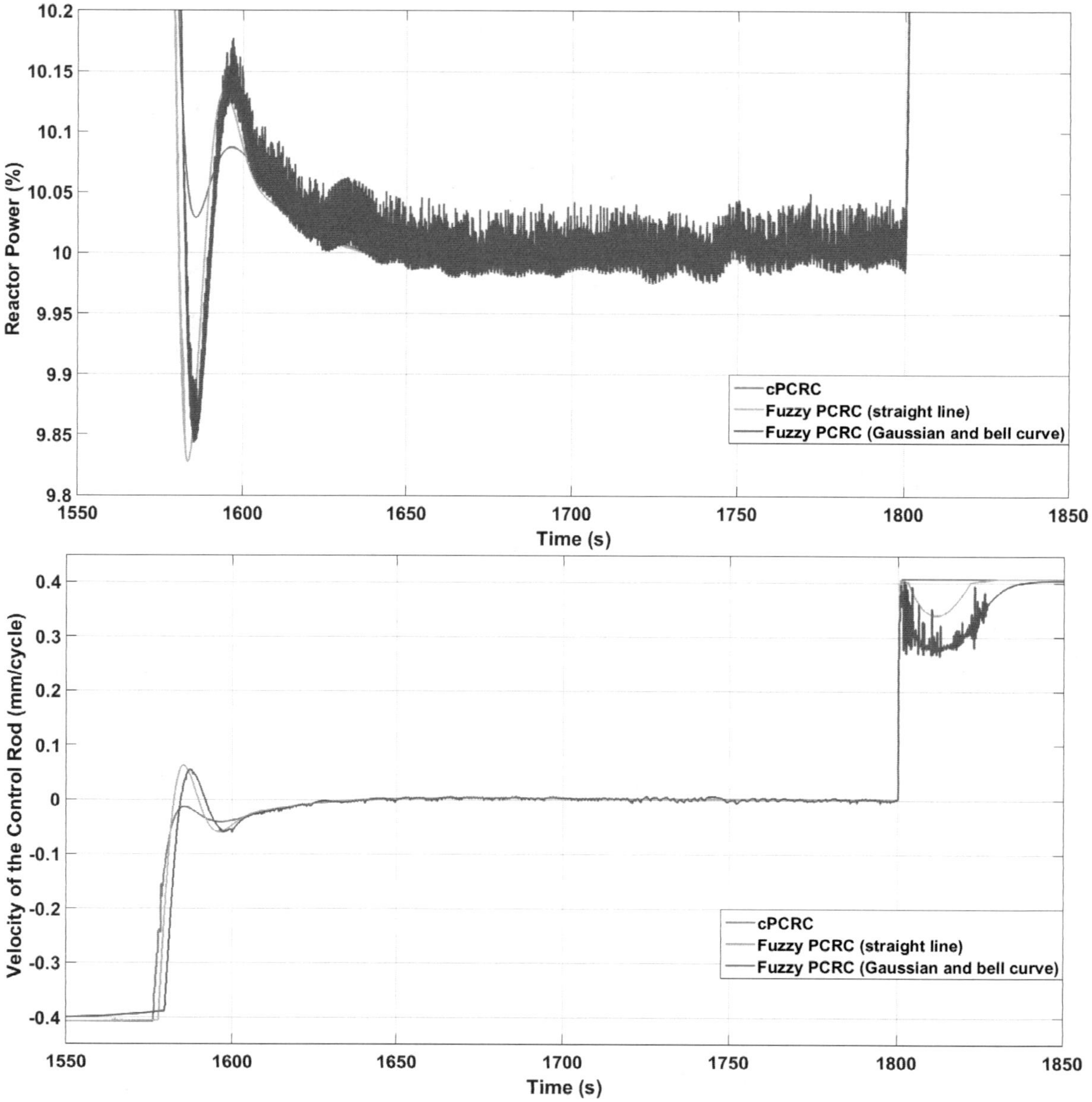

Figure 10. Comparison of different PCRC methods to limit the rate of power increment (by reducing the power to a lower power at 10% FP).

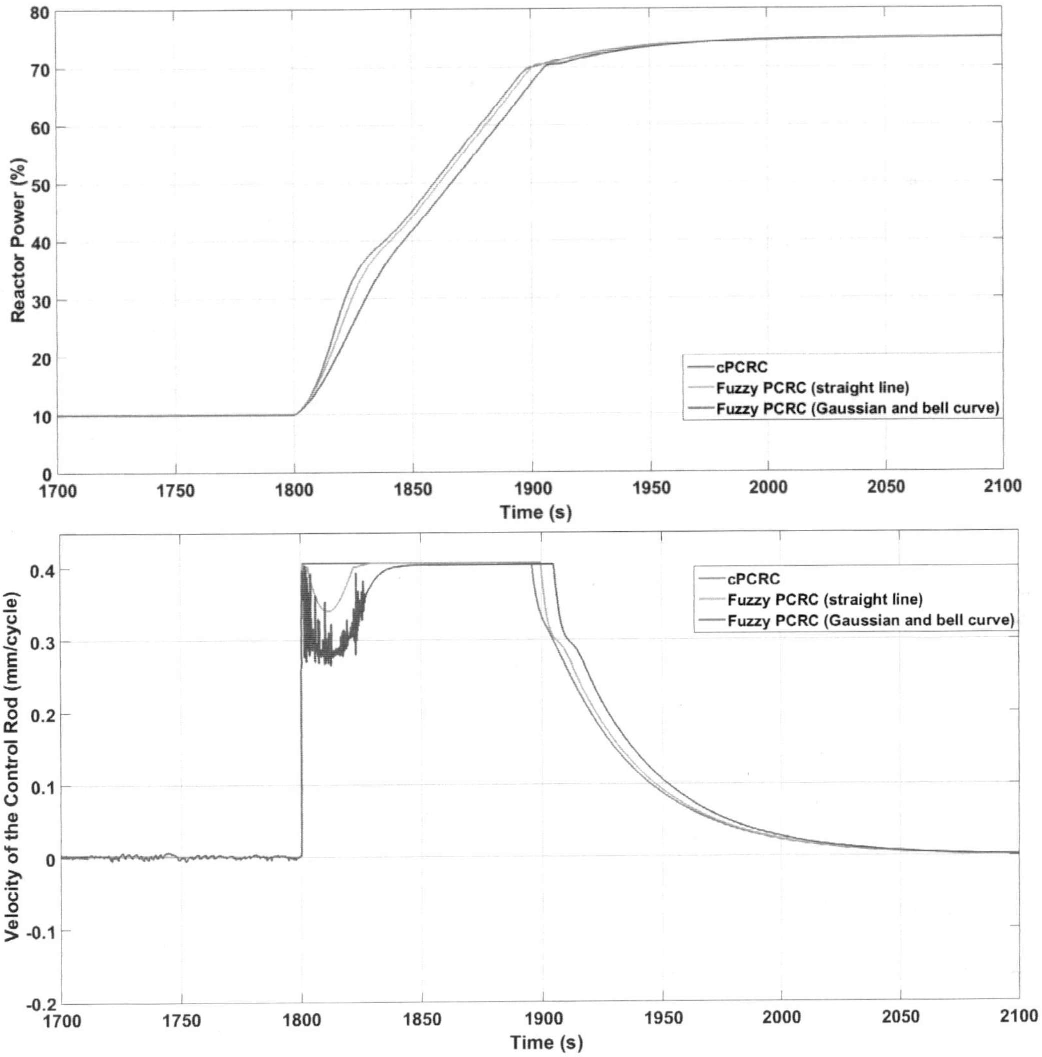

Figure 11. Comparison of different PCRC for limiting the rate of power increment (by increasing the power to 75% FP).

From Figures 8 to 11, it can be seen that the Fuzzy PCRC with straight line membership function has a gently sloping surface which will make the design more robust but with high overshoot at low power. Furthermore, this Fuzzy PCRC also is able to provide almost similar tracking to cPCRC at high and low powers within acceptable limits. On the contrary, the Fuzzy PCRC with Gaussian and bell curves membership function has tighter control with steep surface which would make the system become more oscillatory [24]. Its aggressive signal may increase the possibility of damage in CRDM. Only the Fuzzy PCRC (straight line) can significantly reduce the chattering error in the actuator during steady-state condition while others could not.

From the results above, it can be concluded that the single linear function type or the simplest triangles membership functions can produce a better control performance. In addition, the exponential function is computationally expensive, hard to be applied, non-linear in nature and is not capable of outputting a true zero (i.e., nonzero at all point). In contrast, the linear function is easier to compute and optimize. It converges faster compared to the smooth function by around a factor of six. The consideration of linear and complex membership functions in this work is to investigate the potential of both functions in improving the performance of core power control system.

Performance comparison between cPCRC and Fuzzy PCRC (straight line) at 75% FP is tabulated in Table 2 quantitatively.

Table 2. Transient and steady-state response for core power control at 750kW

Properties	***cPCRC***	***Fuzzy PCRC (straight line)***
Average (%)	74.971%	74.973%
Min (%)	74.635%	74.637%
Max (%)	75.018%	75.001%
Δe_{ce} (%)	0.383%	0.365%
Settling Time (T_s)	118.5 s	119.0 s
Percent Overshoot (%)	0.024%	0.002%
Rise Time (T_r)	84.5 s	86.0 s

From the experimental result obtained to ensure a stable power with small oscillation can be observed by measured the average, minimum and maximum value of power level during steady-state at 75% FP in a short time range; 2000 s to 2500 s. The deviation chattering error Δe_{ce} can be calculated based on the difference between maximum and minimum values. During transient, the power tracking performance can be evaluated based on the minimum value of rise time T_r and settling time T_s, where T_s is time it takes for the error between reactor power and the power demand to fall to within 5%, and T_r is time it takes for the response to rise from 10% to 90% of the steady state response. The percent overshoot is the difference between peak value and power demand as a final value in percentage.

From Table 2, it can be observed that the proposed Fuzzy PCRC (straight line) is still unable to reduce the settling time for a better power tracking performance. Due to this, a new value of control rod speed limiter was investigated for Fuzzy PCRC (straight line). The performance of cPCRC at different control rod speed constraint is shown in Figure 12.

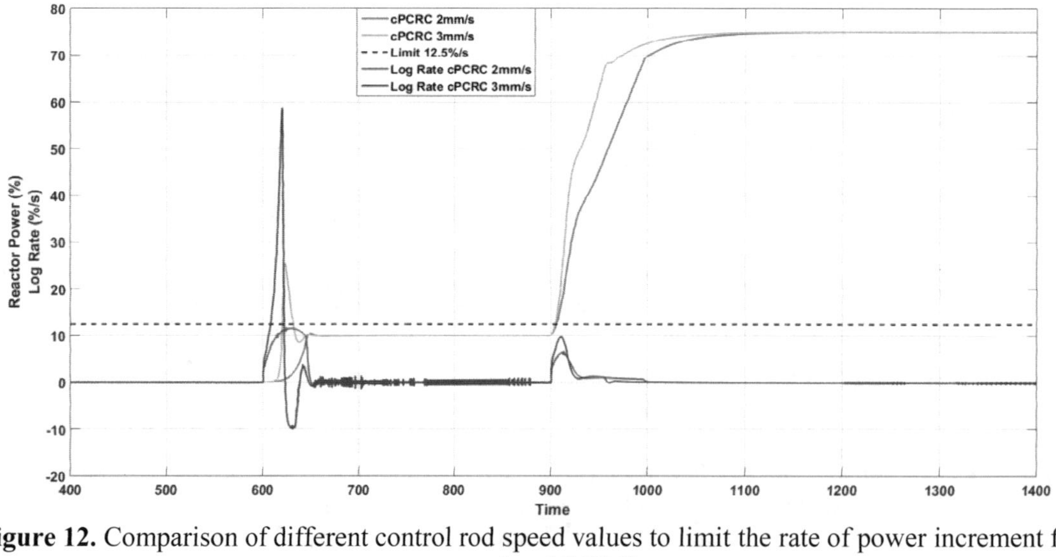

Figure 12. Comparison of different control rod speed values to limit the rate of power increment for conventional PCRC.

By referring to Figure 12, the rise time is too short and the rate of power change is more than 33.3%/s which caused the reactor to trip in real scenario at RTP [18]. The performance of Fuzzy PCRC (straight line) is optimized by using different values of control rod speeds as shown in Figure 13.

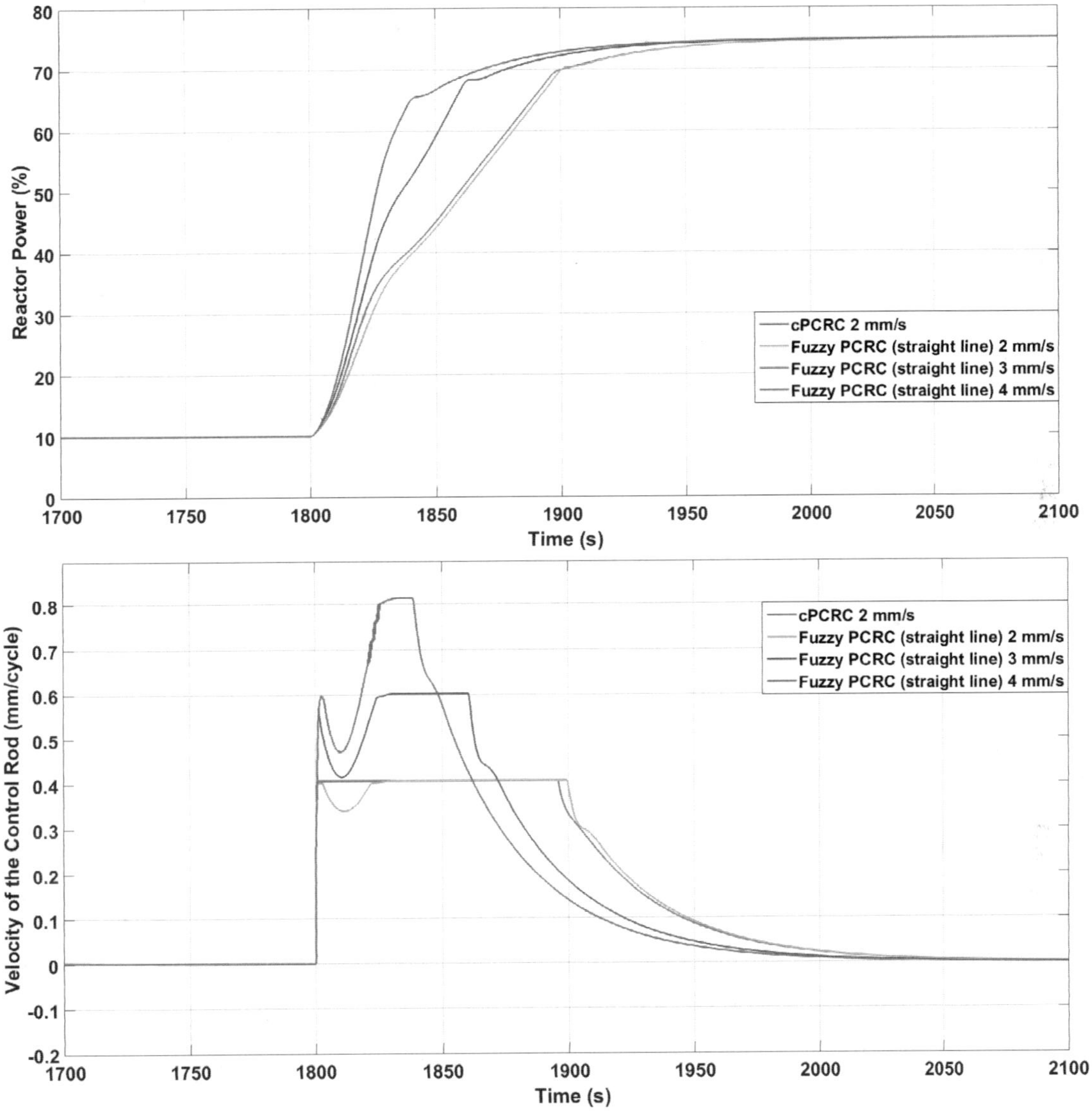

Figure 13. Different control rod speed values for nPCRC (by increasing the power to 75% FP).

In Figure 13, it can be observed that the tracking performance of Fuzzy PCRC (straight line) is improved by eliminating the unsmooth control surface and reducing the settling time when the control rod speed is increased from 2mm/s to 4mm/s. As a comparison in cPCRC, the rate of power change for the increased control rod speed from 3mm/s and 4mm/s is less than 12.5%/s. The performance of cPCRC and Fuzzy PCRC (straight line) with different control rod speed at 750kW is tabulated in Table 3.

Table 3. Performance summary for cPCRC and Fuzzy PCRC (straight line) at 750kW with different control rod speed.

Properties	*cPCRC with 2mm/s*	*Fuzzy PCRC (straight line) with 2mm/s*	*Fuzzy PCRC (straight line) with 3mm/s*	*Fuzzy PCRC (straight line) with 4mm/s*
Settling Time (T_s)	118.5 s	119.0 s	94.0 s	84.0 s
Percent Overshoot (%)	0.024 %	0.002 %	0.001 %	0.017 %
Rise Time (T_r)	84.5 s	86.0 s	58.5 s	50.0 s
Δe_{ce} (%)	0.383 %	0.365 %	0.191 %	0.166 %
Maximum Rate of Power Change (%/s) between 0.1 %FP - 10 %FP	11.580 %/s	11.000 %/s	11.520 %/s	11.800 %/s
Maximum Rate of Power Change (%/s) between 10 %FP - 75%FP	6.240 %/s	6.609 %/s	6.544 %/s	7.577 %/s

Overall, based on Table 3, a Fuzzy PCRC (straight line) with high control rod speed generally offers a better result than cPCRC. In addition, it can remove unwanted line noise for on CRDM when moving control rods and offer small settling time to reach demanded power.

6. Conclusion

In this work, a new design of PCRC is proposed by using fuzzy logic to improve the tracking performance of core power control at RTP. Instead of using a constant gain in cPCRC, Fuzzy PCRC offers generally better results which is able to reduce chattering noise maximum up to 56%. The core power control performance with Fuzzy PCRC can then be further improved by increasing the speed of the control rod. It has been proven in this work that the rate of power change at increased speed does not exceed 33.3 %/s which is the trip parameter.

7. References

[1] Jamil F, Abid M, Haq I, Khan A Q and Iqbal M 2016 Fault diagnosis of Pakistan Research Reactor-2 with data-driven techniques *Ann. Nucl. Energy* **90** p 433–440.

[2] Benítez-Read J S, Ruan D, Najera-Hernandez M, Perez-Clavel B, Pacheco-Sotelo J O and Lopez-Callejas R 2005 Comparison between a continuous and discrete method for the aggregation and defuzzification stages of a TRIGA reactor power fuzzy controller *Prog. Nucl. Energy* **46** p 309–320.

[3] Pérez-Cruz J H and Poznyak A 2008 Neural control for power ascent of a TRIGA reactor *Proc. Am. Control Conf.* p 2190–2195.

[4] Eom M, Chwa D and Baang D 2015 Robust disturbance observer-based feedback linearization control for a research reactor considering a power change rate constraint *IEEE Trans. Nucl. Sci.* **62** p 1301–1312.

[5] Coban R 2014 Power level control of the TRIGA Mark-II research reactor using the multifeedback layer neural network and the particle swarm optimization *Ann. Nucl. Energy* **69** p 260–266.

[6] Minhat M S, Mohd Subha N A, Hassan F and Mohamad Nordin N 2019 An improved control rod selection algorithm for core power control at Reaktor TRIGA PUSPATI *Symp. Control Syst. Signal Procesing* p 1-23.

[7] Minhat M S, Selamat H and Mohd Subha N A 2018 Adaptive control method for core power control in TRIGA Mark II reactor *IOP Conf. Ser. Mater. Sci. Eng.* **298** p 1-15.

[8] Luan X and Young A 2011 Load-following control of nuclear reactors based on Takagi-Sugeno fuzzy model *IFAC Proceedings Volumes* **44** p 8253-8258.

[9] Cavallaro F 2015 A Takagi-Sugeno fuzzy inference system for developing a sustainability index of biomass *Sustainability* **7** p 12359–12371.

[10] Eliasi H, Menhaj M B and Davilu H 2011 Robust nonlinear model predictive control for nuclear power plants in load following operations with bounded xenon oscillations *Nucl. Eng. Des.* v **241** p 533–543.
[11] Eliasi H, Menhaj M B and Davilu H 2012 Robust nonlinear model predictive control for a PWR nuclear power plant *Prog. Nucl. Energy* **54** p 177–185.
[12] Coban R 2010 Computational intelligence-based trajectory scheduling for control of nuclear research reactors *Prog. Nucl. Energy* **52** p 415–424
[13] Rojas-Ramírez E, Benítez-Read J S and Segovia-De-Los Ríos A 2013 A stable adaptive fuzzy control scheme for tracking an optimal power profile in a research nuclear reactor *Ann. Nucl. Energy* **58** p 238–245.
[14] Cammi A, Ponciroli R, Borio Di Tigliole A, Magrotti G, Prata M, Chiesa D and Previtali E 2013 A zero dimensional model for simulation of TRIGA Mark II dynamic response *Prog. Nucl. Energy* **68** p 43–54.
[15] Wang G, Wu J, Zeng B, Xu Z and Wu W 2016 State-space model predictive control method for core power control in pressurized water reactor nuclear power stations *Nucl. Eng. Technol.* **49** p134–140.
[16] Baang D, Suh Y and Park C 2017 Feedback power control for TRIGA-II research reactor *Trans. Korean Nucl. Soc. Spring Meet.* (Jeju, Korea) p 1-2.
[17] Baang D, Suh Y and Kim S H 2017 Power controller design and application to research reactor *Trans. Korean Nucl. Soc. Spring Meet.* (Gyeongju, Korea) p 1-2.
[18] Malaysia Nuclear Agency 2014 Safety analysis report for PUSPATI TRIGA reactor.
[19] Wai T C 2009 *Fuzzy Associative Memory Architecture* (Nanyang Technological University)
[20] Daher A, Hoblos G, Khalil M and Chetouani Y 2018 Parzen window distribution as new membership function for ANFIS algorithm- Application to a distillation column faults prediction *Chemom. Intell. Lab. Syst.* **51** p 241–248
[21] Nwankpa C E, Ijomah W, Gachagan A and Marshall S 2018 Activation functions : comparison of trends in practice and research for deep learning (*ArXiv 2018)* p 1–20
[22] Hatami E, Vosoughi N and Salarieh H 2016 Design of a fault tolerated intelligent control system for load following operation in a nuclear power plant *Electr. Power Energy Syst.* **78** p 864–872.
[23] Tan Y, Shuai C, Jiao L and Shen L 2017 An adaptive neuro-fuzzy inference system (ANFIS) approach for measuring country sustainability performance *Environ. Impact Assess. Rev.* **65** p 29–40.
[24] Jantzen J 1997 A robustness study of fuzzy control rules *Comp. Sci.* p 1–6.

Acknowledgments
This work was supported by a Fundamental Research Grant Scheme (FRGS) grant (FRGS/1/2017/TK04/UTM/02/55) from the Ministry of Higher Education, Malaysia. The author would like to thank all individuals who have contributed to this paper especially to Dr Nurul Adilla Binti Mohd Subha from Universiti Teknologi Malaysia and to the Malaysian Nuclear Agency under the Ministry of Energy, Science, Technology, Environment, and Climate Change, Malaysia (MESTECC) for the support.

Performance of imaging system at Reactor TRIGA Mark II PUSPATI

Khairiah Yazid[1,2,a)], Mohd Zaid Abdullah[1], Muhammad Rawi Mohamed Zin[2], Rafhayudi Jamro[2], Julie Andrianny Murshidi[2] and Faridah Mohamad Idris[2]

[1]School of Electrical and Electronic Engineering, Universiti Sains Malaysia, 14300 Nibong Tebal, Penang, Malaysia.
[2]Malaysian Nuclear Agency, Bangi, Kajang, 43000 Selangor, Malaysia.

[a]khairiah@nm.gov.my

Abstract. The neutron radiography facility in TRIGA MARK II PUSPATI reactor (RTP) was successfully upgraded to obtain higher image quality as well as reducing exposure time. The major improvement is the implementation of a new neutron detector for neutron imaging. In this work, the implementation of a new neutron imaging system based on CCD camera and LiF/ZnS scintillator is presented. This article focuses on the qualitative and quantitative results obtained by using the new imaging system. The characteristics of this new imaging system were investigated in terms of spatial resolution.

1. Introduction

Reactor TRIGA Mark II PUSPATI (RTP) is a swimming pool-type light water research reactor with enriched uranium-zirconium-hydride fuel and graphite reflector (1 MW). The reactor has three radial beam ports, one tangential beam port and one thermal column.

The neutron imaging facility at RTP has been developed since the late 1980s and is located at one of the radial beam ports. The film-based technique was established at this facility. However, this facility produced low neutron intensities at the sample position, and this led to very long exposures. To overcome this limitation, an upgrade plan for the neutron facility was initiated in 2015. This upgrade was essential as there was an urgent need for a modern imaging system at this facility. Three years after the start of the upgrade plan, the newly designed collimator was successfully installed [1] [2]. Recently, extensive experiments using film have been carried out to characterise the new neutron beam setup [3].

A major step in the improvement of the neutron radiography activity at this upgraded facility is the implementation of a digital neutron detector for real time neutron radiography. The CCD camera-based detector is a key component in the new neutron imaging system. The common modern detector for a neutron imaging system is a CCD camera optically coupled to a scintillator screen. This digital detector offers some very explicit advantages over the conventional film-based technique. Generally, the chain of components in a CCD camera scintillator detector consists of: scintillator-mirror-lens-CCD camera-interface-PC-software. The most common scintillation screen for neutron imaging consists of zinc sulphide (ZnS) mixed with lithium fluoride (6LiF), doped with silver (Ag) for blue light emission or copper and gold (Cu, Au) for green light emission.

The objective of this work is to characterise the performance of the new imaging system at RTP in terms of spatial resolution.

2. Experimental

2.1. Neutron imaging system

All experiments in this work were performed at the upgraded neutron radiography facility at RTP. The facility is a new well-shielded concrete construction (bunker). Figure 1 shows a simplified view of the neutron radiography facility at RTP. It consists of a 232 m long divergent collimator, a 3 cm diameter aperture, mechanical beam shutter and an external beam trap. The imaging test station is located 130 m away from the beam port exit. Between the beam port exit and imaging equipment a horizontal moveable beam trap is installed. It is a sandwich construction consisting of several 10 cm thick borated polyethylene and 15 cm thick lead. To perform neutron radiography, the sample is positioned at a 100 cm distance from the beam port exit between the beam trap and imaging equipment (L/D ratio ~110). Due to the limited space in the bunker, the maximum L/D ratio is ~120. The thermal neutron flux at the sample position is 104 n $s^{-1}cm^{-2}$ at 750 kW reactor operating power. This leads to a long exposure time at this low flux position for better image quality.

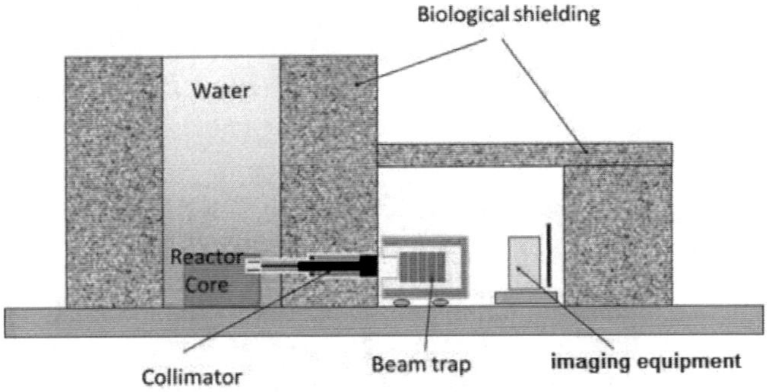

Figure 1. Layout of neutron radiography facility at RTP.

Figure 2 shows the present status of the installed imaging equipment at RTP. The detection system consists of a scintillator converter screen, a mirror and a commercial CCD camera. The scintillator screen is 0.1 mm thick and has a 20 cm*20 cm field of view based on 6LiF/ZnS co-doped with Cu and Au. The camera is a Peltier cooled (-27 ºC) low intensity CCD sensor comprising of 2759 x 2200 pixels (4.54 x 4.54 μm2 size) and a quantum efficiency of ~75%. A 45° front-surfaced aluminium mirror and a 50 mm f/1.2 high-performance manual focus lens is used to project the greenlight from the scintillator onto the CCD sensor. The detector is housed within an L-shaped light-tight camera box and a small additional shielding (lead plate plus a B4C layer) was used to protect the CCD unit from the direct beam. However, this shielding of the detection system was insufficient. High-energy gammas from the radial beam port may hit the CCD chip directly. Hence, an additional 2 cm lead shielding was fabricated to minimise the camera's gamma exposure.

Figure 2. The present setup of the neutron imaging equipment inside the bunker at RTP.

2.2. Modulation transfer function

The modulation transfer function (MTF) is a fundamental tool for assessing the performance of imaging systems. In neutron imaging particularly, MTF is one of the most useful tools to evaluate the performance of an imaging system on a quantitative basis. There are several approaches to obtain the MTF.

The most common method is use a slit to determine the modulation transfer function of the detector system. This approach is less sensitive to noise. However, the disadvantage is that the slit must be polished or extracted with very high precision and must be very well aligned with the beam.

In this work, a simple method for obtaining the MTF of neutron imaging detectors from edge images is implemented. From the digital edge image an edge spread function (ESF) is generated. ESF is measured using an opaque object with a straight edge. Figure 3 shows the test object used to measure the MTF. The cadmium plate was chosen to obtain significant different intensities. The triangle slits were fabricated in order to produce slanted pattern edges between the air and the plate. The test object was positioned at close contact to the scintillator screen and scanned at 750 kW reactor operating power with 5 minutes exposure time. All neutron images were 16-bit grayscale images and all white noise were suppressed in a pre-processing step before the MTF calculation process to avoid noise amplification.

Figure 3. Test object

The calculation of the MTF was implemented in the following steps:

(1) Acquisition of an image of a slanted edge. A small region of interest (ROI) surrounding the edge was manually extracted. The ROI was selected for a high contrast sharp edge that is homogenous on both sides.
(2) Construction of an ESF. The ESF of a system is the image of an ideal step function. ImageJ software was used to extract the step function from the ROI by taking 1D line profile cross section (cross-sectional plots).
(3) In order to avoid noise from differentiation, the edge profile was fitted using curve fitting. [4]. Therefore, in this work a curve fitting using Arctan (Equation 1) was selected in fitting the ESF. The reason for selecting Arctan fitting is due to close contact measurement.

$$f(x) = a \arctan \frac{(x-a)}{\sigma} + \frac{1}{2} \quad (1)$$

where a, μ and σ are the amplitude, position of edge, and width, respectively.

(4) The differentiation of fitted ESF was performed in order to obtain the LSF.

(4) The LSF was then converted to the frequency domain using a fast Fourier transform (FFT) to the LSF to obtain the MTF.

IOP Conf. Series: Materials Science and Engineering **785** (2020) 012023 doi:10.1088/1757-899X/785/1/012023

2.3. Qualitative method

The most rapid way to investigate the performance of a neutron imaging system is by visual inspection with the PSI Siemens Star test pattern. The test device developed by [5] at PSI consists of a metallic gadolinium-based star pattern which measures resolution in microns. Another test device uses the American Society for Testing and Materials (ASTM) standard objects: Beam Purity Indicator (BPI) and Sensitivity Indicator (SI).

3. Results and discussion

3.1. Quantitative results

A close contact radiography of the test object in Figure 3 is shown in Figure 4. As previously explained, the edge was detected using ImageJ software as display in Figure 5(a) and then the ESF was fitted using the Arctan function given in Equation (1). Figure 5(b) display the corresponding ESF using Arctan function. **Error! Reference source not found.** shows the LSF pattern. The LSF, calculated by differentiating the fitted ESF, shows visible 'spikes'. In this work, the real LSF and Gaussian fitted LSF are plotted. To calculate the MTF from the LSF, the fast Fourier transform (FFT) is applied. From the result in Figure 5(c), the measured MTF is 50% at 8 lp/mm and 10% at 14 lp/mm corresponding to MTF resolutions of 6.25 μm and 35.7 μm.

Figure 4. Contact radiography of test object.

IOP Conf. Series: Materials Science and Engineering **785** (2020) 012023 doi:10.1088/1757-899X/785/1/012023

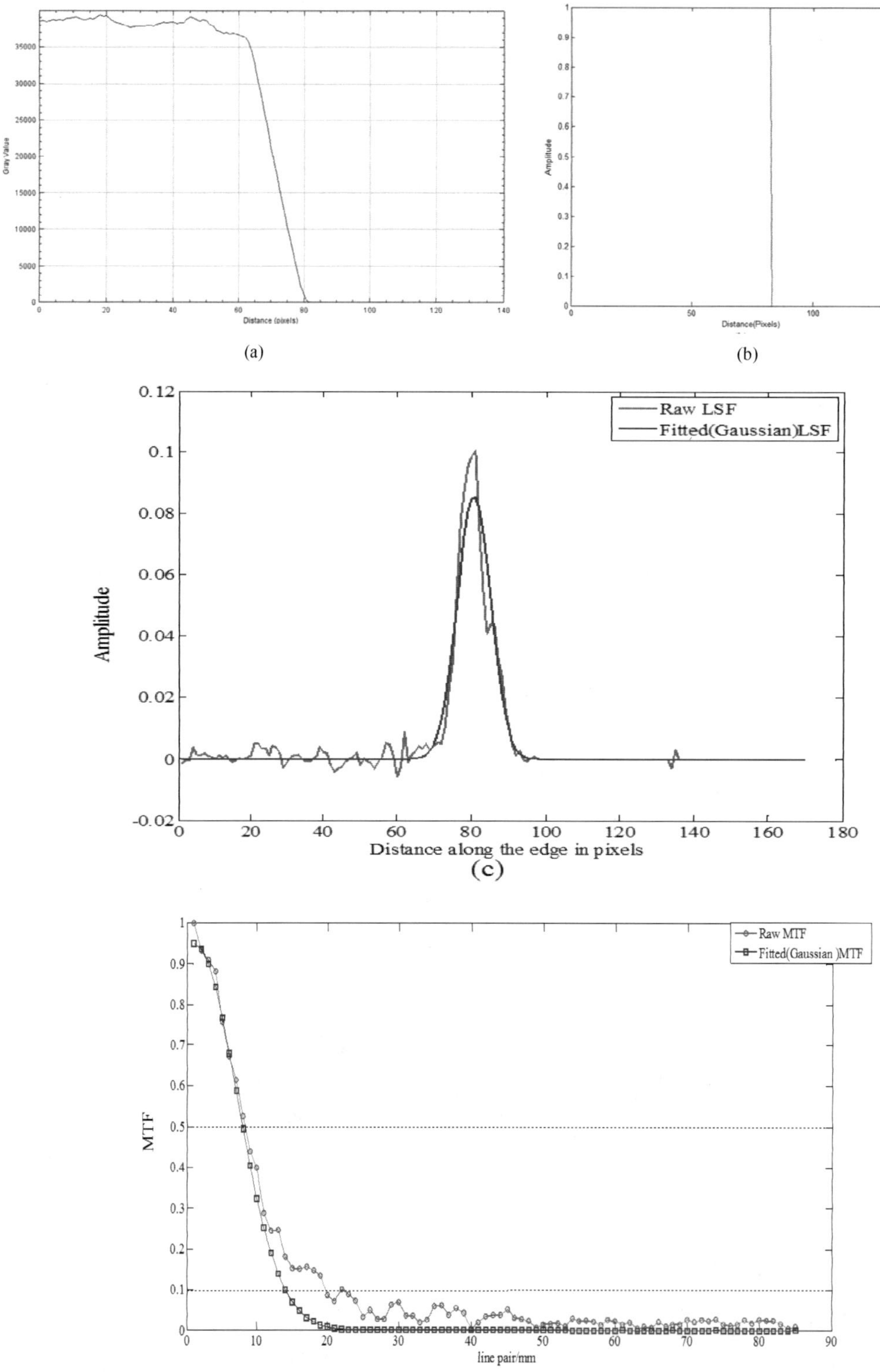

(d)

Figure 5. Measurement of Modulation transfer Function (a) Edge line profile using ImageJ software, (b) the corresponding ESF using Arctan function, (c) The LSF pattern and (d) MTF of the edge.

IOP Conf. Series: Materials Science and Engineering **785** (2020) 012023 doi:10.1088/1757-899X/785/1/012023

3.2. Qualitative Results

Figure 6 shows the image of the Siemens star structure acquired with a 0.1 mm LiF/ZnS scintillator screen. The image of the PSI-Siemens star spatial resolution test object shows that individual features of about 400 μm can be resolved. Figure 7 shows the image of the ASTM standard objects. Clearly visible in the image of BPI are the BN disks, the cadmium wires and the density difference between the Teflon block and the open region in the centre. From the image of SI, only two aluminium shims are clearly visible. However, all the holes are not visible. From the images in Figure 6 and Figure 7, it can be observed that small structures in micrometre scale cannot be detected. In both cases, the L/D ratio becomes the major limiting parameter for the resolution, and blurring can be seen in the image.

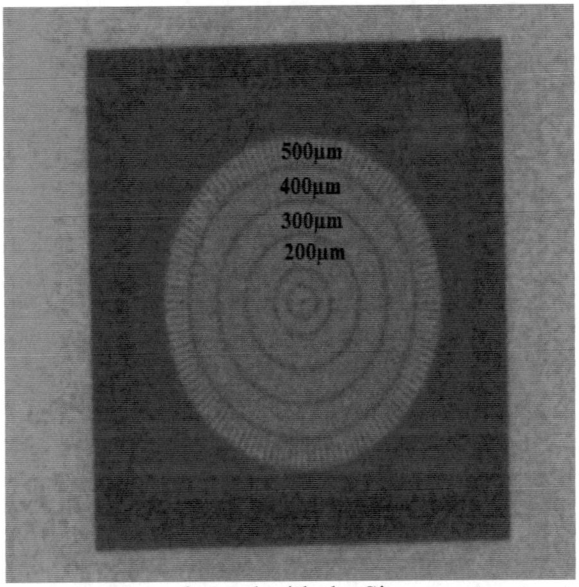

Figure 6. Resolution measurement performed with the Siemens star structures on 0.1mm LiF/ZnS scintillator screen.

Figure 7. (a) Conventional photograph of Purity Indicator (BPI) and Sensitivity Indicator (SI) and (b) Neutron image of (a).

4. Conclusion

A simplified edge-based MTF measurement method to determine the MTF of the neutron imaging system at RTP from an edge image has been described. The measured MTF using the method is 50% at 8 lp/mm and 10% at 14 lp/mm corresponding to MTF resolutions of 6.25 μm and 35.7 μm. However, this edge-based slanted method is prone to error and the MTF results can be improved by means of a slit or pre-sampled MTF machine fabricated edge. The results show that this new imaging system at RTP provides enough resolution for neutron radiography applications where such capability is needed but resources are limited. The implementation of the new neutron digital detector has a high impact on the neutron radiography activities at RTP.

5. References

[1] Zin M R M *et al.* 2015 Simulation of collimator for neutron imaging facility of TRIGA MARK II PUSPATI Reactor. *Phys. Procedia* **69** p 138–142.
[2] Jamro R *et al.* 2017 Monte Carlo Simulation for designing collimator of the neutron radiography facility in Malaysia. *Phys. Procedia* **88** p 361–368.
[3] Sarkawi M S *et al.* 2019 IOP Conf. Ser.: Mater. Sci. Eng. **555** p 012018
[4] Smith E H B PSF estimation by gradient descent fit to the ESF *Proc. SPIE 6059: Image Quality and Sys. Perform. III* p 60590E
[5] C. Grunzweig *et al.* 2007 Highly absorbing ga- dolinium test device to characterize the performance of neutron imaging detector systems *Rev. Sci. Instr.* **78** p 053708–053708-4.

Acknowledgements
The experiment and instruments used in this paper were performed at Malaysian Nuclear Agency, Bangi, Malaysia. The author would like to acknowledge all the contributors in completing this paper.

IOP Conf. Series: Materials Science and Engineering **785** (2020) 012024 doi:10.1088/1757-899X/785/1/012024

Development of industrial single photon emission computed tomography (ISPECT)

Hanafi Ithnin[1,2,a)], Elmy Johana Mohamad[1,*], Nazrul Hizam Yusoff2, Hearie Hassan[2], Norliana Mohd Lip[3]

[1]Instrumentation and Sensing Technology Research Group (InSet), Faculty of Electrical and Electronic Engineering, University Tun Hussein Onn, Parit Raja 86400, Malaysia.
[2]Industrial Technology Division, Malaysian Nuclear Agency, MOSTI, Kajang 43000, Malaysia.
[3]Faculty of Computer and Mathematical Science, Universiti Teknologi Mara, Seremban 70300, Malaysia.

[a]hanafi_i@nm.gov.my

Abstract. Industrial single photon emission tomography system utilized the gamma-ray emission of radioactive tracer to images the cross-section of tracer distribution for industrial process monitoring and diagnosis. In this paper, 36 units of Sodium Iodide, NaI scintillation detectors were used to study flow in a pipelines. The detectors were arranged in two geometrical arrangement, a hexagon and nonagon configuration. The detectors are connected to three 12-channel gamma counter which is controlled by data acquisition and image reconstruction software developed for the system. The images from these two set-ups were also compared with single detector time-average system. Depending on the size of the pipeline used, i.e. for 36 detectors, the nonagon arrangement gives more information on the distribution of the tracer. Thus nonagon set-up is more suitable to be used for representing or demonstrating the image of the tracer distribution in the pipeline.

1. Introduction

Computed tomography (CT) is a computer-aided tomographic process uses irradiation (such as gamma, x-ray, neutron and others) as well as non-irradiation (such as magnetic, electrical capacitance, infrared and others) to produce either two-dimensional or three-dimensional representations of the scanned object both externally and internally. It makes use of computer-processed combinations of many data or images taken from different angles to produce cross-sectional images (virtual "slices") of specific areas of a scanned object, allowing the user to see inside the object without cutting.

In gamma-ray computed tomography system, there are two mode of approach, namely transmission and emission tomography. The different of the two modes are; transmission utilized the attenuation of the gamma ray passing through an object, while in an emission tomography system the interest is on imaging the distribution of the gamma ray itself. Single Photon Emission Computed Tomography, SPECT system is one type of emission tomography where it has been used extensively in medical field but few have reported the uses of this system for industry [1-4]. This is as a result of the applications and problems in industry are diverse from one to another.

Some of previous studies are working on development different types and configuration of industrial SPECT system. One of the earlier groups was Legoupil *et.al.* who developed a parallel-collimation portable SPECT system consist of 36 NaI(Tl) detectors to investigate the flow pattern in a multiphase flow system [5]. The other more current work in industrial SPECT is Park et.al. who developed a system with 12-gonal diverging collimators [6]. Both SPECT system produced information of the flow patterns in different problem of interest. The results of the system have shown the prospect of an industrial SPECT system for industrial process diagnosis.

In the present work, two configuration of an industrial SPECT system, namely ISPECT is studied for imaging liquid-liquid flow in a pipeline. Both ISPECT systems consist of 36 NaI detectors with hexagonal (6x6) and nonagon (9x4) arrangements.

2. Methodology/ ISPECT system

The ISPECT system developed in this study use 36 NaI scintillation detectors, connected to three 12 port single channel analyzer as shown in Figure 1. The three analyzers were then connected to special (in-house) developed data acquisition software. All of detector in this system has 3 cm lead collimator in front of the detector with 0.5 cm collimator opening. The two configuration of ISPECT studied in this work are hexagon and nonagon. In hexagonal system, there are six arrays of detectors with six detectors in each array. For nonagonal system there are four detectors in an array of overall nine arrays in the system. The maximum diameter of both of these systems is set to be 50 cm and the interest region of scanning is 30 cm in diameter. Figure 2 shows the arrangements of the two systems.

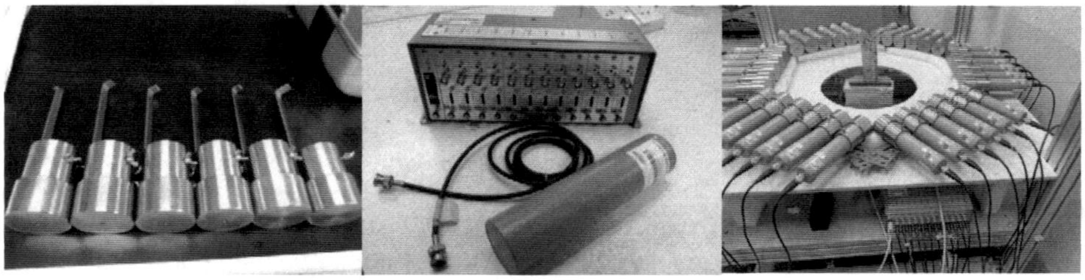

Figure 1. ISPECT hardware system consists of detectors, collimators and gamma counters.

Although the fastest data collection for the system is 50 ms, in this system the counts of gamma-ray were summed up to 1 second for each of 36 detectors. This will allow the system to image the cross section of the tracer flow per seconds. The image for the ISPECT system is reconstructed using Maximum Likelihood-Expectation Maximization method (ML-EM) [7-9] shown in equation 1.

$$y^{(n+1)} = \frac{y_j^{(n)}}{\sum_{i'} h_{i'j}} \sum_i h_{ij} \frac{g_i}{\sum_k h_{ik} y_k^{(n)}} \tag{1}$$

The pixel for the images is simulated using Monte-Carlo, MCNPX code [10]. For the time being in the present work, the pixel is set to be 30x30 for 30 cm region of interest, ROI thus each pixel has 1 cm spacing. The average detection efficiency is also calculated from the MCNPX simulation by averaging the total of 648 point source. For the evaluation on position of tracer in reconstructed image, caesium Cs-137 sealed source were used and moved around inside the ROI. In this work, the source is positioned at the centre, 5 cm, 10 cm and 15 cm from the centre of the ROI.

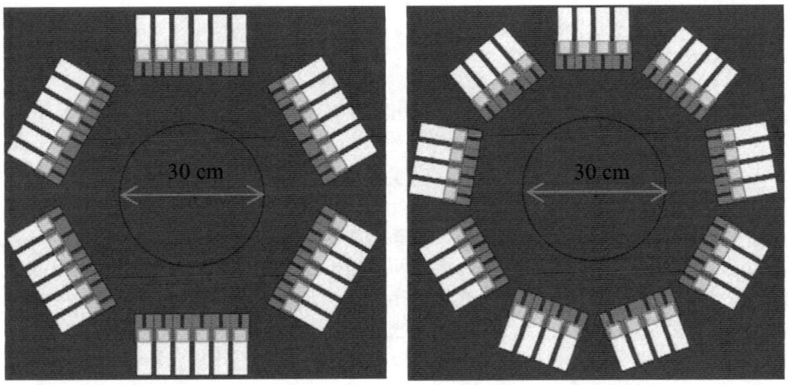

Figure 2. Hexagon and nonagon arrangement of ISPECT detector system.

3. Results

3.1. Detection efficiency map

Figure 3 shows the detection efficiency map from the MCNPX simulation for hexagonal and nonagonal ISPECT system used in this study. The average detection efficiency for hexagonal system is 4.24 x 10^{-5} and nonagon system is 4.53 x 10^{-5}. Both images show there is high detection on the edge near the detectors. Nevertheless, for nonagon configuration, the image (Figure 3b) shows brighter edge effect as compared to hexagonal image (Figure 3a). The image from nonagon system shows more uniform pixel mapping inside 5 cm from the edge whereby hexagon images has more dark pixel which means low detection in around the center of the ROI. The average detection efficiency for the region without the edge effect, 20 cm inside ROI for hexagon system is 1.488 x 10^{-3} while nonagon system is 1.584 x 10^{-3}.

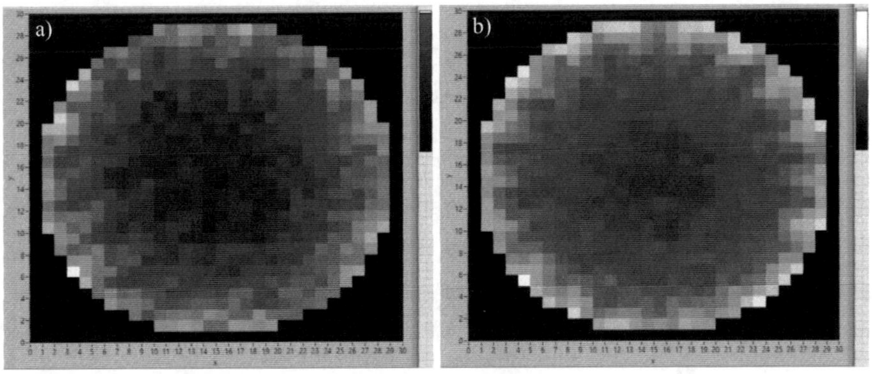

Figure 3. Detection efficiency map of ISPECT system; a) hexagon and b) nonagon.

3.2. Image resolution

The highest image resolution for both systems is mainly depend on the simulated point source mapping which is 1 cm^2. For evaluating and comparing the image resolution of the two systems, Cs-137 sealed source is placed at the center of the ROI and the image is reconstructed using ML-EM method. Figures 4a (i) and 4b (i) respectively shows the reconstructed images obtained from hexagonal and nonagonal system. Figures 4a (ii) and 4b (ii) are the profile plot across the ROI for the two images. It is evidently indication that both of the systems are able to detect the source at the centre of the ROI. The images show that hexagonal system has more blurring effect near the centre of the images with significant artefact in the surrounding image. For nonagonal system, the surrounding image has no artefact and there is less blurring near the center of the images. The average grey value near the center of the source detected for hexagonal system is 224 while for nonagonal system is 219. Both of these elements are clearly seen in the profile plot across the reconstructed images.

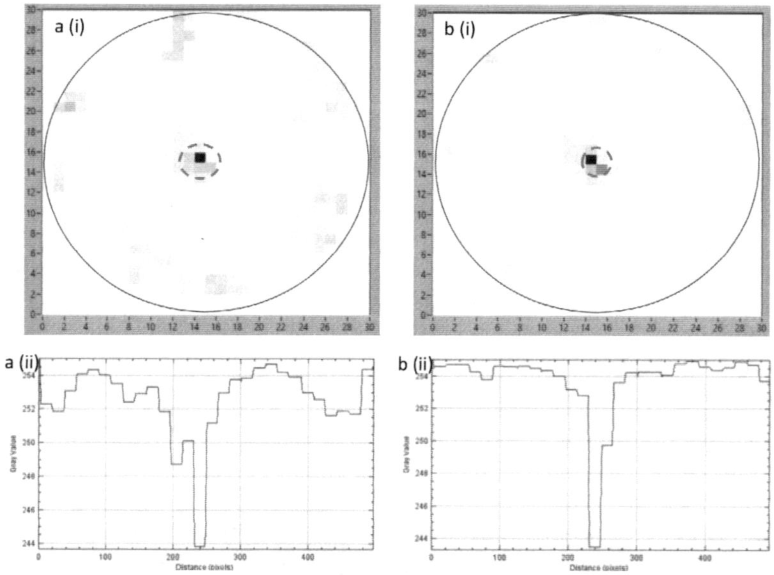

Figure 4. Reconstructed images of ISPECT for source positioned at the centre with profile plot; a) Hexagon system b) Nonagon system

3.3. Image accuracy/position

Figure 5 shows the reconstructed images from three different positions around the ROI for both of hexagonal and nonagonal system. The source is positioned at 5 cm positive y-axis, 10 cm positive x-axis and 15 cm negative x-y-axis. The images clearly demonstrate that, it can be detected at all position, for both of the system. Although the quality of the reconstructed images is differ from one to another. It can be seen near the edge of the ROI which also means near the gamma detectors, the reconstructed images define the position well in one pixel and there are no surrounding artefacts. At source positioned 10 cm from the centre, blurring effect around the detected position for the hexagonal system are more significant compared to nonagonal system. The reconstructed image for source position at 5 cm of the centre has more blurring but only hexagonal system has surrounding artefacts at positive x-axis direction. The nonagonal system has defined the source position at this point relatively well without much blurring effect.

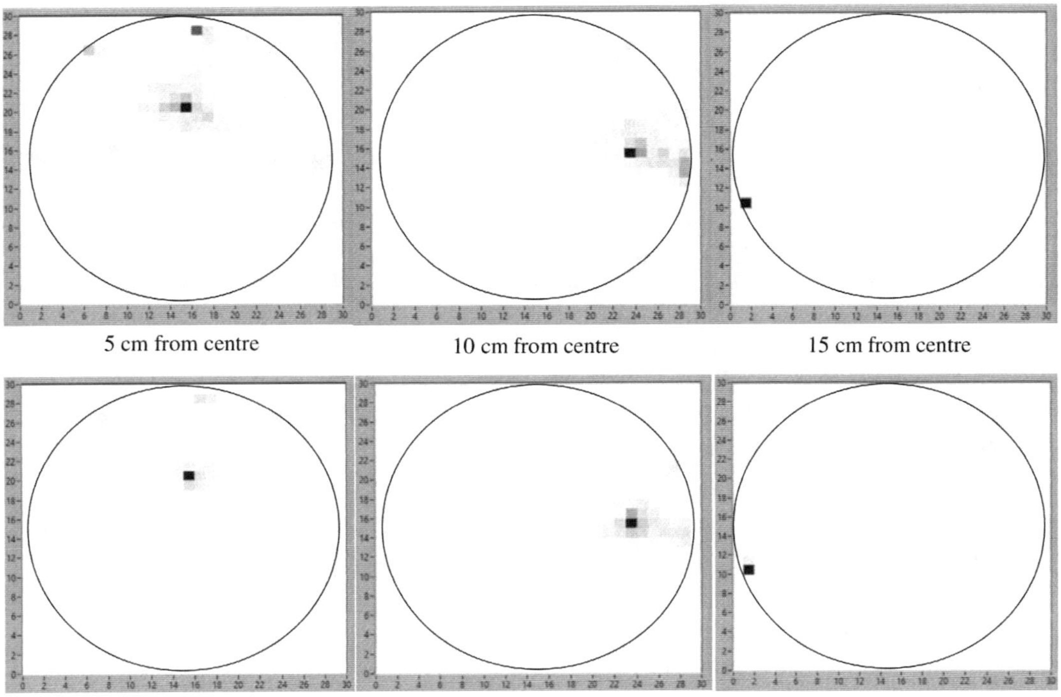

Figure 5. Reconstructed images from three different positions around the ROI; hexagon (top) and nonagon (bottom).

4. Conclusion

Hexagonal and Nonagonal ISPECT system has been developed and the reconstructed image was compared. The result shows that in detection efficiency mapping, nonagon system has higher edge counts but more uniform pixels value inside 20 cm of the region of interest, ROI. The reconstructed images showed that nonagonal system was better than hexagonal system due to its better resolution and less surrounding artefacts. Both systems have shown good image accuracy where both system could detect different position of the source. On the other hand, the quality of reconstructed images for nonagonal system was better due to the images from hexagonal system that has more blurring effect and has artefacts at 5 cm source position. The study has shown for 36 detectors system, the nonagon arrangement could give more information on the distribution of the tracer and better quality of the reconstructed images and thus, more suitable compare to hexagonal system for imaging the distribution of tracer in the pipeline.

5. References

[1] International Atomic Energy Agency 2008 Industrial Process Gamma Tomography *IAEA TECDOC-1589* (International Atomic Energy Agency).

[2] Legoupil S *et al.* 1996 Determination of the detection process in an experimental tomograph for industrial flow visualization using radioactive tracers *IEEE Trans. Nucl. Sci.* **43**(2) p 751–760.

[3] Park J G *et al.* 2010 Experimental test of double-layer method for industrial SPECT. *Nucl. Technol.* **175** p 113–117.
[4] Park J G *et al.* 2011 Double-layer method to improve image quality of industrial SPECT. *J. Instr.* **6** p C12032.
[5] Legoupil S *et al.* 1997 An experimental single photon emission computed tomograph method for dynamic 2D fluid flow analysis. *Appl. Radiat. Isot.* **48** p 1507–1514.
[6] Park J G *et al.* 2014 Development of advanced industrial SPECT system with 12-gonal diverging-collimator *Appl. Radiat. Isot.* **89** p 159–166.
[7] Moon T K 1996 The expectation–maximization algorithm *IEEE Signal Process. Mag.* **13** p 47–60.
[8] Wernick M N, Aarsvold, J N 2004 Emission Tomography *The Fundamentals of PET and SPECT* (California: Elsevier).
[9] Panin V Y *et al.* 1999 Total variation regulated EM algorithm (SPECT reconstruction) *IEEE Trans. Nucl. Sci.* **46** (6) p 2202–2210.
[10] Pelowitz D B 2011 MCNPX™ user's manual ver. 2.7.0 (California: Los Alamos National Laboratory).

Acknowledgement
The authors would also like to extend the appreciation to all personnel of the Plant Assessment Technology (PAT) group, Industrial Technology Division, Malaysian Nuclear Agency.

IOP Conf. Series: Materials Science and Engineering **785** (2020) 012025 doi:10.1088/1757-899X/785/1/012025

Development of data acquisition and control software for gamma-ray process tomography system

N Yussup[1,a)], N A A Rahman[1], M M Ibrahim[1], S C Soh@Shaari[1], N Saad[1], H Hassan[2] and J Abdullah[2]

[1] Technical Support Division, Malaysian Nuclear Agency, 43000 Kajang, Malaysia.
[2] Industrial Technology Division, Malaysian Nuclear Agency, 43000 Kajang, Malaysia.

[a]nolida@nm.gov.my

Abstract. Data acquisition (DAQ) and control software has been developed for gamma-ray process tomography system in Malaysian Nuclear Agency. The gamma-ray computed tomography (CT) system uses Computer Automated Measurement and Control (CAMAC) standard for nuclear counting and data acquisition instead of the common Nuclear Instrumentation Module (NIM) standard. For scanning configuration, the system consists of an array of NaI(Tl) scintillation detectors that moves in a small degree of angle by using a stepper motor. The gamma-ray CT software comprises of graphical user interface (GUI) for the system automation that includes CAMAC modules setup, motor and scanning control, data acquisition from CAMAC crate controller via ethernet interface and data recording. The software is developed by using National Instruments LabView development package.

1. Introduction

Computed tomography (CT) has been widely used for industrial processes such as petrochemical industries in order to improve the design, operation, optimization and troubleshooting [1-2]. There are many types of CT such as x-ray, positron emission tomography (PET), single-photon emission computed tomography (SPECT) and gamma-ray. In Malaysian Nuclear Agency (Nuklear Malaysia), a gamma-ray CT system was developed for imaging and visualizing of industrial multiphase reactors [3].

Gamma-ray CT is a technique that can provide a two-dimensional (2D) concentration, or the hold-up distribution, of the phases in multiphase reactor systems. If several such individual layers, each with their measured distribution are stacked up, a three-dimensional (3D) field can be reconstructed. Gamma CT allows the visualization of failure points in 3D analysis and in sections compares to a well-known technique column scan. The column scan is able to analyze and identify failures in industrial processes but lack of visualization capabilities. A tomography system is an update methodology used to determine the cross-section density profile inside a processing vessel [4].

IOP Conf. Series: Materials Science and Engineering **785** (2020) 012025 doi:10.1088/1757-899X/785/1/012025

2. System Overview

Radiation detectors and a source of radiation are two primary elements of a tomography system. The third generation configuration was adopted for the scanner where the entire hardware arrangement in the scanning configuration which includes the detectors, source and collimators have to be rotated around the test section. Hence, the detector array has enough and high spatial resolution cells to allow measurement of a fan-beam projection of an entire sample cross-section. Penetrating radiation from an isotopic gamma-ray source of Cs-137 and a bank of NaI(Tl) scintillation detectors in combination with a dedicated mechanical gantry set-up were used to construct the gamma-ray CT system in Nuklear Malaysia. The schematic diagram of the developed industrial process gamma-ray CT is shown in Figure 1.

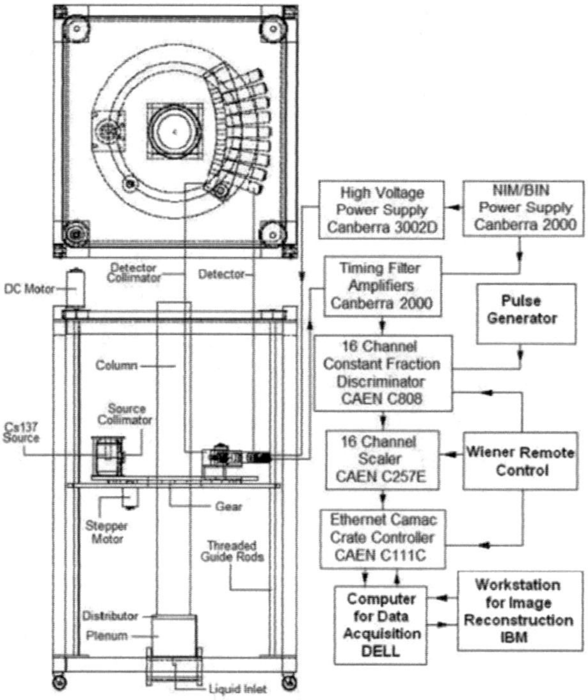

Figure 1. Schematic diagram of industrial process gamma-ray CT [3].

The mathematical methods used for image reconstruction are Expectation-Maximization (EM) and Alternating-Maximization (AM) algorithms. Both algorithms are written in Visual-Fortran programming language. However, this paper will not explained the reconstruction process using the algorithms as it is only focus on the development of data acquisition and motor control software. As shown in Figure 1, the gamma-ray CT hardware can be divided into two parts which are nuclear counting modules and scanning mechanism. The CAMAC modules for nuclear counting will be described in detailed in Section 3.1 while the close-up of scanning mechanism is shown in Figure 2.

The fan beam arrangement consists of an array of nine units of NaI(Tl) scintillation detector with a diameter and length of 2 x 2 inches to accommodate a fan angle of 40° where the detector pitch-to-pitch is 5°. An encapsulated Cs-137 sealed source located opposite from the center of the array of detectors. The detector plane and the sealed source are placed on a projection plate as shown in Figure 2. Since the inertia of the system to be moved in this configuration is large, a temporal resolution of the order of a few seconds cannot be expected.

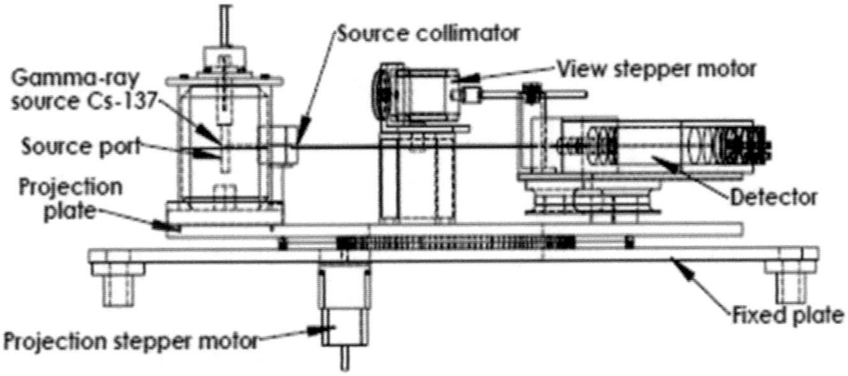

(a) Illustration of industrial process gamma-ray CT mechanism [3].

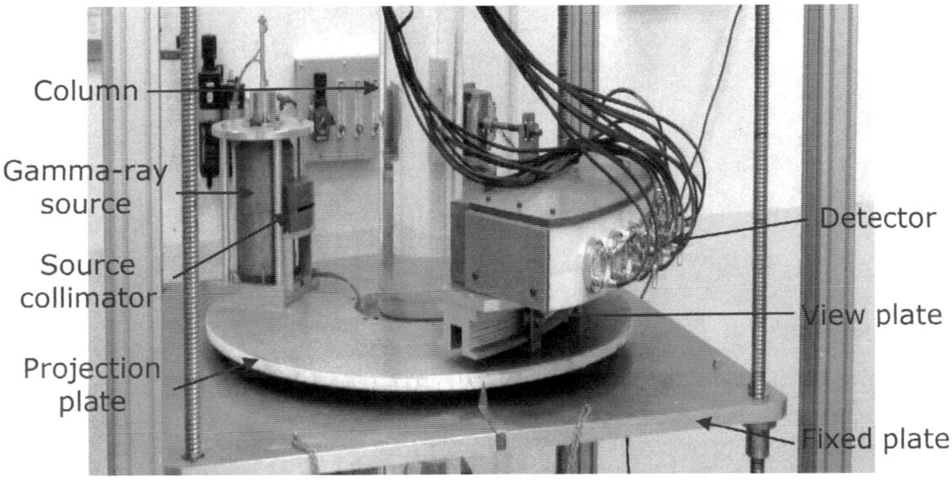

(b) Photograph of industrial process gamma-ray CT.

Figure 2. Side view of industrial process gamma-ray CT.

3. Software Development

The software is designed and developed to control all associated nuclear electronics equipment and to acquire gamma-ray transmission data from the CAMAC counting system. The software is also designed to control the movement of the detector plane and projection plate for tomography scanning process. The software is developed by using an older version of National Instrument LabView package which is LabView 8.6. The flow chart of gamma-ray CT scanning process is shown in Figure 3. The finite-state machine (FSM) is used to write the software programs. FSM is a mathematical model of computation used to represent and control execution flow. The output of a state machine is a function of the input where the status or state changes from one state to another based on inputs or satisfied conditions, resulting the output [5].

The software is divided into two sub modules as follows:

1) System setup and data acquisition
2) Motor control and motion

Both modules are integrated and represented as shown in Figure 3. The software begins with system and data file setup. Then the software will acquire data from the nuclear counting system where the data will be displayed and recorded. This step will be repeated according to the number of samples entered in the system setup. The next step is operating the scanning mechanism. The data acquisition and scanning operation will be repeated until the total scanning angle is reached.

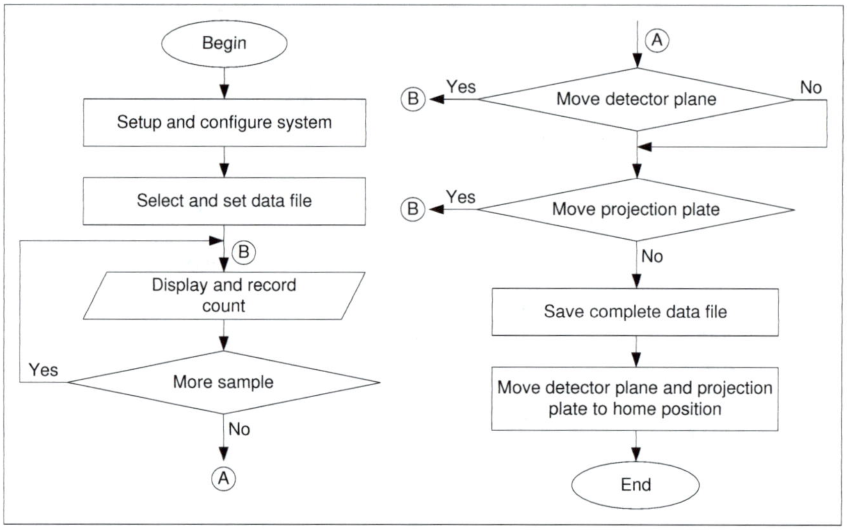

Figure 3. Flow chart of the gamma-ray CT scanning process.

3.1. System Setup and Data Acquisition

Nuclear counting system for the developed gamma-ray CT consists of high voltage power supply, amplifier and CAMAC modules as shown in Figure 1. The CAMAC modules play important role in the software development where the system setup and gamma count are controlled and acquired based on the modules. There are three CAMAC modules involve as hardware. The first one is Ethernet CAMAC crate controller (C111C) which is responsible as the communication interface between CAMAC crate and computer. The communication protocol is Transmission Control Protocol/Internet Protocol (TCP/IP) where the address and port number need to be determined prior to connection establishment. TCP control socket commands used in the software program are listed in Table 1.

Table 1. TCP control socket commands reference [6].

TCP control socket command	Description
`CCCZ`	Generate Dataway Init
`CCCC`	Generate Crate Clear
`CSCAN`	Executes a Crate scan and returns a buit mask with bitn = 1 if in the slot n a CAMAC board is detected
`CFSA <function> <slot> <addr> <data>`	24-bit CAMAC command; returns Q and data

In order to configure the counting system parameters such as discriminator threshold, pulse width and dead time, 16-channel constant fraction discriminator (C808) is incorporated into the CAMAC crate. The parameters can be set up via computer by programming the software based on CAMAC functions listed in Table 2.

Table 2. C808 CAMAC functions [7].

CAMAC Function	Description
F(16) N A(0..15)	Writes Discriminator Threshold on W1..W8.
F(17) N A(0)	Writes Pattern of Inhibit on W1..W16 (1=enabled).
F(18) N A(0)	Writes Output Width on W1-W8 for channels 0 to 7.
F(18) N A(1)	Writes Output Width on W1-W8 for channels 8 to 15.
F(19) N A(0)	Writes Dead Time on W1-W8 for channels 0 to 7.
F(19) N A(1)	Writes Dead Time on W1-W8 for channels 8 to 15.

16-channel scaler (C257) is used to count the pulse generated by the array of detector which consists of nine units of scintillation detector resulted from the CT scanning process based on the counting system setup. The count for each detector then is acquired, displayed and recorded by the computer. Hence, the software is programmed in order to deliver this function. For this purpose, CAMAC functions in Table 3 are utilized.

Table 3. C257 CAMAC functions [8].

CAMAC Function	Description
F(0) N A(0..15)	Reads the value held in the counter addressed by A(0..15). Q is true only if the input is inhibited.
F(2) N A(0..15)	Same as above, but with A(15) reads the 15th channel and if the input is inhibited, resets all the channels and the LAM line.

3.2 Motor Control and Motion

Motor control module for the CT is developed using electro-mechanical and digital electronic components such as stepper motor driver and stepper motor besides a computer unit. The function of the module is to control stepper motor angular movement either in clockwise or counter clockwise from one point to another. The stepper motor is controlled by the computer using parallel communication. It allows digital signals that have been set up from computer to operate the stepper motor. As shown in Figure 2(a), there are two stepper motors incorporated in the development. One stepper motor is for the movement of the detector plane called view stepper motor. Another one is for the rotation of projection plate called projection stepper motor.

Typically, a scan involves obtaining gamma ray counts for 100 projections around the column. Therefore, the projection plate is rotated in increments of 3.6° step angle for a total of 360° scanning angle. In each projection, the source and collimator is stationary while the detector plane which consists of an array of nine detectors sweeps from an angle range from a -2.4° to +2.4° which is from right to left. The default step angle for the detector plane is 0.2° which results in 25 projections for each detector. Since nine detectors are used, a total of 225 projections are scanned for the movement of the detector plane alone. The sweep operation is done such as equiangular fan beam is eventually created where each detector covers an angle of 4.8° [3].

Once the detector plane motion is completed, the stepper motor is programmed to bring the detectors back to their original position. Then, the projection plate where source and detectors are placed, is rotated clockwise by increment of the plate step angle with respect to the center of the column. The next view is scanned in a similar manner. Two proximity sensors are utilized where they are placed at the detector plane and projection plate each in order to move the respective part to home position. The flow chart of the motion is shown in Figure 4.

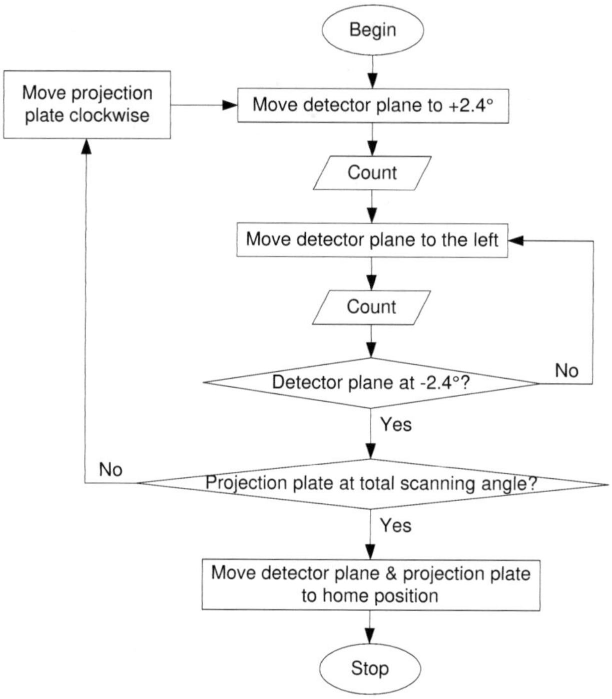

Figure 4. Flow chart of the motor motion.

4. Results and Discussion

The GUI of gamma-ray CT comprises of four panels. Top panel displays the operation status and name of selected menu. The left panel consists of menu and integration time in order to set up and configure the system. The motor status during scanning process is also located at the left panel. There are four menus that can be selected which are as follows:

1) System Setup and Configuration
2) C808 Discriminator Setup
3) Scan
4) Stepper Motor Adjustment

The centre panel which is the GUI main panel will change according to the selected menu above. Hence, there are four displays for the main panel. On the other hand, the right panel consists of setting related to the data files. There are three data files that can be selected which are Back-Projection (BP), Expectation-Maximization (EM) and Densitometry depends on the type of CT and process required by the user. The file is created based on the selected data file where the data header and arrangement will be different from each other. The data file is save in text format (.txt). The GUI of gamma-ray CT software is shown in Figure 5.

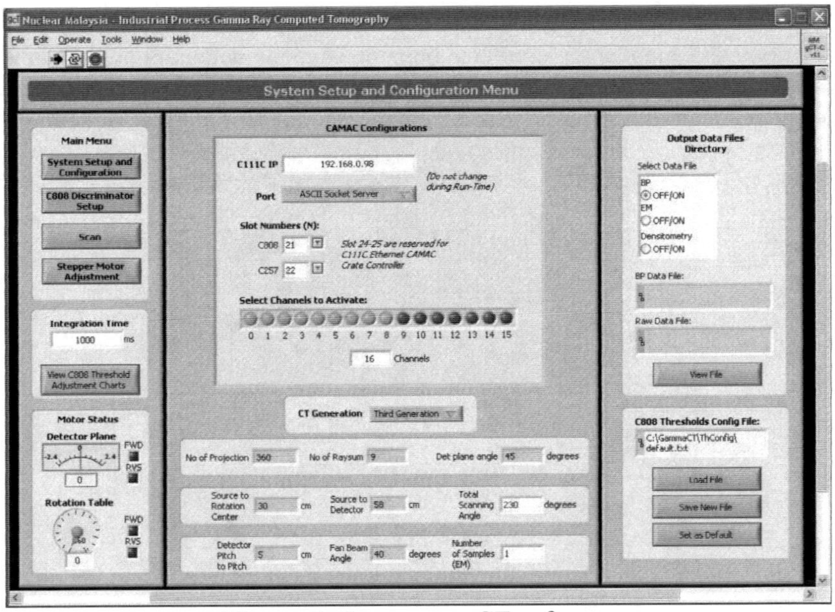

Figure 5. Gamma-ray CT software.

Figure 5 also shows the main panel for System Setup and Configuration menu. CAMAC crate controller and CT scanning parameters are set up under this menu. In order to configure C808 discriminator parameters which are discriminator threshold, pulse width and dead time, the second menu which is C808 Discriminator Setup needs to be selected by the user. The main panel and the LabView program for C808 Discriminator Setup are shown in Figure 6 (a) and 6(b) respectively.

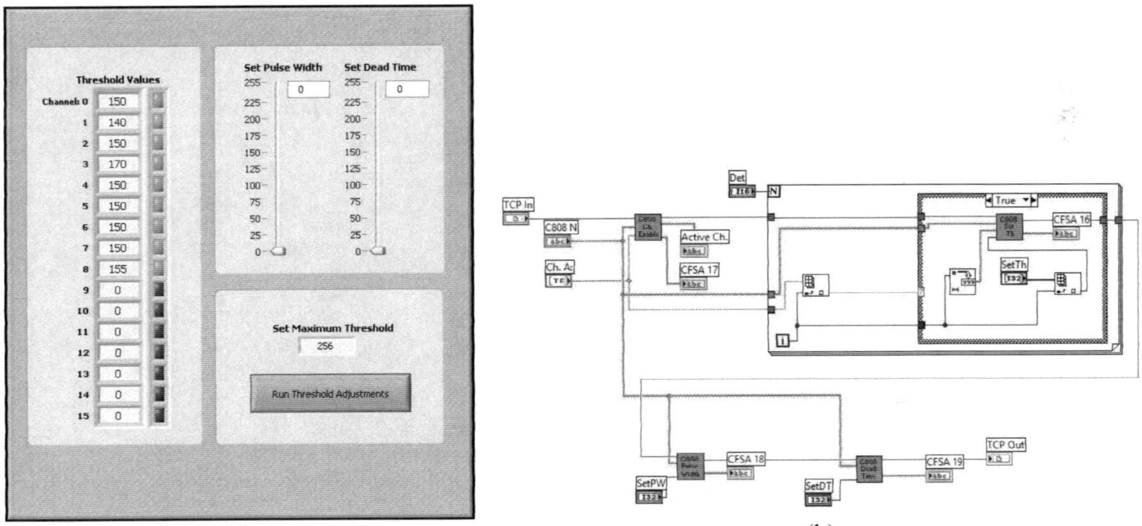

(a) (b)

Figure 6. The main panel and LabView program for C808 Discriminator Setup.

The LabView program as shown in Figure 6(b) is developed based on the C808 CAMAC functions list in the Table 2. The program is written in sequence where the channel is enabled by sending functions CFSA17 followed by other functions in order to set up the threshold value for each detector channel, besides pulse width and dead time of the acquired signals.

The gamma-ray CT scanning process where the count data of detectors are acquired is shown in Figure 7. The data is acquired and recorded periodically based on the integration time set up at the left panel. The status of the scanning process which includes the current number of sample, detector step and rotation step of the scanning process is also displayed in the Scan panel. Therefore, the user is able to know the progress of the on-going CT scanning process.

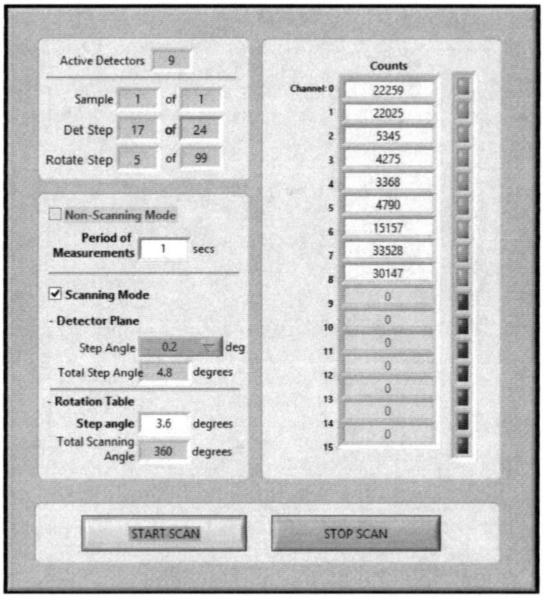

Figure 7. The main panel for scanning process.

The output is saved in text file according to the selected data file. The differences between the outputs of each data file are data header and count arrangement. The output sample is shown in Figure 8. The text file will be transferred to Visual-Fortran based program for image reconstruction.

114, 20 (No. of samples at each position, sample frequency)
0 0 (View number, projection number) → Projection 0 of view 0 begins
21 111 99 10 373 66 246 182 287 (Counts for detectors 1-9, sample one)
124 149 156 146 118 134 137 115 90 (Counts for detectors 1-9, sample two)
114 157 170 168 155 145 142 137 75 (Counts for detectors 1-9, sample three)
⋮ ⋮ ⋮
(Total of 114 samples)
⋮ ⋮ ⋮
0 1 (View number, Projection number) →Projection 1 of view 0 begins
150 194 136 12 486 45 439 315 383 (Counts for detectors 1-9, sample one)
105 143 165 130 117 132 132 161 97 (Counts for detectors 1-9, sample two)
116 191 159 144 162 151 134 126 108 (Counts for detectors 1-9, sample three
⋮ ⋮ ⋮
(Total of 114 samples)
And so on for all the 100 (0-99) views and 25 (0-24) projections for each view

Figure 8. The sample of output for EM data file.

5. Conclusions

Software for gamma-ray CT has been developed and tested in Nuklear Malaysia. All features including data acquisition from CAMAC crate controller via Ethernet interface, scanning control, motor motion and data file generation work accordingly. In term of the scanning operation and automation, there is no difference between using CAMAC and NIM standard. Computer Ethernet and parallel connections need to be stable in order to ensure the reliability of the automatic data acquisition and scanning process. Since the software is developed in-house, it can be modified and improved based on the user requirement.

6. References

[1] Chaouki J, Larachi F and Dudukovic M P 1997 *Ind Eng Chem Res* **36** p 4476–4503.
[2] Chen J W, Gupta P, Sujatha D, Al-Dahhan M H, Dudukovic M P, Toseland B A 1998 *Flow Meas. Instr.* **9**(2) p 91–101.
[3] Abdullah J, Mohd Said M K, Abdul Rahman N A, Mohd S, Shari M R, Yussup N, Che Soh S, Saad N, Hassan H, Mustapha M *et al.* 2011 *Appli. Mech. and Mater.* **83** p 41-47.
[4] Calvo W A P, Hamada M M, Sprenger F E, Vasques P A S, Rela P R, Martins J F T, Matos Pereira J C S, Omi N M and Mesquita C H 2009 *Nukleonika* **54**(2) p 129–133.
[5] Wang J and Tepfenhart W 2019 *Formal meth. in comp. sci.* (CRC Press) pp 33-36
[6] Technical Information Manual 2008 *MOD. C111C Ethernet CAMAC crate controller* rev 7 (C.A.E.N.) p 31.
[7] Technical Information Manual 2001 *MOD. C808 16 channel constant fraction discriminator* rev 2 (C.A.E.N.) p 12
[8] Technical Information Manual 2001 *MOD. C257 16 channel scaler* rev 0 (C.A.E.N.)

Acknowledgements

Authors wishing to acknowledge assistance from personnel of the Plant Assessment Technology (PAT) Group, Malaysian Nuclear Agency for their support during the course of this software development in particular and research project in general.

Preliminary study to determine the maximum settling velocity and model parameter of Cu nanoparticle by settling method

Azraf Azman[1,2,a)], Mohd Zamri Yusoff[1], Mohd Zaid Hassan[2], Hafizal Yazid[2], Prem Gunnasegaran[1], KC Ng[3]

[1]UNITEN, Jalan IKRAM-UNITEN 43000 Kajang Selangor, Malaysia.
[2]Malaysian Nuclear Agensi, Bangi 43000, Kajang Selangor, Malaysia.
[3]University of Nottingham Malaysia Campus, Jalan Broga,43500 Semenyih, Selangor Malaysia.

[a] azraf@nuclearmalaysia.gov.my

Abstract. Nanoparticle suspensions also known as nanofluids are often polydisperse and tend to settle with time. Its settling process is very challenging and quite complexs to understand. In this study, Copper (Cu)nanoparticles were dispersed in base fluid (water) and the settling behaviour were examined to obtain the settlement parameter. A series of experiments adopted from sludge settling methods were conducted and the results were analysed. Based on the experiments, the settling curve was successfully established.

1. Introduction

Nanofluids are colloidal suspensions composed of nanoparticles with the size of less than 100 nm which are dispersed in a carrier fluid. These nanoparticles act as an additive material for carrier fluidto obtain a better thermal conductivity enhancement.Nanofluids are a topic of great interest due to its physical properties. These materials has significant feature in many fields such as aerospace, biomedicine and thermal engineering.

In the field of thermal engineering, nanofluids has shown a significant improvement in flow and heat transfer of the liquids/fluids. However, its practical application remains as a significant issue due to instability of the nanoparticle in the solution/fluid. The nanoparticles tend to agglomerate once in suspension. The agglomeration results in sedimentation and this will decrease the efficiency of nanofluids thermal conductivity.

Sedimentation is the tendency ofparticles in suspension to settle out of the fluid in which they are entrained and come to rest against a barrier. These particle move fromrelatively higher density through a less dense fluid. This is due to their motion through the fluid in response to the forces acting on them such as gravity, centrifugal acceleration, or electromagnetism.This process is relevant for a wide range of natural and industrial processes such as waste water treatment [1], the fouling of heat transfer surfaces [2], stability of dust clouds [3], sedimentation in natural waters, such as in rivers, lakes and oceans [4],and metallurgy [5].

Sedimentation in a field of gravity is well studied for cases where the sedimenting material (dispersed phase) is sufficientlydiluted to make its volume fraction much lower than that of the continuous phase. According to [3], these particles move uniformly downward with their constant terminal sedimentation velocity. This is due to the different nature of the sedimentation components (carrier fluid, nanoparticle, particle properties) and the concentration at which these compounds occur. Buoyancy-gravitational flow instabilities may occur in the continuous phase when the solid particles are settled or when the solid particles settling process is completed [7]. In reality, the volume of nanofluids settling will depend on both its hindered settling and compression behaviour. These two factors, are both influenced by the composition of the nanoparticle size distributions, surface properties, rheology, carrier fluid, etc.

According to [8],instability issues always exist after nanofluids were formed. The settling of nanoparticles can be neglected in natural convection compared to nanofluids forced flow. However, it may not be practical to ignore/neglect these issues in nanofluids natural convection due to temperature driven velocity is too small. In extreme cases, the sedimentation or settling issues of nanofluids need to be treated using stabilizers. It is a significant topic for nanofluids natural convection and nanoparticles sedimentation. To the best of the authors' knowledge, no correlative works have been reported yet.

Therefore,in this work, the method to obtain the settling velocity and settlement parameter for Cu nanoparticles is adopted according to secondary sludge settlersdesign.Generally it involves two steps: (1) settling at low concentrations (including discrete and flocculent particle settling), and (2) settling at high concentrations (including hindered and compression settling) [9]. Detailed information on the settling behaviour of a nanofluids sample can be obtained from a batch of settling and hindered settling curve[10].

2. Materials and methods

2.1. Preparation of nanofluids

The black colour, spherical shape Cu nanoparticles were purchased from the Sigma Aldrich, USA (concentration%),. The physical and chemical properties are summarised in Table 1. The nanoparticles were in a dry state properties in as in Figure 1. In order to study the instability of suspension, five samples with different concentrations were prepared which is 0.5%, 1%, 2%, 3% and 4%.

Table 1. Physical properties of Cu nanoparticles.

Nanoparticle	Size	Density, ρ	Colour
Cu	60 – 80 nm	8933	Black

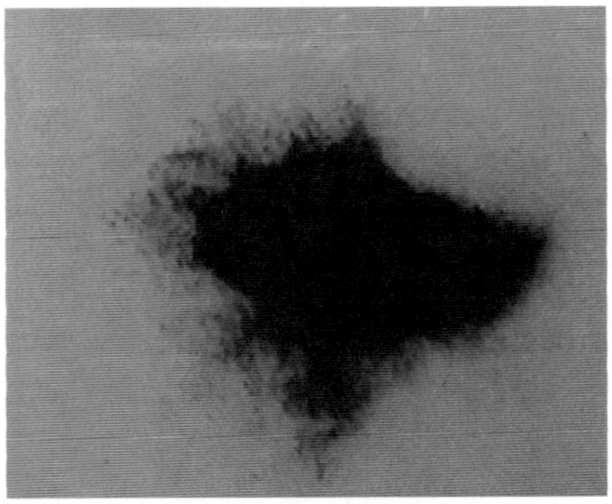

Figure 1. Cu nanoparticle powder.

IOP Conf. Series: Materials Science and Engineering **785** (2020) 012026 doi:10.1088/1757-899X/785/1/012026

Figure 2. Scanning Electron Microscopy (SEM) image of typical Cu nanoparticle morphology.

The mass of the nanoparticles for different concentrations were estimated using the weight percent ratio. The nanoparticles with different weights (x,y,z,g) were mix with 20 mL deionized water(DW). The mixture was ultrasonicated for two hours to break any possible agglomeration of nanoparticles.

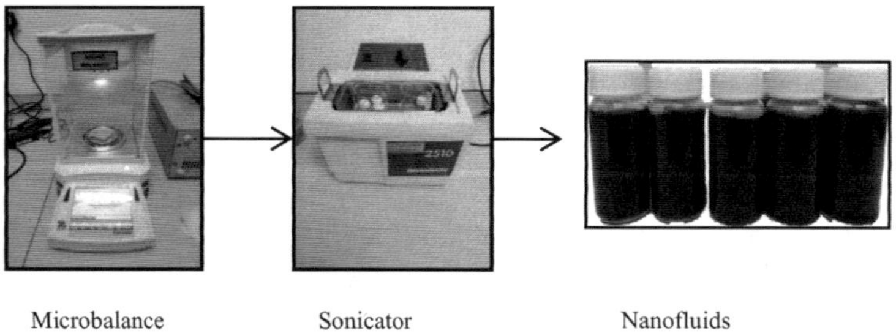

Figure 3. Preparation of Cu nanofluidswith different concentrations.

2.2. Experimental setup

The experiment to determine the gravity settling velocity of the Cu nanoparticle suspension was carried out using a digital web camera. Then the vials were placed in front of a white colored cardboard in an enclosed compartment. A usb led light bar was placed at upper and behind the sample vial. The white background and a led lighting were applied for better photographic effect and contrast. The digital web camera captured the dynamic events of the suspension and showed the movement of the solid-liquid interface with time.

IOP Conf. Series: Materials Science and Engineering **785** (2020) 012026 doi:10.1088/1757-899X/785/1/012026

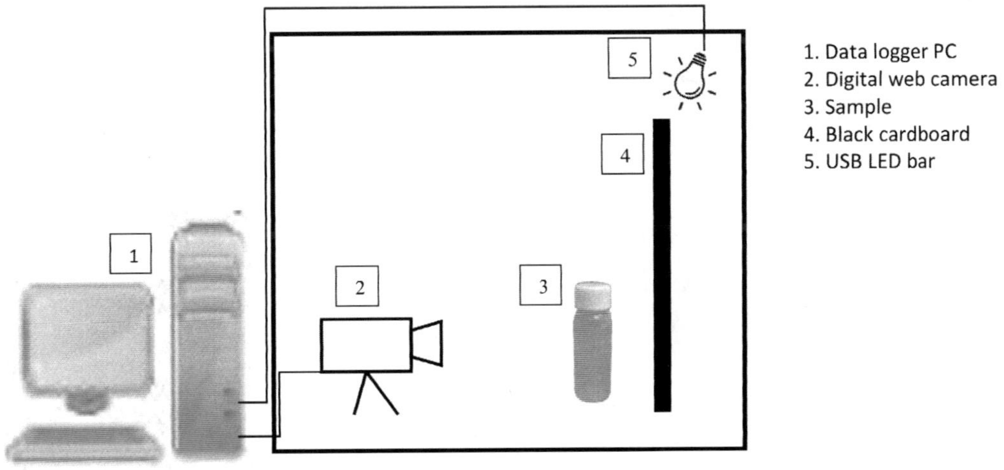

Figure 4. Illustration of the experimental set-up for Cu nanoparticle suspension/nanofluids gravity settling velocity.

3. Result and discussion

The experiments demonstrated that the unstable suspensions settle rapidly during the first minutes of the experiment. The change of the interface height slows down the settling of nanoparticles considerably. The liquid above the interface becomes more transparent whenmost of the larger particles have settled out. However there are still a small particles remains for a period of time as observed from the liquid phase is not completely clear. Based on [10] in Figure 5, there are four different phases in a unhindered settling curve or batch settling curve. However, batch settling curve only indicate information on the settling behaviour at the suspensions interface.

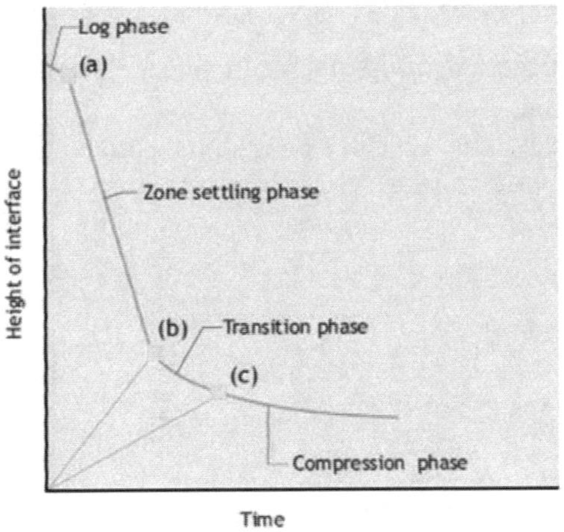

Figure 5. Four different phases in batch settling curve (Experimental Methods in Wastewater Treatment, 2016).

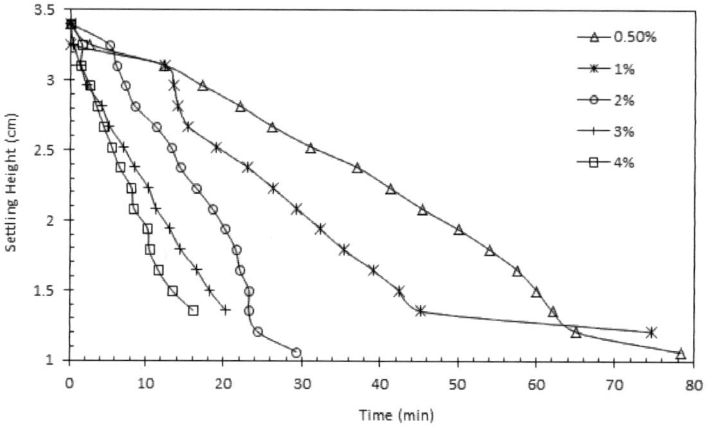

Figure 6. Batch settling curve of various concentration of Cu nanofluids.

In Figure 6, the 0.5% and 4% Cu nanofluid concentrations showed the slowest and fastest settling time respectively. This suggest that there are some large object/particles or clustered forms due to instability of charge present in the suspension. Cu nanofluids with 4% concentrations settles less then 20 minutes where asthe 0.5% concentrations settles over 60 minutes. Figure 7 showed that the hindered settling velocity settles rapidly with higher nanofluid concentrations compared to lower concentrations. Note that the velocity of the particles with concentrations of 4% is ~4.8m/s compared to ~1.2m/s for 0.5% concentrations. Referring to sludge settling test, the relation between the particle concentration and the zone settling velocity can be described as an exponential decaying function as shown by Vesilind,1968.

$$V_{hs}(X) = V_0 \cdot e^{-rV \cdot XTSS} \quad (1)$$

where V_{hs} is the hindered settling velocity, V_o is terminal velocity, -rV is the settling parameter and XTSS is the nanoparticle concentration.

However, from Figure 7, the settling of Cu suspensions shows that the relation between concentration and zone settling velocity can be described as an inverse to exponential decay or a logarithmic function as followed.

$$V_{hs}(X) = V_0 \cdot (1- e^{-rV \cdot XTSS}) \quad (2)$$

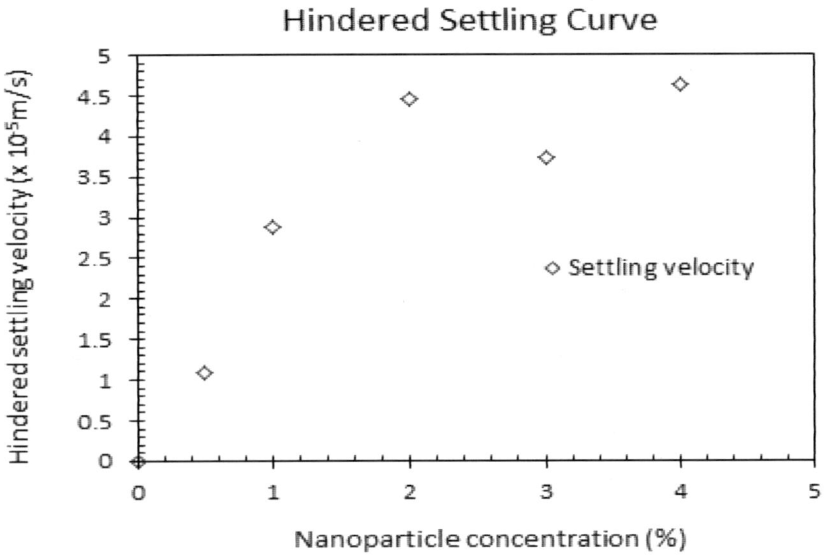

Figure 7. Hindered settling velocity of Cu nanoparticles.

The hindered settling velocity increase at higher concentrations, and it is assume that there are some large object or clustered forms due to instability of charge present in the suspension. Note that in Figure 6, the settling pattern shows the settling time of the nanoparticles became faster with respect to concentrations.

In Figure 8, the circles represent measured settling velocities and the line is the calculated settling velocities logarithmic function as followed in Eq.2. From the calculated equation the terminal velocity of Cu is estimated at V_O, 7.62e-05 m/s from the settling function. The dashline shows the 95% confidence intervals of the estimated parameters for the settling function. In accordance with the results, sedimentation will occur in suspensions after the preparations.

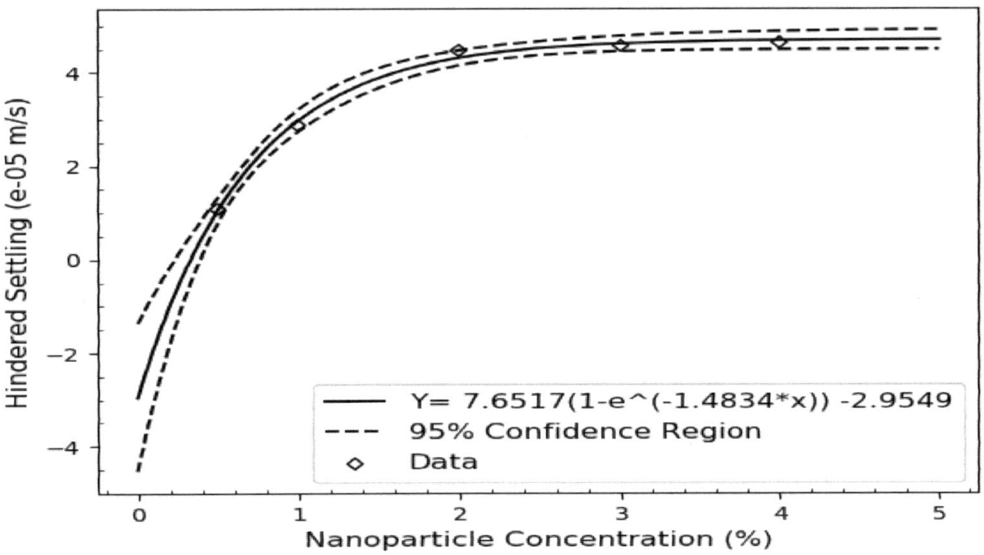

Figure 8. Settling velocity as a function of solid concentration.

4. Conclusions

Nanofluids based on Cu nanoparticles were prepared at various/different concentrations using water as a base fluid. In accordance with the preliminary results, settling of Cu nanoparticles was slower with lower concentrations. Observation within 24 hours of nanofluids settling prepared via the two-step method, revealed that settling and agglomeration depends on the size and concentration of nanoparticles. The higher the particle concentration in the fluid, the smaller the inter-particle distance between the particles., As a result, the probability of agglomeration to occur is higher due to Vander Waals attraction.These phenomena will also exerts a negative influence of the heat transfer performance of nanofluids, which requires complementary studies on the determination of sizes and concentrations of nanoparticles in a suspension.

5. References

[1] Diehl S 1997 Continuous sedimentation of multi-component particles. *Math. Meth. in the Appl. Sci.* **20** p 1345– 1364.

[2] Brahim F, Augustin W, Bohnet M 2003 Numerical simulation of the fouling process. *Int. J. of Thermal Sci.* **42**:323–334.

[3] Goossens D 2006 The granulometrical characteristics of a slowly-moving dust cloud *Earth Surf. Processes and Landforms* **10** p 353–362.

[4] van Maren D S 2007 Grain size and sediment concentration effects on channel patterns of silt-laden rivers *Sedimentary Geol.* **202** p 297–316.

[5] Holbeach J W, Davidson M R 2009 An Eulerian-Eulerian model for the dispersion of a suspension of microscopic particles injected into a quiescent liquid *Eng. Fluid Appl. of Comp. Mech.* **3** p 84–97

[6] Di Felice R 1999 The sedimentation velocity of dilute suspensions of nearly monosized spheres *Int. J. of Multiphase Flow* **25** p 559–574.

[7] Bergantz G W Ni J 1999 A numerical study of sedimentation by dripping instabilities in viscous fluids *Int. .J. of Multiphase Flow* **25** p 307–320.

[8] Meng X , Zhang X, Li Q 2016 Numerical investigation of nanofluid natural convection coupling with nanoparticles sedimentation *Appl. Thermal Eng.* **95** p 411–420

[9] Zhang Y, Wang H, Qi L, Liu G, He Z, Jiang S 2016 Simple model of sludge thickening process in secondary settlers. *Front. Environ. Sci. Eng.* **10**(2) p 319-326.

[10] Experimental Methods in Wastewater Treatment 2016 EISBN: 9781780404752

[11] Akhatov J S, Juraev ET, Juraev TI, Avdievich VN 2018 Study of sedimentation process in nanofluids with various concentrations of SiO_2 and Al_2O^3 nanoparticles, Appl. Solar Ener. **54**(6) (Allerton Press, Inc.) p 428–432

Acknowledgements

The author would like to thank all research officers from Agensi Nuklear Malaysia (ANM) and from Jabatan Perkhidmatan Awam Malaysia for their supports and helps in completing this work.

IOP Conf. Series: Materials Science and Engineering **785** (2020) 012027 doi:10.1088/1757-899X/785/1/012027

Ion irradiation-induced the radiation hardening in Fe and Fe-0.3Si alloys

Phongsakorn Prak Tom[1,2,a)], Kenta Murakami[1], Luu Vu Nhut[1]

[1]Nagaoka University of Technology, 1603-1, Kamitomioka Nagaoka, Niigata, Japan.
[2]Malaysian Nuclear Agency, Bangi, 43000 Kajang, Selangor, Malaysia.

[a]phongsakorn@nm.gov.my

Abstract. Depth profile of Helium irradiation effects in reactor pressure vessel (RPV) model material is identified using a nano-indentation technique in order to develop a new methodology to clarify the solute-vacancy interaction. 2 MeV He ion was irradiated in pure iron at room temperature with and without an energy degrator. The indentation hardness was measured by nanoindentation changing the load at 9 levels. Hardening in irradiated materials was clearly observed in all indentation loads as well as the well-known indentation size effect. It was interesting that the significant increase of hardness was recognized at 0.8 μm depth, which is corresponding to the obstacle of deformation at 3~4 μm depth, that would be the cause of helium implantation. Furthermore, dose effects on 3MeV Si irradiated RPV model materials were confirmed.

1. Introduction

Fast neutron irradiation induces lattice defects in metal such as interstitial atoms, vacancies, and their clusters [1][2][3]. The diffusion of the irradiation-induced defects may cause nano-scale clustering of solute atoms as well as the further clustering of the defects, and these clusters obstacle dislocation motion and lead hardening. In reactor pressure vessel (RPV) made of low-alloy ferrite steels, which is one of the most important components in nuclear power plants suffered by a fast neutron, solute atoms such as Cu, Mn, Ni and Si gather to form solute clusters, and these clusters cause irradiation embrittlement of the RPV. Clustering of these elements is mainly controlled by the diffusion mechanism of irradiation-induced vacancies. Though it is important to observe the interaction between vacancies and solute atoms, experimental techniques to directly observe the vacancy motion is limited.

Helium babble is another important cluster related to the vacancy motion. Helium atoms are generated from nuclear transmutation reactions in materials during neutron irradiation. Transmuted helium gathers with vacancies to form the bubbles. Helium babble formation in ferrite steels is an interesting topic for advanced reactors though is not observed in the RPVs of light water reactor. Furthermore, it is relatively easier to observe helium babbles using a transmission electron microscope (TEM) comparing with the other experimental technique to observe vacancy-type defects, such as residual resistivity method or positron annihilation [4][5].

Ion irradiation is widely used due to well-controlled irradiation condition such as irradiation temperature and dose rate [6][7]. We use ion accelerator to understand the effect of solute elements on vacancy migration by focusing the helium babble development behaviour. However, the weak point of ion irradiation is a large gradient of irradiation defects as well as the injected ions along with the depth direction. Therefore, as the first step, the objective of this study is to develop the instant detection of the depth profile of irradiation effects using a nano-indentation technique using the model materials of RPV.

2. Materials and methods

As the RPV model materials, pure-Fe and Fe-0.3Si, 5 x 10 mm were used in this study. The chemical composition of the materials is shown in Table 1. The samples were heated to 1000 ^{0}C for 1 h and followed by 800 ^{0}C for 1 h before quenching in water. After that, it was kept at 600 ^{0}C for 24 h to reduce the residual stress. After the heat treatment, the samples were polished one-side using polishing paper and lastly buff polishing with a solution of 0.05 μm. Finally, it was electrochemical-polished by ethanol 90% and perchloric acid 10% to reduce stress concentration on the surface may be generated by the polishing process.

Table 1. The chemical composition of Fe based materials (wt %).

	C	Si	Mn	P	S	Ni	Cr	Cu	O	N
Fe	0.0065	<0.01	<0.01	<0.003	0.0021	<0.02	<0.02	<0.02	0.0087	0.0006
Fe-0.3Si	0.0085	0.30	<0.01	<0.003	0.0019	<0.02	<0.02	<0.02	0.0018	0.0006

Stopping and Ranges of Ions in Matter (SRIM) is the Monte Carlo simulation code used for calculating ion deposition profiles in materials by ion implantation [8]. Ion energy plays a significant role in implantation depth of ion into materials. Mazumder *et. al.* [9] performed irradiation many times with different ion energy in order to irradiate the materials in the uniform depth. In this study, the energy degrader material was used to control the irradiation depth.

2 MeV He^{2+} ion implantation by 1.7 MV Tandem accelerator at Nagaoka University of Technology was performed to pure-Fe at room temperature with helium of 5 x 10^{15} ions/cm2 and 1.5 x 10^{16} ions/cm2. These fluences are corresponding to the concentration at a peak of 0.2% and 0.6%, respectively. Titanium foil was used as energy degrader to reduce stopping range of implantation depth. The estimated depth profile of implanted He is about 3.15 μm at the displacement damage peak. Furthermore, the use of 2 μm thickness titanium foil decreases the implantation depth to approximately 1.9 μm at the displacement damage peak. Figure 1 shows the estimation of helium depth distribution calculated by SRIM.

For comparison, the Si^{2+} ion irradiation was conducted by 1.7 MV Tandem Cockcroft-Walton accelerator at the HIT facility of the University of Tokyo [10] up to the fluence of 1.12 x 10^{16} and 3.34 x 10^{16} ions/cm2 at 400 ^{0}C. They are corresponding to the average dose of 5 dpa and 15 dpa, respectively. Figure 2 shows the SRIM calculation displacement depth profile for 3MeV Si ion implantation in Fe. The implantation reaches to approximately 1.3 μm from the surface. Pure-Fe and Fe-0.3Si were irradiated to evaluate the effect of solute material Si to the irradiation hardening.

Nanoindentation tests were performed to determine the mechanical properties of irradiation-induced hardening for both He and Si irradiation samples. Dynamic Ultra Micro Hardness Tester (DUH-211) with Berkovich trigonal pyramid shape indenter tip used in this study. The Oliver and Pharr method was applied to calculate indentation hardness [11]. The schematic diagram of the nanoindentation load applied as shown in Figure 3. For He irradiation samples, nine different loads were performed to analyse the indentation size effect. For Si irradiation sample, the load was set at 5 mN.

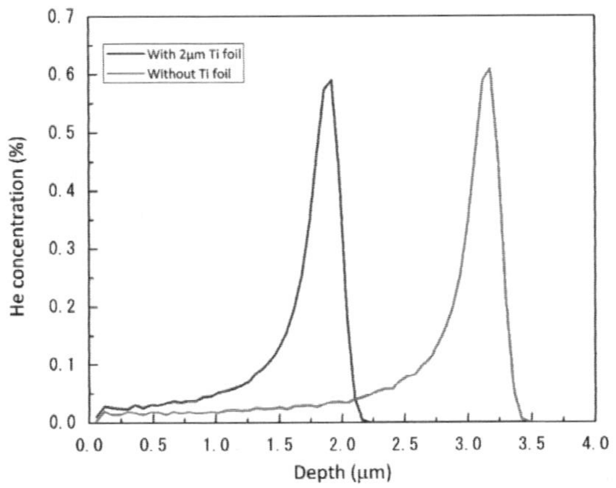

Figure 1. The estimation depth of helium concentration level of 0.6 at% profile calculated by SRIM for He^{2+} irradiation in Fe with energy degrader 2μm Ti foil and without Ti foil.

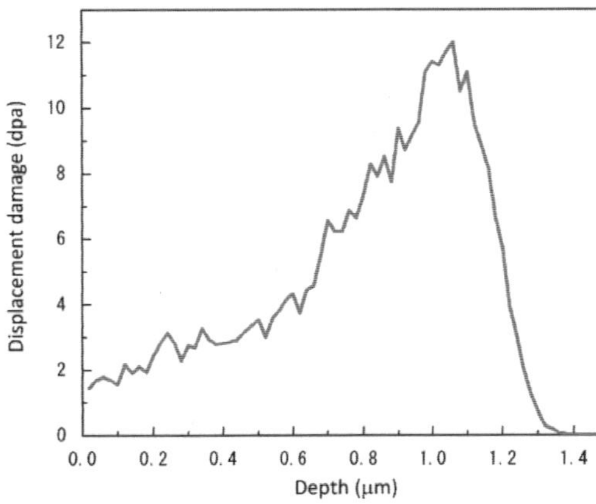

Figure 2. SRIM calculation displacement depth profile for 3MeV Si^{2+} ion implantation in Fe with fluence 1.12×10^{16} n/cm^2.

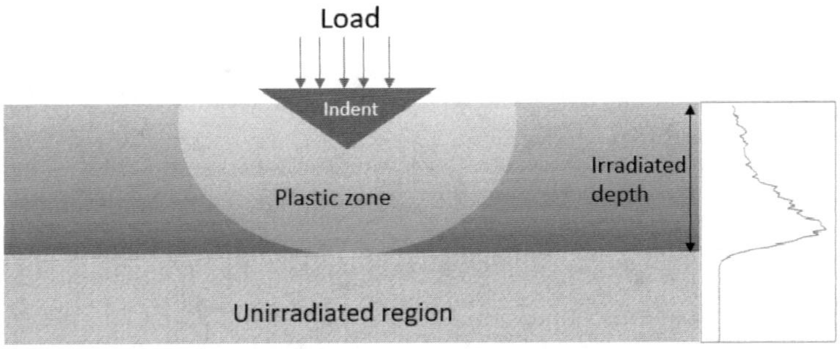

Figure 3. The schematic diagram of the nanoindentation procedure used.

3. Results and discussion

Figure 4 shows the relation between nano-hardness of pure Fe irradiated by He^{2+} ion with indentation applied load to the irradiation depth. We can clearly observe that applying low indentation load resulted in high of nano-hardness. This is due to the indentation size effect (ISE) [12]. The figure also shows irradiation hardening. The irradiated pure Fe by 0.2% and 0.6% of helium concentration showed an increase of nano-hardness compared to unirradiated pure Fe. However, the nano-hardness was slightly decreased when the concentration of He increases from 0.2% to 0.6%. It was reflected to study by Chen et al. [13] that found the nano-hardness decreasing after He concentration increases from 0.1% to 0.2%. They observed the formation of small cavities when increasing the He concentration. In our case, the slight recovery of hardening may also be due to the formation of cavities and voids in the irradiated region. The indentation load also affects the penetration of irradiation depth. It has a trend similar to how hardness changes with increasing depth in the case of non-irradiated Fe. The anomalous bumps observed in the hardening in Figure 4 at approximately 0.8 μm could be representing the region where the hardening is the most obvious. Usually, the deformation region of nano-indentation is 4~5 time deeper than the penetration depth of the indenter. The bumping region represents the obstacle of dislocation at the depth of approximately 3~4 μm. In the region, injected helium would exist at the highest concentration. The depth profile of irradiation-induced hardening represents the microstructure related to helium.

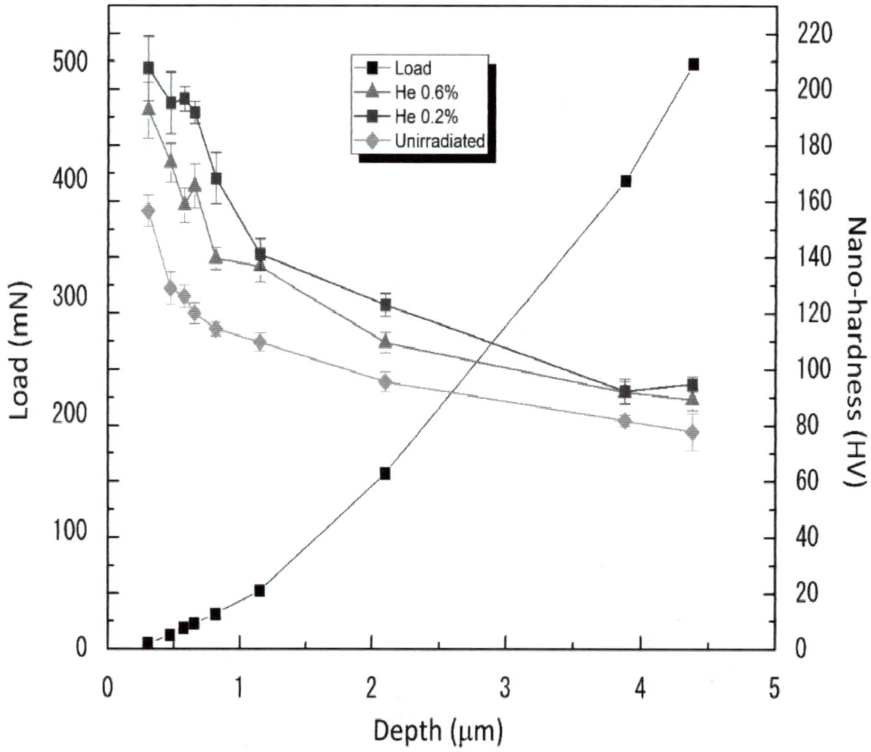

Figure 4. The relation between nano-hardness of He irradiation pure Fe and indentation load in the function of indentation depth.

The 5mN load with maximum indent depth of approximately 0.3 μm was applied to Si2+ ion irradiated specimens. The dose dependence of irradiation hardening for pure Fe and Fe-0.3Si alloys irradiated at 400 0C was shown in Figure 5. Hardness change is the difference of the hardness before and after irradiation; ΔHV = Hirr. – Hunirr. The hardening increase with increasing dose. The results may show the change in hardening as a function of the square root of dose. The solute element Si affect the irradiation hardening. The Si element contained in the alloys resulted in decreases of irradiation hardening compared to pure Fe. Liu et al. [14] also reported the same trend where the presence of Si solute element in Fe-Cu alloys decreases the irradiation hardening. Moreover, the findings are in line with the findings from Murakami et al. [15] which stated the presence of solute element Si enhance the recovery of damage defects.

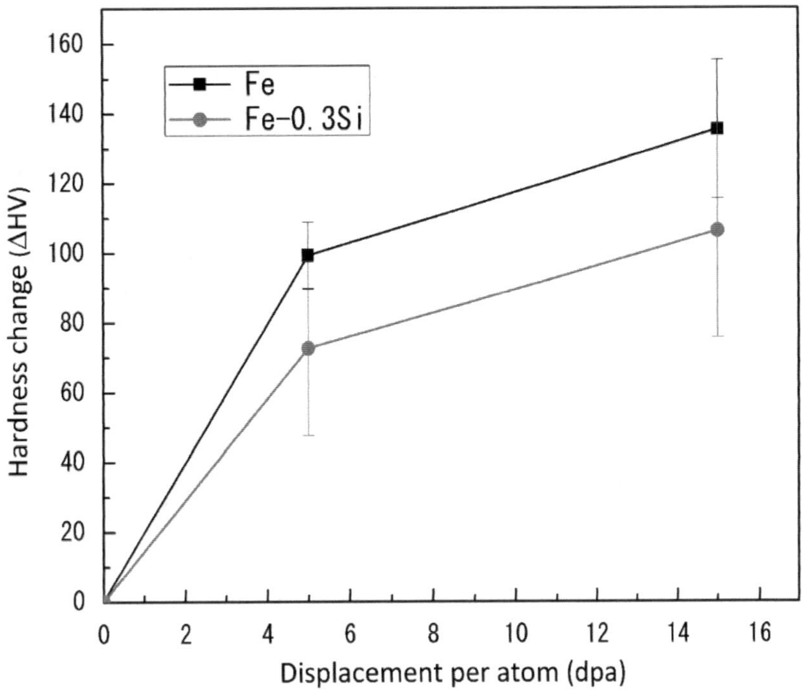

Figure 5. Irradiation hardening of Si^{2+} irradiation pure Fe and Fe-0.3Si for 5dpa and 15dpa at $400^{0}C$.

4. Conclusion

We irradiated pure Fe and Fe-0.3Si with He^{2+} and Si^{2+} ion and found that both ion irradiation enhances the irradiation hardening. He^{2+} ion irradiation performed to pure Fe for 0.2% and 0.6% He concentration at room temperature. Nano-hardness decreases with the increase of He concentration from 0.2% to 0.6% through both are harder than the unirradiated material. For Si^{2+} ion irradiation, we irradiated 5dpa and 15dpa for Fe and Fe-0.3Si at 400 ^{0}C. The irradiation hardening increases with increasing of dose at 400 ^{0}C for both materials, pure Fe and Fe-0.3Si. Further, the addition of solute element Si in Fe resulted to reduce the irradiation hardening. The present of Si may contribute to the enhancement of radiation damage recovery.

5. References

[1] Yu K Y *et al.* 2012 Radiation damage in helium ion irradiated nanocrystalline Fe *J. Nucl. Mater.* **425**(1–3) p 140–146.

[2] Zeman A, Debarberis L, Kupča L, Acosta B, Kytka M, and Degmová J 2007 Study of radiation-induced degradation of RPV steels and model alloys by positron annihilation and Mössbauer spectroscopy *J. Nucl. Mater.* **360**(3) p 272–281,.

[3] Yabuuchi K, Kasada R, and Kimura 2013A Effect of alloying elements on irradiation hardening behavior and microstructure evolution in BCC Fe *J. Nucl. Mater.* **442**(1-3) SUPPL.1 p S790–S795.

[4] Iwai T, Murakami K, Katano Y, Iwata T, Onitsuka T, and Abe H 2010 A positron beam study on vacancy formation in iron by ion beam irradiation at low temperature *J. Phys. Conf. Ser.* **225** p 1–4.

[5] Murakami K. *et al.* 2015 Role of nickel and manganese in recovery of resistivity in iron-based alloys after low-temperature proton irradiation *Philos. Mag.* **95**(15) p 1680–1695.

[6] Fujii K, Fukuya K, and Hojo T 2013 Effects of dose rate change under irradiation on hardening and microstructural evolution in A533B steel *J. Nucl. Sci. Technol.* **50(**2) p 160–168.

IOP Conf. Series: Materials Science and Engineering **785** (2020) 012027 doi:10.1088/1757-899X/785/1/012027

[7] Liu C, Hashimoto N, Ohnuki S, Ando M, and Shiba K 2013 Dependence of dose and he on irradiation-hardening of Fe-ion irradiated Fe-8Cr model alloy *Mater. Trans.* **54**(1) p 96–101.
[8] Ziegler J F, Ziegler M D, and Biersack J P 2010 The stopping and range of ions in matter (2010)," *Nucl. Inst. Methods Phys. Res. B* **268**(11–12) p 1818–1823.
[9] Mazumder B, Bannister M E, Meyer F W, Miller M K, Parish C M and Edmondson P D 2015 Helium trapping in carbide precipitates in a tempered F82H ferritic-martensitic steel *Nucl. Mater. Energy* **1** p 8–12.
[10] Murakami K, Iwai T, Abe H, and Sekimura N 2016 Improvement of the High Fluence Irradiation Facility at the University of Tokyo *Nucl. Inst. Methods Phys. Res. B* **381** p 67–71.
[11] Yan W, Pun C L, and Simon G P 2012Conditions of applying Oliver-Pharr method to the nanoindentation of particles in composites *Compos. Sci. Technol.* **72(**10 p 1147–1152.`
[12] Prasitthipayong A, Vachhani S J, Tumey S J, Minor A M and Hosemann P 2017 Increasing temperature *Acta Mater.*
[13] Chen S, Wang Y, Hashimoto N, and Ohnuki S 2015 Helium irradiation hardening in Fe, F82H-IEA and F82H-ODS steel *J. Plasma Fusion Res.* **11** p 79–82.
[14] Liu L *et al.* 2016 Effects of solute elements on hardening and microstructural evolution in neutron-irradiated and thermally-aged reactor pressure vessel model alloys.
[15] Murakami K, Iwai T, Abe H, and Sekimura N 2016 Interaction between solute atoms and radiation defects in Fe-Ni-Si and Fe-Mn-Si alloys under irradiation with proton ions at low-temperature *J. Nucl. Mater.* **482** p 47–52.

IOP Conf. Series: Materials Science and Engineering **785** (2020) 012028 doi:10.1088/1757-899X/785/1/012028

Automatic winder shielding materials focused on shape of the shielding using Monte Carlo N-particle

Mohd Noorul Ikhsan Mohamed[1,a], Nuri 'Adilah Nashrumillah Idris[1], Nur Solehah Abdul Majid[1]

[a]ikhsan@nuclearmalaysia.gov.my

Abstract. Automatic winder is a device that can be controlled using a remote controller which can be operated from a distance. It is equipped with electronic component like battery, raspberry pi and mechanical component such as gear, motor, and drive cable. This device will be used to perform radiographic testing which highly exposed to the gamma radiation. Front surface of the automatic winder is an area that receive high dose rate, thus proper shielding needs to be considered to protect component inside the automatic winder. In this study, two type of shapes were selected to be installed at the front surface of the winder housing. By using Monte Carlo N-Particle, increase the thickness of depleted Uranium from 1 mm, 3 mm, 5 mm, 7mm, 9 mm and 11 mm will reduce the dose rate received to the automatic winder. When comparing the shape, it shows that hemisphere shape gives better shielding compared with rectangular shielding.

1. Introduction

Gamma radiation is classified as electromagnetic (EM) radiations beside any other EM radiations such as radio waves, infrared radiation, ultraviolet radiation, X-rays and microwaves [1][2]. Gamma radiation has been utilized in various sectors for example, in medicine (radiotherapy), industry (sterilization and disinfection) as well as the nuclear industry. Shielding against gamma radiation is very crucial because it could cause skin, blood and eye disorders to human being who exposed to the radiation [3]. Therefore, different types of shielding materials have been manufactured depending on the applications in order to provide necessary protection to the radiation worker and their surrounding environment [4].

In finding an appropriate shielding material for certain application, the weight, space and cost of the shielding materials have become the main concerns among the researchers around the globe. In principle, materials comprised of high-atomic number (Z) and high-density elements are good candidates to be used as effective gamma shields due to their high probability of interactions and larger energy transfer with gamma rays [1]. Materials used for Gamma radiation should be highly dense such as concrete and Lead [5]. Although the low-density materials can also be considered in reducing the intensity of Gamma radiation, but the higher-density materials are more effective [6]. The sufficient installation of high-quality Gamma radiation shielding material is required to reduce the risks of excessive gamma exposure.

IOP Conf. Series: Materials Science and Engineering **785** (2020) 012028 doi:10.1088/1757-899X/785/1/012028

Automatic Winder is a device that can be controlled using a remote controller which can be operated from a distance. It is equipped with electronic components like battery, raspberry pi and mechanical components such as gear, motor, and drive cable. This device will be used to perform radiographic testing which highly exposed to gamma radiation. In this study, two parameters were considered in order to protect the Automatic Winder from malfunction which is type and thickness of the shielding material. Monte Carlo N-Particle Transport Code (MNCP) has been used to optimize shielding material for industrial radiography Automatic Winder in this study. Three types of shielding used in the study were Lead, Tungsten, and depleted Uranium.

Lead has a good physical property that makes a fantastic shield property against a gamma-ray in nuclear medicine. The main advantage of using a Lead shield is because of its compactness due to its density [7][8]. However, it is dangerous to health because of its toxicity and the dust particles that formed on the surfaces of Lead objects [8]. Moreover, the Lead physical property such as strength and hardness are not suitable for the hard environment. However, Lead is a common radiation shielding material for radiation protection [6]. Tungsten has a high gamma attenuation property, but it does not possess the chemical toxicity associated with lead [9]. Thus, Tungsten has many advantages therefore it is one of the most important materials for nuclear applications. Furthermore, Tungsten composite materials are used in high gamma radiation shielding applications because of their high density and favorable mechanical properties. One-third less Tungsten material has the same effectiveness at absorbing energy compared with lead and it can provide compact shielding without compromising the attenuation characteristics [1]. From [10] there is an equation that provides the relationship between ambient dose equivalents with the thickness of shielding material. A case study by [11] indicates that the measured dose values approximately decrease exponentially with the increase of Lead thickness. Furthermore, the sensitivity measured of the ion chamber decrease with an increasing the Lead thickness as a result of decreasing the dose values, hence, the dose unpredictability increased at low dose values. Lastly, the objective of this study is to compare the ability of three different types of shielding material on gamma radiation and to study the effect of different shapes of shielding material in shielding gamma radiation.

2. Methodology

The basic algorithm of all Monte Carlo methods is the fundamental that is based on a stochastic component. Consequently, there is a wide variety of algorithms under the name Monte Carlo and used for very different applications from materials modelling to risk assessment in finance [12]. The first Monte Carlo method is attributed to Ulam and Metropolis, that was developed while working at the Los Alamos National Laboratory during the Manhattan project[13].

The Monte Carlo method that will be used throughout this study follows the kinetics of a series of events, known as the probabilities of occurrence, hence its name kinetic Monte Carlo (KMC). It was first used in the field of radiation damage by H. Heinisch in the early 90's [14] and since then it has been applied to different types of metals, particularly iron, iron alloys[15][16] and Tungsten [17][18]. Monte Carlo simulations comprises of a branch of experimental mathematics that involves experiments with random numbers [19]. Differing from a physical experiment, Monte Carlo simulation uses random numbers to conduct many virtual experiments on the computer and then analyzes the results of the experiments on a statistical basis to draw the conclusions. The stochastic approach of Monte Carlo method favoured its application to address diffusion problems such as mass and heat transfer problems in material science [20][21].

IOP Conf. Series: Materials Science and Engineering **785** (2020) 012028 doi:10.1088/1757-899X/785/1/012028

The generation of structure for the cell card, surface card and data card were written in the input of MCNP as shown in Figure 1. The details for each equipment that are involved in the experiment are illustrated in code that was generated for MCNP software. The main equipment involved throughout the experiment were Automatic Winder, Gamma projector, radiation detector and the shielding materials. The arrangement of the equipment is shown in Figure 2.

```
c
c -------cell cards-------
c
  1  0   999
  2  2   -0.00125 -999 #10 #11 #20 #21 #22 #41 #60 #70
  20 1 -2.69        -200 201 250
  21 3 -19.1        -201 250
  22 2 -0.00125     -250
c
  60 4 -11.36        600 -601 602 -603 -604 605 -606 #70
  70 2 -0.00125      700 -701 702 -703 -704 705 -706
c
  10 1 -2.69        -100 101 401
  11 2 -0.00125     -101 401
c
  41 4 -11.36        -401
c
c

c -------surface cards-------
c
  100 rpp  -17.5 17.5  -5.25 5.25  -73.5 -43.5 $ Winder Box
  101 rpp  -17.3 17.3  -5.05 5.05  -73.3 -43.7
c
  200 rpp  -6.75 6.75 -9.5 9.5  -23  0          $ Projector Box
  201 rpp  -6.65 6.65 -9.3 9.3  -22.8 -0.2
  250 rpp  -4.05 4.05 -6.8 6.8  -20.3 -2.7      $ Depleted U shield
c
c
  401 rpp  -4.0 -0.6 -1.0 1.0 -58.5 -57.0       $ bleeper inside winder
c
c
c
c
  600 px -17.5                         $ Surface Shielding hemisphere
  601 px  17.5
  602 py  -5.25
  603 py  5.25
  604 pz  -37.5
  605 pz  -43.5
  606 c/x 0 -43.5 5.25
c
  700 px -17.3
  701 px  17.3
  702 py  -5.15
  703 py  5.15
  704 pz  -37.6
  705 pz  -43.4
  706 c/x 0 -43.5 5.15
c
  999 rcc 0 0 -100  0 0 280  50
c
```

```
c --------data cards--------
c
mode p
imp:p 0 1 2 4 8 8 8 8 8 8
c
sdef POS=0 0 150 PAR=2 ERG=0.38
c
c
m1   13027              1       $Aluminium
c
m2   7014               0.755268 $Air
     8016               0.231781
     18000              0.012827
     6012               0.000124
m3   92238              1       $ DU
c
m4   82207              1       $ Lead
c
c
c
c
f4:p 41
df4 IU=2       $ Sv/h/particle
fm4 1.48e12    $ convert to Sv/h
c
c
c
nps 2000000000
```

Figure 1. MCNP input file.

2.1 Studies of gamma radiation level

The initial test conducted in this study was to look at the radiation levels at several different locations during the radiography test where it has been simulated using MCNP software. This test is important for identifying and comparing radiation levels at specific places to help build effective shielding. Although the test is a basic test in which the results could be anticipated beforehand, it is important to verify that the MCNP model produced is correct and working properly. Figure 2 shows the experimental set up where the radiation levels at locations a, b and c have been studied. The detector is located inside the automatic winder as shown in Figure 2 (b). It was predicted that the readings at location a will be

IOP Conf. Series: Materials Science and Engineering **785** (2020) 012028 doi:10.1088/1757-899X/785/1/012028

zero while the readings at location b will be higher than location c due to the distance with the detector is shorter.

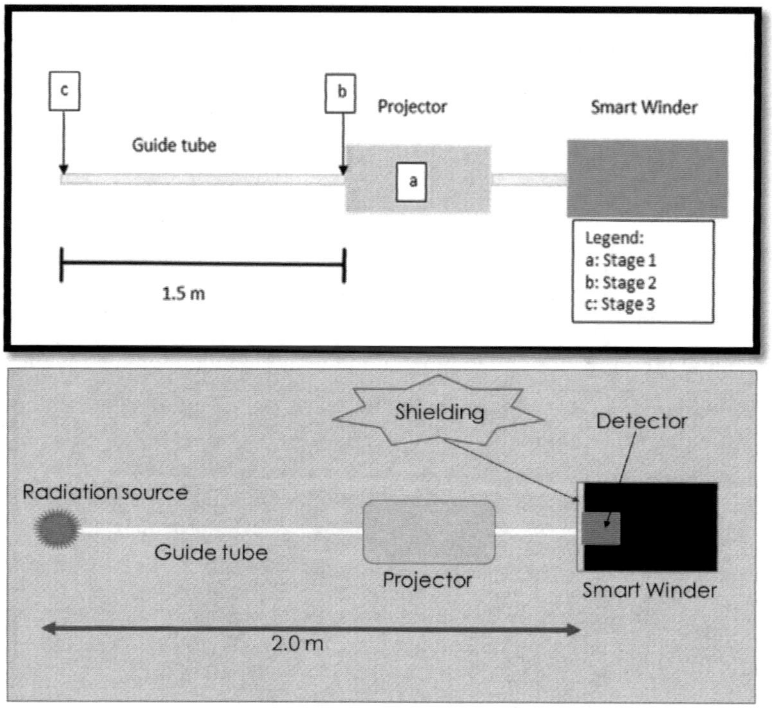

Figure 2. Equipment arrangement.

2.2 Studies of different types and thicknesses

The MCNP model from the initial test has been used in this second test where a layer of shielding material was placed between the radiation source and the detector located at the inside of the automatic winder wall. In this study, the radiation location was only at position c as it was the longest exposure position during radiographic testing. Radiation level readings without shielding material that were carried out in the preliminary tests were used as controls. Then the Lead shielding material with a thickness of 1 mm was placed in the position as shown in Figure 2 and the readings of the detector with the radiation source at position c were recorded. The test was repeated with the thickness of the Lead shielding material raised to 11 mm with intervals of 2 mm. The tests were then repeated with the Tungsten and depleted Uranium shielding materials at the same thicknesses as the tests conducted on the Lead. All the readings were recorded and plotted in a graph.

2.3 Studies of shape of shielding material
Figure 3 shows the three shapes of shielding material that have been studied using the MCNP simulation method. The purpose of this study is to observe if the shape of the shielding material affects the ability to withstand the Gamma radiation. The shape 1 is a flat form. It is a common form which used as a control and a comparison of the other two forms that were tested. The shape 2 is vertically curved and the third shape is horizontally curved as shown in Figure 3.

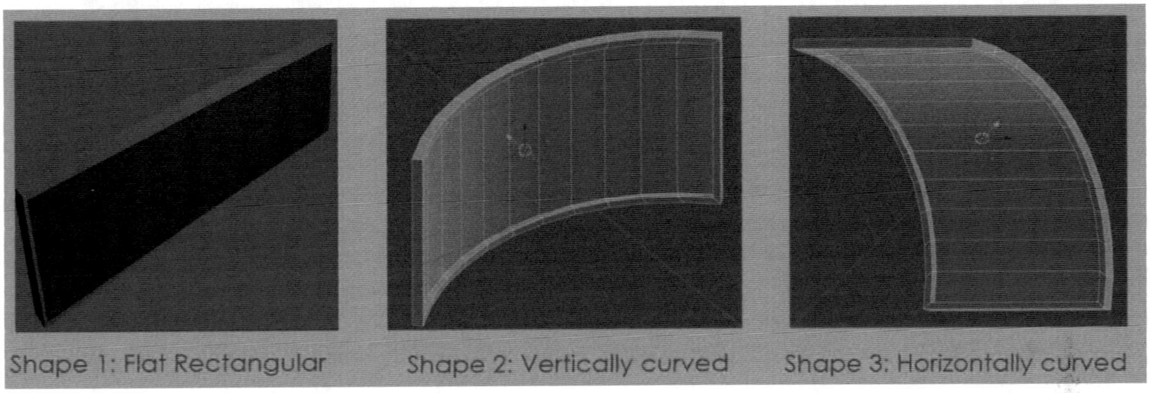

Figure 3. Three shape of shielding material studied.

3. Result and discussion

In order to achieve two objectives of this study, MCNP has been used. From Figure 4. stage 1 is where the source secure in gamma projector equal to 0 Sv/hr. At stage 2, the source in the guide tube was 42 μSv/hr that just outside the gamma projector. The source at the end of the guide tube is 58 μSv/hr (stage 3).

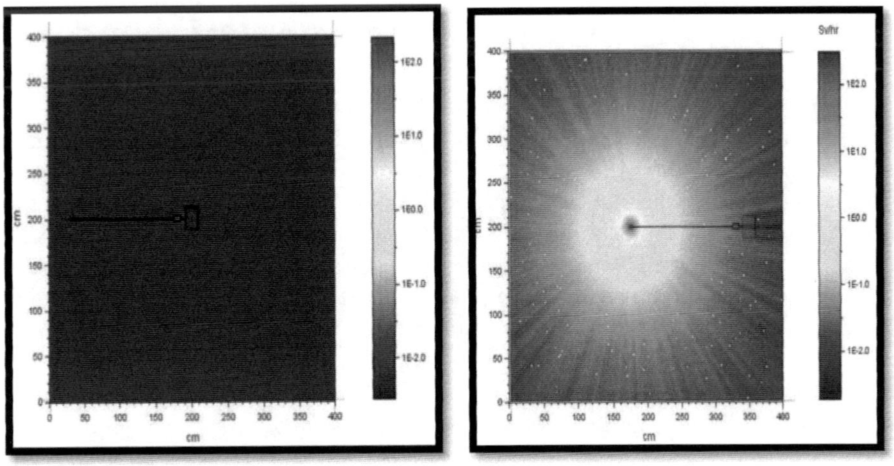

Figure 4. Intensity at different location during exposure.

3.1 Studies of Gamma Radiation Level in Radiographic Testing.
There are three types of shielding used to find out this objective which are Lead, Tungsten, and depleted Uranium. Figure 5 shows that depleted Uranium is the best shielding material compared to Tungsten and Lead. The dose rate (μSv/hr) is decreasing due to the increase of shielding thickness, hence dose

rate (µSv/hr) is inversely proportional to shielding thickness. As for depleted Uranium, the dose rate is decreasing from 35 (µSv/hr) to 10 (µSv/hr) following the increasing of the shielding thickness. In conclusion, depleted Uranium is the most effective shielding material on gamma radiation.

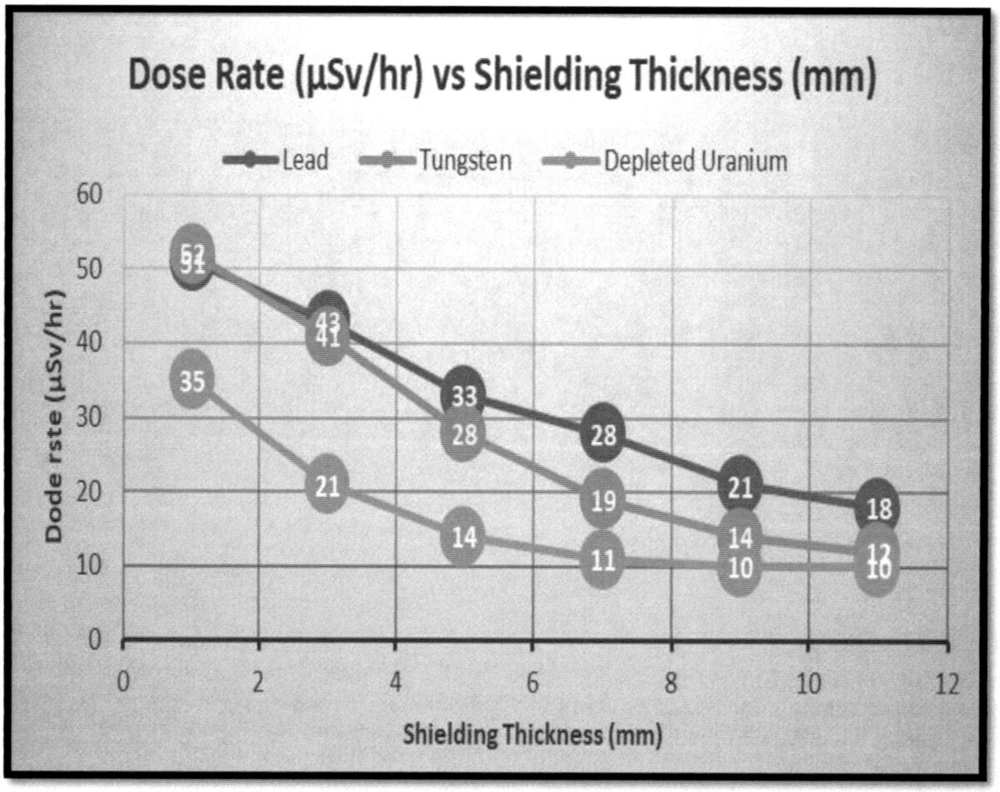

Figure 5. Dose rate (µSv/hr) versus the shielding thickness (mm) by different types of material shielding.

3.2 Studies of shielding material of different types and thicknesses

Based on the result of objective one, depleted Uranium is the most effective shielding material on gamma radiation compared to Lead and Tungsten. Therefore, for this objective, the researcher tested the effect of different shapes of shielding using depleted Uranium. The shaped of shielding material tested for this objective are flat rectangular, vertically curve, and horizontal curve. Figure 6 shows that the vertically curved material shielded more gamma radiation compared to the horizontally curved shape. Flat rectangular is the less effective shape in material shielding compared with the other two. In conclusion, vertically curved is the most efficacy shape of the shielding material.

IOP Conf. Series: Materials Science and Engineering **785** (2020) 012028 doi:10.1088/1757-899X/785/1/012028

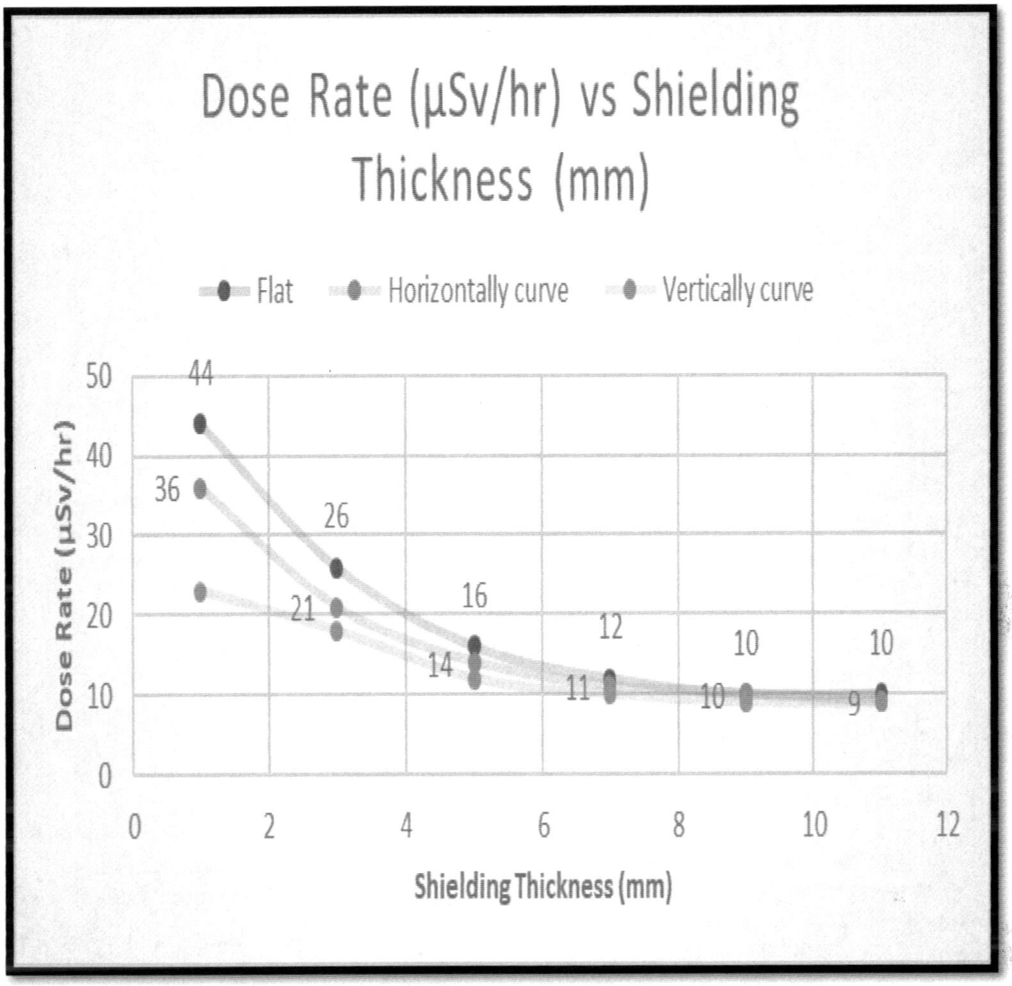

Figure 6. Dose Rate (µSv/hr) versus the shielding thickness (mm) for depleted Uranium.

3.3 Studies of shielding material of different types and thicknesses

The results show that the shape of a shielding materials affects the ability. Figure 7 shows that curved shapes are more effective for filtering Gamma radiation, which have given lower dose rate readings as compared to the flat shape. Meanwhile, the comparison between shape 2 (vertically curved) and Shape 3 (horizontally curved) shows that shape 2 is more effective in filtering Gamma content.

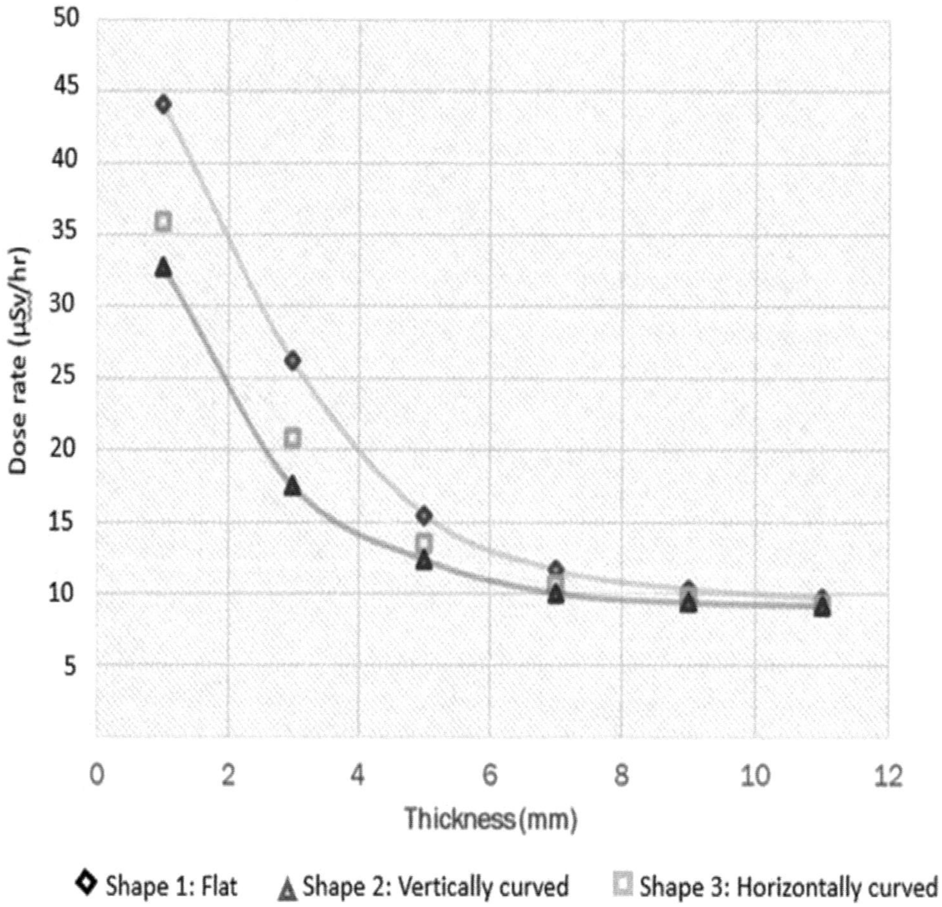

Figure 7. Thickness of a shielding materials vs dose rate.

4. Conclusion
This study concluded that among the three tested materials, depleted uranium proved to be the most effective shielding material of IR-192 gamma radiation, followed by tungsten and lead. For the second objective, it concluded that horizontally curved is the most efficacy shape of the shielding material. Followed that the vertically curved material shielded more gamma radiation compared to the flat rectangular shape.

5. References
[1] Poltabtim W, Wimolmala E, Saenboonruang K 2018 Properties of Lead-free gamma-ray shielding materials from metal oxide / EPDM rubber composites *Rad. Phys. and Chem.* **153** p 1–9. https://doi.org/10.1016/j.radphyschem.2018.08.036.

[2] Lucas J What are gamma rays? Live science contributor (2018) Retrieved https://www.livescience.com/50215-gamma-rays.html.

[3] Erramli. H, Asri, J E 2019 Gamma Rays: Applications in environmental gamma dosimetry and determination samples gamma-activities induced by neutrons, use of gamma radiation techniques in peaceful applications, ed. Almayah B A (IntechOpen), doi: 10.5772/intechopen.85503.

[4] Oto B. *et al.* 2015 Investigation of gamma radiation shielding properties of various ores *Prog. Nucl. Energy*, **85**(11) pp. 391-403, 10.1016/j.pnucene.2015.07.016.

[5] Luković J. *et al.* 2015 Synthesis and characterization of Tungsten carbide fine powders *Ceram. Int.* **41**(1) 10.1016/j.ceramint.2014.09.057 p1271-1277,

[6] Jamal N, Aida N, Amin B 2019 Conventional and new lead-free radiation shielding materials for radiation protection in nuclear medicine : a review, p 165.

[7] Mirji R, LoboRadiation B 2017 Shielding Materials : A brief review on methods , scope and significance *Nat. Conf. on Adv. in VLS and Microelect.* ed Jabin P C (, (Huballi, India:Science College).

[8] AbuAlRoos N J, Amin N A B, Zainon R 2019 Conventional and new lead-free radiation shielding materials for radiation protection in nuclear medicine: A review. *Rad. Phys.and Chem. p* 108439.

[9] Tekin HO, Sayyed M I, Altunsoy E E, Manici T Shielding properties and effects of WO3 and PbOhttps://doi.org/10.1016/j.apradiso.2016.12.040 p861-867.

[10] Costa P R, Caldas L V E 2002 Evaluation of protective shielding thickness for diagnostic radiology rooms: Theory and computer simulation, httpsmass attenuation coefficients by using MCNPX code *Dig. J. Nanomater. Biostruct.* **12** ://doi.org/10.1118/1.1427309 p 73–85.

[11] Yücel H, Güllüoğlu E, Çubukçu Ş, Üncü Y A 2016 Measurement of the attenuation properties of the protective materials used as a thyroid guard and apron for personnel protection against diagnostic medical x-rays **27**(1) p 111–128.

[12] Kroese D P, Brereton T, Taimre T, Botev Z I 2014 Why the Monte Carlo method is so important today. *WIREs Comp. Stat.* **6**(6) https://onlinelibrary.wiley.com/doi/abs/10.1002/wics.1314. doi:10.1002/wics.1314 p 386-392.

[13] Metropolis N, Ulam S 1949 The Monte Carlo method *J. of the American Stat. Assoc.* **44**(247) p 335-341.

[14] Heinisch H L 1990 Computer simulation of high energy displacement cascades *Rad. Effects and Defects in Solids* **113**(1-3) p 53-73.

Effects of natural rubber latex fraction to glycidyl methacrylate on grafting polymerization and its application as Th^{4+} and UO_2^{2+} ions bio adsorbent

Nurrul Assyikeen Md. Jaffary[1,2,a)], Nor Hasimah Mohamed[2], Khoo Kok Siong[1], Sofian bin Ibrahim[2], Siti Nor Syuhada Bt Hamdan[2], Ungku Aida Farzana Binti Ungku Azly[2], Syazwani Mohd Fadzil[1]

[1]Nuclear Science Program, Nuclear Technology Research Center, Faculty of Science and Technology National University of Malaysia, 43600 UKM Bangi, Selangor, Malaysia
[2]Malaysian Nuclear Agency, Bangi, 43000 Kajang, Selangor, Malaysia

[a]assyikeen@nuclearmalaysia.gov.my

Abstract. Natural rubber (NR) latex was modified by in situ gamma radiation-induced grafting copolymerization of glycidyl methacrylate (GMA). NR latex to GMA emulsion ratio of 100:0, 50:50, 40:60, 30:70 and 20:80 have been investigated at the absorbed dose of 150kGy. The grafted GMA natural rubber-based follow the subsequent chemical modification to anchor phosphate group by open up the epoxy ring. The modified NR was characterized by Fourier transform infrared. Demonstration of modified NR on Th^{4+} and UO_2^{2+} ions adsorption in 25ml of 10ppm standard solution gave 95% and 28% percentage removal respectively.

1. Introduction

Actinides such as thorium and uranium are radioactive materials that occur naturally in our environment and has a significant deposit in monazite [1]. These radioactive materials have upgraded to the elevated level under the chemical processing of monazite ore in order to recover rare earth elements (REEs) and hence, it is called technologically enhanced radioactive material (TENORM). Stream waters close to the rare earth processing operated near Ipoh, Malaysia plant site had revealed to have anthropogenically influenced considering the chemical mobility of radioactive elements in the environment [2]. Separation of these radioactive elements from wastewater before discharging into the stream water using conventional method such as solvent extraction and chemical precipitation are extremely difficult since the present of the radioactive materials is relatively low in concentration and have significant amounts of co-existed elements [3]. Furthermore, these methods required high capital cost, high energy or high volume of reagent, generation of toxic sludge and waste products that need further disposal. Therefore, many studies focused on the development of alternative separation methods with low capital cost and reduce the environmental burden.

Recently, a novel chitosan/clinoptilolite (CS/CPL) adsorbent has been synthesized in the beads form, and the maximum experimental adsorption capacities of 328.32mgTh^{4+}/g and 408.62mg UO_2^{2+}/g were recorded [4]. A silicon dioxide nano-balls (nano-SiO2) was also prepared by Kaynar and his team with adsorption capacity of Th^{4+} achieved was 188.2 mg/g [5]. An agriculture waste base adsorbent was among the most studied by researchers for Th^{4+} ion removal and recovery due to economic reason. De-oiled karanja seed cake has also been used for Th^{4+} sorption, and the study showed that the bio sorbent suited with pseudo-second order kinetic model with spontaneous and endothermic in nature [6]. Other biomaterials such as rice bran, rice husk, and wheat bran have evaluated where extraction efficiencies of 95.4 and 74.6 % achieved from 4 M nitric acid of 200-4000 ppm Th^{4+} for RB and WB respectively [7]. Two different groups of adsorptive materials, which are activated carbons and zeolites have been explored as Th^{4+} adsorbent as well [8]. The activated carbons were prepared from solvent extracted olive pulp (SEOP) and olive stone (OS) by a two-step physical activation method with steam. The study showed that synthetic zeolites have highest removal ability for Th^{4+}.

Other researchers introduced groups like carboxylate on titanium dioxide-densified cellulose [9] and octyl(phenyl)-N,N-diisobutylcarbamoyl-methylphosphine oxide (CMPO) ligand on merrifield chloromethylated resin with high adsorption uranium capacity [10]. The strong chelators with target metal can be shown as pKchelators/pKnatural ligands. The greater this ratio, the stronger this chelator will be. Various suitable chelators (sodium bicarbonate, citric acid, two-stage acid leaching procedure) has been used following techniques of natural attenuation, physicalmethods, chemical processes for uranium remediation from contaminated soil [11].

Natural rubber (NR) is one of the leading commodity in Malaysia and interesting material for chemical modification for actinides removal. Natural rubber possesses a double bond on its polymeric backbone prone to various chemical modifications. Recently, NR latex has successfully functionalized with sulfonic through grafting polymerization for heavy metal removal application [12]. Therefore, in this study we will demonstrate a modification of NR latex by radiation-induced graft polymerization as actinides bio adsorbent. This study will focus on the effect of GMA concentration on the degree of grafting and their effect on adsorption of Th^{4+} and UO_2^{2+} from acidic solution.

These adsorbents with chelate effect can be introduced by grafting polymerization for heavy metal removal [13][14]. Study on grafting glycidyl methacrylate (GMA) through electron beam irradiation technique on water hyacinth fibers (*Eichhornia crassipes*) [15] and banana trunk based fibrous [16] has successfully done for wastewater treatment application.

2. Materials and methods

2.1. Materials
High ammonia natural rubber, HANR latex (60% DSC) used in the present work supplied by Titilatex Sdn Bhd, Malaysia. Glycidyl methacrylate monomer (purity 97%) and Tween 20 (polyethylene glycol sorbitan monolaurate) supplied by SIGMA. Phosphoric acid (85%) was supplied by R&M. UO_2^{2+} and Th^{4+} standard solution in 2% nitric acid were supplied by Perkin Elmer. The dose rate of gamma loop in SINAGAMA facilities was determined to be 2.14 kGy/hr by Fricke dosimetry before grafting experiments.

2.2. Radiation-induced graft polymerization
Grafting polymerization was carried by adding NR latex to monomer emulsion containing 5% GMA, 0.5% Tween 20 and 94.5% deionized water at a weight ratio of 100:0, 50:50, 40:60, 30:70 and 20:80 (%wt:%wt) in 60ml screw-capped plastic bottles. All samples were purged with nitrogen gas for 20 minutes before irradiated with gamma ray at a dose of 150 kGy. The GMA grafted natural rubber (GMA-g-NR) were washed using acetone and distilled water in order to remove glycidyl methacrylate homopolymer (PGMA). The residue was dried in drying oven at 50°C for several days until a constant weight was reached. The following Eq. **Error! Reference source not found.** measured the degree of grafting for grafted NR:

$$\% \, \boldsymbol{Grafting} = \frac{\boldsymbol{Treated - Untreated}}{\boldsymbol{Untreated}} \boldsymbol{x100\%}, \tag{1}$$

where, treated and untreated are the weights of the dry natural rubber (g) after and before grafting, respectively.

2.3. Chemical modification

GMA-g-NR were placed into phosphoric acid solution in the presence of solvent at 80°C for 3 hours. Finally, the functionalized NR refer as Ph-g-NR was washed with distilled water and dried in an oven overnight.

2.4. FT-IR analysis

FTIR spectra were recorded on a Bruker Platinum ATR TENSOR II. NR, GMA–g-NR and Ph-g-NR were sliced and directly characterized using ATR cell. FTIR spectra were recorded in the range from 500 to 4000 cm^{-1}, with a resolution of 4 cm^{-1} and averaged over 50 scans.

2.5. Water sorption test

Water uptake was determined by immersing 0.5gram of Ph-g-NR in excess of deionized water at room temperature [12]. Samples weight were recorded at time 0, 4, 12 and 36 hours. The water uptake was calculated as follows Eq.2:

$$\boldsymbol{Water\ uptake}\ (\%) = \frac{(\boldsymbol{m_t} - \boldsymbol{m_o})}{\boldsymbol{m_o}} \tag{2}$$

where m_o and m_t are the mass of the modified rubber at initial and time t, respectively.

2.6. ICPMS analysis for adsorption efficiency

The sorption efficiencies of Ph-g-NR toward representative actinide ions were carried out using 10ppm concentration of Th^{4+} and UO_2^{2+} ions in 25ml 0.1M HNO_3 acid medium. 0.5g of modified NR were used and the sorption was done using in Wisd shaker bath at 25°C and shaking were set at 200 RPMs for 4hrs. The concentration of the solutions was measured by using a Perkin-Elmer Nexion 350x model Inductively Coupled Plasma Mass Spectrometry (ICPMS).

3. Results and discussion

3.1. Grafting degree (GD)

Effect GMA concentration on the grafting yield is shown in Figure 1. The grafting degree for formulation with NR latex to GMA emulsion ratio of 100:0, 50:50, 40:60, 30:70 and 20:80 has been measured to be 0, 5, 10, 20 and 40 % respectively. It can be noted that grafting degree increases with increased of GMA content. The increased of monomer concentration provides enough contact area between monomer and thermally activated radicals of NR latex particles which will benefit to the grafting polymerization. [16, 17]. It is showed that the graph still increasing dramatically with the increase in GMA concentration. The study at higher GMA monomer concentration shall be conducted and discuss in the future study.

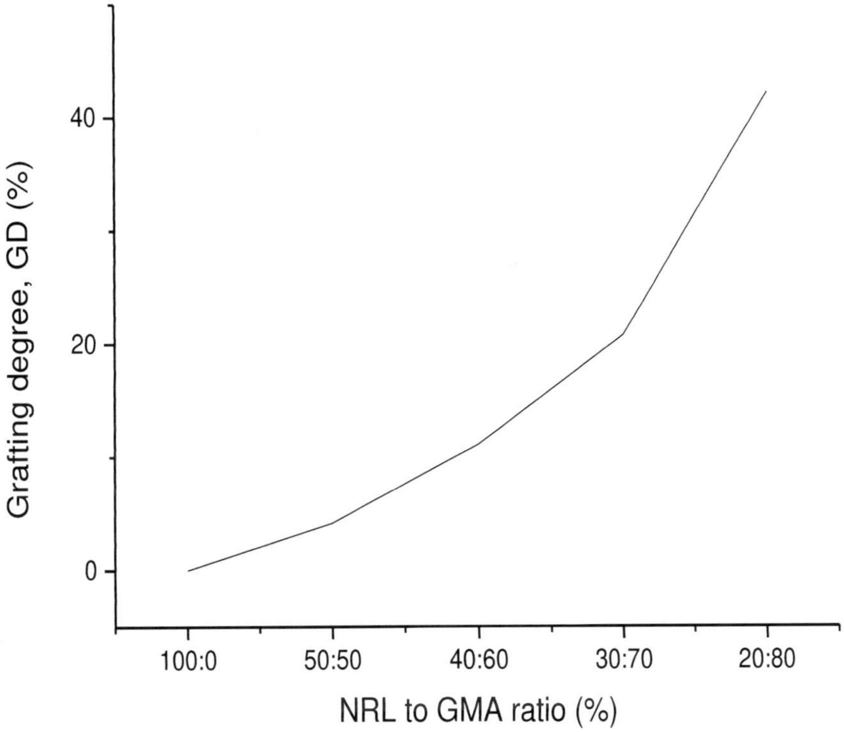

Figure 1. Grafting degree at different ratios between latex to GMA.

3.2. FTIR analysis on grafted natural rubber, GMA-g-NR

FTIR spectra in Figure 2 shows that the grafting polymerization of GMA onto NR was confirmed by the presence of absorption band at 1728cm^{-1} assigned to C-O stretching of the ester group from GMA and symmetrical stretching of the epoxy ring near 1251cm^{-1}. The band at 993cm^{-1} and 908cm^{-1} was attributed to asymmetrical epoxy ring. The successful of GMA grafting on NR based is clearly seen on the formulation NR latex to GMA ratio (%,%) of 20:80 and 30:70 by the existence of 993cm^{-1} and 908cm^{-1} bands. Based on the FTIR spectra, clearly showed that increasing of GMA concentration had increases the grafting degree of modified NR, thus giving the higher integrated area at 993cm^{-1} and 908cm^{-1}.

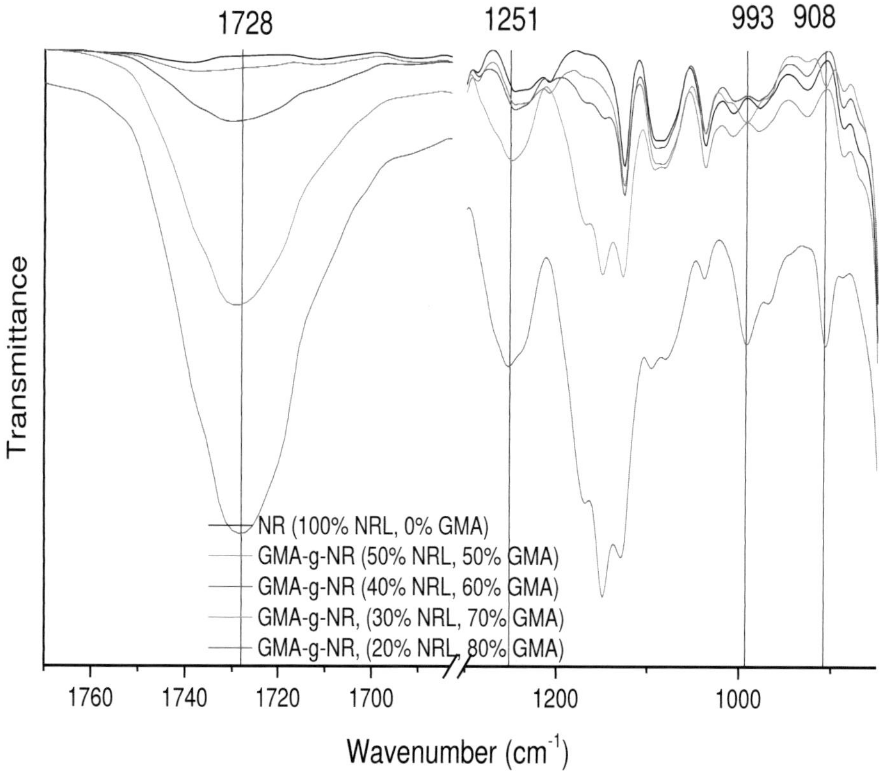

Figure 2. FTIR analysis on dry grafted NR at various fraction of latex to GMA ratio.

3.3. FTIR analysis on modified natural rubber, Ph-g-NR

FTIR spectra in Figure 3 shows the effect of phosphate modification on grafted GMA-g-NR at various fraction. The contraction of epoxy ring to anchor phosphate group was established by appearance of wide band in the region 900–1080 cm^{-1} which was assigned to P–OH stretching vibration. The wideband around 3100-3600cm^{-1} attributed to -OH band that verifies the hydrophilic properties of Ph-g-NR. GMA-g-NR has transformed into Ph-g-NR showed by reducing of the C-O stretching at 1728cm^{-1}.

The successful of Ph anchored on the opening epoxide can be observed on NR latex to GMA ratio (%,%) of 20:80 and 30:70 where these formulations are chosen for water sorption test and actinides removal application.

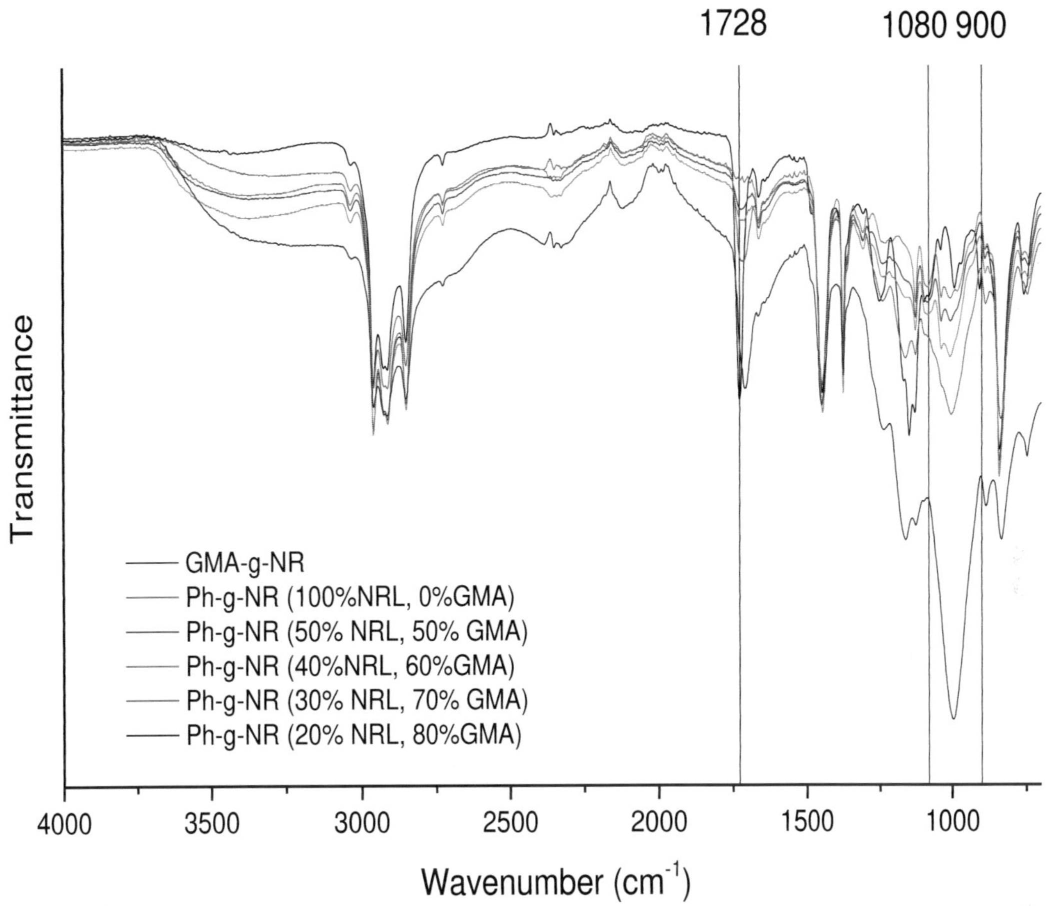

Figure 3. FTIR analysis on dry phosphate modified NR at various fraction of NR latex to GMA.

3.4. Water sorption test

Water uptake study conducted using sample formulation NR latex to GMA ratio (%,%) of 20:80 and 30:70 with 20% and 40% grafting degree respectively in Figure 4 showed that the water uptake of the modified rubber is increasing with higher grafting degree.

The GMA-g-NR modification using phosphoric acid leads to the hydrolysis of the epoxide. Figure 3 show that the phosphate group functionalized on the epoxide has increased according to the grafting degree while the phosphate group come with -OH group functionality that will increase hydrophilic properties

IOP Conf. Series: Materials Science and Engineering **785** (2020) 012029 doi:10.1088/1757-899X/785/1/012029

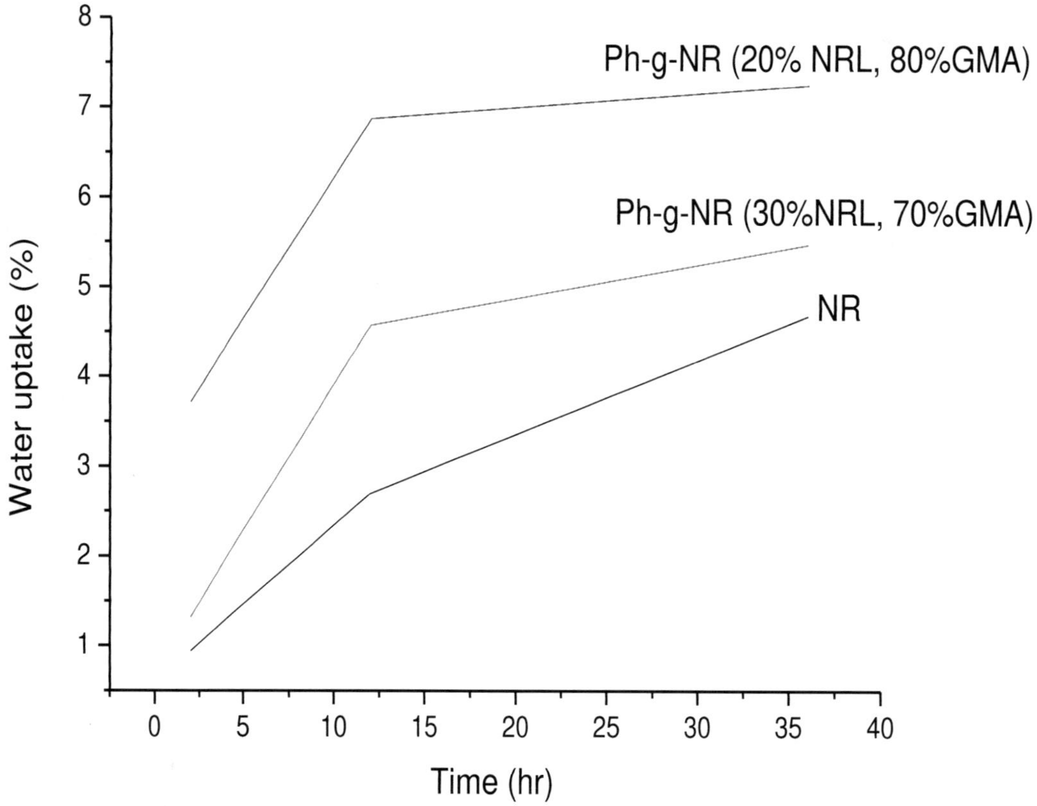

Figure 4. Effect of grafting degree on water uptake.

3.5. ICPMS analysis for adsorption efficiency

In order to demonstrate the effect of grafted GMA onto NR, actinide adsorption test performed on Ph-g-NR. Figure 5 shows the Th^{4+} and UO_2^{2+} removal using modified NR at 20% and 40% grafting degree. The high percentage removal of thorium was observed using 40% GD compared to the 20% GD. This is due to the higher density of phosphate functional group anchored that will form the ligand binding with Th^{4+}. Acidic phosphates are known to have an affinity toward actinide ions [18].

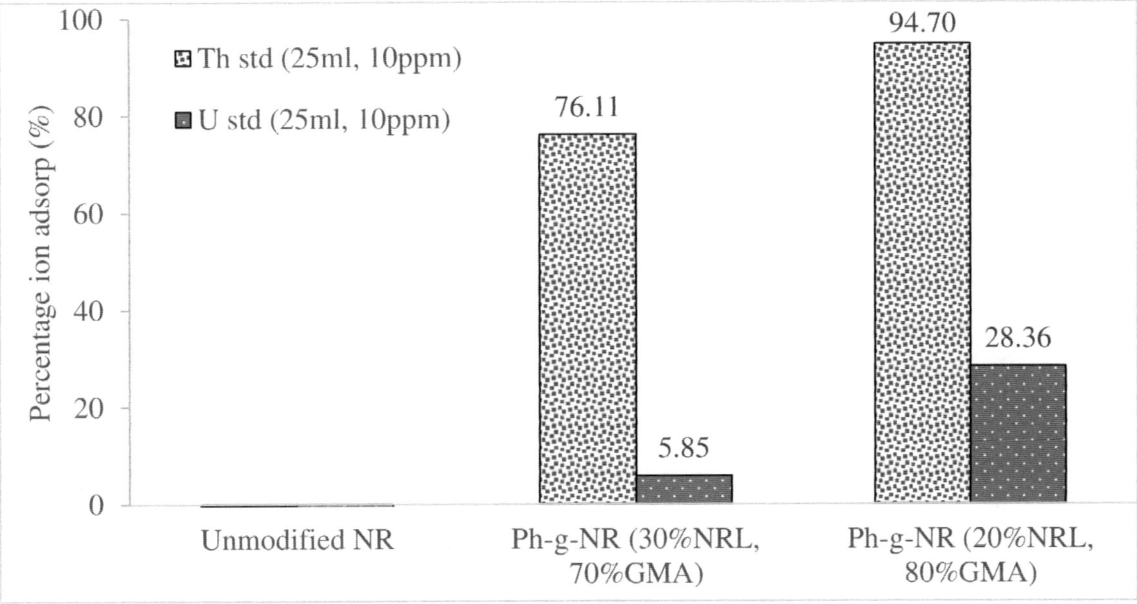

Figure 5. Effect of grafting degree on Th^{4+} and UO_2^{2+} ion removal.

It is also observed that the adsorbent tends to remove Th^{4+} ion compared to UO_2^{2+} ion. When the chelate has modified with adding the R-OH groups, it contributes to the addition of the neutral oxygen donors and gives rise to a rule of ligand design [19, 20]. Addition of groups containing neutral oxygen donor leads to an increase in selectivity of the ligand due to the higher magnitude of charge/radius (nm) ratio for large metal ions of Th^{4+} (4/0.100=40) over smaller metal ions of UO_2^{2+} (2/0.2807)[20]. The results suggested that the neutral oxygen donors in Ph-g-NR prefer undergoing complexation with Th^{4+} than UO_2^{2+}.

4. Conclusion

Grafting polymerization of GMA onto NR in the latex form by gamma irradiation as actinides bio adsorbents were successfully developed. Grafting degree has been showed increases with increasing of monomer. FTIR analysis has confirmed the success of grafting by the presence ester group and an epoxy ring of GMA in the grafted NR. The fraction of NR latex to GMA gave effect on the grafting degree and does affect the removal efficiency of actinide ion. It was found that the modified NR has a higher tendency for Th^{4+} removal compared to UO_2^{2+} ion.

5. References

[1] Al-areqi W M, Majid A A, Sarmani S, Nor C 2015 Thorium : Issues and prospects in Malaysia p 040005.

[2] Ito A, Otake T, Shin K-C, Ariffin KS, Yeoh F-Y, Sato T 2017 Geochemical signatures and processes in a stream contaminated by heavy mineral processing near Ipoh city, Malaysia. *Appl. Geochem* **82** p 89–101.

[3] Zhang X, Gu P, Liu Y 2018 Decontamination of radioactive wastewater : state of the art and challenges forward *Chemosphere.*

[4] Humelnicu D, Dinu M V, Drăgan E S 2011 Adsorption characteristics of UO22+ and Th4+ ions from simulated radioactive solutions onto chitosan/clinoptilolite sorbents. Journal of Hazardous Materials **185**(1) p 447–55.

[5] Kaynar Ü H, Ayvacıklı M, Hiçsönmez Ü, Çam Kaynar S 2015 Removal of thorium (IV) ions from aqueous solution by a novel nanoporous ZnO: Isotherms, kinetic and thermodynamic studies *J. of Env. Radioactivity* **150**(Supplement C) p 145–51.

[6] Varala S, Kumari A, Dharanija B, Bhargava SK, Parthasarathy R, Satyavathi B 2016 Removal of thorium (IV) from aqueous solutions by deoiled karanja seed cake: Optimization using Taguchi method, equilibrium, kinetic and thermodynamic studies *J. of Env. Chemical Eng.* **4**(1) p 405–17.

[7] Boveiri Monji A, Ghoulipour V, Mallah M H, Maraghe-Mianji B 2015 Selective sorption of thorium (IV) from highly acidic aqueous solutions by rice and wheat bran *J. of Radioanalytical and Nucl. Chem.* **303**(1) p 949–58.

[8] Metaxas M, Kasselouri-Rigopoulou V, Galiatsatou P, Konstantopoulou C, Oikonomou D 2003 Thorium removal by different adsorbents *J. of Hazardous Mater.* **97**(1–3) p 71–82.

[9] Anirudhan T S, Sreekumari S S, Jalajamony S 2013 An investigation into the adsorption of thorium(IV) from aqueous solutions by a carboxylate-functionalised graft copolymer derived from titanium dioxide-densified cellulose *J. of Env. Radioactivity* **116**(Supplement C) p 141–7.

[10] Kesava Raju C S, Subramanian M S 2007 Sequential separation of lanthanides, thorium and uranium using novel solid phase extraction method from high acidic nuclear wastes. *J. of hazardous mater.* **145**(1–2) p 315—322.

[11] Gavrilescu M, Pavel LV, Cretescu I 2009 Characterization and remediation of soils contaminated with uranium *J. of hazardous mater.* **163**(2–3) p 475–510.

[12] Phetphaisit C W, Yuanyang S, Chaiyasith W C 2016 Polyacrylamido-2-methyl-1-propane sulfonic acid-grafted-natural rubber as bio-adsorbent for heavy metal removal from aqueous standard solution and industrial wastewater *J. of Hazardous Mater.* **301** p 163–71.

[13] Seko N, Ninh N T Y, Tamada M 2010 Emulsion grafting of glycidyl methacrylate onto polyethylene fiber *Rad. Phys. and Chem.* **79**(1) p 22–6.

[14] Sharif J, Mohamad S F, Fatimah Othman N A, Bakaruddin N A, Osman H N, Güven O 2013 Graft copolymerization of glycidyl methacrylate onto delignified kenaf fibers through pre-irradiation technique *Rad. Phys. and Chem.* **91** p 125–31.

[15] Madrid J F, Nuesca G M, Abad L V 2013 Gamma radiation-induced grafting of glycidyl methacrylate (GMA) onto water hyacinth fibers. *Rad. Phys. and Chem.* **85** p 182–8.

[16] Selambakkannu S, Othman N A F, Bakar K A, Shukor S A, Karim Z A A 2018 Kinetic and mechanistic study of adsorptive removal of metal ions by imidazole-functionalized polymer graft banana fiber. *Rad. Phys. and Chem.* **153** p 58–69.

[17] Mohamed N H, Ripin A 2016 Modification of flax fibres by radiation induced emulsion graft copolymerization of glycidyl methacrylate. *Rad. Phys. and Chem.*

[18] Mhatre A M, Chappa S, Chaudhari C V, Bhardwaj Y K, Pandey A K 2018 Phosphate functionalized radiation grafted Teflon for capturing and quantifications of U(VI) and Pu(IV) ions at ultra-trace concentration in aqueous samples *J. of Radioanal. and Nucl. Chem.* **317**(2) p 1141–9.

[19] Hancock R D, Martell A E 1989 Ligand design for selective complexation of metal ions in aqueous solution *Chem. Rev.* **89**(8) p 1875–914.

[20] Pan N, Li L, Ding J, Li S, Wang R, Jin Y *et al.* 2016 Preparation of graphene oxide-manganese dioxide for highly efficient adsorption and separation of Th(IV)/U(VI) *J. of Hazardous Mater.* **309** p 107–15.

Acknowledgments

Appreciation to RAYMINTEX, Nuclear Malaysia for providing required sample study, to SINAGAMA, Nuclear Malaysia for providing irradiation facilities and other Nuclear Malaysia divisions who were giving support through sample analysis using required instruments.

IOP Conf. Series: Materials Science and Engineering **785** (2020) 012030 doi:10.1088/1757-899X/785/1/012030

Radiation vulcanization of natural rubber latex by Caesium-137 source

C K Chai[1,a)] and M N Mat Lazim[1]

[1]Malaysian Nuclear Agency, Bangi, 43000 Kajang, Selangor.

[a]chai@nuclearmalaysia.gov.my

Abstract. Radiation vulcanized natural rubber latex is a rubber latex vulcanized by using ionizing radiation such as gamma radiation or electron beam. Gamma radiation is preferred over electron beam due to its higher penetration power, and Cobalt-60 is commonly used as a gamma radiation source for the irradiation process. Caesium-137 is another source for gamma radiation but it is not commonly used for RVNRL preparation despite its longer half-life. Caesium-137 retains its source activity much longer than Cobalt-60 and a Caesium-137 based irradiation facility would require less replenishment compares with the one with Cobalt-60. One of the reasons preventing the application of Caesium-137 for RVNRL preparation is its lower gamma radiation energy. This study aimed to analyze RVNRL prepared by using Caesium-137 source from a gamma cell and to compare its tensile properties with RVNRL prepared by using Cobalt-60 source. Results showed evidence of radiation induced crosslinking in RVNRL prepared by Caesium-137 source. At radiation dose of 12 kGy, the tensile properties of RVNRL prepared by Caesium-137 source is comparable to those by Cobalt-60. At radiation dose below 12 kGy, the effectiveness of Caesium-137 in radiation vulcanization is found to be better than that of Cobalt-60.

1. Introduction

Radiation vulcanization of natural rubber latex (RVNRL) is a technology for vulcanizing natural rubber latex by using high energy ionizing radiation such as gamma radiation or electron beam [1-3]. The most significant advantage of this technology when comparing to conventional sulphur vulcanization of natural rubber is the absence of sulphur and its accelerators in RVNRL. Radiation vulcanization does not require sulphur to crosslink latex as appose to sulphur vulcanization. Products made from RVNRL are free from sulphur and sulphur vulcanization accelerators residue, and as a result, these products are free from carcinogenic nitrosamine and SO_x pollutants, having low ash residue upon incineration and very low cytotoxicity [4]. In short, they are more user and environmental friendly compared to products made from sulphur vulcanized latex. In radiation vulcanization, rubber chains in latex are crosslinked via free radicals mechanism. The free radicals involved are produced from the interaction between water and ionizing radiation, a process called radiolysis [5]. The use of monomer acrylate based sensitizer such as n-butyl acrylate in radiation vulcanization has reduced the amount of required radiation from 250 kGy to less than 15 kGy [6-8], which has remarkably reduces vulcanization cost and promises commerciality of RVNRL.

Malaysian Nuclear Agency possesses the only RVNRL facility in the country. It is a pilot plant designed for demonstrating the viability of RVNRL production for industry application and technology

IOP Conf. Series: Materials Science and Engineering **785** (2020) 012030 doi:10.1088/1757-899X/785/1/012030

transfer. It is based on Cobalt-60 source for gamma radiation and capable of producing 6000 ton of RVNRL annually when it is loaded with 1 MCi of Cobalt-60. Cobalt-60 has a half-life of 5.26 years [9]. Hence, the source needs replenishment in a regular manner in order to maintain its activity and production capacity, and the cost is considerably high. Caesium-137 is a potential replacement for Cobalt-60 as it is also a gamma radiation emitter [10]. It is better than Cobalt-60 in reducing frequency of source replenishment as it has longer half-life of 30.17 years. However, radiation energy of Caesium-137 is almost 50% lower than Cobalt-60's. This leads to lower penetration power and ability to interact with medium it's travelling in. It may also affect its effectiveness as a gamma radiation emitter for RVNRL. Therefore, it is crucial to investigate the suitability and effectiveness of Caesium-137 for RVNRL preparation. The objective of this paper is to report the tensile properties of RVNRL produced by Caesium-137 gamma irradiator and to compare the properties with RVNRL irradiated by Cobalt-60.

2. Experimental

2.1. Materials
Raw latex used in this work was high ammonia latex concentrate supplied by TT Latex, Jelebu, Negeri Sembilan. Sensitizer used was n-butyl acrylate (nBA) purchased from Sigma-Aldrich, Co., USA. Stabilizer used was Octosperse KL20, a 20% fatty acid soup produced by Tiarco Chemical (M) Sdn. Bhd., Ipoh, Perak. All the stated materials were used as received.

2.2. Sample preparation
RVNRL formulation were prepared in accordance to Table 1. Sensitizer, stabilizer and water were made into emulsion, added to raw latex and stirred slowly for at least 2 hours. Formulated latex was filled into capped PE bottles of 500 ml capacity. The latex samples were irradiated by using research loop at MINTech-Sinagama and gamma cell Biobeam GM8000 for 4, 8 and 12 kGy respectively. Both research loop at MINTech-Sinagama and camma cell Biobeam Gm8000 are gamma irradiators available in Malaysian Nuclear Agency. The research loop uses Cobalt-60 while the gamma cell uses Caesium-137 as radiation source. The dose rate of the research loop and gamma cell were 2.11 kGy/hour and 0.73 kGy/hour respectively. Irradiated latex samples, now called RVNRL, were used to prepare latex dipped films for tensile tests. The films were conditioned in desiccator for at least 16 hours prior to testing.

Table 1. RVNRL formulation.

Ingredient	Amount, part per hundred rubber (phr)
Latex concentrate	100
nBA	5
Stabilizer	0.06
Water	For dilution to 52% total solid content

2.3. Tensile tests
Tensile tests were conducted in accordance with ASTM D412 standard using a Shimadzu universal testing machine model AG-XDplus, with a crosshead speed and controlled temperature of 500 mm/minute and 25±2 °C respectively. Tensile properties measured included modulus at 700 % elongation, tensile strength and elongation at break.

IOP Conf. Series: Materials Science and Engineering **785** (2020) 012030 doi:10.1088/1757-899X/785/1/012030

3. Results and discussion

Tensile properties of a latex vulcanizate such as tensile strength, modulus and elongation at break are the most basic properties referred by latex product manufacturers when judging the quality of a latex product. In the case of radiation vulcanization, changes in tensile properties under different vulcanization condition such as type of radiation, radiation dose and sensitizer are good indicators of the effects of such vulcanization condition. In this regards, the effects of gamma radiation from Caesium-137 source are investigated by studying tensile properties of RVNRL irradiated at various radiation dose.

Graphical representation of the result of modulus (at 700% elongation), tensile strength and elongation at break of RVNRL irradiated by Caesium-137 and Cobalt-60 sources at radiation dose ranged from 0 – 12 kGy are shown in Figure 1, 2 and 3 respectively. By referring to Figure 1, it is obvious that the modulus of RVNRL irradiated by Caesium-137 increases with radiation dose. Modulus of a latex article is reflected on its degree of stiffness, and it is directly proportional to crosslink density as a result of vulcanization [11]. Hence, the increase of modulus in RVNRL is an evidence of crosslinking in the latex upon exposure to gamma radiation from Caesium-137. This claim is supported by tensile strength and elongation at break of RVNRL irradiated by Caesium-137. In Figure 2, the trend of tensile strength against radiation dose of Caesium-137 is similar to modulus, which is due to the same reason, i.e. radiation induced crosslink. According to Figure 3, the change of elongation at break shows reverse trend compares to tensile strength and modulus. Elongation at break of RVNRL showed an instant increment upon irradiation but then reduced with the increase of radiation dose due to increase of embrittlement of RVNRL. This phenomenon is a consequence of increase in rigidness of rubber network as a results of radiation induced crosslink among rubber chains in irradiated latex [12]. The tensile test results of the RVNRL samples have elucidated the capability of gamma radiation from Caesium-137 in vulcanizing natural rubber latex.

Comparison of tensile properties between RVNRL prepared by using Caesium-137 and Cobalt-60 sources are also showed in Figure 1, 2 and 3. It is evident that tensile properties of RVNRL irradiated by Caesium-137 up to 8 kGy are better than RVNRL by Cobalt-60. There is no significant difference in tensile properties between both RVNRL samples at radiation dose of 12 kGy. The effectiveness of Caesium-137 gamma radiation is relatively higher than Cobalt-60 in RVNRL preparation when the latex is irradiated at radiation dose below 12 kGy. This finding is unexpected and rather on the contrary with the fact that the radiation energy and dose rate of Caesium-137 are about half and one-third of Cobalt-60's respectively. Theoretically, the rate of radiation polymerization is proportional to dose rate and monomer concentration [13]. Research also showed that vulcanization dose, i.e. the radiation dose that produces RVNRL with optimum tensile strength, of sensitized natural rubber latex irradiated with low energy electron beam decreases by increasing dose rate [14]. Obviously, besides dose rate, there is possibility of other factor or combination of factors such as radiation energy, irradiation condition and interaction between radiation and sensitizer that affect the outcome of irradiation process in a positive manner, that require further investigation.

IOP Conf. Series: Materials Science and Engineering **785** (2020) 012030 doi:10.1088/1757-899X/785/1/012030

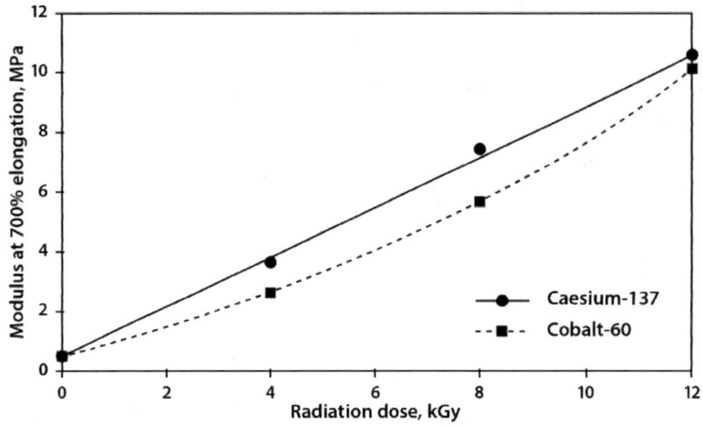

Figure 1. Modulus at 700% elongation of RVNRL irradiated by Caesium-137 and Cobalt-60 source.

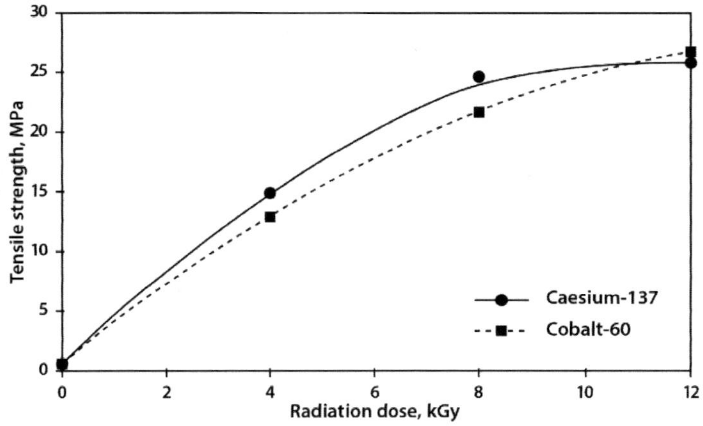

Figure 2. Tensile strength of RVNRL irradiated by Caesium-137 and Cobalt-60 source.

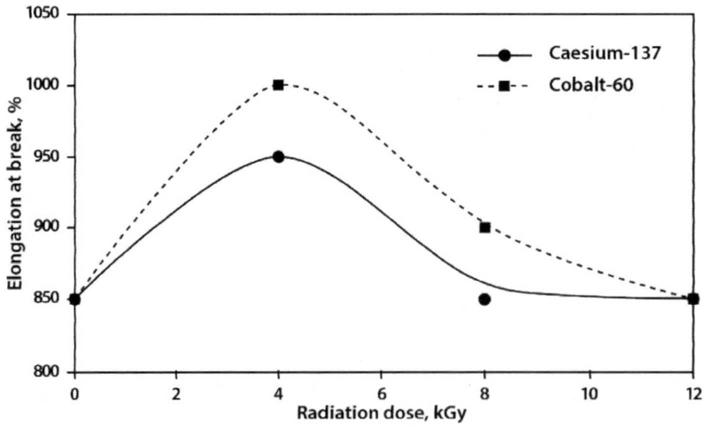

Figure 3. Elongation at break of RVNRL irradiated by Caesium-137 and Cobalt-60 source.

4. Conclusion

RVNRL can be prepared by using gamma radiation from Caesium-137 source. Caesium-137 gamma radiation is capable of inducing crosslink in rubber polymers. Despite its lower dose rate and radiation energy, the effectiveness of Caesium-137 is higher than Cobalt-60 when the source is used to vulcanize latex at radiation dose below 12 kGy. RVNRL irradiated at 12 kGy by both radiation source did not show significant difference in term of tensile properties.

5. References

[1] Minoura Y and Asao M 1961 Studies on the γ-irradiation of natural rubber latex: the effects of organic halogen compounds on crosslinking by γ-irradiation *J. of Appl. Polymer Sci.* **5**(16) p 401-407.

[2] Feroza Akhtar, Yoshii F and Makuuchi K 1990 Radiation vulcanization of natural rubber latex (NRL) using low energy electron beam accelerator *Proc. of the Int. Symp. on Rad. Vulcanization of Nat. Rubber Latex* p 210-217.

[3] Makuuchi K and Yoshii F 1995 Radiation vulcanization of NR latex with low energy electron beams *Rad. PhysiCaesium and Chem.* **46**(4-6) p 979-982.

[4] Kalyani Liyanage N M V 1999 Radiation vulcanized natural rubber latex (RVNRL) – a material with less or no protein allergens? *Bul. of the Rubber Research Inst. of Sri Lanka* **39** 33-39.

[5] Makuuchi K and Cheng S 2012 Radiation processing of polymeric materials and its industrial applications (United States: John Wiley & Sons, Inc)

[6] Chen Z H and Makuuchi K 1990 Radiation vulcanization of natural rubber latex with 3 MeV electron beams (part 1) *Proc. of the Int. Symp. on Rad. Vulcanization of Nat. Rubber Latex* p 358-367.

[7] Chen Z H & Makuuchi K 1990 n-Butyl acrylate as a sensitizer for radiation vulcanization of natural rubber latex *Proc. of the Int. Symp. on Rad. Vulcanization of Nat. Rubber Latex* p 326-335.

[8] Haque M E, Dafeder N C, Akhtar F and Ahmad M U 1996 Radiation dose required for the vulcanization of natural rubber latex *Rad. PhysiCaesium and Chem.* **48**(4) p 505-510.

[9] Zin W M W 1998 Semi industrial scale RVNRL preparation, products manufacturing and properties *Rad. PhysiCaesium and Chem* **52**(1-6) p 611-616.

[10] International Atomic Energy Agency 2006 *Gamma Irradiators for Rad. Processing* (Viena: IAEA).

[11] Alliger G and Sjothun I J 1978 *Vulcanization of Elastomers* (New York: Robert E. Krieger Publishing Company).

[12] Ratnam C T, Nasir M, Baharin A and Khairul Zaman 2000 Electron beam irradiation of epoxidized natural rubber *Nucl. Inst. and Meth. in PhysiCaesium Res. B* **171** p 455-464.

[13] Makuuchi K 2003 An introduction to radiation vulcanization of natural rubber latex (Bangkok: T.R.I. Global Co., Ltd).

[14] Chirinos H, Yoshii F, Makuuchi K and Lugao A 2003 Radiation vulcanization of natural rubber latex using 250 keV electron beam machine *Nucl. Inst. and Meth. in PhysiCaesium Res. B* **208** 256-259

Acknowledgments

Authors wish to thank operators of gamma cell and MINTec-SINAGAMA research loop for their professional assistance during the irradiation of latex samples.

IOP Conf. Series: Materials Science and Engineering **785** (2020) 012031 doi:10.1088/1757-899X/785/1/012031

Effect of low dose gamma irradiation on the tensile and thermal properties of natural rubber-based blend

N H Lazim[1], and S A Shamsudin[1,a)]

[1] Centre for Frontier Sciences, Faculty of Science and Technology, Universiti Kebangsaan Malaysia, 43600 Bangi, Selangor, Malaysia

[a]aisyah@ukm.edu.my

Abstract. Natural rubber (NR) is a renewable natural resource that has been utilized in many applications. Due to its low tensile strength, NR is commonly blended with another polymer to improve the properties of NR. However, most of the blended NR has miscibility issue, leading to poor mechanical properties. Gamma irradiation (GR) is a safe and green alternative than conventional chemical method for inducing crosslinking in blended NR. This study was aimed at investigating the mechanical and thermal properties of NR/liquid natural rubber/polystyrene-*block*-polyisoprene block copolymer (NR/LNR/PS-*b*-PI) blend using GR method. The polymer blend is prepared according to the ratio of NR:LNR:PS-*b*-PI of 6:1:3 using an internal mixer. The NR/LNR/PS-*b*-PI blends are irradiated with gamma ray at doses of 0, 5, 6, 7, 8 and 9 kGy. Tensile and thermal tests are conducted to evaluate the properties of unirradiated and irradiated polymer blend. The tensile strength and modulus of the irradiated blend are increased with increasing irradiation dose before decrease at 9 kGy. Lower melting point of blend indicates that PS-*b*-PI are compatible in NR/LNR blend. The improvement in tensile properties of irradiated NR/LNR/PS-*b*-PI blend is related to the radiation-induced crosslinking.

1. Introduction

Demand for natural rubber (NR) in various sectors has increased owing to its properties such as high elasticity, low cost, good wear and weather resistance. However, NR has limited temperature stability and low tensile strength. Polymer blending technique is an alternative to improve NR performance in which two or more polymers are combined to produce new polymer blend with desired properties. The blending of natural rubber with other elastomers and polymer such as liquid natural rubber (NR/LNR) [1], polylactic acid (NR/PLA) [2], and styrene-butadiene rubber (NR/SBR) [3] have shown improvement in tensile properties compared to neat NR.

It is known that the main problem associated with polymer blend is the immiscibility between polymer pairs that led to poor mechanical properties. Hence, many studies have been carried out to improve the miscibility of NR by adding NR with chemical compatibilizer such as maleic anhydride, methyl methacrylate and glycidyl methacrylate [4-6]. Although the properties of blended polymer may be improved, the high cost of compatibilizer materials increase production costs [7]. The addition of block copolymers (BCP) in the NR-based polymer blend has been widely studied [8-9]. Polystyrene-*block*-polyisoprene (PS-*b*-PI) is a type of BCP consists of a glassy domain (PS) and rubbery part (PI). Chattopadhyay S. et. al (2001) blended natural rubber with polystyrene and reported that the addition of PS-*b*-PI has increased mechanical properties of the blend [8].

Gamma irradiation (GR) is a green and fast modification technology to induce crosslinking in polymer blend. GR has short curing time, low cytotoxicity, high productivity, and no residue in final products [10]. GR can modify the structure of materials by causing several reactions to occur in the system during irradiation such as chain crosslinking, scission and branching [11]. The property improvement of the polymer blend, however, depends on the irradiation dose of γ-irradiation treatment.

The main objective of this study is to evaluate the effect of the low gamma irradiation doses on the tensile and thermal properties of natural rubber/liquid natural rubber/polystyrene-*b*-polyisoprene (NR/LNR/PS-*b*-PI) polymer blend.

2. Description of the experiment and research methods

In this work, method of polymer blending and gamma irradiation have been chosen to fabricate irradiated NR/LNR/PS-*b*-PI blend. As experimental samples, the natural rubber (NR), liquid natural rubber (LNR) and polystyrene-block-polyisoprene (PS-*b*-PI) have been used. NR, in 62% solid content is purchased from Mardec Sdn Bhd and has been kept in ammonia to prevent degradation. The LNR, made through depolymerization of NR has been supplied by the Polymer Processing Laboratory, National University of Malaysia. The PS-*b*-PI block copolymer has been supplied by Takenaka Laboratory, Kyoto University, Japan. The preparation of the blended NR/LNR/PS-*b*-PI consist of blending NR: LNR: PS-*b*-PI at a ratio of 6:1:3 by weight percentage (wt.%) by using an internal mixer machine (Plasti-Corder, Brabender) at temperature of 190°C at a screw speed of 100 rpm for 15 minutes. The polymer blend has been kept overnight to achieve equilibrium expansion. Samples for the tensile test have been prepared using the hot press machine (LP 50, LabTech). The gamma irradiation has been carried out at a radiation facility in Malaysian Nuclear Agency, Selangor, Malaysia using Iridium-192 source. The samples have been irradiated at different doses of 5,6,7,8 and 9 kGy in room temperature. Experimental studies are carried out to determine the optimum irradiation dose on the mechanical and thermal properties of irradiated NR/LNR/PS-*b*-PI blend. Tensile test has been carried out using the Universal Testing Machine (Model M350-10CT) in a three-point bending mode with a cell load of 5 kN. For thermal test, Differential Scanning Calorimetry (Mettler Toledo DSC 882e) has been used to determine the melting point of the samples. 5 mg of samples are heated from -50°C to 300°C with 10°C/min of heating rate. All data are then analyzed for discussion.

3. Results and discussion

Based on the results of the experimental studies, the effect of irradiation dose on the mechanical and thermal properties of NR/LNR/PS-*b*-PI polymer blend has been analyzed.

As shown in Figure 1 (a), it is observed that the tensile strength of NR/LNR blend is improved with the addition of block copolymer, PS-*b*-PI. The primary cause for the improvement in tensile strength could be due to the PS-*b*-PI has enhanced the compatibility of the polymer blend and help to increase the strength of the rubber. Similar effect of PS-*b*-PI as compatibilizer has been found in NR/PS blend [8]. The addition of PS-*b*-PI also increases the Young's modulus of NR/LNR blend from 0.17 MPa to 0.91 MPa, as shown in Figure 1 (b). Upon gamma irradiation on the polymer blend and increasing the irradiation doses, a remarkable increase in tensile strength and modulus of irradiated samples can be observed. The maximum tensile strength and Young's modulus is obtained in the sample irradiated with dose of 8 kGy, having values of 0.25 MPa and 3.4 MPa, respectively. Apparently, the Young's modulus of 8 kGy-irradiated sample is 273% higher than unirradiated NR/LNR/PS-*b*-PI blend. This improvement is associated to the radiation-induced crosslinking [12].

It is discovered that the tensile strength and Young's modulus decreased to 0.21 MPa and 2.98 MPa, respectively, when the polymer blend has been irradiated with a 9 kGy dose. The poor tensile properties suggest that the 9 kGy irradiation dose has promoted chain scission and eventually degrades the NR/LNR/PS-*b*-PI polymer blend.

IOP Conf. Series: Materials Science and Engineering **785** (2020) 012031 doi:10.1088/1757-899X/785/1/012031

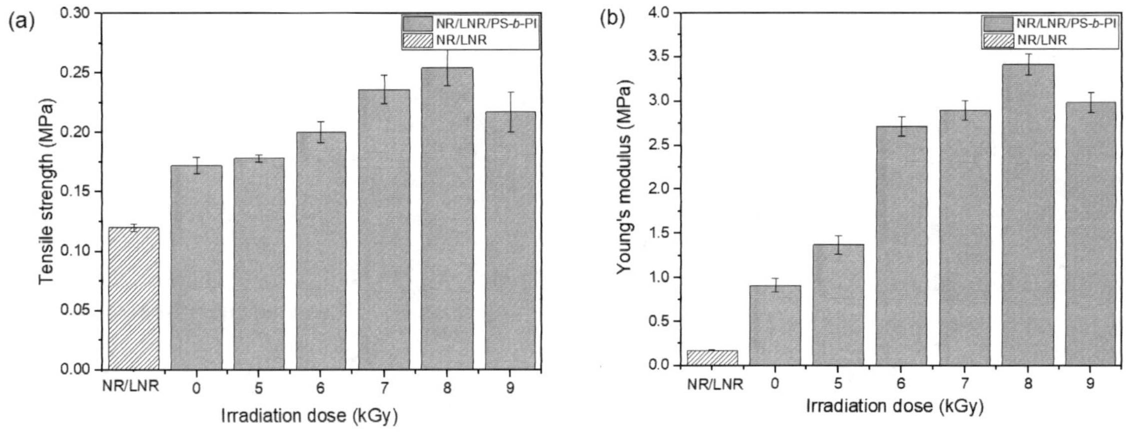

Figure 1. (a) Tensile strength and, (b) Young's modulus of NR/LNR blend and NR/LNR/PS-*b*-PI at various irradiation doses.

The DSC melting thermograms of the neat PS-*b*-PI and NR/LNR/PS-*b*-PI polymer blend is shown in Figure 2. The melting temperature (T_m) has been determined at the base of the melting endothermic peak. The T_m of the neat NR/LNR cannot be determined due to its T_m is found lower than -60 °C [13]. The T_m for PS-*b*-PI and NR/LNR/PS-*b*-PI polymer blend is at 152.67 °C and 68.33°C, respectively. The decrease of the T_m of the polymer blend explains that the PS-*b*-PI is miscible in the NR/LNR blend during the blending process. For the γ-irradiated polymer blend, a slight increase in the T_m is observed, except for 9 kGy-irradiated blend sample. The increase in T_m of irradiated samples implied that the chain crosslinking has been generated by the irradiation procedure. However, an excessive irradiation dose of 9 kGy decreases the T_m of NR/LNR/PS-*b*-PI to 57.33°C.

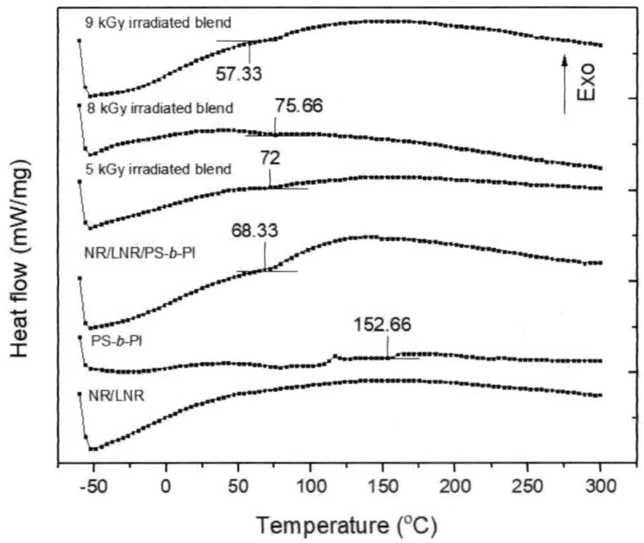

Figure 2. DSC thermograms of the neat NR/LNR, PS-*b*-PI and NR/LNR/PS-*b*-PI polymer blend at various irradiation doses.

4. Conclusion

This research study has been done to investigate the effect of various irradiation doses on the mechanical and thermal properties of the natural rubber/liquid natural rubber/polystyrene-*block*-polyisoprene polymer blend using the gamma irradiation method. It is shown that adding 3 wt% of block copolymer PS-*b*-PI and low irradiation doses (5,6,7, and 8 kGy) have increased the tensile strength and Young's modulus of polymer blends. The low doses of gamma irradiation have enhanced the chain crosslinking of the NR/LNR/PS-*b*-PI polymer blend. It has been shown that 9 kGy irradiation dose decreases the tensile properties due to degradation process. From thermal test, a lower melting point found in NR/LNR/PS-*b*-PI implied that PS-*b*-PI is compatible in the NR/LNR blend. From this study, it is possible to determine the optimal irradiation dose which will allow maximum tensile and thermal properties improvement in NR/LNR/PS-*b*-PI polymer blend.

5. References

[1] Rosli N A, Ahmad I, Anuar F H and Abdullah I 2016 *Polym. Test.* doi: 10.1016/j.polymertesting.2016.07.021.
[2] Buys Y F, Aznan A N A and Anuar H 2018 *IOP Conf. Ser.: Mater. Sci. and Eng.* **290** p 012077.
[3] Jovanović S, Samaržija-Jovanović S, Marković, G, Jovanović V, Adamović T and Marinović-Cincović M 2018 *J. of Thermo. Comp. Mat.* **31** p 265-287.
[4] Rosli N A, Ahmad I, Anuar F H and Abdullah I 2019 *Carbo. Polym.*
[5] Bakar R A and Fauzi M S 2012 *J. Chem. Eng.* **6** p 962-966.
[6] Juntuek P, Ruksakulpiwat C, Chumsamrong P and Ruksakulpiwat Y 2012 *J. Appl.Polym. Sci.* **125** p 745-754.
[7] Lazim N H, Samat N 2017 *Procedia En.g* **184** p 538-543.
[8] Chattopadhyay S and Sivaram S 2001 *Polym. Int.* **5** p 67-75.
[9] Wang H, Zhang X, Nie H, Wang R and He A 2019 *Comp. Part A: Appl. Sc. and Mfg.* **116** p 197-205
[10] Manaila E, Stelescu M D and Craciun G 2012 *Adv. Elast.—Tech., Prop. and App.* p 3–34
[11] Stelescu M D, Airinei A, Manaila E, Craciun G, Varganici N F C, Pamfil D and Doroftei F 2018 Polymers **10** p 1206.
[12] Pongsathit S and Pattamaprom C 2018 *Rad. Phys. and Chem.* **144** p 13-20
[13] Ibrahim P, Daik R and Zin W W M 2016 *Rad. Eff. and Def. in Sol* **171**(11-12) p 1006-1015

Acknowledgements

The authors wish to thank the Ministry of Higher Education, Malaysia for the grant FRGS1/2018/TK05/UKM/02/13, which has made this research work possible, the Universiti Kebangsaan Malaysia (UKM) for the grant GUP-2017-109 for their support of this research and Agensi Nuklear Malaysia for the gamma irradiation facility.

IOP Conf. Series: Materials Science and Engineering **785** (2020) 012032 doi:10.1088/1757-899X/785/1/012032

Determination of WUBHS speed range for electron beam irradiation of wire and cable

Siti Zulaiha Hairaldin[1,a)], Ruzalina Baharin[1], Shahrina Akma Mansur[1], Mohd Suhaimi Jusoh @ Yusoff[1], Mohamad Zulhailmee Mohd Zainal[1], Azmi Ali[1]

[1]Electron Beam Irradiation Centre (ALURTRON), Malaysian Nuclear Agency, Bangi, 43000 Kajang, Selangor, Malaysia.

[a]sitizulaiha@nm.gov.my

Abstract. The electron beam machine was installed in 1991 at Nuclear Malaysia Agency. There are about one million meters of wire and cable irradiated in the facility every year. Electron beam irradiation introduces to crosslink the polymer material for high-performance wire and cable for the automobile industry. ALURTRON is facing few challenges while of irradiating wire and cable to determine the speed and to comply with the product quality (QC) test. The uncertainty of Wire Under Beam Handling System (WUBHS) speeds while irradiating wire and cable due to different situations. In this study, Polyethylene (PE) and Polyvinylchloride (PVC) wire and cable irradiated with electron beam irradiation at suitable speed based on the material, size and calculation etc. The sample analyse with hot set and gel content test. The speed decided at range 220-232, 180-196 and 180-194 m/min for 6.4 mm tube, 9.5 mm tube and PVC wire respectively.

1. Introduction

The EPS3000 providing irradiation services for various product treatment and purposes. Few type of product irradiated in the facility such as silicon wafer, wire, tubing, polymer, hydrogel, electronic component, gemstone [1], waste water treatment etc. ALURTRON is certified to ISO 9001 Quality Management System (QMS) since 2003. With the implementation of ISO 9001:2015 at the facility the data record becomes more systematic and structured. Quality assurance ensures the intended process (machine, material, man method) is available to be carried out and compliance with the standard. Meanwhile, quality control is a process to control the output of the process to minimize the variation of quality, avoid reject or undelivered products.

Radiation for an insulated wire and heat shrinkable tube has been widely used for the last 50 years, starting in 1950s [2]. In ALURTRON, we started to irradiate for wire & cable since the installation of Wire Under Beam Handling System in 1994 [3]. Usually, polymeric material with high molecular weight is used for electron beam irradiation to improve the material properties. Few types of polymer widely used electron beam accelerator in commercial industry: Polypropylene (PP), Polyethylene (PE), Polyvinylchloride (PVC), nylon, Ethylene-propylene-diene (EPDM), etc. The 3.0 MeV EPS3000 equipped with product handling system. The Product handling system is mechanical equipment that moves products from one location to the irradiation room. Two different types of product handling available in ALURTRON: 1) toe conveyer and 2) Wire Under Beam Handling System (WUBHS). The toe conveyer is used for flat product while the WUBHS is for wire & cable handling.

The product needs to pass the test at a certain range. In this study, the gel test will be matched with the hot set result and will be used as a reference by the electron beam (EB) machine operator. However, the problem with irradiation of wire and cable is the uncertainty of speed while irradiating different batch of samples. The speed keeps changing and not fixed at only one speed. ALURTRON is committed to ISO standard and the implementation. Therefore any change must be recorded and shall be evaluated occasionally. This paper aims to determine the speed range of WUBHS processing parameter for data record improvement

2. Material and Methods

Samples of wire and cable diameter size irradiated with electron beam machine (EPS 3000) according to irradiation process procedure in Quality Manual. Material type of wire and tube is polyvinylchloride (PVC) and polyethylene (PE) respectively (Table 1).

Table 1. Samples of wire and tube irradiation.

Samples	Material	Diameter size (mm)
Wire	PVC	0.5, 0.85, 1.25
Tube	PE	6.4
Tube	PE	9.5

The parameter of the electron beam machine is decided by considering a few factors which is material, size, hot set test, and gel content result, etc. The speed of WUBHS estimated by the equation:

$$\text{Speed (m/min)} = \frac{\text{Beam Current x No. Of Turns x K Factor}}{\text{Dose (Mrad)}} \quad (1)$$

The speed calculated based on the required dose and quality test data. Two types of test available in the QA/QC laboratory for wire and cable testing: 1) hot set and 2) gel content. The hot set test measured the heat expansion behaviour of the material. The test conducted as a preliminary test and used as a preference due to quick and reproducible results. The gel content test measures the crosslinked degree amount of a polymeric material and it is a compulsory test required by the company. However, the gel test takes about 5 days to complete. The hot set test takes a shorter time compared to the gel content test, which is about 25-40 minutes per sample only. Therefore, the hot set test used as a preliminary reference for QC process due to the shorter time required.

The diameter of the sample measured using the micrometer caliper (MITUTOYO/CD-15APX, Japan). UV-VIS spectrophotometer machine (UV-VIS Spectronic Genesis 5, Thermo Electron Corporation) to measure absorbed dose by the product. Cellulose triacetate (CTA) dosimeter film (FTR-60, Fujifilm, Japan) used as a dosimeter due to wide range of dose 5-300 kGy and provide reliable data and suitable for our facility application.

Load mass for the hot set test was determined by balance (Sartorius/BSA 6202S) Germany. The dumbell cutter used to cut the 9.5 mm tube. Samples were put in Kendro oven at a specific temperature. Stopwatch to measure 15 minutes hot set test heating timing following procedure of hot test test (Quality manual). Elongation of the sample measured by laser teleheight equipped with an oven. Tetrahydrofuran (THF) and Xylene used as a solvent for PVC wire and PE tube respectively following procedure. All data recorded, compiled and analyzed.

IOP Conf. Series: Materials Science and Engineering **785** (2020) 012032 doi:10.1088/1757-899X/785/1/012032

3. Material and Methods

Samples of wire and cable diameter size irradiated with electron beam machine (EPS 3000) according to irradiation process procedure. Material type of wire and tube is polyvinylchloride (PVC) and polyethylene (PE) respectively (Table 2).

Table 2. Samples of wire and tube irradiation.

Samples	Material	Diameter size (mm)
Wire	PVC	0.5, 0.85, 1.25
Tube	PE	6.4
Tube	PE	9.5

The diameter of the sample measured using the micrometer caliper (MITUTOYO/CD-15APX, Japan). UV-VIS spectrophotometer machine (UV-VIS Spectronic Genesis 5, Thermo Electron Corporation) to measure absorbed dose by the product. Cellulose triacetate (CTA) dosimeter film (FTR-60, Fujifilm, Japan) used as a dosimeter due to wide range of dose 5-300 kGy and provide reliable data and suitable for our facility application.

Load mass for the hot set test was determined by balance (Sartorius/BSA 6202S) Germany. The dumbell cutter used to cut the 9.5 mm tube. Samples were put in Kendro oven at a specific temperature. Stopwatch to measure 15 minutes hot set test heating timing following the Hot Set procedure (Quality manual). Elongation of the sample measured by laser teleheight equipped with an oven. Tetrahydrofuran (THF) and Xylene used as a solvent for PVC wire and PE tube respectively. All data recorded, compiled and analyzed.

4. Results

The Speed data of each sample indicate in Figure 2. The lot no refers to the batch delivery of the product in the current month. The uncertainty of WUBHS speeds while irradiating wire and cable due to several reasons. The WUBHS Speed is fluctuating. It is not convenient for the operator to fill in the process change form each time the speed change. The variation of speed may be due to 1) different batch of product and involves different blending process. Another crucial factor also contributes to changing of speed which is 2) different supplier of raw material and 3) change of formula etc. Situation 2) and 3) is in control, the customer will inform us if they do change the formula or supplier. However, for situation 1) is very subjective which is no reference and indication of record available/provided by customer to refer to.

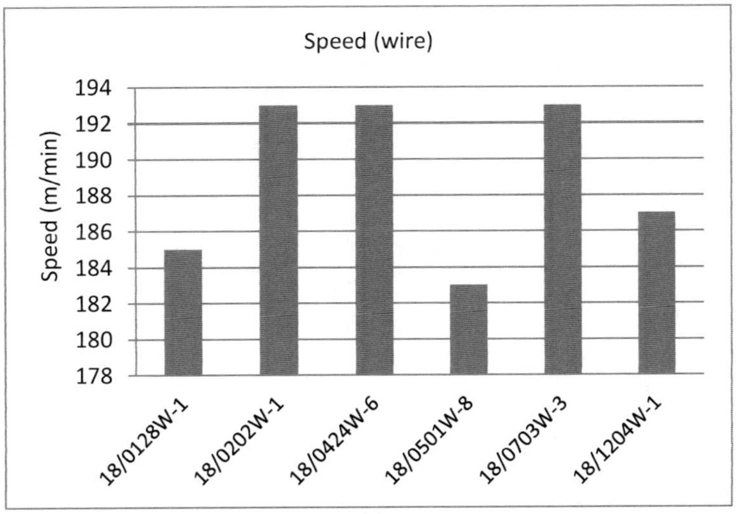

(a)

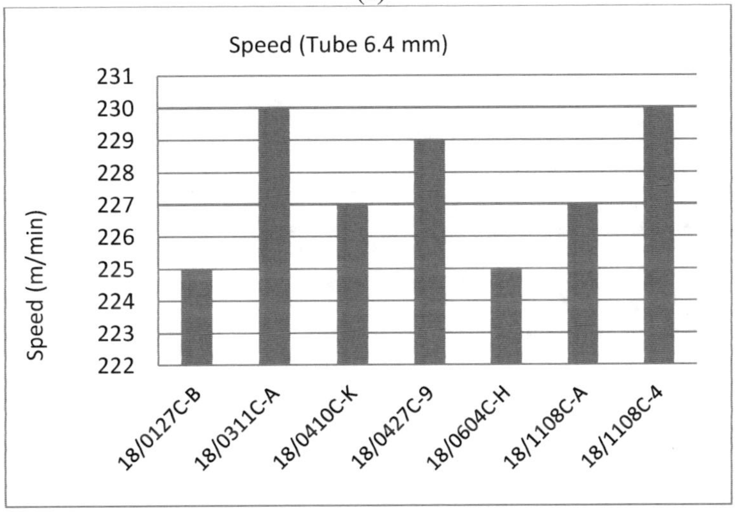

(b)

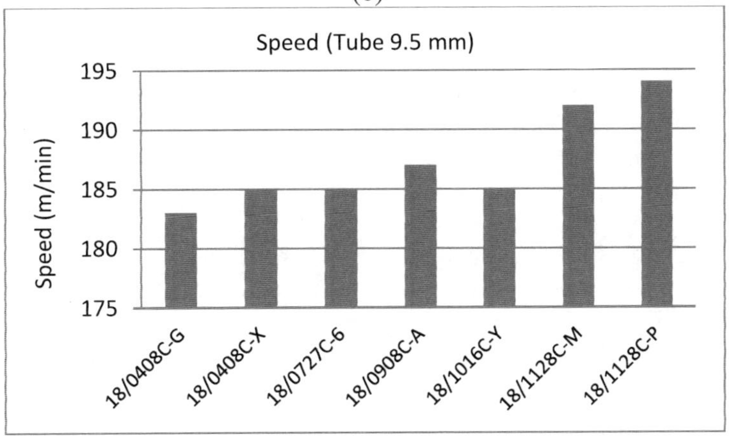

(c)

Figure 1 (a-c). WUBHS speeds data analysis for wire and cable.

Parameter for electron beam processing for wire decided in Table 2. The word "Before" refers to the last parameter used (single speed data record) and "After" is refer to the flexible speed with minimum and maximum speed. This data recorded in the irradiation process control form (EBM/FO/OP/02) in Table 3 and 4. Therefore, the process of data recording and filing is more effective by applying the range of speed data record instead of a single speed.

Table 3. The Parameter of electron beam processing for wire.

Product		Before	After
Wire (all size)	Parameter/size (mm)	PVC	PVC
	Acc. Voltage (MeV)	1.0	1.0
	Beam Current (mA)	30	30
	No. turns	54	54
	Dose (kGy)	90	90
	Speed (m/min)	187	180-194

Parameter of tubes decided as indicate in Table 4:

Table 4. The Parameter of electron beam processing for 6.4 mm tube and 9.5 mm tube

Product		Before		After	
Tube	Parameter/sample size (mm)	6.4	9.5	6.4	9.5
	Acc. Voltage (MeV)	2.5	2.5	2.5	2.5
	Beam Current (mA)	30	30	30	30
	No. turns	64	64	64	64
	Dose (kGy)	160	190	160	190
	Speed (m/min)	230	194	220-232	180-196

5. Conclusion

Tube and wire product speed are decided at range 220-232, 180-196 and 180-194 m/min for 6.4 mm tube, 9.5 mm tube and PVC wire respectively. Speed suggested being recorded as a range type instead of a single data speed parameter.

6. References

[1] Idris S, Hairaldin S Z, Tajau R, Karim J, Jusoh S, and Ghazali Z, Ahmad S 2015 An experience of electron beam (EB) irradiated gemstones in Malaysian Nuclear Agency *AIP Conf. Proc.* **1584**(1) p 170-174.

[2] Cleland M R 2006 Industrial applications of electron accelerators *CERN Accelerator School:Small Accelerators*, *Proc.* p 383-416.

[3] Hashim S A, Jahar S, Muhamad A 2011 Electron beam cross linking service for tubular products: Malaysian Nuclear Agency's experience *Nucl. Tech. Conv. 2011* (Malaysia).

[4] Quality Manual 2019 Electron Beam Irradiation Centre (ALURTRON) (Bangi: Malaysian Nuclear Agency)

Acknowledgment
The authors wish to express their sincere thanks to Electron beam Irradiation Centre (ALURTRON) facility and personnel who have worked and contributed to carrying this experiment.

Development of marine grade radiation cross-linked high density polyethylene (HDPE) floater pontoon material

Muhammad Hazim Muhammad Sayuti[1], Meor Yahaya Razali[1], Mohd Faizal Abd Rahman[1], Siti Salwa Mohammad Shirajuddin[1], Mohd Roslie Ali[1], Ahmad Zuhdi Mohd On[1],

[1]Radiation Processing Technology Division, Malaysian Nuclear Agency, 43000, Kajang, Selangor

hazim@nuclearmalaysia.go.my

Abstract. Crosslinked High Density Polyethylene (HDPE) was developed using electron beam radiation techniques. The crosslinked HDPE can be used as marine grade cross-linked HDPE floater pontoon material. Thermoplastic resin, HDPE blended with Ethylene Vinyl Acetate (EVA) in various weight percentage compositions before being radiate with electron beams irradiation. These composites showed significant changes in mechanical properties and temperature stability after the irradiations. From this study, the optimum dose for HDPE and HDPE/EVA blends will be obtain. This study also will provide the optimum irradiation dose in order to have a good, high strength and impact for the production of pontoon material. The present of EVA in the blend will act as impact modifier also being investigated in this study. From the study, the result showed some significant changes in thermal and mechanical properties after being irradiated with electron beam. The comparison of the result between these composite composition of irradiated and non-irradiated HDPE composite used for floater pontoon material is presented in this paper

1. Introduction

High Density Polyethylene (HDPE) is a thermoplastic that provide a high strength-to-density ratio, and has been widely used in the production of plastic bottles, pipes, geomembranes, and plastic lumber. HDPE is known for its large strength to density ratio. The density of HDPE can range from 930 to 970 kg/m^3. Although the density of HDPE is only marginally higher than that of low-density polyethylene, HDPE has little branching, giving it stronger intermolecular forces and tensile strength than Low Density Polyethylene (LDPE) [1]. The difference in strength exceeds the difference in density, giving HDPE a higher specific strength. It is also harder and able to withstand somewhat higher temperatures (120 °C/ 248 °F for short periods) [2].

In order to obtain more excellent material properties of HDPE, Ethylene vinyl acetate copolymer (EVA) can be used to strengthen HDPE. Ethylene vinyl acetate copolymer (EVA) being used in this project are as an impact modifier for improving impact properties due to the presence of olefinic segments and interactive vinyl acetate groups in EVA[3]. In addition, the vinyl acetate concentration (VA %) may provide an additional variable to EVA for its role as impact modifier of HDPE. Furthermore, the glass transition temperature of EVA is quite low, which also indicates the possibility of achieving better low temperature impact properties for the HDPE/EVA blend

Radiation technology used is based on the use of high energy ionizing radiation, mainly high energy electrons (0.2-10 MeV) from electron accelerators as a source of energy in different industrial applications. The scientific basis of radiation technology is the reactive chemical species produced by absorbed radiation resulting in changes of chemical, physical and biological properties.

In many industrial applications radiation processing has proven to be a technology of choice either because of its economic competitiveness or its technical superiority. Although the chemical effects of ionizing radiation have been known for more than a century its industrial applications become possible only after the availability of reliable gamma source and powerful electron accelerator during the last couple of decades. Worldwide there are approximately 200 Co-60 gamma irradiators and 100 electron beam accelerators being used in major fields of industry namely health-care, polymers, food and environmental applications [4]. This project is intended to develop irradiated HDPE and of High Density Polyethylene (HDPE) and Ethylene vinyl acetate (EVA) (HDPE/EVA) blends for pontoon material purposes. The main objective of blending is to enhance the material properties that will meet the sufficient material requirement for HDPE/EVA resin. Pontoon or floats are the airtight hollow structures, similar to pressure vessel, designed to provide buoyancy in water. Their principal applications are in marine engineering applications such as watercraft hulls, aircraft floats, floating pier and pontoon bridge construction. So, in order to meet the standard of pontoon production, it must have a high strength and impact properties and also high temperature resistance.

For the blending of HDPE/EVA, the blends will be processed using different formulation in high speed mixer. The process then will be followed using electron beam in order to expose the blends onto various kinds of doses before characterized. In order to characterize and investigate the properties of this polymer blend, the testing will be involved in aspect of mechanical, thermal and morphological. The blends will undergo Gel Content, DSC, tensile, flexural, hardness and impact testing.

2. Experimental

2.1. Materials

High Density Polyethylene (HDPE), Ethyl Vinyl Acetate (EVA) and Trimethylolpropane triacrylate (TMPTA) was used in this project. HDPE that been used in the study is HDPE HD5201AA as it is for extrusion grade material.

2.2. Preparation of crosslinked HDPE

Crosslinking of HDPE was carried out by electron beam in various doses 50, 100, 150 and 200 kGy respectively. HDPE pellet been dried first at 60°C for 24 hours to eliminate moisture. The HDPE pellets were blends with the internal mixer melting for about 10 minutes before the 3 weight percent of TMPTA was added. The mixture compounded using internal mixer with a 190°C in temperatures and screw speed of 50 rpm. The compound been allowed to cool in room temperature for a few minutes before subsequently been hot pressed into a 1mm and 4mm thickness sheet using hot press machine. The hot press machine operated at 170°C for about 15 minutes. The press samples undergo the electron beam irradiation at various doses which are 50, 100, 150 and 200 kGy respectively. The prepared samples are being cut before being used for the characterizations. The cross-linked HDPE samples being prepared to the standard and requirements of each characterization respectively. The process flow details showed in the figure 1 below.

IOP Conf. Series: Materials Science and Engineering **785** (2020) 012033 doi:10.1088/1757-899X/785/1/012033

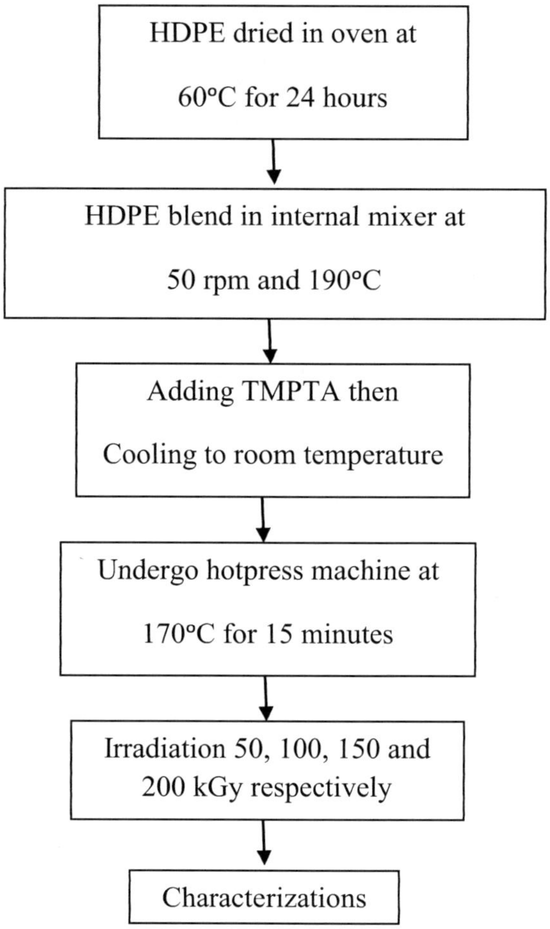

Figure 1. Process flow of preparation of crosslinked HDPE.

2.3. Preparation of HDPE/EVA blends with crosslinker agent

The project continued with blending of HDPE and EVA blend with present of TMPTA. Firstly, the HDPE and EVA pellets put into the internal mixer and let it to melt before the TMPTA was added. The compositions of TMPTA that used for the mixing with HDPE/EVA is 3 weight percent meanwhile the HDPE/EVA is 90/10 weight percent. The mixture compounded using internal mixer with a 190°C in temperatures and screw speed of 50 rpm. The compound was allowed to cool in room temperature for a few minutes before subsequently been hot pressed into a 1mm and 4mm thickness sheet using hot press machine at 170°C for about 15 minutes before being irradiated with 50, 100, 150 and 200 kGy. The prepared sample then being characterized and the optimum composition of HDPE/EVA were collected. The process flow details showed in the Figure 2 below.

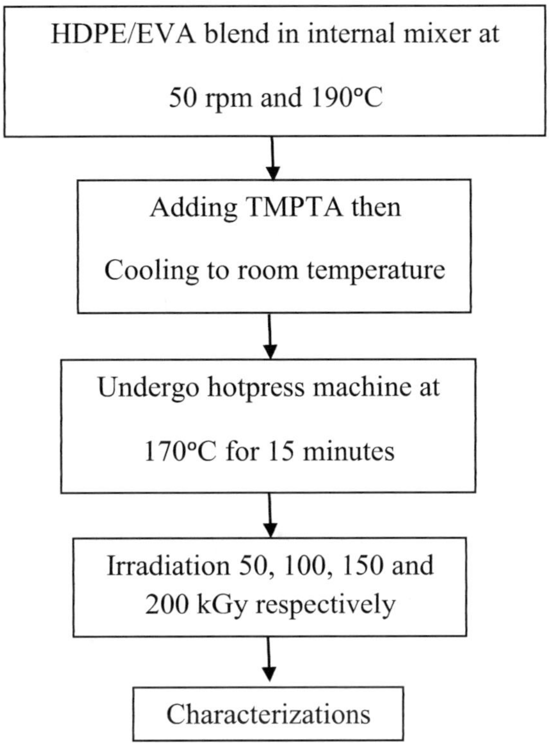

Figure 2. Process flow of preparation of HDPE/EVA blends with crosslinker agent.

2.4 Gel Content

The crosslink degree of HDPE and HDPE/EVA was determined by a gel content method. The products obtained from electron beam were cut into 0.2 cm^2 with three time repetitions before being put into a wire mesh. Then, the samples prepared were being boiled in 40 mL xylene, at 60°C. The solution was being left for about 24 hours. Then, the sample were collected and washed by ethanol to get rid the excessive ethanol after being boiled. The samples then were being dried at 70°C in a vacuum oven for 4 hours.

To measure the crosslink degree, the dried sample then collected and measured it final weight in order to obtain the mass of sample that do not dissolve, that indicate the crosslink part of the samples. The crosslink degree was calculated using the following equation. Each sample were tested in triplicate and average values were reported.

$$\text{Gel Content} = (\text{Wt. of sample after boiling}) / (\text{Wt. of sample}) \times 100\% \quad (1)$$

2.5 Differential Scanning Calorimetry (DSC)
The miscibility of the blends was investigated by means of differential scanning calorimetry in inert atmosphere. The sample of 4 to 7 mg will be pressed in Aluminium pan and were being heated at a rate of 10°C/min from 25°C to 200°C. The crystallization (T_c) and melting temperature (T_m) had being taken as those corresponding to the peak values of the crystallization exotherms and melting endotherms, respectively. The percentage of crystallinity X_c was calculated using the following equation:

$$\%\text{Crystallinity} = \frac{\Delta Hm}{f\Delta Hm\ ideal} \times 100\% \quad (2)$$

where, ΔH_m is measured heat fusion, f is the weight fraction of the component in equation, and ΔH_m ideal is the entalphy of fusion for 100% crystalline (ΔH_m (PLA) =288 J/g).

2.6 Tensile Test
Tensile test was carried out according to ASTM D638 (Standard Test Method for Tensile Properties). The tensile properties were measured at environment temperature and humidity on the universal testing machine. The testing was being performed under ambient conditions. The cross head speed was fixed in a value of 50 mm min^{-1}. Five specimens for each formulation were tested and average values will be recorded.

2.7 Hardness Test
Hardness test were done according to ASTM D2785 by using Shore D Rockwell Hardness tester. The test was done under ambient conditions. The samples standard dimensions are 6mm. Hardness values were being calculated. Five specimens were tested and the average value was being calculated.

2.8 Impact Test
The Izod impact test was carried out according to the ASTM D256 (A) standard test method to determine the pendulum impact resistance of notched specimens for the composites. The testing machine, Izod impact tester consists of a heavy base with a vise for clamping the specimen in place during the test. The samples is in rectangular bar which was held vertically at half of it length and notch according to standard. Five specimens of each formulation will be tested and the average impact strength was calculated.

3. Results and discussion

3.1. Characterization of HDPE

3.1.1. Gel Content. From the graph shown in Figure 3, the degrees of crosslinking were increased with increasing of irradiation dose. From 0% of gel content degree, it is increased up to 70% when being irradiated with electron beam with dose of 50, 100, 150 and 200kGy respectively. This showed that HDPE blends undergo the crosslinking process throughout the irradiation using electron beam.

IOP Conf. Series: Materials Science and Engineering **785** (2020) 012033 doi:10.1088/1757-899X/785/1/012033

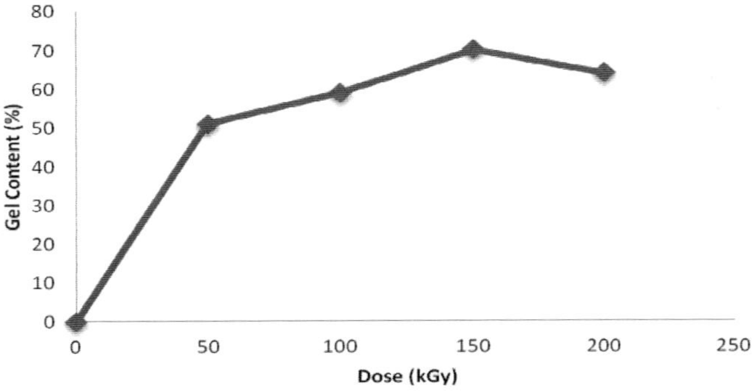

Figure 3. The crosslink degree of HDPE.

Crosslink with 3 phr TMPTA with 50kGy irradiation dose have a 51% of crosslink degree and increase slightly when the sample bombarded with 100kGy. With the presence of 100kGy of electron beam irradiation, the crosslinking degree is about 59%. For dose 150kGy, the result also showed the increment in crosslinking degree from 59% to 70% and it decrease slightly with the 200kGy as shows in the Figure 1. This shows the irradiation dose of 150kGy of HDPE is the optimum dose for HDPE crosslinking. Increasing the dose would enhance the chances of crosslinking process occurs in the backbone of HDPE, hence the crosslink degree increased [5]. Since the available sites on the HDPE backbone are limited, the higher dose of electron beam irradiation will degrade the rather than crosslink.

3.1.2. Differential Scanning Calorimetry (DSC). The thermal properties of samples were determined by DSC analysis. Figure 4, 5, 6, 7 and 8 shows the DSC curve of heat flux versus temperature at different dose. The value of melting temperature, Tm, glass transition temperature, Tg and degree of crystallinity, %Xc of each samples were summarized in Table 1.

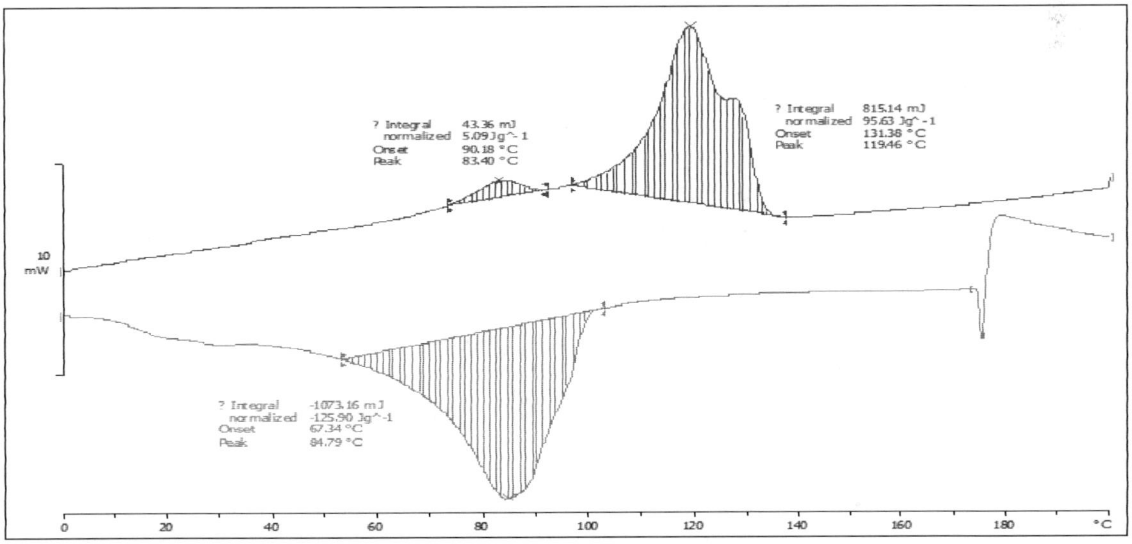

Figure 4. DSC thermogram at of HDPE pure.

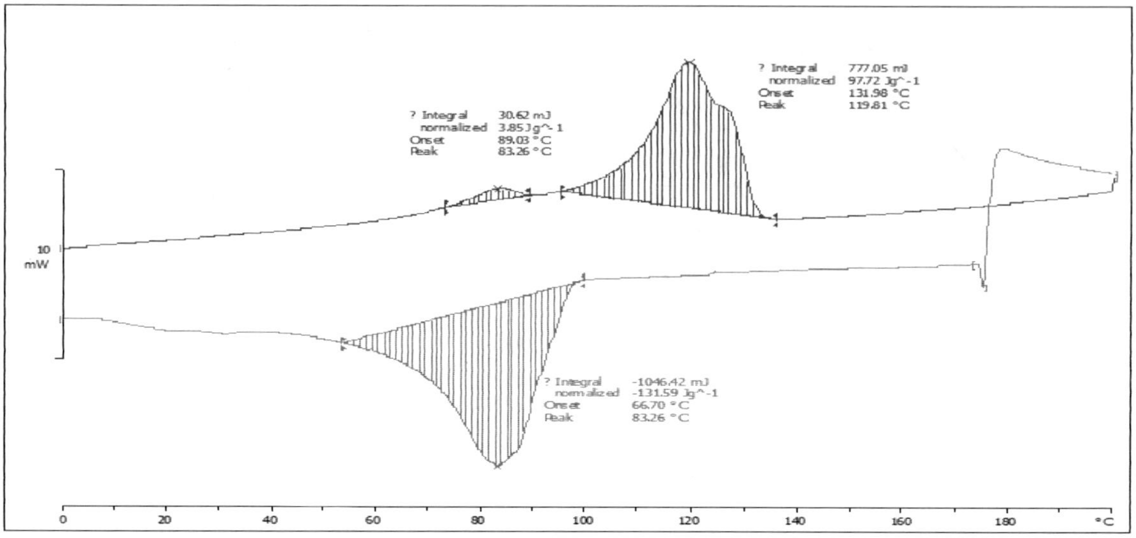

Figure 5. DSC thermogram at of HDPE50.

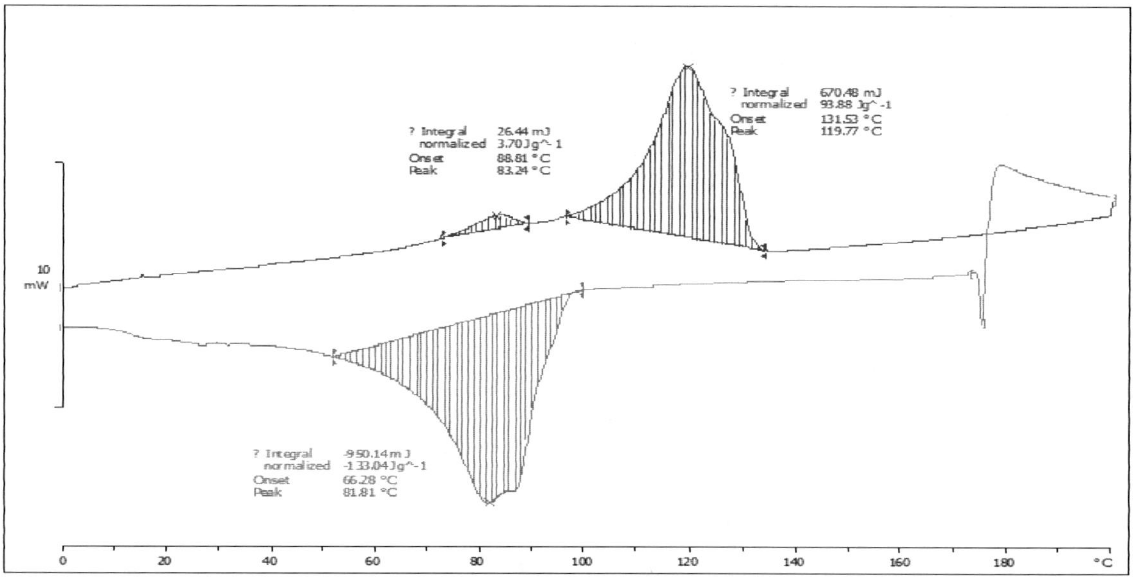

Figure 6. DSC thermogram at of HDPE100.

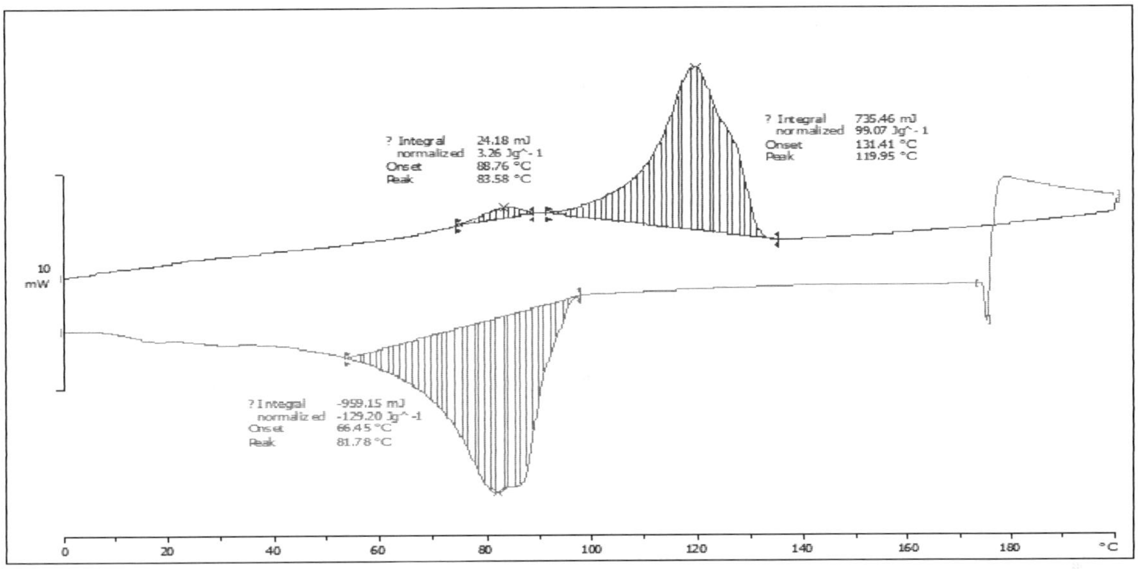

Figure 7. DSC thermogram at of HDPE150.

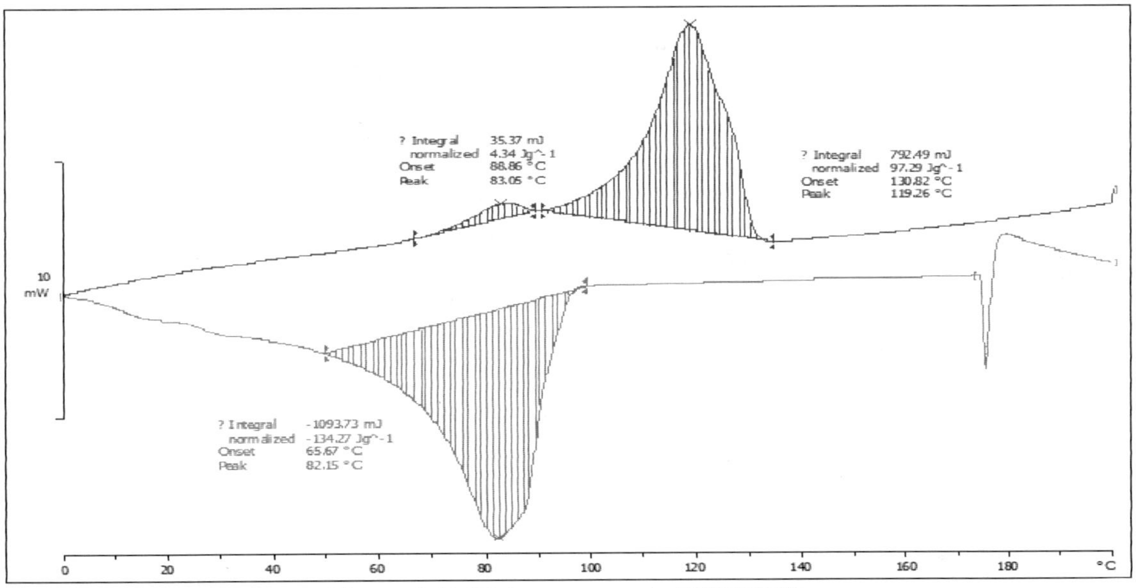

Figure 8. DSC thermogram at of HDPE200.

Table 1. Tm, Tg and Xc value for different irradiation dose (HDPE)

Sample Description	Tm (°C)	Tg (°C)	Xc (%)
HDPE pure	131.38	67.34	54.41
HDPE50	131.98	66.70	55.62
HDPE100	131.98	66.28	55.51

HDPE150	132.41	66.45	56.78
HDPE200	131.82	65.67	55.97

The Tm value of the HDPE pure is increase in increasing of the irradiation dose but decreasing when being bombarded with 200kGy. For the Tg, the Tg for the HDPE pure is decreased when bombarded with electron beam. The reduction of Tg could be attributed to chain branching due to a crosslinking reaction between HDPE backbones. Researcher also reviewed that the endothermic peaks at Tg could be attributed to the enthalpy relaxation effects coming from the thermal history of the samples [(6)]. The Xc showed in table were increasing when the HDPE is been irradiate with electron beam. From the result, this is indicated that by exposed the blend to the electron beam, the Tm of HDPE increase but decreasing their Tg.

3.1.3. Tensile Test. Tensile strength with different dose was summarized in Figure 9 below. Tensile strength of the HDPE increased with increasing of irradiation dose. Tensile strength of pure HDPE was 30.1 MPa. When undergo the irradiation dose from 50 up to 200kGy, the tensile strength increasing to 30.3, 31.3, 32.2 and 32.1 MPa respectively. This occurred due to the cross linking of the HDPE. The highest tensile strength recorded is 32.2 MPa with irradiation dose of 150 kGy.

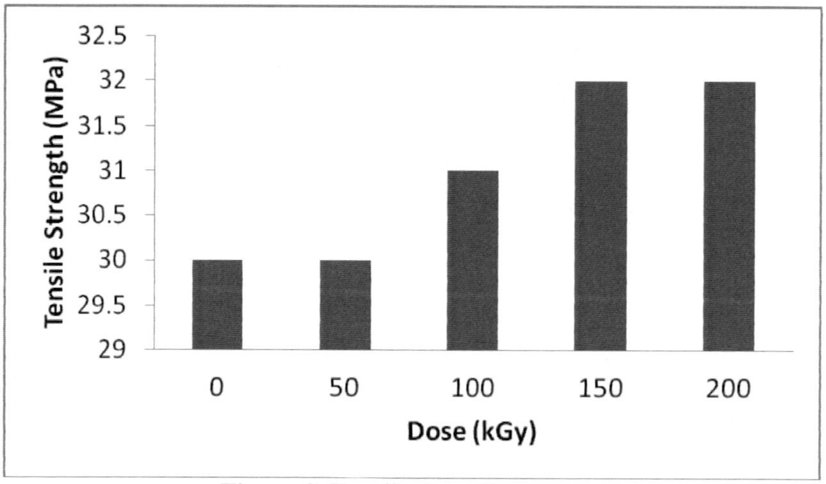

Figure 9. Tensile strength.

3.1.4.Impact. The impact strength with the different dose of irradiation is shown in Figure 10. It can be seen that with increasing of irradiation dose, the impact toughness increase for prepared sample.

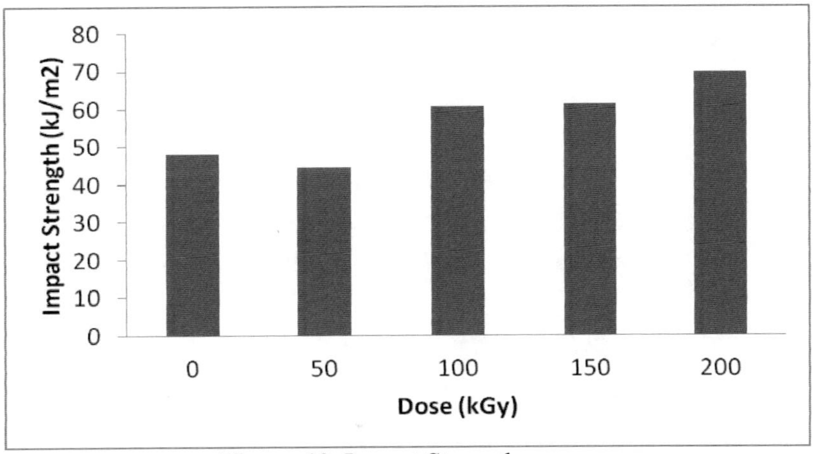

Figure 10. Impact Strength.

3.1.5. Hardness. The hardness test was carried out and the results shown in the Figure 11 below. From the figure, the hardness of cross linked HDPE showed at 150 kGy achieved the highest value and decreasing when being radiated with 200 kGy.

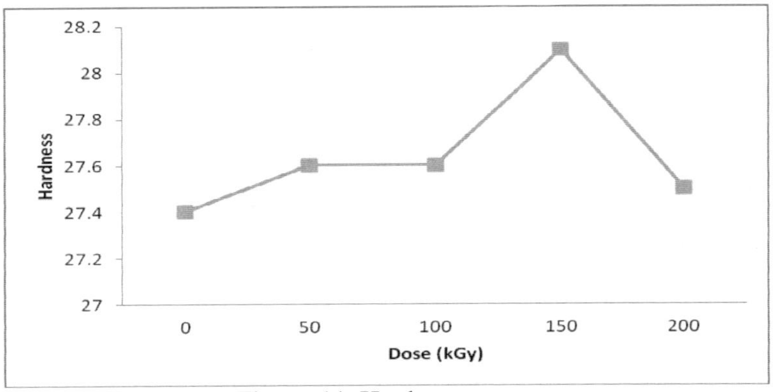

Figure 11. Hardness

3.2. Characterization of HDPE/EVA blends

3.2.1. Gel Content. From the graph shown, the degree of crosslinking was increased with increasing of irradiation dose. From 0% of gel content degree, it is increased up to 42% when being irradiated with electron beam with dose of 50, 100, 150 and 200kGy respectively. This showed HDPE/EVA blends undergo the crosslinking process throughout the irradiation using electron beam.

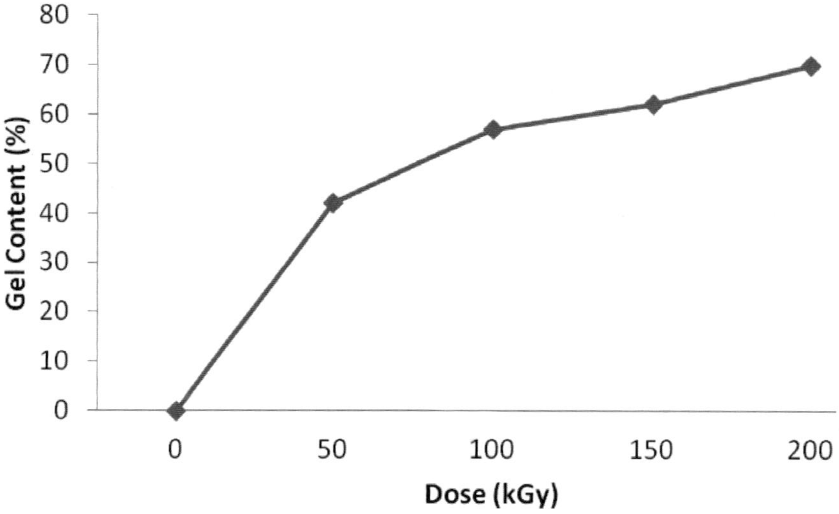

Figure 12. The crosslink degree of HDPE/EVA.

Crosslink with 50kgy irradiation dose have a 42% of crosslink degree and increase slightly when the sample bombarded with 100kGy. With the presence of 100kGy of electron beam irradiation, the crosslinking degree is about 57%. For dose 150kGy, the result also showed the increment in crosslinking degree from 57% to 62% and it increase slightly with the 200kGy as shows in the Figure 11. Increasing the dose would enhance the chances of crosslinking process occurs in the backbone of HDPE/EVA, hence the crosslink degree increased.

3.2.2. DSC. The thermal properties of samples were determined by DSC analysis. Figure 3 and 4 shows the DSC curve of heat flux versus temperature at different dose. The value of melting temperature, Tm, glass transition temperature, Tg and degree of crystallinity, %Xc of each samples were summarized in Table 2.

Table 2. Tm, Tg and Xc value for different dose of HDPE/EVA.

Sample Description	Tm (°C)	Tg (°C)	Xc (%)
HDPE/EVA	131.38	67.34	54.33
HDPE/EVA50	132.78	65.69	55.87
HDPE/EVA100	132.40	67.22	54.32
HDPE/EVA150	131.86	66.30	55.97
HDPE/EVA200	131.08	66.16	55.23

The Tm value of the HDPE/EVA is increase in increasing of the irradiation dose but decreasing when being bombarded with 150 and 200kGy. For the Tg, the Tg for the HDPE/EVA is decreased when bombarded with electron beam. The reduction of Tg could be attributed to chain branching due to a crosslinking reaction between HDPE and EVA. The Xc showed in table were increasing when the HDPE/EVA is being irradiate with electron beam. It seems that, by exposed to electron beam, had increase it Tm and but decreasing their Tg.

3.2.3. Tensile Test. Tensile strength with different dose was summarized in Figure 13 below. When undergo the irradiation dose from 50 up to 200kGy, the tensile strength does not have significant changes but decreasing from lower dose to higher, the tensile strengths are 28, 25, 25, 25 and 24 MPa respectively. This occurred due to the presence of EVA onto HDPE. The tensile strength of HDPE/EVA blend is decreasing significantly with increasing of radiation dose. This is due to the formation of voids and poor interaction between HDPE/EVA blends caused the poor in tensile strength. These showed that higher dose applied on the HDPE/EVA blends will reduce it tensile strength.

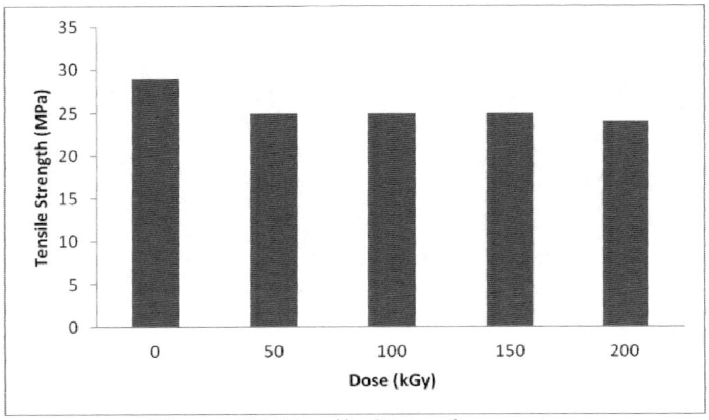

Figure 13. Tensile Strength.

3.2.4. Impact. The impact strength with the different dose of irradiation is shown in Figure 14 below. It can be seen that with increasing of irradiation dose, the impact toughness increase for prepared sample, and from the figure, the impact strength for HDPE/EVA blend showed an increment in impact strength value compared to pure cross linked HDPE. This is due to the present of EVA in HDPE blends as an impact modifier.

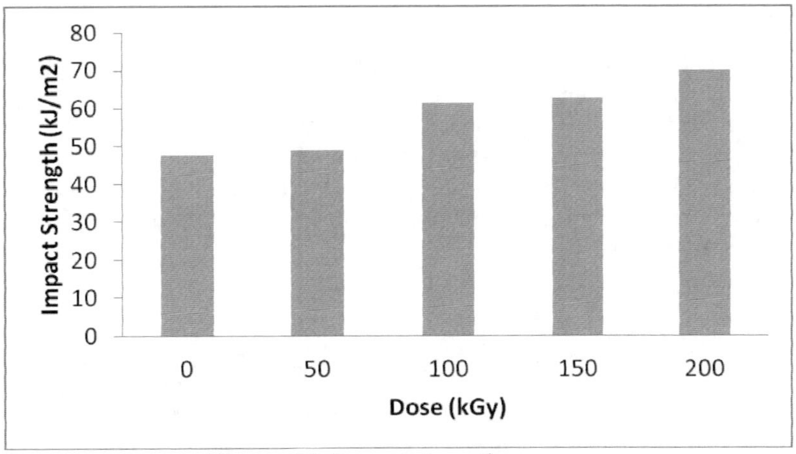

Figure 14. Impact strength.

3.2.5. Hardness. The hardness test was carried out and the results shown in the Figure 15 below. From the figure, the hardness of cross linked HDPE/EVA blend showed at 100 kGy achieved the highest value and uniform when being irradiated further. This showed that 100 kGy of radiation dose is the optimum dose for the blend.

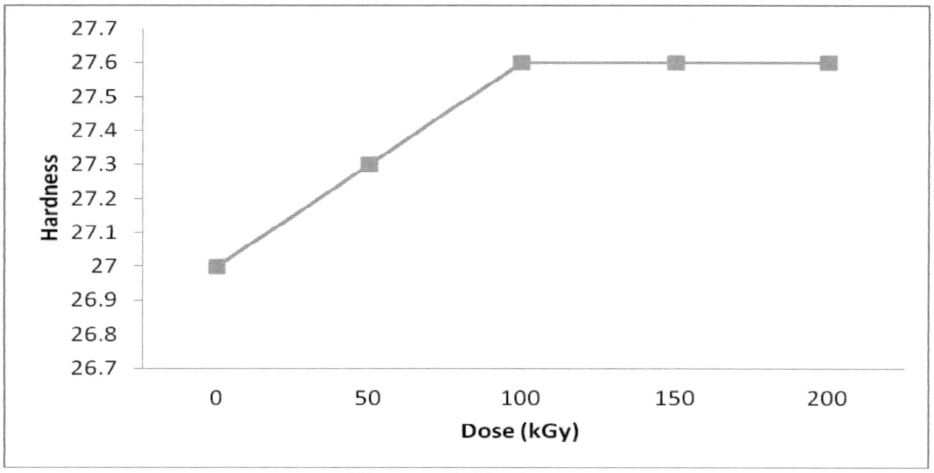

Figure 15. Hardness.

4. Conclusions

From the results of the study, the crosslinking degree of the HDPE with TMPTA as crosslinker showed the optimum grafting degree is at 150 kGy of irradiation dose with degree of crosslinking is 70%. For the thermal test, DSC showed that the Tm value increased and the Tg is decreased with increasing of %Xc. This result also supported by tensile and hardness testing that indicated that 150kGy is the optimum dose in order to have a good, high strength and impact for the production of pontoon material as the normal pontoon used only HDPE. And this property is really critical for every production of pontoon supplier as it is the main issues in making the pontoon. On the other hand, the blends of HDPE/EVA with the presence of 3 phr of TMPTA were being tested. From the result, the tensile strength HDPE/EVA blends decreased gradually with increasing irradiation dose. These results implied that the samples became less stiff and had lower resistance to deformation within these compositions and indicated that higher dose is not suitable for the blends. For the impact strength of HDPE/EVA blends, it showed that with increasing irradiation dose, the impact toughness increase from 40 to 50kJ/m^2. This is due to the present of EVA in HDPE blends as an impact modifier.

5. References

[1] Kumar G S, Sunil C, James J and Student 2014 Fabrication of glass fiber helmet. *Proc. of The Intl. Conf. on Info., Eng., Manag. and Security 2014.*

[2] Alamin N F A 2018 Modeling and analysis of film blowing process, PhD diss. (Sudan University of Sciences and Technology).

[3] Jun Y, Jun Zhang, Yu Yao 2006 Melt grafting of poly(Ethylene-Vinyl Acetate) copolymer with Maleic Anhydride *J. of Appl. Polym. Sci.* **102** p 841-846 10.1002/app.24320.

[4] Sarma K S S 2004 Prospects and development of radiation technologies in developing countries. *Emerging appl. of Rad. Proc. IAEATECDOC-1386* (Vienna: IAEA) p 14-20.

[5] Bhattacharya A 2000 Radiation and industrial polymers *Prog. in Polym. Sci.* **25**(3) p 371-401.

[6] Sepe M P 1997 *Thermal anal. of polym.* 95 iSmithers Rapra Publishing.

[7] Yousef J. (2015). *The effect of high-energy electron beam on drawn and undrawn high density polyethylene fibers*.

[8] Mizera A, Manas M, Manas D, Stanek M, Cerny J, Bednarik M, Ovsik M *Modifi. of Polym. Prop. by Irrad. Prop.s of HDPE after Rad. Cross-linking* ISBN: 97811618041081.

[9] Celli A 1992 Thermal Properties and Physical Ageing of Poly (L-Lactic Acid) *Polym.* **33** p 2699-2703.

[10] Drobny J G, 2003 *Radiat. Tech. for Polym.* (Boca Raton: CRC Press) ISBN 1587161087.

[11] Woods R J 1994 *Appl. Radiat. Chem. Radiat. Proc.* (New York:Wiley-Interscience), ISBN 0471544523.

[12] Manas M, Stanek M, Manas D *et al.* Temperature stability of irradiated polymers *Chemicke listy* **105**(S) p 254-256 ISSN 00092770.

[13] M Stanek *et al.* 2011 Optimization of Injection molding process *Int. J. of Math. and Comp. in Sim.* **5** p 413-421.

[14] M. Stanek *et al.* 2011 Simulation of injection molding process by cadmould rubber *Int. J. of Math. and Comp. in Sim.* **5** p 422- 429.

[15] Manas D *et al.* 2011 Thermal effects on steels at different methods of separation *Chemicke Listy*, 105 p 713-715.

[16] Manas D *et al.* 2011. Improvement of mechanical properties of the TPE by Irradiation *Chemicke Listy* **105** p 828-829.

[17] Stanek M *et al.* 2009 Influence of surface roughness on fluidity of thermoplastics materials, Chemicke Listy, **103** p 91-95.

[18] Manas D *et al.* 2009 Influence of mechanical properties on wear of heavily stressed rubber parts (Kautschuk Gummi Kunststoffe) **62**, p 240-245.

[19] Manas M *et al.* 2009 Modification of Polyamides properties by irradiation *Chemicke Listy* **103** p 24-26.

[20] Manas D *et al.* 2009 Wear of multipurpose tire treads, *Chemicke Listy* **103** p 72-74.

[21] Sanda S *et al.* Injection mold cooling system by DMLS *Chemicke Listy* **103**, pp. 140-142.

[22] M. Stanek et al. 2009 Plastics parts design supported by reverse engineering and rapid prototyping *Chemicke Listy* **103** p 88-91.

[23] Sanda S *et al.* 2011 Gate effect on quality of injected part *Chemicke Listy* **105** p 301-303.

[24] Stanek M *et al.* 2011 How the filler influence the fluidity of polymer *Chemicke Listy* **105** p 303-305.

[25] Pata V *et al.* 2011 Visualization of the wear test of rubber materials *Chemicke Listy* **105** p 290-292.

[26] Kyas K, Stanek M, Krumal M, Holik Z 2011 Simulation of ruber injection holding process, *Chemicke listy* **105**(15) p S354-S356

[27] Javorik J *et al.* 2011 The Shape Optimization of the pneumatic valve diaphragms *Int. J. of Math. and Comp. in Simulation* **5** p 361-369.

[28] Pata V *et al.* 2011 Visualization of the wear test of rubber materials *Chemicke Listy* **105** p 290-292.

[29] Tamboli S M, Mhaske S T, Kale D D 2004 Cross-linked polyethylene *Indian J. of Chem. Tech.* **11** p 853-864.

[30] Khalaf M N2015 Mechanical properties of filled high density polyethylene J. of Saudi Chemical Soc. **19** 88-91.

IOP Conf. Series: Materials Science and Engineering **785** (2020) 012034 doi:10.1088/1757-899X/785/1/012034

Microstructure and leaching characteristics of steel slag and CRT glass

Ahmad Yusri Mohamad[1,a)], Maslina Jamil[2,b)], Nur Izzi Md Yusoff[1] and Syazwani Mohd Fadzil[4]

[1]Smart and Sustainable Township Research (SUTRA), Faculty of Engineering and Built Environment, Universiti Kebangsaan Malaysia, 43600 UKM Bangi, Selangor Darul Ehsan, Malaysia.
[2]Centre for Innovative Architure and Built Environment (SErAMBI), Faculty of Engineering and Built Environment, Universiti Kebangsaan Malaysia, 43600 UKM Bangi, Selangor Darul Ehsan, Malaysia.
[3]Nuclear Technology Research Centre, Faculty of Science and Technology, Universiti Kebangsaan Malaysia, 43600 UKM Bangi, Selangor Darul Ehsan, Malaysia.

[a]P91989@siswa.ukm.edu.my
[b]maslinajamil@ukm.edu.my

Abstract. In view of the large amount and types of waste materials generated, an extensive suite of environmental tests is carried out to assess the potential of reusing these materials. This study investigates the morphology and leaching characteristics of steel slag and Cathode Ray Tube (CRT) glass. The morphology and elemental composition of the samples are examined using field emission scanning electron microscope (FESEM) and energy dispersive X-Ray (EDX). The result of the leaching test is analysed and compared with the toxicity characteristics leaching procedure (TCLP) limit for hazardous elements. Results of the FESEM tests show that each sample has different morphological characteristics and texture. The surface of steel slag is rough, uneven, irregular and porous compared to the surface of granite and CRT glass. The outcome of the TCLP test shows that none of the identified hazardous elements exceed the standard limits, which indicates these materials may be suitable for use as construction materials.

1. Introduction

In line with the global awareness of the need to preserve and conserve the environment, many studies have been carried out to explore the possibility of reusing by-products (steel slag) and waste products (CRT glass), for instance as a construction material. In addition to preserving and conserving the environment, the use of by-products and waste products will help to reduce the need for large landfills. [1][2]. With respect to the environmental assessment, the aggregate samples used in this study are investigated for their toxic compound concentration due to the concern for the leaching of heavy metals. Galvin *et al.* [3] stated that the potential risk of environmental contamination is due to the leaching of toxic wastes into the soil. The problem with leaching is not due to the amount of contaminants (heavy metal element) but the amount of water that seeps into surface water or subsurface water. Other factors influencing the rate of leaching as noted by Tiruta-Barna and Geankoplis studies in Law and Evans [4] are the volume to surface area ratio. Oluwasola *et al.* [5] studied the potential of utilizing Electric Arc Furnace (EAF) steel slag and copper mine tailing as asphalt paving. Testing was conducted on four different asphalt mixtures using both types of waste. The result of leaching analysis shows that the utilization of mixes containing either EAF steel slag or copper mine tailing or both did not produce hazardous elements exceeding the standard TCLP limits.

This study investigated morphology and leaching characteristics of waste material by-product; steel slag and waste product. Morphology sample test was examined by using FESEM and EDX. The leaching test was analysed and compared with TCLP procedures.

2. Materials and method

The natural aggregates used in this study were obtained from Kajang Rock (M) Sdn. Bhd., Selangor. The steel slag was supplied by Lion Titco Resources Sdn. Bhd., Banting, Selangor and is categorized as EAF. Steel slag is a residual material produced during the manufacture of stainless steel through EAF and BOF (Basic Oxygen Furnace) processing. [6][2]. The CRT glass samples were obtained from Nippon Electric Glass (NEGM) in Shah Alam, Selangor. The CRT glass was subjected to several preparation processes. All components of the CRT glass were separated using a laser cutting method, and the glass components were processed by crushing and grading in order to obtain angular glass particles.

The morphology and elemental composition of the aggregate samples, i.e. granite, steel slag and CRT glass, were examined using FESEM (Supra 55vp-ZEISS (Merlin-Zeiss, Germany). The outer surface of the aggregate samples was coated with a thin layer of platinum using the k550 sputter coater before being mounted in the equipment. The samples were then examined at a resolution of 15kV. The same samples were subjected to EDX analysis to determine the elemental distribution on the surface of the aggregate sample.

The TCLP developed by the United States Environmental Protection Agency US EPA (1992) was used in this study to assess the leaching of heavy metals from the aggregates. Leaching test is conducted to analyse the leached hazardous elements when a material is exposed to leaching fluids. The sample is subjected to leaching test before performing any physical and mechanical tests because the tests able to evaluate the environmental safety as a suitable road construction material [8]. The leachate of the aggregate samples (steel slag and CRT glass) were prepared by putting a representative solid portion of the materials in an extraction bottle, and leaching fluid was then added to the bottle. The bottle was sealed tightly and placed on the rotator and rotated at 300 rpm for 18 hours at laboratory ambient temperature of 25 °C. After 18 hours agitation the leachate was removed from the bottle and filtered through a 0·45 μm membrane filter to remove the suspended solids. The metal contaminants in the leachate solution were characterised by using inductively coupled plasma optical emission spectrometry (ICP-MS). Table 1 show the conditions of leaching test.

Table 1. Conditions of leaching tests.

Test conditions	TCLP (EPA Test Method 1311)
Leaching solution	2.5% acetic acid
Liquid to solid ratio (cm^3/g)	20:1
Leaching time (h)	18
pH of leaching solution	4.0 +/- 0.1
Temperature (^0C)	19 – 25

Sources : [7][5]

3. Results and discussion

3.1 Microstructure analysis

Figure 1 shows the FESEM micrographs and the result of EDX elemental analysis of the aggregate samples The FESEM images for each sample shows a different morphology and texture. Unlike the surface of granite and CRT glass, steel slag has a porous, rough, uneven, and irregular surface texture. This result is consistent with those obtained by Oluwasola [5]. Granite has a rougher surface texture.

Kosior-Kazberuk [9] examined the surface of steel slag and granite, and found that the rough surface texture of both materials gives them the ability to improve internal friction, thereby strengthening the bond between the materials in the mixes. The benefits of using steel slag in road construction is that it provides better interlocking and friction, thus resulting in better stability, rutting resistance and skid resistance in comparison to the utilization of natural materials [2]. CRT glass, on the other hand, has a smooth surface texture. In a related work, Ling and Poon [10] and Romero et al. [11] have shown that CRT glass has a smooth surface and an angular grain shape.

EDX spectrum of the aggregate samples were used to compare the elemental composition of the samples. Figure 1 shows that granite and CRT glass have a high weight percentage of silica of 41.3 wt% and 24.2 wt%, respectively. The presence of silica can increase the strength of a material by virtue of its pozzolanic properties [12][13]. The high silica content in CRT glass makes it a suitable substitute for the fine aggregates used in various building material products, such as foam glass, glass ceramic bricks and concrete materials [14][15].

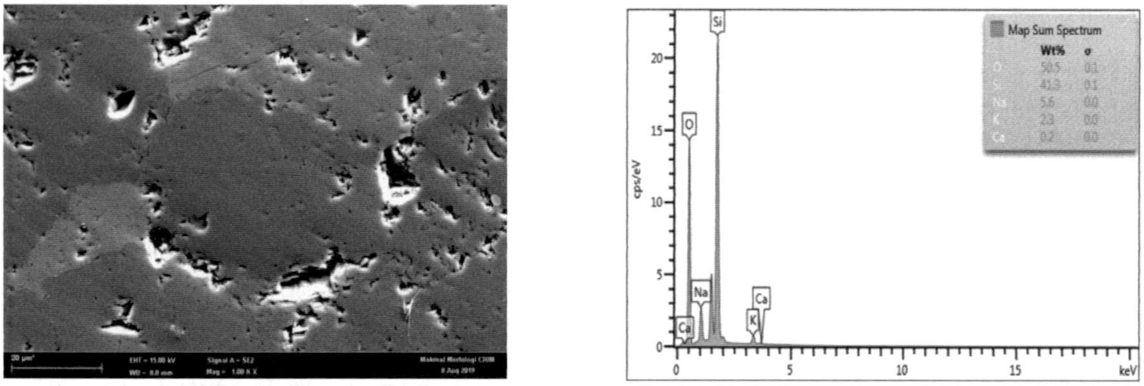

(a) Granite

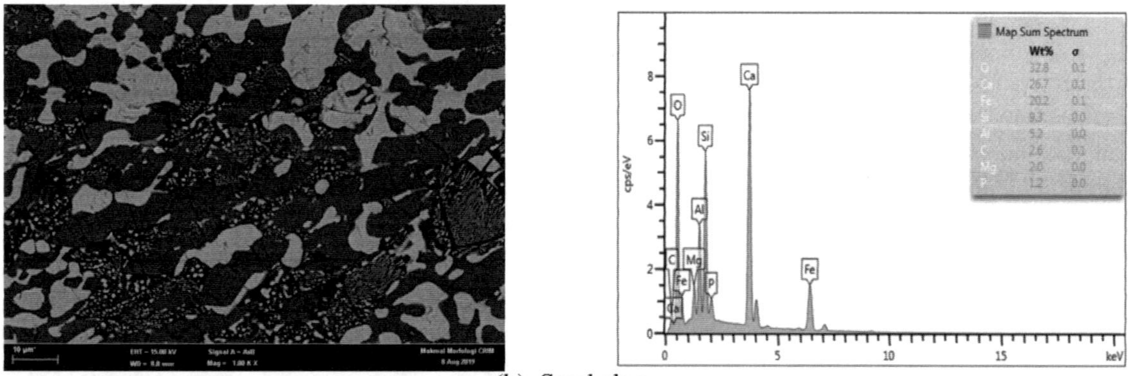

(b) Steel slag

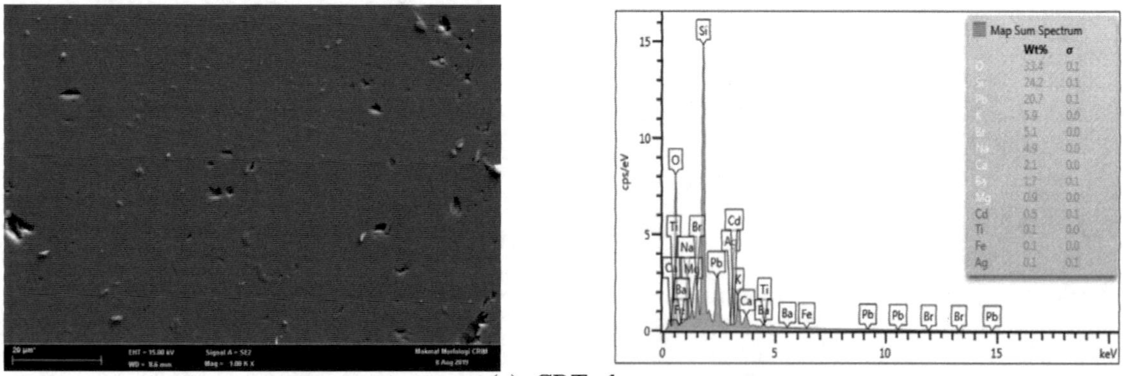

(c) CRT glass

Figure 1.FESEM micrograph and EDX of sample aggregate.

3.2 Result of TCLP

The leaching test measures the concentrations of seven heavy metals, i.e. Arsenic (As), Barium (Ba), Cadmium (Cd), lead (Pb), silver (Ag), Selenium (Se) and Mercury (Hg), in the steel slag and CRT aggregates, and the results are presented in Table 2. It shows that the heavy metals with the highest concentrations in steel slag and CRT glass are Pb and Ba, respectively. However, the concentrations of all metals did not exceed the regulatory limit of TCLP test. Therefore, the waste aggregates are non-hazardous and do not present toxicological problems.

Table 2. Concentration of heavy metals in leachates.

Parameters	Results (mg/L)		
	Steel slag	CRT Glass	TCLP limits
As	0.001	0.001	5.000
Ba	0.033	0.628	100.000
Cd	0.0003	0.0005	1.000
Pb	0.005	4.694	5.000
Ag	-0.0002	0.0001	5.000
Se	0.0001	-0.0004	1.000
Hg	0.0004	-0.00003	0.200

4. Conclusions

The results of this study suggest that among the benefits of utilising steel slag and CRT glass aggregates are safety for leaching problem and preservation of the environment. The following conclusions can be drawn based on the results of this research:

- The microstructure analysis of the granite, steel slag and CRT glass samples showed that steel slag is more porous and rougher than granite and CRT glass. The roughness of steel slag provides better interlocking and friction, thus resulting in better stability, rutting resistance and skid resistance when compare with natural materials.
- The results of TCLP test revealed that the concentrations of the seven heavy metals were below the TCLP regulatory limit for waste material aggregates containing steel slag and CRT glass. This shows that the leaching that occurs when these materials are used in field conditions does not present a threat to the environment.

5. References

[1] Autelitano F, Giuliani F 2016 Electric arc furnace slags in cement-treated materials for road construction: Mechanical and durability properties, *Constr. Build. Mater.* doi:10.1016/j.conbuildmat.2016.03.054.

[2] Hainin M R, Aziz M M A, Ali Z, Putra Jaya R, El-Sergany M M, Yaacob H 2015 Steel Slag as A Road Construction Material, *J. Teknol.* **73** p 33–38. doi:10.11113/jt.v73.4282.

[3] Galvín A P, Ayuso J, Agrela F, Barbudo A, Jiménez J R, Analysis of leaching procedures for environmental risk assessment of recycled aggregate use in unpaved roads 2013 *Constr. Build. Mater.* **40** p 1207–1214. doi:10.1016/j.conbuildmat.2011.12.091.

[4] Law D W, Evans J 2013 Effect of leaching on ph of surrounding water, *ACI Mater. J.* **110** p 291–296.

[5] Oluwasola E A, Hainin M R, Aziz M M A, Warid M N M 2016Volumetric properties and leaching effect of asphalt mixes with electric arc furnace steel slag and copper mine tailings, *Sains Malaysiana* **45** p 279–287.

[6] Oluwasola E A, Hainin M R, Aziz M M A 2014 Characteristics and utilization of steel slag in road construction, *J. Teknol.* **70** p 117–123 doi:10.11113/jt.v70.3591.

[7] Kalembkiewicz J, Sitarz-Palczak E 2015 Efficiency of leaching tests in the context of the influence of the fly ash on the environment *J. Ecol. Eng.* **16** p 67–80 doi:10.12911/22998993/589.
[8] Pasetto M, Baldo N 2016 Recycling of waste aggregate in cement bound mixtures for road pavement bases and sub-bases, Constr. Build. Mater. **108** 112–118 oi:10.1016/j.conbuildmat.2016.01.023.
[9] Kosior-Kazberuk M, Grzywa M 2014 Recycled aggregate concrete as material for reinforced concrete structures *J. Sustain. Archit. Civ. Eng.* **7** p 60–66 doi:10.5755/j01.sace.7.2.7135.
[10] Ling T C, Poon C S 2014 Use of CRT funnel glass in concrete blocks prepared with different aggregate-to-cement ratios, Green Mater. **2** p 43–51.
[11] Romero D, James J, Mora R 2013 Hays CD Study on the mechanical and environmental properties of concrete containing cathode ray tube glass aggregate, *Waste Manag.* **33** p 1659–1666. doi:10.1016/j.wasman.2013.03.018.
[12] Walczak P, Małolepszy ., Reben M, Rzepa K 2015 Mechanical properties of concrete mortar based on mixture of CRT glass cullet and fluidized fly ash, *Procedia Eng.* **108** p 453–458. doi:10.1016/j.proeng.2015.06.170.
[13] Pauzi N N M , Jamil M, Hamid R, Abdin A Z, Zain M F M 2019Influence of spherical and crushed waste cathode-ray tube (CRT) glass on lead (Pb) leaching and mechanical properties of concrete *J. Build. Eng.* **21** p 421–428. doi:10.1016/j.jobe.2018.10.024.
[14] Meng W, Wang X, Yuan W, Wang J, Song G 2016The Recycling of leaded glass in cathode ray tube (CRT) *Procedia Env. Sci.* **31** p 954–960. doi:10.1016/j.proenv.2016.02.120.
[15] Yu-Gong, Tian X M, Wu Y F, Tan Z 2016 Recent development of recycling lead from scrap CRTs: A technological review *Waste Manag.* **57** p 176–186. doi:10.1016/j.wasman.2015.09.004.

IOP Conf. Series: Materials Science and Engineering **785** (2020) 012035 doi:10.1088/1757-899X/785/1/012035

Preliminary study of prompt gamma neutron activation analysis for Melamine

Muhamad Noor Izwan Ishak[1, a)], Susan Maria Sipaun[1], Normanshah Dahing[1], Haerie Hassan[1], Nurliyana Abdullah[1]

[1]Plant Assessment Technology (PAT) Group, Industrial Technology Division, Malaysian Nuclear Agency, Bangi, 43000 Kajang, Selangor, Malaysia

[a]noorizwan@nm.gov.my

Abstract. Prompt gamma neutron activation (PGNA) technique can be applied for detecting explosives in hidden places by using spontaneous fission source 252Cf source or neutron generator that emits thermal and cold neutrons. Some of emitted neutrons will be captured by nuclei and results in production of prompt gamma rays. Explosive materials can be identified by observing prompt gamma energy spectrum using high purity germanium (HPGe) detector. Gamma energy peaks indicate the presence of corresponding explosive nuclei in scanned materials. This paper describes the experiments conducted to investigate the possibility of explosive detection by using PGNA with 241Am/Be neutron source and limitations that come with it..

1. Introduction

The threat of explosive devices on military and civilian is one of the most important issues in many countries. The detection of explosives in hidden places has become an important aspect of the problem. Prompt gamma neutron activation analysis is capable of detecting elements in explosive item. It also offers the possibility of measuring the elemental density of most elements in inspecting materials. Melamine ($C_3H_6N_6$) contains 67% nitrogen by mass. Melamine is used as dummy explosives sample in this experiment due to its chemical formula. Nitrogen (N) in melamine is a key element of chemical explosives as shown in Table 1 below.

Table 1. Some Representative Common High Explosives and Their Compositions.

Explosives	Formula	wt % C	wt % H	wt % N	wt % O
Ammonium nitrate (AN)	$H_4N_2O_3$	**0**	**5.04**	**35.01**	**59.97**
Ammonium picrate(Expl D)	$C_6H_6N_4O_7$	**29.28**	**2.46**	**22.76**	**45.5**
Cyclonite (RDX)	$C_3H_6N_6O_6$	**16.22**	**2.72**	**37.84**	**43.22**
Pentaerythritol tetranitrate (PETN)	$C_5H_8N_4O_{12}$	**19.00**	**2.55**	**17.72**	**60.73**
Trinitrotoluene (TNT)	$C_7H_5N_3O_6$	**37.02**	**2.22**	**18.5**	**42.26**

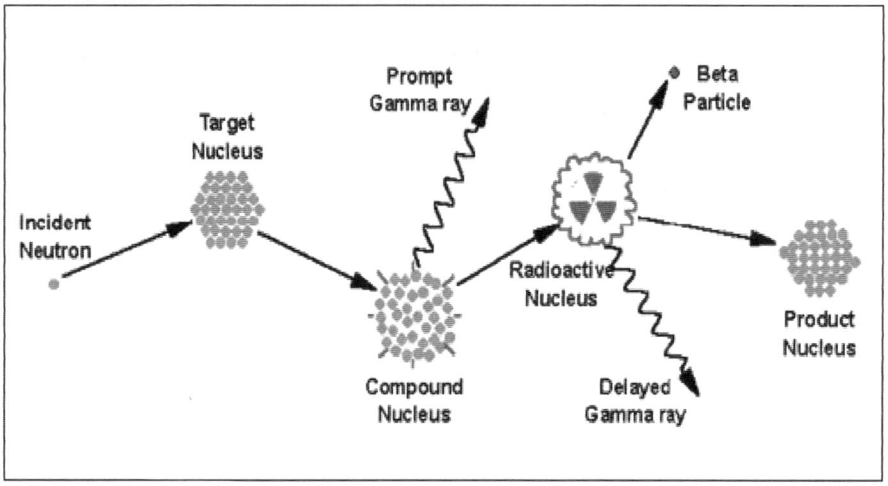

Figure 1. The process of neutron capture by target nucleus followed by the emission of prompt gamma ray, delayed gamma and beta particle.

The basic principle of prompt gamma activation analysis techniques is illustrated in Figure 1 [1]. When neutron emitted by a neutron source (i.e. ^{252}Cf) enter a medium of the target, depending on the type of material, neutrons can pass through with long distance before interacting with the nucleus. Neutrons can interact with the nuclei of materials in two ways to yield characteristic prompt gamma-ray emission. First, fast neutron (> 1 MeV) can scatter in-elastically from a nucleus, losing energy in the process and exciting the nucleus to a higher energy state. Immediate excitation of the nucleus produces a gamma-ray characteristic of the excited nucleus. Next, neutrons with energy around the thermal value (0.025 eV) can be captured by a nucleus. A compound nucleus is formed in an excited state and the prompt de-excitation of this capture nucleus causes the emission of a characteristic capture gamma-ray. The presence of a gamma-ray line with a particular energy is used for qualitative analysis. In addition, the intensity of the line is used for quantitative analysis. The range of gamma-ray in a typical bulk medium is approximately 20 to 25 cm.

2. Materials and method

The main components of PGNAA used in this research are High Purity Germanium (HPGe) semiconductor detector, neutron beam source (usually Cf-252 or Am Be) and multichannel analyzer system (MCA). MCA is calibrated using Co-60 and C-137 energy peaks as the reference energy. Sometimes, MCA is calibrated using hydrogen peaks especially on site. Prompt gamma energy peaks formation depends on: (1) The concentration of the elements near the detector, (2) Neutron absorption cross section (unit: barns) of elements. The high cross section of H nucleus (0.3326 barn) causes the energy peak 2223 eV to be detected easier than other energy peaks, (3) Thermal or cold neutron flux emitted from neutron source, and (3) Geometrical alignment of the 3 components and sample. PGNAA system is also simulated using Particle and Heavy Ion Transport System (PHITS) code by using pure Si, C, and N_2 sample in ideal environment.

The multi-element samples will produce multi-energy prompt gamma rays. This is a complex spectrum, which typically contains several hundreds of peak energies. Therefore, if the background is omitted, the most easily determined element is the one which has highest abundance, cross-sections and the intensity of gamma ray. Detection of prompt gamma with certain energy is used for qualitative analysis and the intensity of gamma rays is used for quantitative analysis. Among the advantages of this technique compared with conventional techniques is that there is no damage to the sample, preparation is simple, in-situ measurement, portable and uses only low radioactivity of radioactive sources in the range millicurie (mCi). The prompt gamma spectrum is obtained in neutron flux quantity and compared with experimental results.

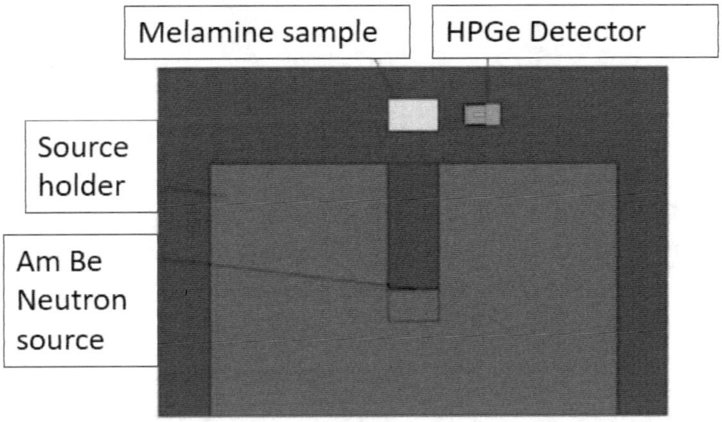

Figure 2 . Basic schematic diagram of PGNAA experimental system.

Figure 3. Experimental setup with HPGe (High Purity Germanium) detector and Am Be neutron source.

3. Results and discussion

Figure 4 and 5 show results from the simulation of mono energetic 0.025 eV neutron beam bombarded in a direction toward melamine cube melamine sample with dimension 2x2x2 cm. It was observed that neutron scattered in all directions as a result of elastic collision with the nuclei contained in melamine. Figure 5 shows the production of prompt gamma rays as a result of neutron capture. The ratio of the number of prompt gamma produced to the number of neutrons collided with melamine is proportional to the neutron capture cross section of nuclei in melamine.

Prompt gamma spectrum from both PHITS simulation and PGNAA experiment are shown in figure 6 and 7 for silicon sample. More energy peaks for silicon nuclei are observed in simulation compared to experiment due to low neutron flux and short scanning time (30 minutes). Other energy peaks in the spectrum are formed due to other elements in the surrounding and background radiation.

Results for simulation with carbon and nitrogen samples are shown in figure 8 and 9. The simulation with these two elements are implemented for comparison with melamine sample due to its chemical formula. It can observe that energy peaks for carbon are easier to obtain compared to nitrogen, which needs longer scanning time or higher neutron flux. Hydrogen energy peak (2223 eV) from melamine sample in figure 10 caused other peaks to be not apparent enough to be observed.

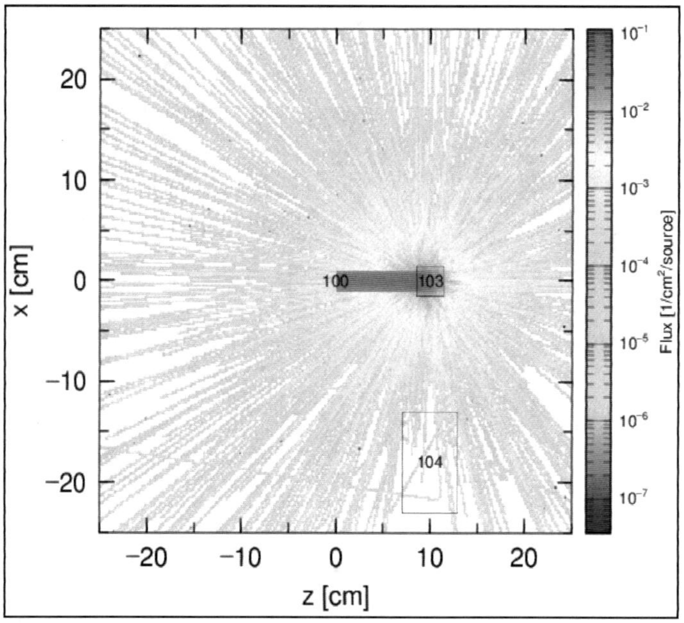

Figure 4. Monoenergetic thermal neutron of 0.025 eV bombarded toward pure melamine cube 103 sample in vacuum space 100. Neutrons scattered across all direction after collision with melamine.

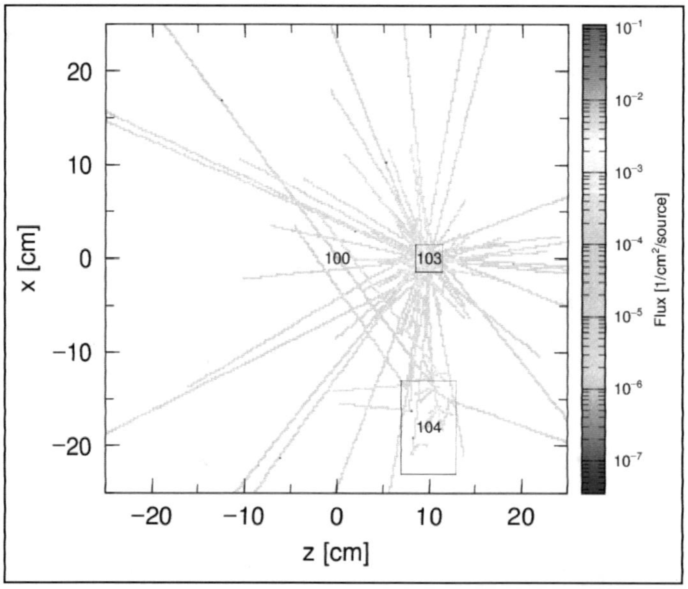

Figure 5. Prompt gamma produced from melamine sample and hitting Ge detector 104 as a result of simulation setup in Figure 4.

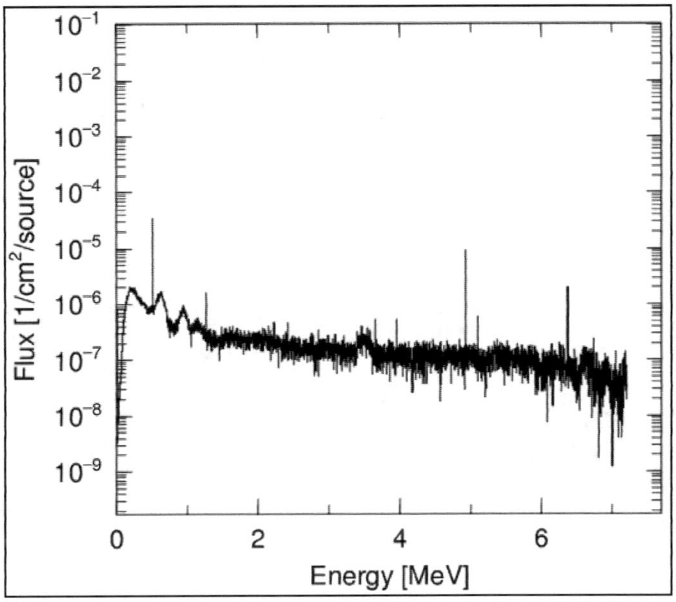

Figure 6. Prompt Gamma Spectrum from Si sample from PHITS simulation.

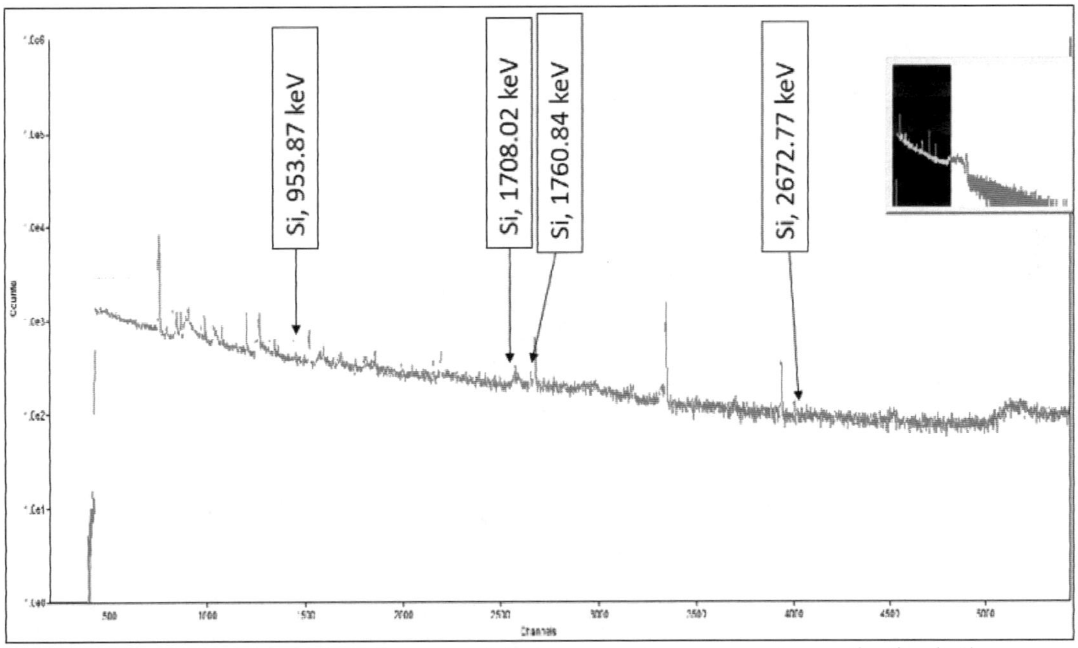

Figure 7. Gamma spectrum of sand measured during prompt gamma neutron activation in the energy range from 0 to 3700 keV for 30 minutes

IOP Conf. Series: Materials Science and Engineering **785** (2020) 012035 doi:10.1088/1757-899X/785/1/012035

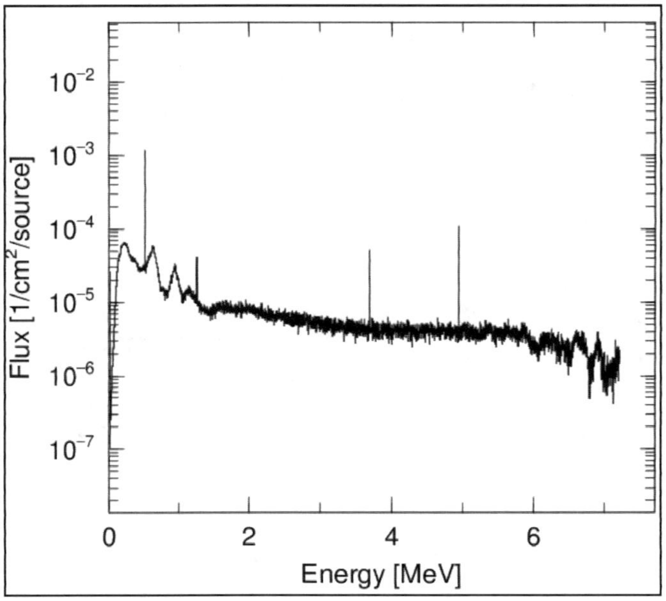

Figure 8. Prompt Gamma Spectrum from C sample from PHITS simulation.

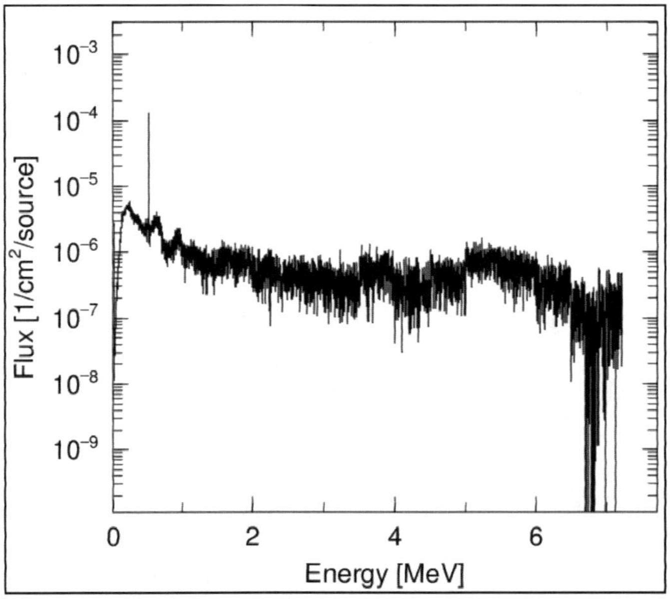

Figure 9. Prompt Gamma Spectrum from N_2 sample from PHITS simulation.

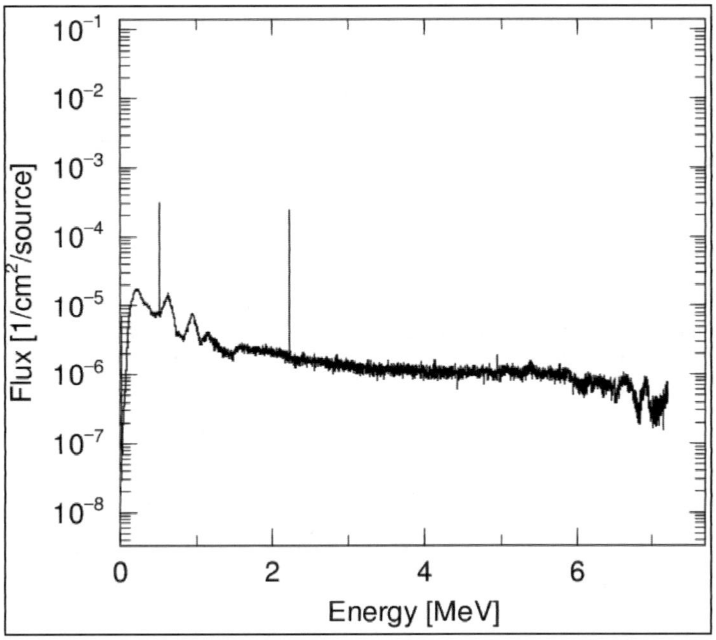

Figure 10. Prompt Gamma Spectrum from $C_3H_6N_6$ (melamine) sample from PHITS simulation.

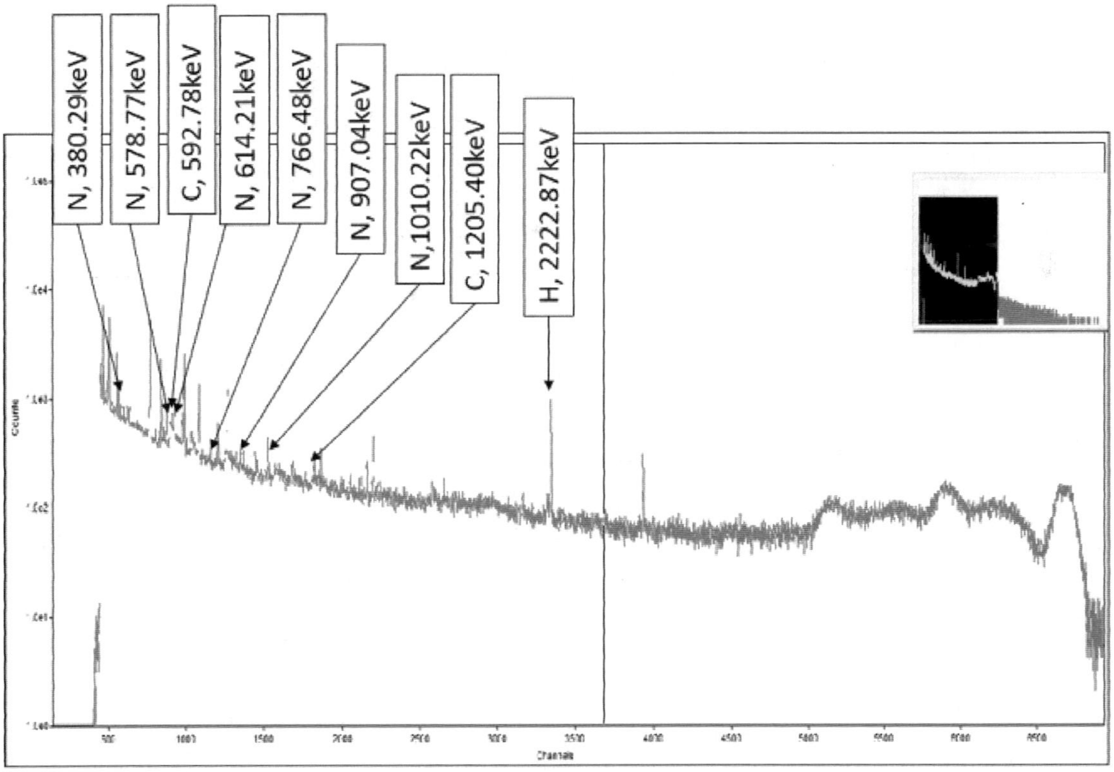

Figure 11. Gamma spectrum of melamine measured during prompt gamma neutron activation in the energy range from 0 to 3700 keV for 30 minutes.

Six energy peak spectrum for N are obtained in Figure 11 representing nuclear reaction: $^{14}N(n,n')$. The gamma peak observed at $E\gamma$ = 2223 keV results from the $1H(n,\gamma)$ reaction. It is difficult to obtain precise estimation of concentration of individual elements in sample considering considering low thermal and cold neutron flux from $^{241}Am/Be$ neutron source and much longer detection time is needed to obtain more prompt gamma energy peaks having smaller neutron absorption cross section. Background gamma radiation and surrounding materials can produce prompt gamma from neutron capture forming additional energy peaks. For the further research, improvisation of various parameters such as detection time, distance between source, detector and sample and source activities need to be taken into account. Other techniques like CHNS analyzer and XRF spectrometer will be considered to obtain mass composition of elements in the melamine sample before investigating the capability of PGNAA make the same assessment.

4. Conclusion

From the preliminary result, it can be concluded that PGNAA technique can be utilized to detect the presence of specific elements in melamine sample. Based on the application of PGNAA in research and industries, PGNAA can also use to study the elemental composition of a sample. This technique is nondestructive, with simple sample preparation, in-situ measurement, portable and use very low activity of radioactive. PGNAA instruments are capable to detect the presence explosive materials once the composition of elements of scanned sample (ex. C/H or N/C ratios) are determined. Although this sample need to be bombarded with high thermal neutron flux. Neutron generator or laser induced neutron source can be utilized for this purpose.

5. References

[1] Dahing L 2014 Non-destructive multi elemental analysis using prompt gamma neutron activation analysis techniques: Preliminary results for concrete sample *AIP Conf. Proc.* **1614** p 20.

[2] Viestia G, Pesentea S, Nebbiaa G, Lunardona M, Sudacb D, Nađb K, Blagusb S, Valkovićb V 2015 Detection of hidden explosives by using tagged neutron beams: Status and perspectives *Nucl. Instr. and Meth. in Phys. Res. B: Beam Interactions with Mater. and Atoms* **241**(1–4) p 748-752

[3] Seabury E H, Caffrey A J 2006 Explosives detection and identification by PGNAA (Idaho Falls, Idaho:Idaho National Laboratory) p 83415.

[4] Lee S, Lee K, Cha H 2014 Activation analysis of indium, KCl, and melamine by using a laser-induced neutron source *J. of the Korean Phys. Soc.* **64**(7) p 982-986.

[5] International Atomic Energy Agency 2006 Database of prompt gamma rays from slow neutron capture for elemental analysis (Vienna: International Atomic Energy Agency).

[6] Molnar Gabor L 2004 Handbook of prompt gamma activation analysis with neutron beams (Kluwer Academic Publishers) ISBN 978-1-4757-0997-1

[7] Ithnin H 2016 Modeling and simulation of portable prompt gamma neutron activation analysis (PGNAA) system using MCNPX code *Int. Nucl. Sci. Tech. and Eng. Conf. 2018 iNuSTEC 2018* (Universiti Teknologi Mara).

IOP Conf. Series: Materials Science and Engineering **785** (2020) 012036 doi:10.1088/1757-899X/785/1/012036

Development on advanced electrophoretic deposition system for fabrication and processing of irradiated hybrid nanocomposites

Mohd Roslie Ali[1,a)], Meor Yahaya Razali[1], Mohd Hamzah Harun[1], Ahmad Zuhdi Mohd On[1], Muhammad Hazim Muhd Sayuti[1], Khairul Azhar Abdul Halim[1]

[1]Radiation Technology Division, Malaysian Nuclear Agency, Bangi, 43000 Kajang, Selangor, Malaysia.

[a]roslie@nuclearmalaysia.gov.my

Abstract. Electrophoretic deposition (EPD) is attracting increasing interest as a materials processing technique for a wide range of technical applications. This technique enables the production of unique microstructures and nanostructures as well as novel and complex material combinations such as hybrid nanocomposites, in a variety of macroscopic shapes, dimensions and arrangements starting from micron-sized or nanosized particles. In the standard EPD process (lab scale), two electrodes, the working electrode (substrate material) and the counter electrode are immersed in a suspension. In pilot scale, the fully automated system consists suspension compartment with ultrasonic transducer, electrodes in-line system, programmable AC/DC power supply, washing system, drying/ annealing system and others. The advantages of EPD are rather simple equipment, the flexibility in substrate shape/dimension and coating materials choice as well as the ability to easily control the homogeneity and thickness of the coatings. Pilot scale deposition facilities or the advanced EPD system, involves semi-automatic operation with accurate deposit thickness control. This promising technique is a better choice than other techniques which are more expensive, less efficient and harmful to environment. The hybrid nanocomposites have been proven to exhibit superior properties than the conventional materials. To make full use of the properties derived from particles, it is important to control the arrangement of particles with high packing density in hybrid nanocomposites. The Ultrasonic irradiation assisted in this system has been found to enable desired particle arrangement and packing density with various shapes. Besides, with the implementation of Gamma irradiation (Pretreatment) as replacement for chemical treatment and EB irradiation for enhancement of conductivity and uniformity of the coatings (Post treatment) can distribute efficiently. It is expected that the system is enable to be employed in variety functions and applications. This paper will discuss the development of the system, the advantages and the relevancy of this system in variety applications.

1. Introduction

Nanocomposite is a composite for which at least one of the dimensions of the dispersed phase is in the nano-meter range. Even if the term "nanomaterials" or "nanocomposites" appeared at the end of the decade eighties, the practical use of this type of materials is much older. The tremendous potential filler matrix interface, combined with the same size of the components of the dispersed phase, makes this type of nano-composites are very attractive, since they are expected to require much lesser amounts of filler to obtain similar or even better levels of performance (with regards to mechanical, thermal and electrical properties)

One very promising manipulation technique in manipulation and fabrication of nanocomposite is electrophoretic deposition (EPD), which has recently been reviewed in the context of nanocomposites. EPD fundamentally is a combination of two processes, electrophoresis and deposition. In the first step, materials (particles) were suspended in a liquid are forced to move towards an electrode by applying an electric field. In the second step, the particles are collected at the electrode and formed a coherent deposit. EPD has a number of advantages when using particulate suspensions to generate films and coatings. The attraction of this method lies in its simplicity. EPD is a cost-effective method usually requiring simple equipment and yet it offers the possibility of forming monolithic or composite coatings with complex shapes and surface patterns. In addition, EPD requires only short processing time. All of these advantages can be exploited for forming films and coatings of nanomaterials.

Normally the multilayered films or coatings prepared by EPD are just in laboratory scale. It is involved with complicated procedures including inserting/ remaining electrode substrates into/from EPD bath suspensions, electric field applying (output voltage), washing and drying system are manually controlled. Therefore, the multilayered films or coatings obtained from manual or batch process are nearly unrepeatable because of wide range of the procedures

EPD is an important technique usually for colloidal or microstructures coating and deposition. It has the advantages of rapid, large-area deposition of films with high packing density and controllable thickness at a low-cost. It uses applied electric fields, to move electrically charged colloids or particles in suspension towards electrodes and finally to form films deposition it. The fabrication of multilayered coatings is significant technique in the producing energy devices for solar application. There also many targeted application rising up such biomedical, automotive industry, environment and military.

The conventional techniques like dip coating and painting for fabrication and processing of composites have drawbacks especially inability to obtain large area with uniform quality and limited to planar surface, due to difficulties in controlling the microstructures and the thickness of the coatings[1]. Furthermore, current techniques available in market are more expensive and consume more energy which requires specialized equipment. The highly demands for new simple system is more preferable.

In this paper, we proposed to introduce the EPD system which can be developed for fabrication and processing of hybrid nanocomposites. The review of this system includes an overview of building up the system of EPD from lab method to fully utilized system in pilot plant. The details on key parameters, benefits and also the potential of this system will be discussed.

2. Experimental

2.1. Concept of EPD System

Electrophoretic deposition (EPD) is a two-step process for depositing coatings from suspended particles and polymer molecules. In the EPD process, the substrate itself is immersed as an electrode in a suspension of fine particles. Basically, any electrically conductive substrate is applicable for EPD. EPD requires the homogeneous dispersion of charged particles in a suspension medium. The suspension should be stable over the deposition period to prevent uncontrolled microstructure of the coatings. In the standard lab process, two electrodes, the working electrode (substrate/targeted materials) and the counter electrode are immersed in the suspension. When applying an electric field, the particles move towards the oppositely charged substrate material, where in the next step they deposit to form coatings.

Figure 1 shows the schematic diagram of EPD and overall setup of the EPD in lab scale. Our purpose is to upgrade the system in pilot plant in order to make them full utilized in fabrication and processing of Hybrid Nanocomposites. The traditional or conventional of electrophoretic in the lab

IOP Conf. Series: Materials Science and Engineering **785** (2020) 012036 doi:10.1088/1757-899X/785/1/012036

will be enhanced and the development of a new system in the pilot plant will be setup systematically. In the pilot plant, the system of EPD will be assisted with Ultrasonic irradiation as pre-treatment and *in-situ* technique. The sequences of this system will be featured in combination of multi components.

Figure 1. Setup of the electrophoretic deposition in lab scale

Points of Understanding regarding EPD as follows:

- **i)** All composites / colloidal particles that can be used to form stable suspensions and that can carry a charge can be used in EPD such as polymers, pigments, dyes, ceramics and metals
- ii) It differs from electrolytic in several ways (No redox reaction occurs)
- iii) The types of EPD depend on which electrode the deposition occurs (Cathodic or Anodic EPD)
- iv) The deposits /coatings are not necessarily electrically conductive
- v) Useful for applying materials to any electrically conductive surface

2.2 System components

Figure **2** shows the designated of sequences of EPD system in a pilot scale. In these batches system, there are four (4) functions:

- i) Deposition / Coating: In this function, EPD process can be operated under variable conditions. For instance, the deposition time, the voltage applied or output potential. This *in-situ* system also can be synchronizing with ultrasonic irradiation by manipulating the amplitude of the vibration applied.
- ii) Washing: the fabricated materials can be repeatedly dipped or immersed into washing solvent and then withdrawn until they are well-cleaned. Any solvent can be filled in and drained out by controlling the inlet and outlet of the washing container.
- iii) Drying: During this process, the fabricated materials can be dried in the oven with temperature controlled or blown with heated air. The blower also can be installed in the container after the washing process.
- iv) EB irradiation: This post treatment can be applied separately after the fabricated materials have been fully deposited, cleaned and dried. The thickness of the affected material can be varied by changing the beam power. The purpose of Electron-Beam irradiation treatment is to densify the electrophoretic deposit and to make sure it more adherent to substrates. Thus it will assist in uniformity of the deposition or the coatings

IOP Conf. Series: Materials Science and Engineering **785** (2020) 012036 doi:10.1088/1757-899X/785/1/012036

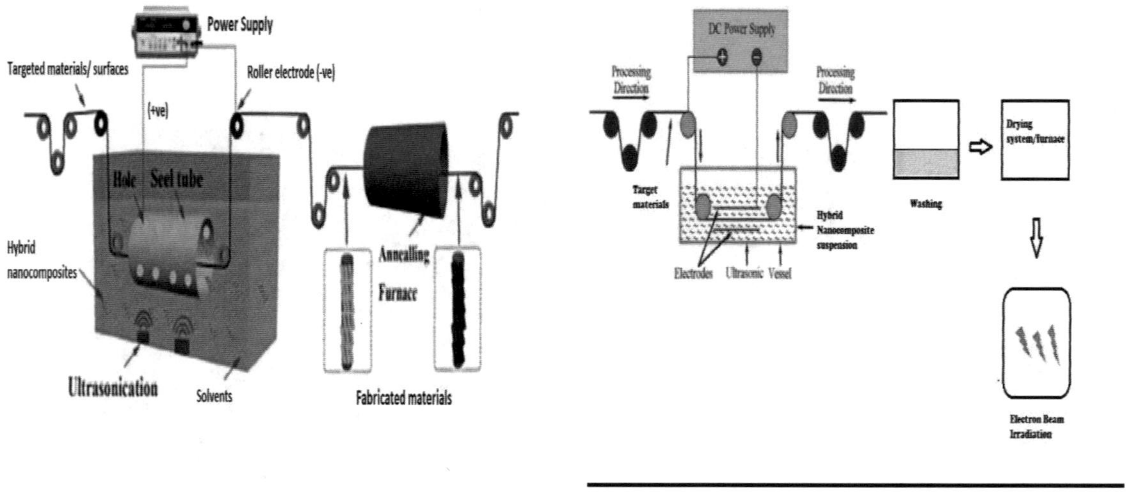

Figure 2. The designated model of EPD in pilot plant.

2.3. Preparation of Hybrid Nanocomposites

There are many types of method that could be used to prepare the hybrid nanocomposites. In this study, the combination of the polymers and other materials will be chosen based on their effectiveness and targeted applications. For example, conductive polymer like polypyrrole can be combined with carbon nanotubes (CNT) or graphene and other advanced materials like titanium dioxide (TiO2) to form nanocomposites. The manipulation of ratios of the mixtures is vital in this stage. The process can be achieved by conventional polymerization or by electropolymerization. As pre-treatment, functionalization by acid reflux or gamma irradiation can be used for achieving the better dispersion in suspension. The stability of the suspension can be achieved by assisting the ultrasonic irradiation as pre-treatment before EPD technique.

3. Discussions

3.1. What happens in EPD process?

As in Figure 3, after migration of the charged particles from the bulk to the electrode of the opposite charged electrode, the de position occurs through the double layer (lyosphere) distortion and thinning followed by coagulation of the particles on the electrode. When the charged particles that is surrounded by the diffuse double layer is subjected to an electric field, it undergoes a distortion and becomes thinner ahead and wider/thicker behind. During the migration of the charged particles, ions of similar charge will move in the same direction of the particles and can react with the counter ions accompanying the charged particle. The latter induces a thinning of the double layer of the particles. Because the double layer of the particles is now thinner, it can easily approach other particles with also a thinner double layer and, if the contact is close enough to favor Van der Waals attractive forces domination, hence the deposition occurs.

Conventionally for this technique, the substrates or targeted materials was needed to be electrically conducted. However, the latest research that was newly investigated, found that the EPD can be applied even on non-conducting substrates or targeted materials. This can happen by isolating the influence of porosity effect.

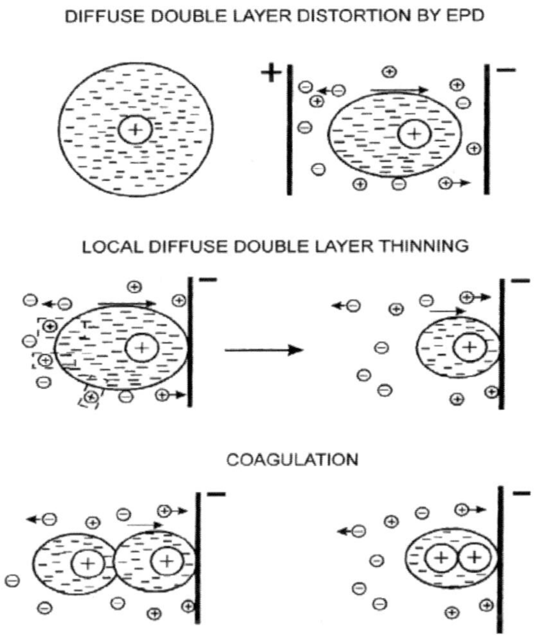

Figure 3. The mechanism of EPD.

The mechanism of EPD involve charged particles in a suspension being deposited onto an electrode under the influence of an applied electric field. Two groups of parameters or factors determine the characteristics of this process; (i) those related to the suspension (Figure 4) and, (ii) those related to the process including the physical parameters such as the electrical nature of the electrodes, the electrical conditions (voltage/intensity relationship, deposition time, and others)

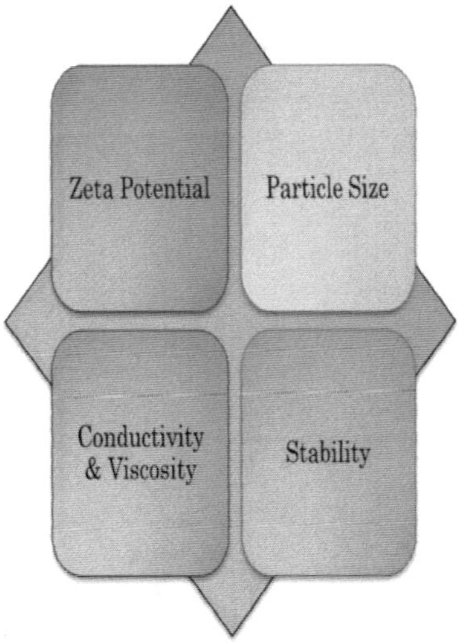

Figure 4. The factors influencing EPD related to suspension.

EPD mainly contains aqueous, organic polar solvents and nonpolar solvents. In general, low voltage range below 20V applicable to aqueous solvents. For medium range of voltage below 100V suitable for polar solvents. As for nonpolar solvents, it requires voltages above 100V. Utilizing nonpolar solvents is preferable for the uniform deposition of materials, but the required high voltage is unfavorable for the chemical stability of the EPD system. For aqueous solvents or electrolytes, even though with a low voltage, bubbles are readily formed due to the production of hydrogen and

IOP Conf. Series: Materials Science and Engineering **785** (2020) 012036 doi:10.1088/1757-899X/785/1/012036

oxygen by water electrolysis, which leads to low-quality EPD deposition or coatings. In addition, highly active hydrogen and oxygen atoms also give rise to side reactions in EPD. In particular, the applied voltage impacts substantially the morphology of EPD materials, in this case, hybrid nanocomposites, as well as the performance of the fabricated materials. Therefore, polar solvents like ethanol and acetone are widely used in EPD.

3.2. Why use ultrasonic irradiation in EPD?

The preparation of stable suspension/dispersion of hybrid nanocomposites in suitable liquids /solvents necessarily prerequisite for successful deposition by EPD. The individual particles are held together by attraction forces of various physical and chemical nature, including van der Waals forces and liquid surface tension. This effect is stronger for higher viscosity liquids, such as polymers or resins. The attraction forces must be overcome on order to deagglomerate and disperse the particles into liquid media. The application of mechanical stress in ultrasonication of the media resulting in alternating high pressure (compression) and low pressure (rarefaction) cycles on the attracting electrostatic forces (Figure 5). Hence the pressure between the particles will separate and make effective means for dispersing and deagglomeration especially in pilot plant system of EPD.

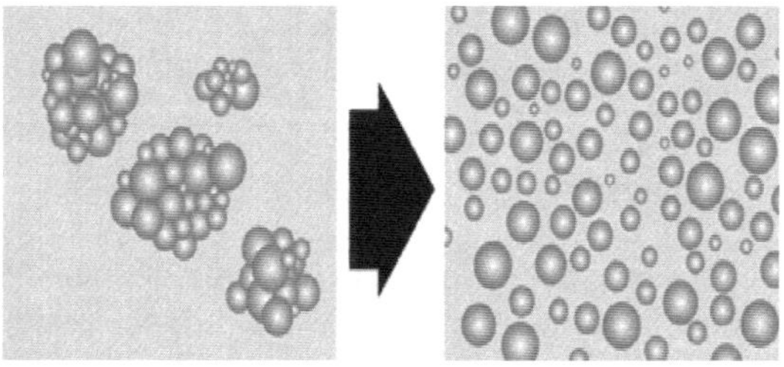

Figure 5. The particles condition before and after ultrasonication.

3.3. Electron-Beam Irradiation (Ionizing) Effect on EPD Coatings

As in Figure 6 on EB Treatment, the reflectivity by the irradiated material is lower compared to laser energy, therefore the beam efficiency is higher. The thickness of affected material can be varied by changing the beam power. This alternative method was successfully used to densify the electrophoretic deposits/ coatings and to make it more adherent to substrates or targeted materials. Thus it will assist in uniformity of the coatings

IOP Conf. Series: Materials Science and Engineering **785** (2020) 012036 doi:10.1088/1757-899X/785/1/012036

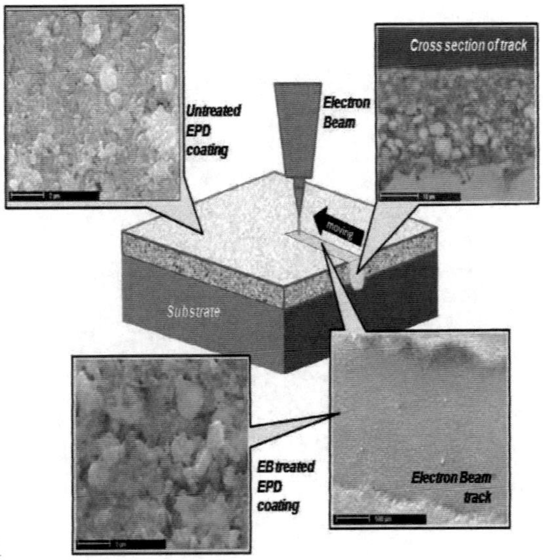

1

Figure 6. The effect of EB irradiation on the EPD coatings.

3.4. Techniques for characterization of Deposited Hybrid Nanocomposites

There were several experimental techniques used for the characterization of deposited hybrid nanocomposites on the substrates or targeted materials.

3.4.1. Characterization of deposition suspension. The stability of the suspensions can be analyzed by Particle Size analyzer and Zeta Potential. The suspensions can be compare between suspensions with and without ultrasonic irradiation applied. The suspensions were ultrasonically treated to ensure its homogeneous distribution before test. Other characterization such as Transmission electron microscope (TEM) can be used to analyze composition of hybrid nanocomposites which synthesized from electropolymerization and functionalization.

3.4.2 Surface morphology and microstructure analysis. The deposits or coating morphology can be observed by Scanning Electron Microscope (SEM). The microscopic Fourier Transform Infrared Analysis (FTIR) and Raman Spectrometer can be utilized to investigate the functional groups of the deposits or coatings. For topography of hybrid nanocomposites surfaces and interfaces, or the surface roughness, can be analyze by Atomic Force Microscope (AFM). The thermal stability of nanocomposites can be evaluated by thermo gravimetric analysis (TGA). As for Differential Scanning Calorimeter (DSC), the nature of crystallization in matrix of the nanocomposites can be studied. X-Ray Diffraction (XRD) can be used to ensure the availability of the deposition and coatings on targeted materials.

Figure 7 and 8 show the images of the MWCNT films deposited on Aluminum plate/substrate by Electrophoretic deposition (EPD) technique and the Scanning Electron Microscope (SEM) micrographs. The SEM micrographs indicate that some cracks in the film surface which are attributed to the tension stress induced by the mismatch of the thermal expansion coefficient between deposition layer and substrate [6]. When EPD was carried out at the optimal parameters, very homogeneous MWCNTs deposits were obtained, about 2μm thick MWCNT coatings were produced.

IOP Conf. Series: Materials Science and Engineering **785** (2020) 012036 doi:10.1088/1757-899X/785/1/012036

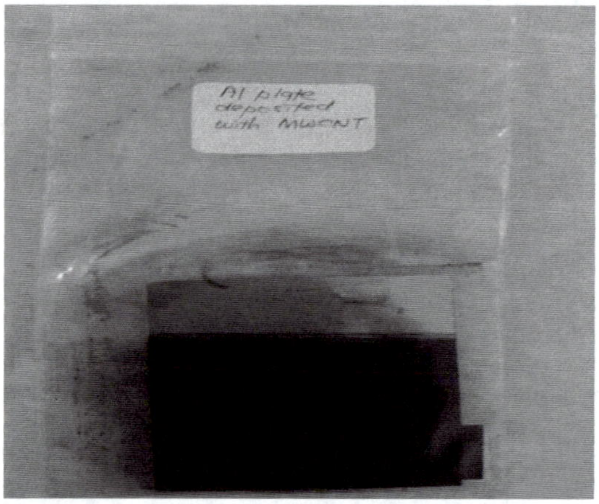

Figure 7. Image of deposited of MWCNTs film onto the Aluminum plate/substrate.

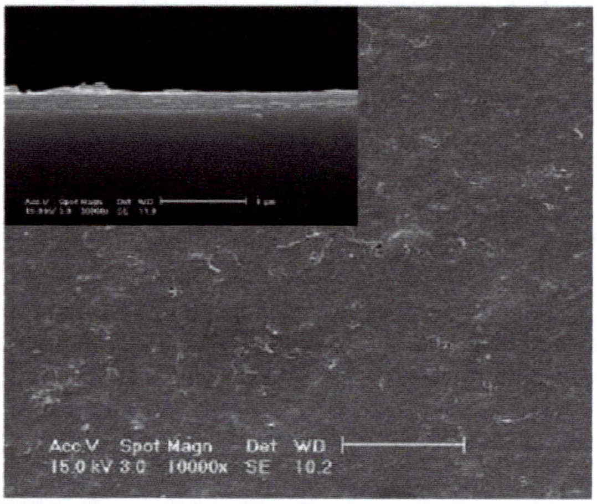

Figure`8. SEM micrographs of surface morphology of MWCNTs film onto the Al-plate, (inset: cross-sectional image of MWCNT films, scale bar: 500 nm).

3.5. Potential applications of EPD system

In recent years, there has been a rising interest in nano and micro-structured materials because they usually exhibit novel and improved properties suitable for various applications. Compared to other deposition and coating process such as grafting, lithography, self-assembly, dip-coating, spin coating and the advanced technique like CVD and PVD, EPD technique offers the advantage of being versatile, cost saving and more importantly, allows deposition of high rates of particles in a controlled structural manner.

The applications of deposits or coatings by EPD can be applied in advanced ceramic materials, coatings for thin and thick films, multilayered composites, functionally graded materials, hybrid materials and nanotechnology. These fabrication and coatings can cover in many potential applications such as shielding for electromagnetic radiation (Power plants, reactors and telecommunication). In reinforcement composites, this technique can have covered materials for construction, automotive and aerospace. EPD under modulated electric fields can offer new application perspectives in biotechnology such as implants, prostheses, drug delivery and medical devices.

EPD is an automated process, a variety of parameters such as voltage, frequency, concentration of dispersed species and deposition time can be controlled to yield a better response under the optimized conditions. With regards to this, large number of fabricated microstructures with improved characteristics have recently been designed and manufactured. In energy and electronic devices, EPD can be future prospect in solar cell technology, energy storage, fuel cells, super capacitor and super electrodes devices.

In environmental field, EPD can be developing for building up the system of desalination separation membranes, air purification, wastewater treatment and the removal of ionic and heavy metals. There also electrochemical system which have been done for water purifications and the development of adsorbents from polymer nanocomposites by this technique. Others promising applications of this versatile technique can be gained in painting technologies, alloy metals processing, corrosion protection and as tools for supporting another processing technique.

4. Conclusions

The investigation confirms that Electrophoretic deposition (EPD) represents a very convenient processing technique for the deposition of hybrid nanocomposites onto various substrates and targeted materials that may be relevant to a variety of applications. Due to high efficiency, scaling up capability, cost effectiveness and easily control in assembly, EPD system exhibits advantages in realizing nanostructured and composite films or coatings for many applications such as energy conversion, fuel cells, biomedical devices, environment, automotive and aerospace. The deposited materials from hybrid nanocomposites exhibited excellent macroscopic homogeneity and uniformity of thickness throughout. A further improvement in the various properties of the conductive or nonconductive substrates, such as its electrical conductivity and optical transmission, should be possible by enhancing the degree of dispersion of the hybrid composites in the well stable suspension. The fully utilized and automated EPD system can promising us future advanced applications in many aspects.

5. References

[1] Azuma S, Yamada H, Kawamura G, Muto H, Mizushima T and Matsuda 2017 A Development of multilayer coating system based on electrophoretic deposition process. J. of the Ceramic Soc. of Japan **125**(4) p 317-321.

[2] Bhardwaj S, Pal A, Chatterjee K, Chowdhury P, Saha S, Barman A, Rana H, Sharma G D, Biswas S 2017 Electrophoretic deposition of plasmonic nanocomposite for the fabrication of dye-sensitized solar cells. Indian J, of Pure and Appl. Phys. **55** p 73-80

[3] Jiang J, Xu C, Su Y, Guo Q, Liu F, Deng C, Yao X and Zhou L 2016 Influence of carbon nanotube coatings on carbon fiber by ultrasonically assisted electrophoretic deposition on its composite interfacial property. Polymers **8** p 302.

[4] Wang C, Li, Sun S, Li X, Zhao F, Jiang B, Huang Y 2016 Electrophoretic deposition of graphene oxide on continuous carbon fibers for reinforcement of both tensile and interfacial strength. Composites Sci. and Tech. **135** 46-53.

[5] Shi Y Y, Li M, Liu Q, Jia Z J, Xu XC, Cheng Y, Zheng Y F **2016** Electrophoretic deposition of graphene oxide reinforced chitosan-hydroxyapatite nanocomposite coatings on Ti substrate. J Mater Sci: Mater Med **27** p 48.

[6] Seuss S, Lehmann M, Boccaccini A R 2014 Alternating Current Electrophoretic Deposition of Antibacterial Bioactive Glass-Chitosan Composite Coatings *Int. J. Mol.* Sci. **15** p 12231-12242.

[7] Shi K, Zhitomirsky I 2013 Electrophoretic nanotechnology of graphene-carbon nanotube and graphene-polypyrrole nanofiber composites for electrochemical supercapacitors J. of Colloid and Interface Sci. **407** p 474-481

[8] Ammam M 2012 Electrophoretic deposition under modulated electric fields: a review. RSC Advances **2** p 7633-7646.

[9] Cho J, Konopka K, Rozniatowski K, Garcia-Lecina E, Shaffer M S P, Boccaccini A R **2009** Characterization of carbon nanotube films deposited by electrophoretic deposition. Carbon **47** p 58-67.

[10] Wang S C, Huang B C 2008 Field emission properties of Ag/SiO2/carbon nanotube films by pulsed voltage co-electrophoretic deposition. Thin Solid Films **517** p 1245-1250.
[11] Jung S M, Jung Y, Suh J S 2008 Horizontally aligned carbon nanotube field emitters fabricated on ITO glass substrates. Carbon **46** p 1973-1977.
[12] Havel M, Behler K, Korneva G, Gogotsi Y 2008 Transparent Thin Films of Multiwalled Carbon Nanotubes Self-Assembled on Polyamide 11 Nanofibers. Adv. Funct. Mater. **18** p 2322-2327
[13] Besra L, Liu M 2007 A review on fundamentals and applications of electrophoretic deposition (EPD). Progress in Materials Science **52** p 1-61.
[14] Besra L, Compson C, Liu M 2007 Electrophoretic deposition on non-conducting substrates: The case of YSZ film on NiO-YSZ composite substrates for solid oxide fuel cell application. J. of Power Sources **173** p 130-136.
[15] Chow Winnie W Y, Wong M K, Li W J, Wong K W 2007 Rapid fabrication of CNT sensors using electro-chemical deposition of functionalized CNTs. Proc. of the 2nd IEEE Int.l Conf. on Nano/Micro Eng. and Molecular Sys.
[16] Boccaccini A R, Cho J, Roether J A, Thomas B J C, Minay E J, Shaffer Milo S P 2006 Electrophoretic deposition of carbon nanotubes. Carbon **44** p 3149-3160.
[17] Moniruzzaman M and Winey K I 2006 Polymer Nanocomposites Containing Carbon Nanotubes. Macromolecules **39** p 5194-5205
[18] Amerio E, Sangermano M, Malucelli G, Priola A, Voit B 2005 Preparation and characterization of hybrid nanocomposite coatings by photopolymerization and sol-gel process. Polymer **46** p 11241-11246.
[19] Kamat P V, Thomas K G, Barazzouk S, Girishkumar G, Vinodgopal K, Meisel D 2004 Self-assembled linear bundles of single wall carbon nanotubes and their alignment and deposition as a film in a dc field. Jacs Art.
[20] Kamada K, Mukai M, Yasumichi M 2003 Electrophoretic deposition assisted by soluble anode. Materials Letters **57** p 2348-2351.

Acknowledgment
The authors would like to thank to Malaysian Nuclear Agency (NUKLEAR MALAYSIA) for highly support in this work. This study was done as a part of Non-funding of internal project (Projek Dalaman) in Nuclear Malaysia.

IOP Conf. Series: Materials Science and Engineering **785** (2020) 012037 doi:10.1088/1757-899X/785/1/012037

Development of the scanning power supply for 300 keV electron accelerator

Leo Kwee Wah[1,a)], M. Azhar[1], R.M. Chulan[1], S.A. Hashim[1], M. Mokhtar[1], Khaidawaton[1], H. Baijan[1], R.M. Sabri[1]

[1]Malaysian Nuclear Agency, Bangi 43000 Kajang, Selangor, Malaysia

[a]leo@nm.gov.my

Abstract. This paper describes the development and study of the scanning power supply. This locally designed low energy electron accelerator with the former energy of 140 keV will be upgraded to 300 keV. As the setup, scanning power supply is required to excite the scanning coils and deflect the energetic electron beam across a titanium foil window in horizontal and vertical direction respectively. The excitation current of the scanning coils is determined by the energy of the electron beam and the deflected length. Therefore, the scanning power supply must be functioning automatically and continuously to ensure the uniform dose distribution and unavoidable heat issue. As the result, the scanning power supply using the combination of signal generator and power amplifier has been developed and functioning as expected.

1. Introduction

Accelerator is a promising and advanced technology with the capability to support various applications such as flue gases treatment and cable cross linking [1], medical applications such as cancer therapy and isotope production [2][3], as a collider for high energy physic experiment and study [4][5] and agriculture for mutation breeding [6]. In Nuclear Malaysia, a localized own developed accelerator technology so-called the low energy electron accelerator [7][8][9] has been upgraded from 140 keV to 300 keV as shown in Figure 1. For such improvement, other auxiliary systems must be upgraded as well such as insulating gas system [8], window cooling system [9], filament power supply [10] and scanning power supply etc. In this paper, the principle and the development of the scanning power supply will be described and discussed.

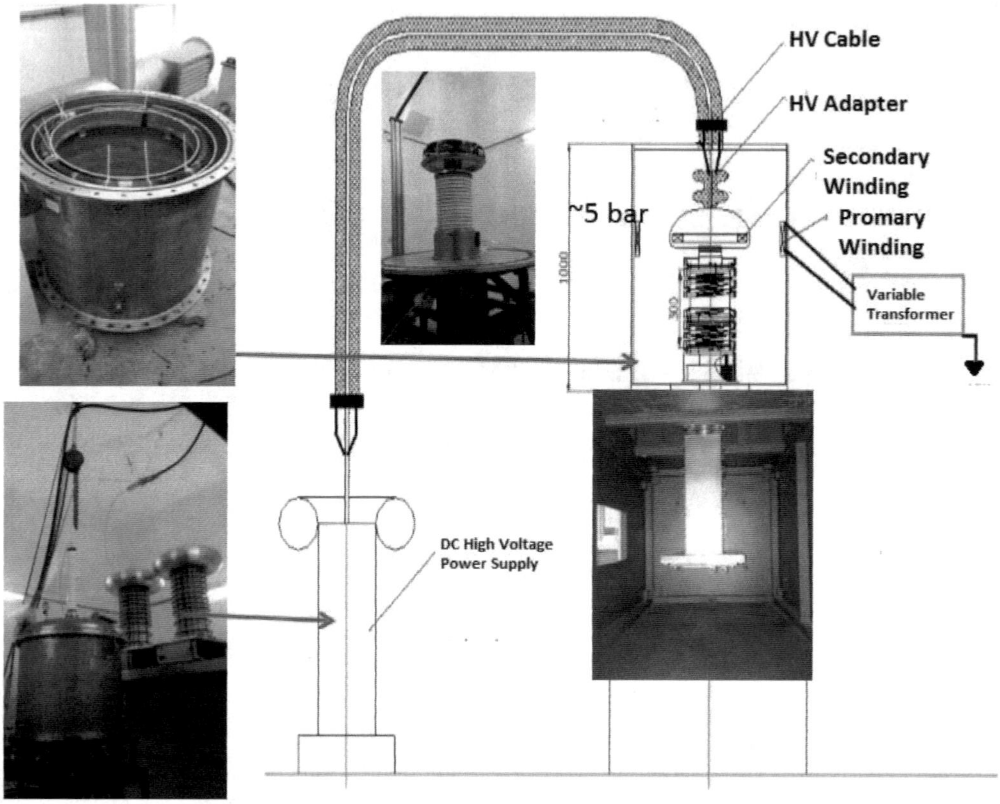

Figure 1. Low energy electron accelerator

2. Parameters of the scanning power supply

Basically, an electron accelerator is a machine to accelerate the electron beam with the required energy and beam current. Whereby, the electron is produced by heating-up the filament or cathode. The heated filament will produce the electron on the surface and with a small applied electric field the electron is extracted and propagate along the accelerating tube. Finally, the energetic electron beam is scanning along the titanium foil window by two pairs of scanning coils as X-axis and Y-axis coil with optimize frequencies respectively [11]. In this low energy electron accelerator, the scanning coils are embedded at the bottom of the accelerating tube as shown in Figure 2, and the specifications of the scanning coils are shown in Table 1.

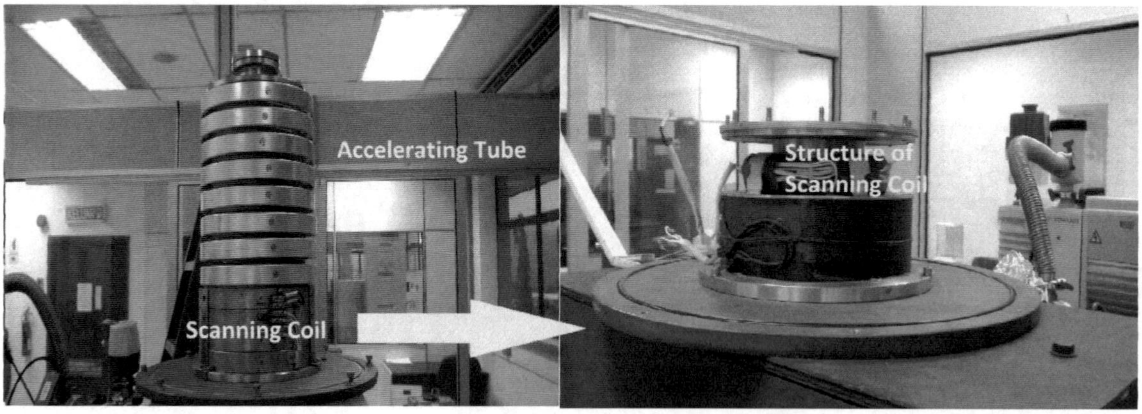

Figure 2. Side view of the accelerating tube with the scanning coils embedded position.

Table 1. Specifications of the scanning coils.

Parameters	Figure
Materials	Copper
Diameter	3 mm
Maximum excitation current	3 A
Number of turns for X-axis coil	200 turns
Number of turns for Y-axis coil	200 turns

Basically, the required scanning excitation current and frequency is determined by the energy of the electron beam and the size of the scanning window [11]. As shown in Figure 3, the window size with the length of 192. 5 mm and width as 97.5 mm.

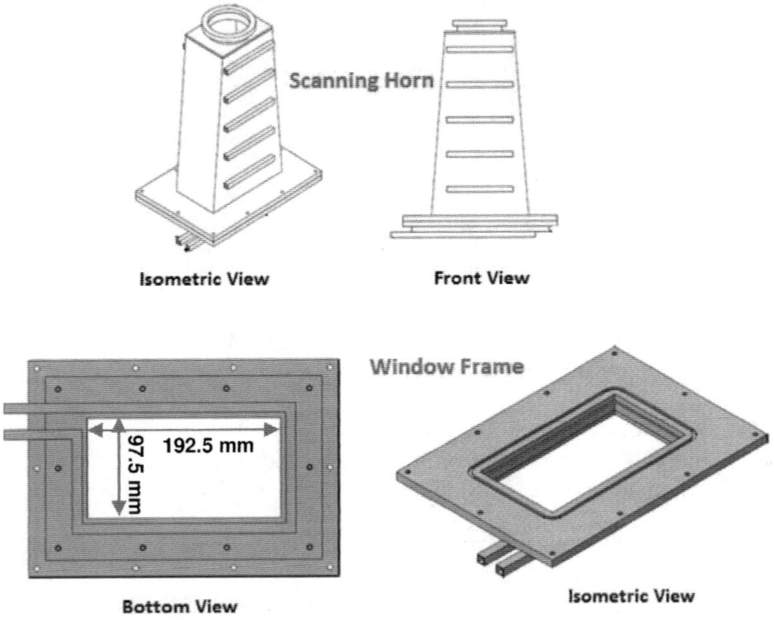

Figure 3. Overview of the scanning horn [11].

At first the desired or optimum frequency must be determined by referring to the physical geometry of the window and the electron beam diameter. According to expression as below,

$$\eta = \frac{x}{2d} \quad (1)$$

$$\eta = 2K \quad (2)$$

$$K = \frac{f_{HF}}{f_{LF}} \quad (3)$$

where:

η is the window length coefficient,
K is the factor of high frequency and low frequency ratio,
x is the length of the scanning window,
d is the electron beam diameter,
f_{HF} is the scanning frequency of Y-axis coil and,
f_{LF} is the scanning frequency of X-axis coil.

IOP Conf. Series: Materials Science and Engineering **785** (2020) 012037 doi:10.1088/1757-899X/785/1/012037

From Figure 4, the electron beam diameter, d is measured by using a photosensitivity paper as 9 mm. Therefore, with the length as 192.5 mm and from (5), (6) η is 10.6944, K is 5.3472.

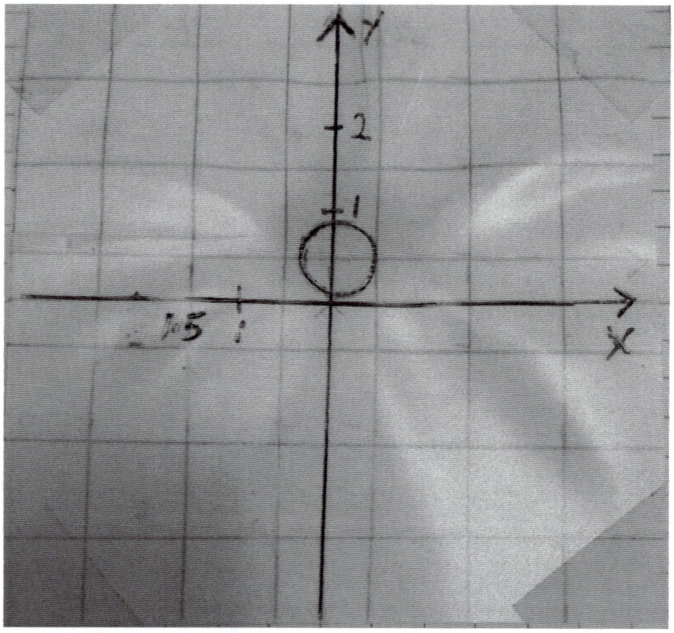

Figure 4. Electron beam size measured by photosensitivity paper.

For defined of those frequencies, the high frequency is defined as 900 Hz (< 1 kHz due to the power of the scanning power supply issue). From Equation (3), the low frequency magnitude is defined as 170 Hz.

After the frequency parameters have been identified. The second stage is to identify the required excitation current for the scanning coil. Therefore, the maximum desired magnetic field by as shown in Figure 5 must be identified, by the relationship between the magnetic field density and the required excitation current, the expression as shown in equations below [11].

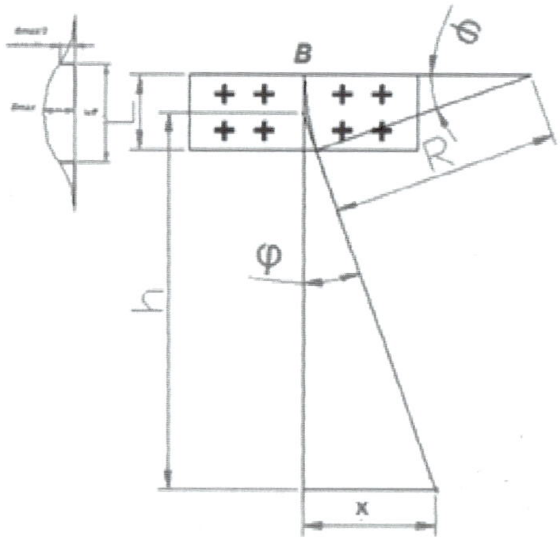

Figure 5. Electron beam trajectory with the interaction of magnetic field (h= 840 mm, x=192. 5mm and y=97.5mm) [11]

$$I_x = \frac{B_{max,x} - 0.1476}{14.2835} \quad \textbf{(4)}$$

$$I_y = \frac{B_{max,\ y} - 0.0399}{14.7297} \quad \textbf{(5)}$$

whereby,

$$B_{max} = \sin \varphi \left(\frac{BR}{l_{eff}} \right) \quad \textbf{(6)}$$

where [11], $l_{eff,x}$ as 0.1198 m and $l_{eff,x}$ as 0.1134 m.

$$BR = \frac{\sqrt{T^2 + 2TE_0}}{300} [MeV] \quad \textbf{(7)}$$

For the scanning pattern, the triangle waveform as shown in the Figure 6 is required [11],

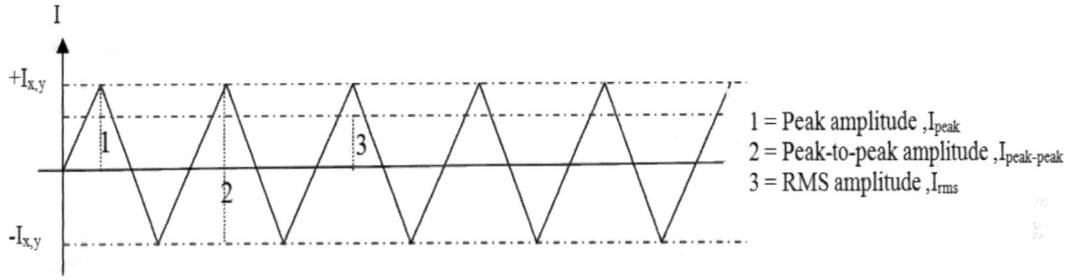

Figure 6. Triangular signal of coil current source

whereby:

$$I_{rms-\ half(x,y)} = \frac{I_{peak(x,y)}}{\sqrt{3}} \quad \textbf{(8)}$$

$$I_{rms(x,y)} = 2X \ \frac{I_{peak(x,y)}}{\sqrt{3}} = \frac{I_{peak-\ peak(x,y)}}{\sqrt{3}} \quad \textbf{(9)}$$

where:

$I_{peak\text{-}peak}$ = Required excitation current of scanning coil
I_{rms}= Observed current at the current meter or digital multimeter

According to [11] and from Equation (4) to (9), the required excitation current for energy up to 300 keV is plotted as shown in Figure 7. At energy of 300 keV, the reading or observed current must be 1.60 A and 0.84 A for X-axis and Y-axis respectively.

IOP Conf. Series: Materials Science and Engineering **785** (2020) 012037 doi:10.1088/1757-899X/785/1/012037

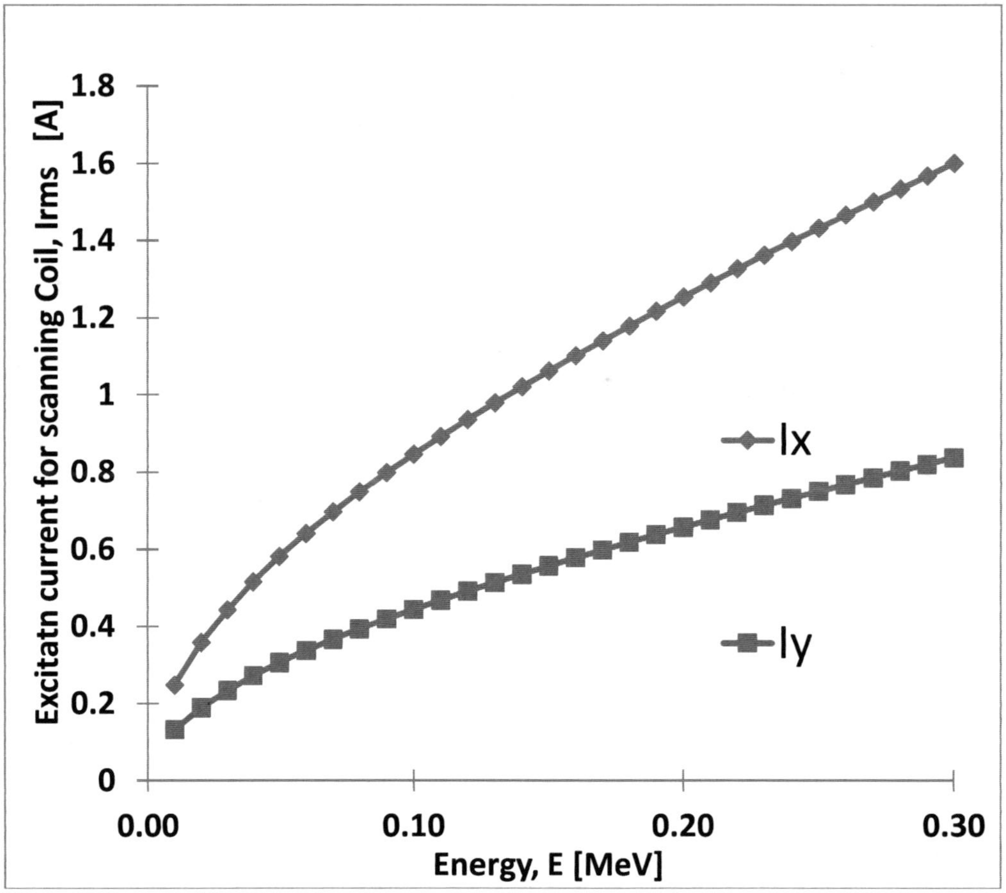

Figure 7. Required horizontal and vertical excitation current for scanning coil Vs beam energy.

3. Specs and configuration of the scanning power supply

Basically, the scanning power supply of the low energy electron accelerator is required to scan the electron beam along the titanium foil window horizontally and vertically. Whereby the frequencies have been identified as 170 Hz for X-axis and 900 Hz for Y-axis. By the simple configuration of the scanning power supply, the signal generator with triangle waveform is required and this small magnitude input must be amplified through an amplifier to supplying the desired excitation current to the scanning coils respectively as shown in Figure 8.

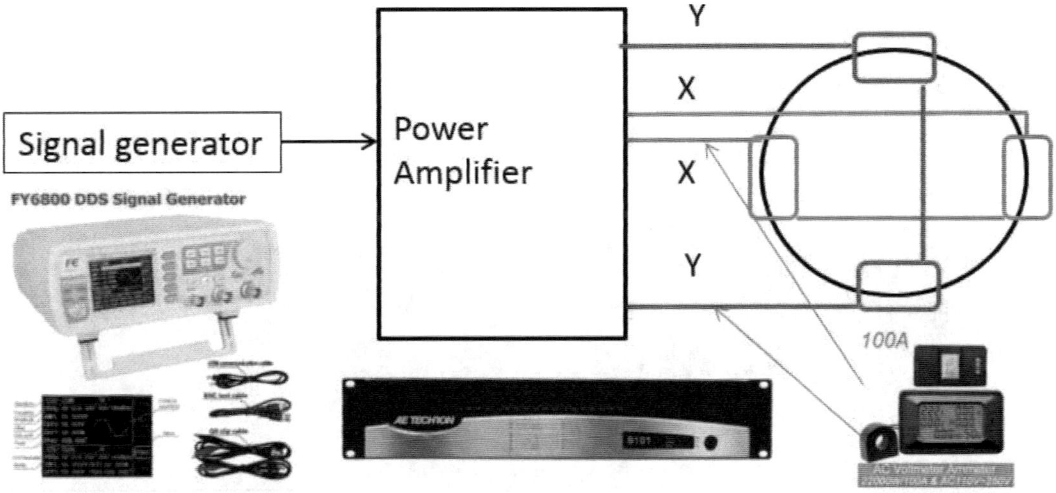

Figure 8. Configuration of the scanning power supply.

The specification of the signal generator FY6800 and power amplifier of AE 8101 as shown in Table 2 and Table 3.

Table 2. Specification of the signal generator FY6800 [12].

Parameters	Figure
Waveforms	Sine, Square (Duty Cycle adjustable), Pulse (Pulse width and cycle time can be set accurately), Triangle/Ramp, Sawtooth Wave, CMOS, Four channels TTL, DC, Half wave, Full wave, Positive Step, Inverse Step, Positive Exponent, Inverse Exponent, Lorenz Pulse, Multitone, Noise, ECG, Trapezoidal Pulse, Sinc Pulse, Narrow Pulse, Gauss White Noise, AM, FM, and other 64 sets customer-defined waveform
Waveform Length	8192 points * 14bits
Sampling Rate	250MSa/s
Vertical Resolution	14 bits
Amplitude (VPP)	Frequency≤10MHz; 1mVpp~20Vpp; 10MHz20MHz: 1mVpp~5Vpp;
Impedance	50Ω±10% (Typical)

Table 3. Specs of the power amplifier AE 8101 [13].

Parameters	Figure
Frequency Response	± 0.25 dB from 10 Hz to 20 kHz at 1 watt
Input Impedance (nominally balanced, nominally unbalanced)	10 k ohms, 5 k ohms.
Maximum Input Voltage	± 10 V balanced or unbalanced
Load Impedance	Dual: 2, 4, 8, 16 Ohm Bridge Mono: 4, 8, 16 Ohm

For this configuration, the input signal is controllable and the gain controller of the power amplifier is fixed. Whereby the excitation current is adjusted by controlling the input signal (manually) of the signal generator. The expected results have been obtained as discussed in next section.

4. Result and discussion

For the configuration as shown in Fig. 8, the results have been recorded by an oscilloscope as shown in Figure 9 and Figure 10. Whereby the parameter for Y-axis is 900 Hz and X-axis 170 Hz, input signal as 12 V and the gain factor is fixed as 10 dB.

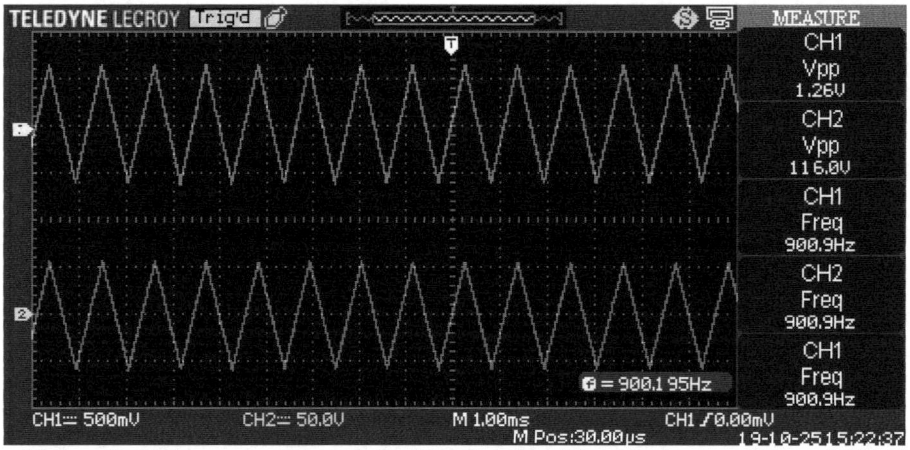

Figure 9. Voltage supplying to the Y-axis coil with triangle waveform of 900 Hz (Probe ratio is 10:1).

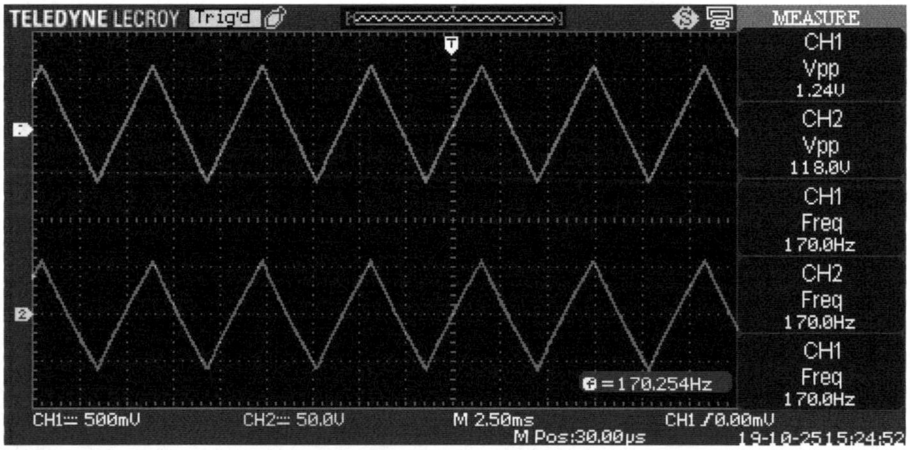

Figure 10. Voltage supplying to the X-axis coil with triangle waveform of 170 Hz (Probe ratio is 10:1).

As the result, the Y-axis coils with the frequency of 900 Hz is capable of supplying 0.23 A (rms value). Meanwhile the X-axis with 170 Hz is capable of supplying 1.02 A (rms value). Both condition with the gain of 10 dB (maximum capability of power amplifier 20 dB), but the excitation current is different and this discrepancy is caused by the impedance of the coil that is directly dependent on the input signal frequency. Therefore, we could estimate the higher power is required with higher frequency. But in the physical geometry of the window, the width of the window is much shorter than the length as shown in Figure 3 Therefore, the desired scanning path along the Y-axis is much shorter than the X-axis. It means the excitation current of the Y-axis is smaller than the X-axis as well. Although, the scanning power supply is compatible with our needs but the specification of the power amplifier should be improved to scan the electron beam along the full size of the window. Although the expected current is still a bit lower as required for 300 keV electron beam, but at the current situation this accelerator is operated at 200 keV, therefore the current specs of the scanning power supply could satisfy our need.

5. Conclusion

In this paper, parameter of the scanning power supply has been identified and developed. Furthermore, the performance of the scanning power supply has been tested and expected excitation current is observed. In near future, the scanning power supply controlled by an automated PC-based centralize control system will be studied and developed as well.

6. References

[1] Machi S 2011 Trends for Electron Beam Accelerator Applications in Industry *Rev. of Accel. Sci. and Tech.* (World Scientific) **4** p 1–10
[2] Dorman E and Emery R 2013 The University of Washington clinical cyclotron a summary of current particles and energies used in therapy, isotope production, and clinical research *Proc. of 20th Int. Conf. on Cyclotrons and Their App. 2013* (Vancouver, BC, Canada).
[3] Tsutsui H 2014 *KEK-Nuclear Malaysia-Sokendai Accel. School in Nuclear Malaysia.*
[4] Edwards D A and Edwards H T 2008 Particle colliders for high energy physics *Rev. of Accel. Sci. and Tech.* (World Scientific) **1** p 99–120
[5] Wilson P B 2008 Electron linacs for high energy physics *Rev. of Accel. Sci. and Tech.* (World Scientific) **1** p 7–41
[6] Hase Y 2014 *KEK-Nuclear Malaysia-Sokendai Accel. School in Nuclear Malaysia.*
[7] Ghazali A B, Rizal M, Rosli D, Rizal M C., Lee C H, Leo K W and Zahidee M T 2005 Development of electron beam machine at Nuclear Malaysia *Tech. and App. Accel. Meeting VIII* (Yogyakarta: *P3TM-BATAN*).
[8] Leo K W *et al.* 2017 Study on the insulation gas system of the 300 keV electron accelerator *AIP Conf. Proc.* **1799** p 050008-1–050008-8.
[9] Leo K W *et al.* 2018 Study on the window cooling system of 300 keV Electron Accelerator *IOP Conf. Series: Mater. Sci. and Eng.* **298** p 012045
[10] Leo K W *et al.* 2019 Study on the filament power supply of the 300 keV electron accelerator *IOP Conf. Series: Mater. Sci. and Eng.* **555** p 012022.
[11] Leo K W *et al.* 2016 Study on the parameters of the scanning system for the 300 keV electron accelerator *AIP Conf. Proc.* **1704** p 020010-1–020010-12.
[12] FY6800Series Fully Numerical control dual channel function/arbitrary waveform generator *User's Manual.*
[13] AE Techron 8101 datasheet

Acknowledgments

We would like to express our gratitude to Dr. Fadeev, from the Budker Institute of Nuclear Physics, Russia, for his guidance and willingness to join us at the Workshop on Electron Accelerator Technology during which he had shared his experience and knowledge with us theoretically and practically.

IOP Conf. Series: Materials Science and Engineering **785** (2020) 012038 doi:10.1088/1757-899X/785/1/012038

Study of the magnet for 1 MeV cyclotron

Maslina Mohd Ibrahim[1,a)], Leo Kwee Wah[2], Noor Farhana Husna A Aziz[1]

[1]Instrumentation and Automation Centre, Technical Support Division, Malaysian Nuclear Agency, Bangi, 43000 Kajang, Malaysia.
[2]Accelarator Development Center, Technical Support Division Malaysian Nuclear Agency, Bangi, 43000 Kajang, Malaysia.

[a]maslina_i@nm.gov.my

Abstract. The cyclotron is a circular particle accelerator. It was invented by Lawrence and Livingston in 1932. Currently, at least 4 cyclotrons are operating in Malaysia with the energy range of 8.5 MeV to 18 MeV. Typically, those cyclotron facilities are used for the medical radioisotope production. The potential of cyclotron with its applications has embarked us to the conceptual design of 1 MeV Cyclotron. Magnet as an essential component of Cyclotron to guide the beam into the circular desired orbit. Therefore, this paper will discuss the design of this magnet for 1 MeV Cyclotron by using different structure of magnet yoke with the aim to achieve a uniform fringe magnetic field of 1.7 Tesla. As the result, the profile at the median plane of the air gap and the characteristic of edge focusing is simulated by using the 2D Finite Element Method Magnetics (FEMM 4.2) software.

1. Introduction

Cyclotron is a circular type of accelerator, which the beam is injected or produced at the center and propagates inside the guiding magnetic field within vacuum region. When the beam is passing through the acceleration components so-called "DEE" (Dee electrode is a hollow structure as acceleration medium), the beam must be synchronized will the RF frequency within the DEE gap. Such synchronization is essential for acceleration. Overall view and principle of the cyclotron is shown in Figure 1. To keep the beam into the ideal orbit, a desired and uniform magnetic field is important. Based on the previous study, the 30 cm Dees with a thickness of 5.07cm was designed. The air gap of the magnet yoke was decided between 5.5 cm to 6 cm. In this study, we chose 6 cm air gap from the magnetic yoke as shown in Figure 2. Two difference design was shown in Figure 2 where (a) is the C-structure and (b) is the H-structure of the core.

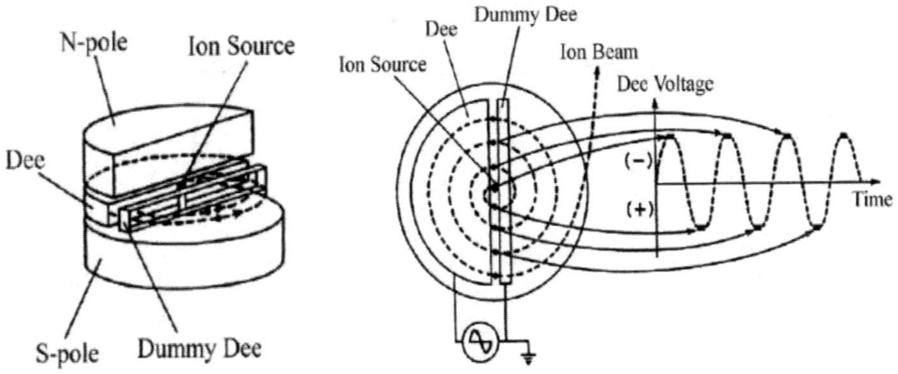

Figure 1. Overview of cyclotron with the acceleration mechanism.

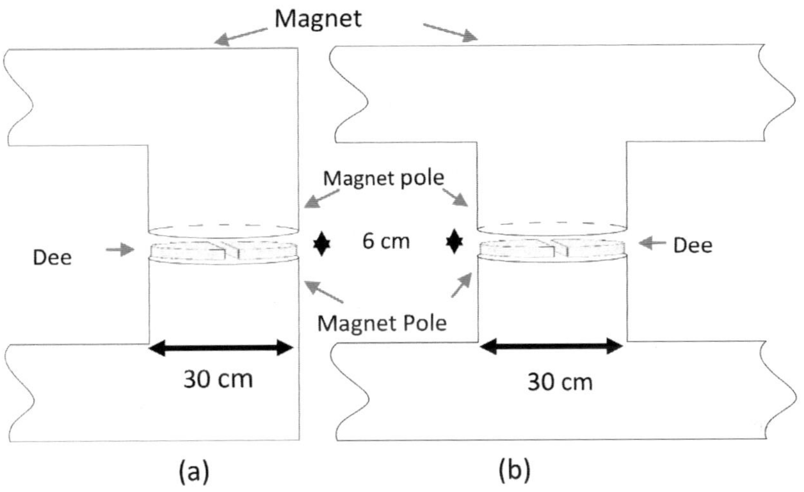

Figure 2. Parameter of magnetic yoke design for 1 MeV cyclotron.

2. Design and simulation

To begin the process of designing the magnetic circuit, several factors should be taken into consideration such as; the nonlinear magnetic properties of steel, fringing of magnetic flux, possible losses at junctions and boundaries, limits on voltage, current, and/or inductance. The ferromagnetic material used for this design is 1018 steel and by calculation, NI = Hsls + Hglg where H is magnetic field intensity and l is the length of the magnetic field in a medium. Whereas s is steel and g is gap, the turns chose for this simulation is 2610 turns. To achieve 1.7 Tesla at the gap area, the current applied for C design and H design are 54 Amps and 40 Amps respectively. The simulation was performed by using a finite element analysis. Finite element analysis calculates all the non-ideal effects as well as the flux density, field intensity, and inductance using the numerical solution of Maxwell's equations and explicitly treating nonlinear materials. In this study, called Finite Element Method Magnetic Version 2 (FEMM 4.2), open source software by David Meeker was used. Circuit properties is as Table 1 below.

Table 1. Circuit properties of C and H structures.

Circuit Properties	C	H
Total Current	54 Amps	40 Amps
Voltage Drop	84.564 V	62.64 V
Flux Linkage	491.917 Wb	492.474 Wb
Flux / Current	9.10957 H	12.3118 H
Voltage / Current	1.566 Ohms	1.566 Ohms
Power	4566.46 W	2505.6 W

Figure 3 below shows the meshed domain of the model for C-structure and H-structure respectively by using the circuit properties at Table 1 in FEMM 4.2.

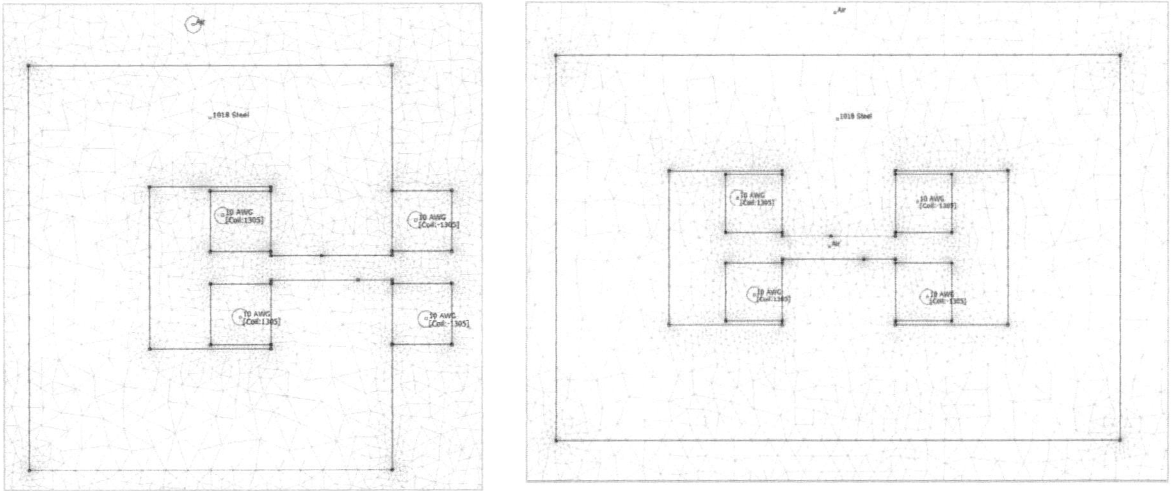

Figure 3. The meshed domain of the model for (a) C and (b) H magnet yoke structure.

3. Result and discussion

Both designs told us that the beam stability condition such as 0< n<1 is fulfilled, whereby:

$$\partial B/_{\partial R} = -0.0043\ (C - strucutre)\ and\ -0.0053\ (H - structure) \quad (1)$$

Further optimization of the turns number of excitation coil, excitation current and heat issue must be considered with more detail study by using 3D simulation code. Figure 4, Figure 5 and Figure 6 below show the result of the simulation using FEMM 4.2 which fulfilled the equation above.

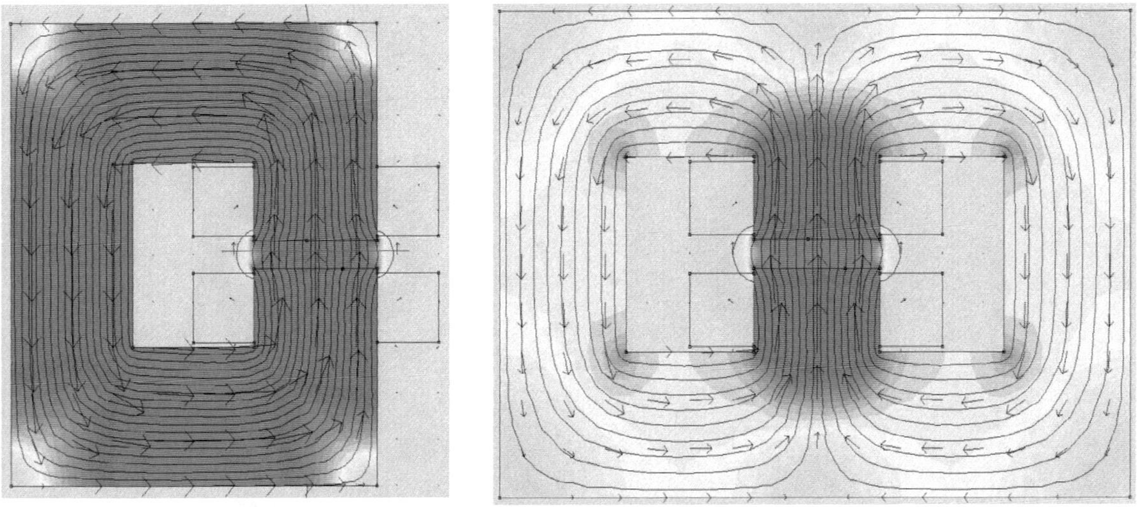

Figure 4. The simulation result of B vector plot.

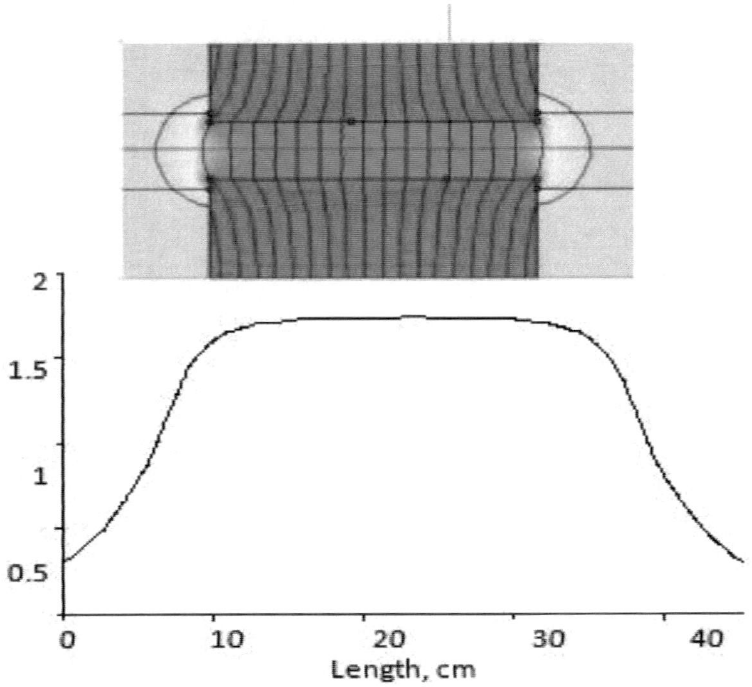

Figure 5. B-value profile at the center of the gap

IOP Conf. Series: Materials Science and Engineering **785** (2020) 012038 doi:10.1088/1757-899X/785/1/012038

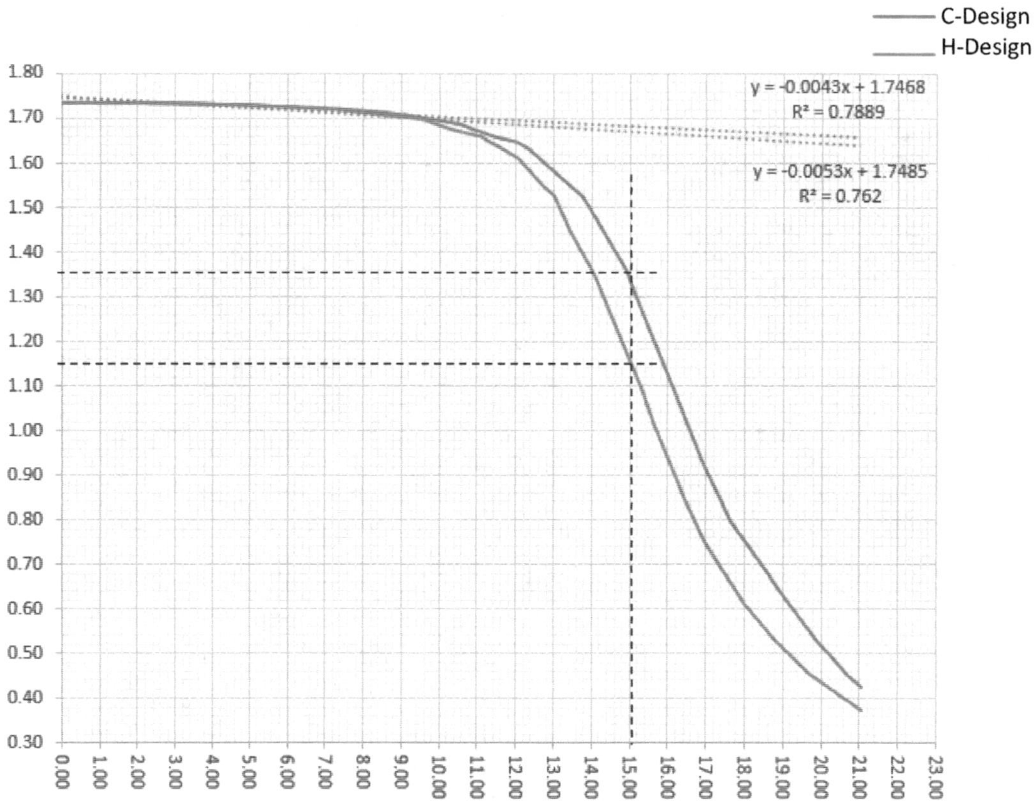

Figure 6. The B-value profile at the edge of magnet for both design.

4. Conclusion

The design of the magnet for 1 MeV cyclotron has been study by the 2D simulation code. The preliminary results indicated that this designed magnet is applicable for 1 MeV Cyclotron. For ideal case, the 3D simulation of magnetic distribution along the median plane is required.

5. References

[1] Koeth T 2005 Report on the 12-inch cyclotron magnet study: measurements, modeling, and future plans http://www.physics.rutgers.edu/cyclotron/papers/12_inch_magnet_studies 4.pdf,.

[2] Koeth T W, Hine G A, Hoffman D E, Krutzler J E, Ponter T S, Rosenberg A J, Ruisard K J, Schneider W S 2011 Comparison of azimuthally varying with constant gradient magnetic fields with the rutgers 12-inch cyclotron http://www.physics.rutgers.edu/cyclotron/papers/AVF_study_Dec_31_2011.pdf, December.

[3] D Meeker 2019 FEMM http://www.femm.info/wiki/HomePage.

[4] Zaremba S and Kleeven W 2015 Cyclotrons magnetic design and beam dynamics *CERN Accel. School, Accel. for Med. App.* (Vösendorf, Austria).

IOP Conf. Series: Materials Science and Engineering **785** (2020) 012039 doi:10.1088/1757-899X/785/1/012039

Examination of corrosion under insulation using gamma ray computed tomography

S Sipaun[1,a)], M F Ab Rahman[1], H Hasan[1], M R Shari[1], A Mahmood[1], N Abdullah[1], M Mustapha[1]

[1]Malaysian Nuclear Agency, Plant Assessment Technology Group (PAT), Industrial Technology Division, 43000 Kajang, Bangi Selangor, Malaysia.

[a]susan@nuclearmalaysia.gov.my

Abstract. In this paper, we propose a screening and scanning approach using gamma ray transmission measurements and computed tomography (CT) to detect and locate corrosion areas in a pipeline. Screening obtains interval measurements of collimated gamma ray along the length of pipe, while scanning gives cross sectional images at the investigated area. The CT scanner consists of a clamped on mechanical gantry, a 1" x 1" NaI(Tl) scintillation detector, 11mCi Ba-133 source, 15mm x 5mm lead collimators and Ludlum scaler ratemeter. This technique was shown to be an effective way to determine the wall loss and erosion areas of at 2.5mm thickness. Several locations of an insulated pipe were screened and scanned. Using a portable gamma ray CT scanner we present the results, where the cross sectional image can indicate the occurrence of corrosion under insulator.

1. Introduction

Corrosion is deterioration of metal that weakens its structure and form. It can happen when the metal is exposed to oxygen and water. This condition allows rust to form through electrochemical reaction/oxidation. Corrosion in pipelines can be difficult to detect especially when the pipes are insulated and its surface appears to be in good condition. Abdullah et. al. suggested that the CT technique can also be used to determine the extent of corrosion under insulation (CUI), to detect blockages and to measure the thickness of deposit/materials built-up on the walls [1]. X-ray CT was used to trace cracks in a ceramic pot, and differentiate ceramic and metal material through the contrasting shades of black and white [2]. Corrosion detection through micro tomography and limited angle tomography has also been suggested by Van Steen and Ewert [2][3]. Recent works show that CT is also used in dimensional metrology and engineering inspection [4][5].

This paper discusses about a two-step process, i.e. screening and scanning of pipes for corrosion detection using gamma ray transmission technique. The first-step is to narrow down the location of suspected corroded area along the length of pipe via interval gamma ray measurements along the investigated pipe. The second-step is to perform a localized computed tomography (CT) scan of that pipe section where structural anomalies are suspected to occur. CT image reveals the pipe's inner structure and enables the analysis of wall loss and erosion without opening the pipe's insulation.

1.1. Computed tomography
Gamma ray computed tomography is an imaging technique that uses the penetrative gamma rays to obtain a two-dimensional view (2-D) cross sectional view of an object. When a beam of gamma ray penetrates a material, the intensity of the beam will be reduced through photoelectric effect and Compton scattering. Material of different densities and composition attenuates the incident beam, and this interaction is described in Equation (1).

$$I = I_o e^{-\mu x} \tag{1}$$

where,
I_o = intensity of incident beam
I = intensity of transmitted beam
μ = linear attenuation coefficient
x = sample thickness.

Figure 1 illustrates gamma ray beam going through an inhomogeneous material and Equation 2 gives the relation of the measured transmission ratio. The ratio corresponds to the sum of the attenuation coefficients multiplied by the corresponding path length of all the elements where the ray traverses. The linear attenuation coefficient, μ for gamma rays is dependent on the given energy, material density and its atomic number. These transmission data are essentially a density distribution map of the object. Information along the known paths is reconstructed from a series of projections and is transformed into a cross sectional image. This cross sectional view allows the viewing of the objects internal non-destructively.

$$\ln (I_o / I) = \mu(x-x') + \mu' x' \tag{2}$$

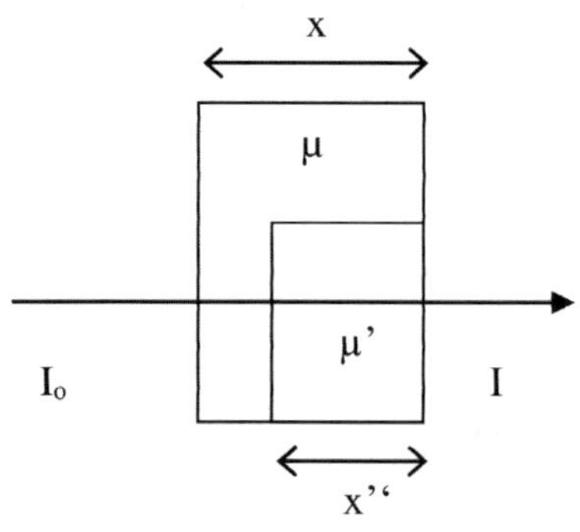

Figure 1. γ-ray transmission through material.

2. Gamma ray pipe inspection
A portable gamma ray CT scanner was developed at the Malaysian Nuclear Agency to inspect industrial pipeline for pipe diameters not more than 50cm [2]. It is a clamp-on system where its mechanical gantry can be fixed onto a vertical or horizontal pipe. The CT scanner consists of a clamped on mechanical gantry, a 1" x 1" NaI(Tl) scintillation detector, 11mCi Ba-133 source, 15mm x 5mm lead collimators and Ludlum scaler ratemeter. The scanner is equipped with a dedicated data acquisition and image reconstruction software that uses filtered back projection algorithm. The pipe sample was placed in the gantry, where the pipe is in the CT systems' field of view. Scan parameters were set according to the sample sizes. Gamma ray pipe inspection involves positioning a gamma ray

source and a radiation detector on opposite sides of the medium of interest and relating changes in the transmitted intensity to changes in the mass per unit area of the material.

2.1. Step1: screening

The insulated 4" pipe was inspected along its length at every 5 cm using 11 mCi Ba-133. The collimated ray passes through the insulation and pipe section and the transmitted ray is recorded by the 1" x 1" NaI(Tl) detector (Figure 2).

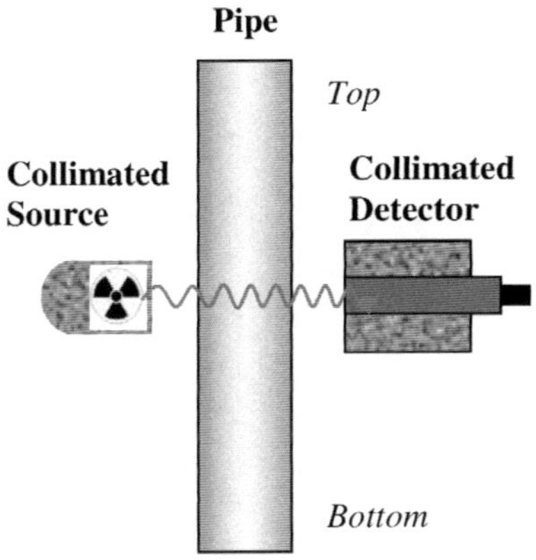

Figure 2. Screening: pipe is inspected along the pipe's length to locate suspected corroded section.

2.2 Step 2: scanning

CT scan was performed at the pipe section (20cm from the top) where suspected wall loss has occurred using a portable gamma ray computed tomography machine. Figure 2 shows the scanning in progress using 11mCi Ba-133. Pipe thickness is 5mm and the insulation thickness is 4cm. At the circumference of the pipe, we added a 1mm copper plate (Plate 1), 2mm (Plate 2) and 3mm (Plate 3) steel so that the rays 'slice' through the plates, insulation and pipe.

Figure 3. CT scanning setup at pipe location 20cm from the top. Three metal plates were added on the insulation surface.

IOP Conf. Series: Materials Science and Engineering **785** (2020) 012039 doi:10.1088/1757-899X/785/1/012039

3. Screening and scanning results

Screening is an interval checking of the pipe for possible locations of suspected wall loss or pipe structure anomalies. In a pipe with intact structure, the density of pipe along its length should remain the same. Figure 4 shows three points (in red) where higher transmitted gamma rays were detected, which indicates the thinning of pipe wall. Figure 5 shows the area around the pipe circumference where there is wall thinning at locations 20cm and 35cm from the top of the pipe, and further narrows down the suspected CUI location. Figure 6 shows the image taken at 20cm position.

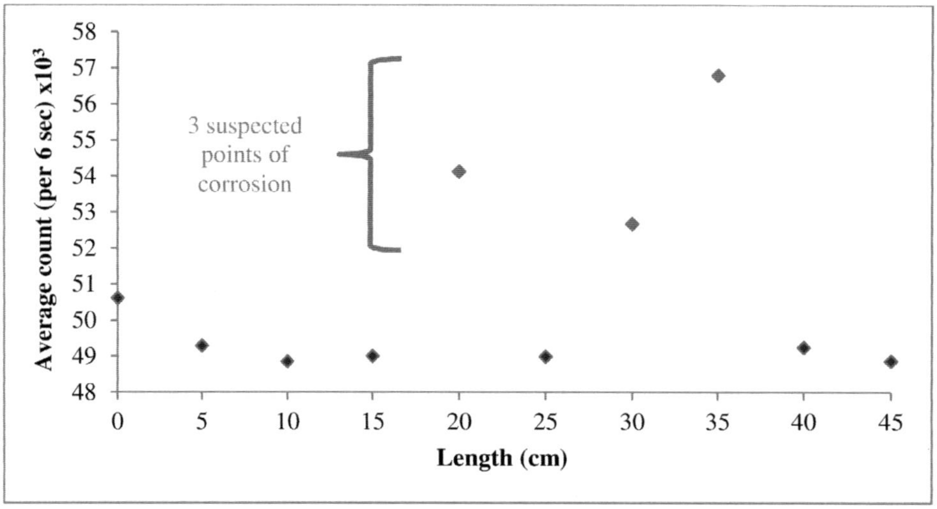

Figure 4. Screening Data: Suspected points of corrosion.

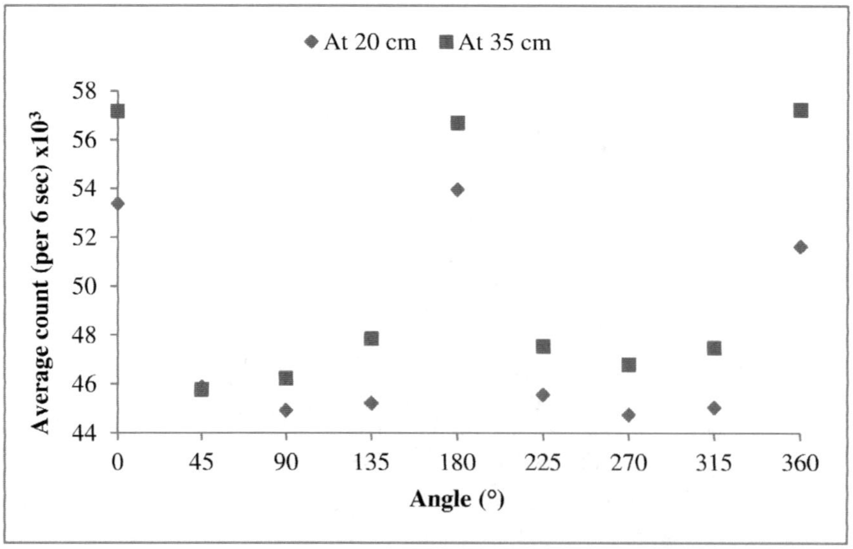

Figure 5. Screening Data: Measurements around the pipe section at 20cm and 35cm from the top

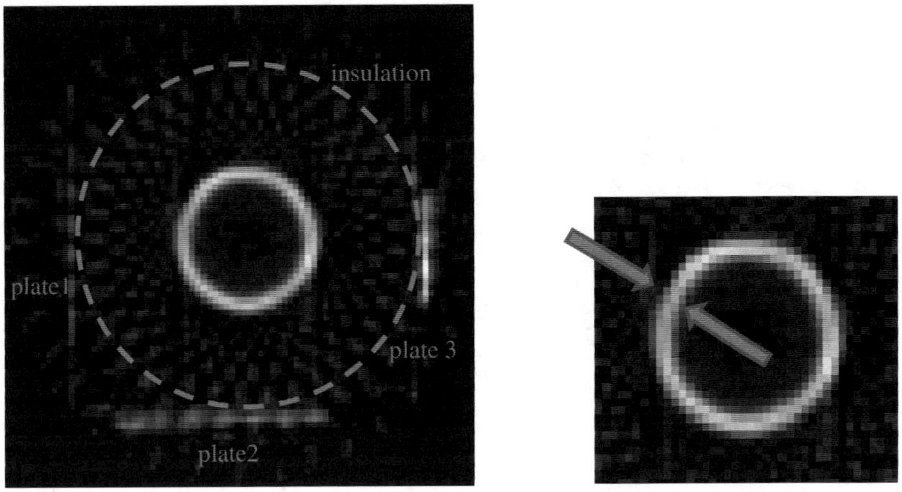

(a) (b)

Figure 6. Scanning: Cross sectional CT image at suspected wall loss section. (a) Image shows the 3 plates, insulation area and the pipe. (b) Zoom in to the pipe. Its wall loss is indicated by the deterioration of image pixels and analysis of the projection data.

In Figure 6, the dark shades represent air and the white/grey shade represent the metal pipe. Plate 1 is a faint shade of grey whereas Plate 2 is greyer and thicker, and Plate 3 is white. Although copper (8.86 gcm^{-3}) is more dense compared to steel (8.05 gcm^{-3}), however its thickness is 1mm compared to 2mm and 3mm steel respectively and therefore absorbs lesser gamma rays. Some artifacts are present in the image as well and it is due to data averaging and effects from the 3 metal plates. Further analysis of the projection data can estimate the thickness of pipe that is lost.

4. Summary

This project proposed the use of gamma ray transmission technique to detect and locate corrosion areas in a pipeline as a two-step process. Step 1 of the 'Screening-Scanning' process obtains measurements along the pipe's length to narrow down suspected areas of pipe anomaly. Step 2 scans the pipe section to reveal the wall loss through CT images. This technique was shown to be an effective way to determine the wall loss and can be used to identify coroded sections.

5. References

[1] Abdullah J, Hassan H, Shari M R, Mohd Ibrahim M, Yussup N, Ithnin H, Mahmood A A 2015 Development and implementation of a portable nucleonic computed tomography system with clamp-on features for engineering inspection *Jurnal Teknologi Special Issue on Sensor Tech. and Cont. Syst. Appl.* **77**(17) p3.

[2] Sipaun S M and Abdullah J 2006 Aspects of X-ray tomographic imaging. Young Researchers Conference on Applied Sciences **1** *Phys. and Mater. Sci.* p 123-127.

[3] Van Steen C, Wevers M and Verstrynge E 2017 X-ray computed tomography for the detection of corrosion-induced damage at the reinforcement-concrete interface 7th *Conf. on Indust. Comp. Tomog iCT 2017* (Leuven, Belgium) accessed from http://www.ndt.net/?id=20802

[4] Ewert U, Tschaikner M, Hohendorf S, Bellon C, Haith M I, Huthwaite P, Lowe M J S 2016 Corrosion monitoring with tangential radiography and limited view computed tomography, *AIP Conf. Proc.* **1706** doi: 10.1063/1.4940574 p 110003
[5] Saengchantr D, Srisatit S and Chankow N 2019 Development of gamma ray scanning coupled with computed tomographic technique to inspect a broken pipe structure inside laboratory scale vessel *Nucl. Eng. and Tech.* **51** p 800-806
[6] Villarraga-Gomez H, Herazob E L and Smith S T 2019 *X*-ray computed tomography: from medical imaging to dimensional metrology *Precision Eng.* **60** p 544-569
[7] Kak A C and Slaney M 1988 Principles of CT imaging (IEEE Press) ISBN 0-87942-198-3

Acknowledgments
This project is part of preliminary work on "Petronas Technology Challenge 4: Effective Field Inspection Tool to Detect and Assess Corrosion Under Insulator (CUI) For Nozzles and Piping Systems". The authors thank the staff of Bahagian Kejuruteraan (BKJ) for the mechanical parts support, and Bahagian Sokongan Technical (BST) for data acquisition GUI work.

IOP Conf. Series: Materials Science and Engineering **785** (2020) 012040 doi:10.1088/1757-899X/785/1/012040

Geometrodynamics of spacetime "surfing"

Anuar Alias[1,a)]

[1]National Center for Particle Physics, University of Malaya, 50603 Kuala Lumpur, Malaysia

[a]aliasanuar37@gmail.com

Abstract. The idea of hyper-fast travel in the framework of General Relativity has firstly been introduced by Miguel Alcubierre. It was interesting, as it shows the possibility to *locally modified spacetime* surrounding an intended vehicle to move at huge speed which is rather different than the method of *spacetime topological breaking* as in a wormhole theory. Instead of *puncturing through* a hole in spacetime as in a wormhole, this method of spacetime geometrical modification, manipulates the geometrodynamics contraction and expansion of spacetime curvature so that the intended object's hypersurface *surfs on* the surrounding spacetime hyperspace. In this paper we will explore in more detail mathematically, how the concept has been derived in the framework of General Relativity. Starting from developing the metric to the expression of energy density that governs the spacetime "surfing" geometrodynamics in relation with the necessary parameters describing the spacetime warp bubble.

1. Introduction

To be a spacefaring civilization with current rocket propulsion technologies that is utilizing Newtonian orbital mechanics, humankind could only travel within the solar system and yet still acquire years of travel to surpass the outermost regions of the solar system. Thus, not to mention the sensible of humanity's dreams for interstellar or perhaps intra-galactic travel, if just by using Newtonian mechanics-base propulsion technologies, because even if such endeavour may approach up to 99% the speed of light [1] which is an almost impossible feat by itself, the journey will still take several years, even of tens to hundreds of thousands years [2]. Thus the speed of light, tremendously huge as it may seems, but nevertheless, if the astronomical scale of the universe is to be considered, light speed is still slow. Therefore, the idea of hyper-fast travel that could surpass many folds the speed of light is very much an interesting subject to study [3]

Surpassing the speed of light seems to violate the law of physics. Yes this notion is true if Special Relativity (SR) is taken into consideration where SR tells that any subject of travel that holds any amount of mass could never even achieve the light speed, not to mention exceeding it. This is due to increase in relativistic mass asymptotically toward infinity as the subject's speed approaching the speed of light, consequently require infinite amount of energy. However, in General Relativity (GR) the concept of spacetime geometrodynamics [4] seems to show that the dynamics of spacetime itself does not constrained toward any limit. These notion has been proven at least in the cosmological scale level when we consider the signatures of the inflationary period of the universe [5] and the currently observable phenomenon of accelerating expansion of the universe [6]. General Relativity has shown that the dynamics of stretching and folding of spacetime may exceed the speed of light by either short cutting the spacetime as in black hole and wormhole [7] or superluminally stretching the spacetime via *expansion* as during the inflationary period of post big bang or via *contraction* as during the theoretically possible scenario of the big crunch period [8].

IOP Conf. Series: Materials Science and Engineering **785** (2020) 012040 doi:10.1088/1757-899X/785/1/012040

Thus, the idea of superluminal travel is not about violating SR but rather a geometrodynamically possible scenario in various GR frameworks. Essentially there are two concepts of superluminal travel that are of "short cutting" and "surfing" on the spacetime. Short cutting concept is about reducing tremendously the distance between two different coordinates as in the wormhole theory which require topological breaking of the spacetime itself of either naturally (from subatomic quantum foam) [9] or induce (through entanglement) [10]. The surfing concept, that is the subject of this paper is about embedding a bubble of warped spacetime hypersurface onto the surrounding hyperspace and using the transition region of warped spacetime properties between the bubble and the surrounding hyperspace that defines the bubble's extrinsic curvature to govern its dynamic's character of expansion and contraction of spacetime. In this paper we investigate the idea of spacetime "surfing" which was firstly introduced by Alqubierre [3] through exploring the mathematical derivation underlying the subject in a rather elaborate details.

2. The metric property of warp bubble

The concept of spacetime "surf" is very much about manipulating spacetime locally, thus we may start by considering a flat spacetime metric that not only describes the hyperspace surrounding the "surfing" spacetime region but also the "interior" of the surfing spacetime region that is by itself a hypersurface [11] with respect to the surrounding region.

$$ds^2 = -dt^2 + dx^2 + dy^2 + dz^2 \tag{1}$$

Imagine the "surfing" spacetime region as a warp bubble that moves along the x axis with a velocity of v,

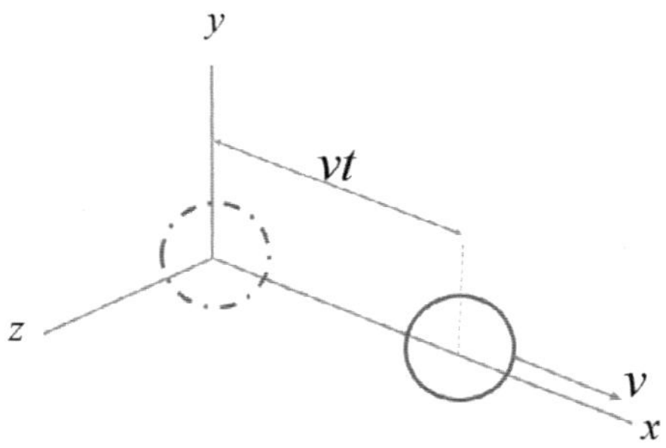

Figure 1. Spacetime warp bubble "surf" (moves) along x axis with velocity of v.

so we may rewrite the metric (1) as:

$$ds^2 = -dt^2 + (dx - vdt)^2 + dy^2 + dz^2 \ , \tag{2}$$

which preserved the essence of equation (1) whereby, if the bubble doesn't move, the equation (2) will becomes (1) again. For representing the characteristic of the bubble where the center is located at $r=0$ and the radius of the bubble can be assigned as R we may consider a function $f = f(r)$ where $f=1$ always within the bubble's radii $|r| = R$,

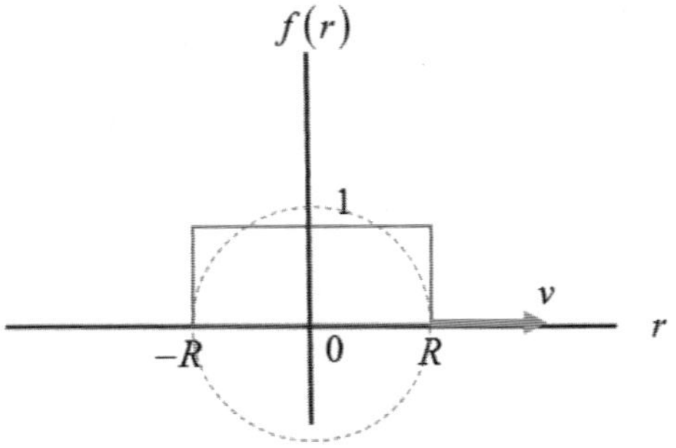

Figure 2. Function $f=f(r)$ characteristic with bubble radii.

Thus, the essence of equation (2) is always preserved and we may rewrite equation (2) with the inclusion of the function $f=f(r)$ as:

$$ds^2=-dt^2+\left(dx-vf(r)dt\right)^2+dy^2+dz^2 \ . \tag{3}$$

If we consider condition at the center of the warp bubble where the spatial coordinate is $(vf(r)t,0,0)$ which then obviously shows equation (3) reduce to $ds^2=-dt^2$. Since the different of proper time is defined as minus of the line element thus $d\tau^2=-ds^2$ therefore $dt^2=-ds^2=d\tau^2$ which obviously implies the *null-time dilation* characteristic of the bubble as proper time is the same as coordinate time $\tau=t$. Inside the bubble, spacetime condition is as normal as in a stationary non relativistic condition. We may now expand this notion to find the geodesic and free fall characteristic of the warp bubble. By relationship of proper time and line element metric, we may derive the extremum [12] of the proper time as below:

$$\begin{aligned}
&d\tau^2=-ds^2=-g_{\mu\nu}dx^\mu dx^\nu \\
&\frac{d\tau^2}{d\lambda^2}=\left(\frac{d\tau}{d\lambda}\right)^2=-g_{\mu\nu}\frac{dx^\mu}{d\lambda}\frac{dx^\nu}{d\lambda}, \\
&\frac{d\tau}{d\lambda}=\sqrt{-g_{\mu\nu}\frac{dx^\mu}{d\lambda}\frac{dx^\nu}{d\lambda}}, \\
&\tau=\int\sqrt{-g_{\mu\nu}\frac{dx^\mu}{d\lambda}\frac{dx^\nu}{d\lambda}}d\lambda, \\
&\tau=\int\sqrt{-g_{\mu\nu}\dot{x}^\mu\dot{x}^\nu}d\lambda \ ,
\end{aligned} \tag{4}$$

which is the extremum expression of the proper time. We may obtain the Lagrangian from Equation (4). But before expanding (4) to obtain the Lagrangian we may have to expand and rearrange metric equation (3) as below

$$\begin{aligned}
ds^2&=-dt^2+dx^2-2vfdxdt+v^2f(r)^2dt^2+dy^2+dz^2, \\
&=-\left(1-v^2f(r)^2\right)dt^2-2vf(r)dxdt+dx^2+dy^2+dz^2 \ , \\
&=-\left(1-v^2f(r)^2\right)dt^2-vf(r)dxdt-vf(r)dtdx+dx^2+dy^2+dz^2,
\end{aligned} \tag{5}$$

so we identify the metric components which are:

$$g_{00}=v^2f(r)^2-1,\quad g_{01}=g_{10}=-vf(r),\quad g_{11}=g_{22}=g_{33}=1. \tag{6}$$

By (4): $\tau=\int\sqrt{-g_{00}\left(x^0\right)^2-2g_{10}x^1x^0-g_{11}\left(x^1\right)^2-g_{22}\left(x^2\right)^2-g_{33}\left(x^2\right)^2}d\lambda$,

$$=\int\sqrt{\left(1-v^2f^2\right)t^2+2vfxt-x^2-y^2-z^2}d\lambda \ ,$$

$$\Rightarrow\left(\frac{d\tau}{d\lambda}\right)^2=\frac{d\tau^2}{d\lambda^2}=\left(1-v^2f^2\right)t^2+2vfxt-x^2-y^2-z^2=-g_{\mu\nu}x^\mu x^\nu \ . \tag{7}$$

From the pure definition of Lagrangian $L=\frac{1}{2}q^2$ where $q=s(x(t))$, as a coordinate position of a metric, thus:

$$q^2=s^2=(-\tau)^2=\frac{-d\tau^2}{d\lambda^2}=\frac{ds^2}{d\lambda^2}=2L \quad \text{since } -ds^2=d\tau^2;\ ds^2=-d\tau^2 \tag{8}$$

and from (7) and (8) $-g_{\mu\nu}x^\mu x^\nu=\left(1-v^2f^2\right)t^2+2vfxt-x^2-y^2-z^2=\frac{d\tau^2}{d\lambda^2}=\frac{-ds^2}{d\lambda^2}$ therefore:

$\left(\frac{ds}{d\lambda}\right)^2=\frac{ds^2}{d\lambda^2}=\frac{-d\tau^2}{d\lambda^2}=q^2=g_{\mu\nu}x^\mu x^\nu=-\left(1-v^2f^2\right)t^2-2vfxt+x^2+y^2+z^2=2L$, thus the Lagrangian is:

$$L=\frac{1}{2}\left(-\left(1-v^2f^2\right)t^2-2vfxt+x^2+y^2+z^2\right), \tag{9}$$

and rearrange as,

$$L=\frac{1}{2}\left(-t^2+y^2+z^2+\left(x-vft\right)^2\right). \tag{10}$$

We may parameterize the geodesic using proper time itself taking $\lambda=\tau$ thus the dot indicates the derivative with respect to τ . By Euler Lagrange equation $\frac{\partial}{\partial\tau}\left(\frac{\partial L}{\partial x^\mu}\right)-\frac{\partial L}{\partial x^\mu}=0$ and from (10) we have 4 equations as follow

$$\frac{\partial}{\partial\tau}\left(\frac{\partial\left(-t^2+y^2+z^2+\left(x-vft\right)^2\right)}{\partial t}\right)-\frac{\partial\left(-t^2+y^2+z^2+\left(x-vft\right)^2\right)}{\partial t}=0 \ ,$$

$$\Rightarrow\frac{\partial}{\partial\tau}\left(-t-vf\left(x-vft\right)\right)+\left(x-vft\right)t\frac{d(vf)}{dt}=0 \ , \tag{11}$$

$$\frac{\partial}{\partial\tau}\left(\frac{\partial\left(-t^2+y^2+z^2+\left(x-vft\right)^2\right)}{\partial x}\right)-\frac{\partial\left(-t^2+y^2+z^2+\left(x-vft\right)^2\right)}{\partial x}=0 \ ,$$

$$\Rightarrow\frac{\partial}{\partial\tau}\left(x-vft\right)+\left(x-vft\right)t\frac{\partial(vf)}{\partial x}=0 \ , \tag{12}$$

$$\frac{\partial}{\partial\tau}\left(\frac{\partial\left(-t^2+y^2+z^2+\left(x-vft\right)^2\right)}{\partial y}\right)-\frac{\partial\left(-t^2+y^2+z^2+\left(x-vft\right)^2\right)}{\partial y}=0,$$

$$\Rightarrow\frac{\partial y}{\partial\tau}+\left(x-vft\right)t\frac{\partial\left(vf\right)}{\partial y}=0\ , \tag{13}$$

$$\frac{\partial}{\partial\tau}\left(\frac{\partial\left(-t^2+y^2+z^2+\left(x-vft\right)^2\right)}{\partial z}\right)-\frac{\partial\left(-t^2+y^2+z^2+\left(x-vft\right)^2\right)}{\partial z}=0,$$

$$\Rightarrow\frac{\partial z}{\partial\tau}+\left(x-vft\right)t\frac{\partial\left(vf\right)}{\partial z}=0\ , \tag{14}$$

where obviously the solution shall be

$$x^{\mu}=\left(x^0,x^1,x^2,x^3\right)=\left(t,x,y,z\right)=\left(1,vft,0,0\right)\ , \tag{15}$$

which are 4 velocities for Eulerian observers that follow timelike geodesics. Since $x=v(t)f\left(r(t)\right)t$, as at $r=0$, $f=1$ and moreover $\lambda=\tau$ so that $t=dt/d\lambda=dt/d\tau=1$ (since $t=\tau$ in the warp bubble), thus the spatial velocity of the warp bubble is at v which does not constrained by speed limit of light speed. This notion of timelike characteristic in the warp bubble can also be proven more by considering the following by referring to the metric equations (6)

$$\begin{aligned}
x_{\mu}=g_{\mu\nu}x^{\nu}\quad\text{for } x_0&=g_{0\nu}x^{\nu}=g_{00}x^0+g_{01}x^1+g_{02}x^2+g_{03}x^3\\
&=g_{00}x^0+g_{01}x^1\quad\text{since } g_{02}=g_{03}=0,\\
&=\left(v^2f^2-1\right)\frac{\partial t}{\partial\tau}+\left(-vf\right)\frac{\partial x}{\partial\tau},\\
&=\left(v^2f^2-1\right)-\left(vf\right)v\quad\text{since }\frac{\partial t}{\partial\tau}=1,\ \frac{\partial x}{\partial\tau}=\frac{\partial x}{\partial t}=v\ ,\\
&=\left(v^2-1\right)-\left(v\right)v\quad\text{since } f=1,\\
x_0&=-1\ ,
\end{aligned}\tag{16}$$

$$\begin{aligned}
\text{for } x_1&=g_{1\nu}x^{\nu}=g_{10}x^0+g_{11}x^1+g_{12}x^2+g_{13}x^3,\\
&=g_{10}x^0+g_{11}x^1\quad\text{since } g_{12}=g_{13}=0,\\
&=\left(-vf\right)\frac{\partial t}{\partial\tau}+\frac{\partial x}{\partial\tau},\\
&=\left(-v\right)+v,\\
x_1&=0.
\end{aligned}\tag{17}$$

The rest, that is $x_2=x_3=0$ obviously, which therefore:

$$x_{\mu}=\left(-1,0,0,0\right) \tag{18}$$

IOP Conf. Series: Materials Science and Engineering **785** (2020) 012040 doi:10.1088/1757-899X/785/1/012040

is defined as covector that is normal to the hypersurface [12] given by $dt=0$, and by equation (15) $x_\mu x^\mu=-1$ confirming the timelike characteristic of the warp bubble.

Defining $\beta=-vf(r)$ we may rewrite equation (5) as below:

$$ds^2=-\left(1-\beta^2\right)dt^2+2\beta dxdt+dx^2+dy^2+dz^2. \tag{19}$$

In tensorial form:

$$\beta^2 = \beta^i \beta_i ,$$
$$dx^2 + dy^2 + dz^2 = dx_j dx^j = \delta_{ij} dx^i dx^j ,$$
$$\beta dx = \beta_i dx^i , \tag{20}$$

where $i, j = 1, 2, 3$, which by equation (5) obviously shows that $\beta = -v$ is only in the direction of the bubble's motion thus otherwise it is zero. Therefore for this case where the bubble is "surfing" along the x axis only $\beta_x = \beta_1 = -v$ while others $\beta_y, \beta_z = \beta_2, \beta_3 = 0$. Now equation (19) in general tensorial form, can be rewritten as

$$ds^2 = -\left(1 - \beta^i \beta_i\right) dt^2 + 2\beta_i dx^i dt + \delta_{ij} dx^i dx^j \ . \tag{21}$$

For even more generalized term to suit the metric tensor g_{00} of equation (21) the term $\alpha = \sqrt{\alpha^i \alpha_i} = 1$ is introduced, thus precisely the equation (21) should be written as

$$ds^2 = -\left(\alpha^2 - \beta^i \beta_i\right) dt^2 + 2\beta_i dx^i dt + \delta_{ij} dx^i dx^j \tag{22}$$

which is the metric as firstly proposed by Alqubierre [3] describing the spacetime metric of warp bubble hypersurface.

3. Shape function

The function $f = f(r)$ is considered as the shape function of the bubble where the conditions required are as in Figure 2 where the values of $f(r)\big|_{|r|<R} = 1$ in the bubble's interior and $f(r)\big|_{|r|>R} = 0$ at the exterior of the bubble. Alqubierre [3] has shown that the function may be written as

$$f(r) = \frac{\tanh\left(\sigma(r+R)\right) - \tanh\left(\sigma(r-R)\right)}{2\tanh\left(\sigma R\right)} \ , \tag{23}$$

where σ is the bubble's thickness factor. We may investigate this function in four conditions of warp bubble radii which are as the following when:-

3.1. The radii is at the positive bubble radius (toward the direction of travel) $r = R$

At the bubble radius, the function value transition between $f = 1$ and $f = 0$ is supposedly to be abrupt however equation (23) will ensure some delay in the transition but controllable. This transition 1 to 0 is also representing spacetime contraction in front of the warp bubble toward the direction of travel. As $r = R$ equation (23) can be shown to reduce to:

$$f(r)\big|_{r=R} = \frac{\tanh\left(\sigma(2R)\right)}{2\tanh\left(\sigma R\right)}$$
$$= \frac{\tanh\left(\sigma(R)\right) + \tanh\left(\sigma(R)\right)}{2\tanh\left(\sigma R\right)\left(1 + \tanh^2\left(\sigma(R)\right)\right)},$$

and thus,

$$f(r)\big|_{r=R} = \frac{1}{1 + \tanh^2\left(\sigma R\right)} \ . \tag{24}$$

IOP Conf. Series: Materials Science and Engineering **785** (2020) 012040 doi:10.1088/1757-899X/785/1/012040

The equation (24) implies that the function's transition is dependence on the value of the bubble thickness factor. As the bubble thickness factor reduce toward zero the transition will be abrupt from $f(r)\big|_{|r|<R}=1$ toward $f(r)\big|_{|r|>R}=0$.

3.2. The radii is at the center of the bubble $r=0$

The center inside the bubble is where any physical object, test particle or even a vehicle stay stationary. As $r=0$equation (23) can be shown deduced to 1 as the following

$$f(r)\big|_{r=0}=\frac{\tanh(\sigma(R))-\tanh(\sigma(-R))}{2\tanh(\sigma R)} \text{ and since } \tanh(\sigma(-R))=-\tanh(\sigma(R)),$$

$$f(r)\big|_{r=0}=\frac{\tanh(\sigma(R))+\tanh(\sigma(R))}{2\tanh(\sigma R)}=1 \quad . \tag{25}$$

3.3. The radii is at the negative bubble radius (toward the opposite of the direction of travel) $r=-R$

Similar to the case in *3.1* the transition from 0 to 1 at the rear of the bubble radius is supposedly to be abrupt however the equation (23) will ensure controllable delay in the transition. This transition of 0 to 1 is also representing spacetime expansion at the rear of the warp bubble as it moves away toward the frontal spacetime contraction in the direction of travel. As $r=-R$ equation (23) can be shown to reduce similarly as in the case of *3.1:*

$$f(r)\big|_{r=-R}=\frac{-\tanh(\sigma(-2R))}{2\tanh(\sigma R)} \text{ and since } \tanh(\sigma(-2R))=-\tanh(\sigma(2R)),$$

$$=\frac{\tanh(\sigma(2R))}{2\tanh(\sigma R)}=\frac{\tanh(\sigma(R))+\tanh(\sigma(R))}{2\tanh(\sigma R)\left(1+\tanh^2(\sigma(R))\right)} \quad ,$$

thus similarly,

$$f(r)\big|_{r=-R}=\frac{1}{1+\tanh^2(\sigma R)} \quad . \tag{26}$$

3.4 The radii is at the exterior region of the bubble $|r|>R$

The exterior region will display flatness of spacetime after the shape function transitions of rear spacetime expansion $(f=0\rightarrow f=1)$and of frontal spacetime contraction $(f=1\rightarrow f=0)$. From equation (23) as $r<-R$, $\left(r\big|_{r<-R}+R\right)<0$ and $\left(r\big|_{r<-R}-R\right)<-2R$ thus

$$f(r)\big|_{r<0}=\frac{\tanh\left(\sigma\left(\left(r\big|_{r<-R}+R\right)<0\right)\right)-\tanh\left(\sigma\left(\left(r\big|_{r<-R}-R\right)<-2R\right)\right)}{2\tanh(\sigma R)},$$

since $\tanh\left(\sigma\left(\left(r\big|_{r<-R}+R\right)<0\right)\right)\approx-1$ and $\tanh\left(\sigma\left(\left(r\big|_{r<-R}-R\right)<-2R\right)\right)\approx-1$ thus $f(r)\big|_{r<-R}\approx0$

as $r>R$, $\left(r\big|_{r>R}+R\right)>2R$ and $\left(r\big|_{r>R}-R\right)>0$, thus:

$$f(r)\big|_{r>R}=\frac{\tanh\left(\sigma\left(\left(r\big|_{r>R}+R\right)>2R\right)\right)-\tanh\left(\sigma\left(\left(r\big|_{r>R}-R\right)>0\right)\right)}{2\tanh(\sigma R)},$$

since $\tanh\left(\sigma\left(\left(r\big|_{r>R}+R\right)>2R\right)\right)\approx1$ and $\tanh\left(\sigma\left(r\big|_{r>R}-R\right)>0\right)\approx1$ thus $f(r)\big|_{r>R}\approx0$, thus:

$$f(r)\big|_{|r|>R} \approx 0 \quad . \tag{27}$$

By Alqubierre warp bubble shape function equation (23) sharp transition of shape function as in Figure 2 becoming rather gradual as depicted in the following figure

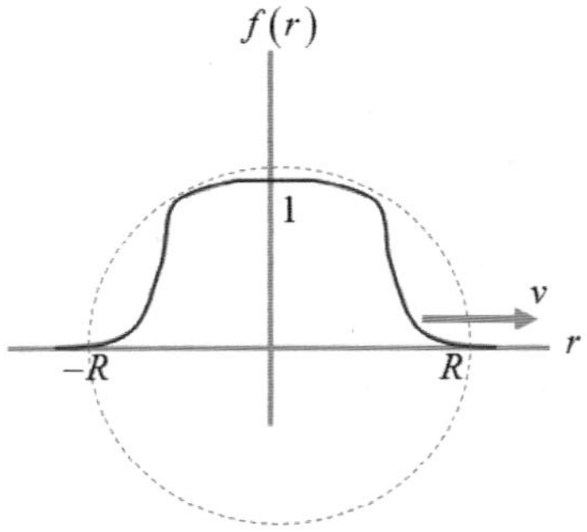

Figure 3. Function $f = f(r)$ characteristic with bubble radii with gradual transition.

The transition region between the warp bubble interior and the surrounding spacetime is actually the governing factor for the dynamics characteristic of the warp bubble.

4. Dynamic of expansion and contraction

The transition region is best described by deriving the extrinsic curvature. The relationship between the extrinsic curvature of the warp bubble and the surrounding spacetime is depicted as below.

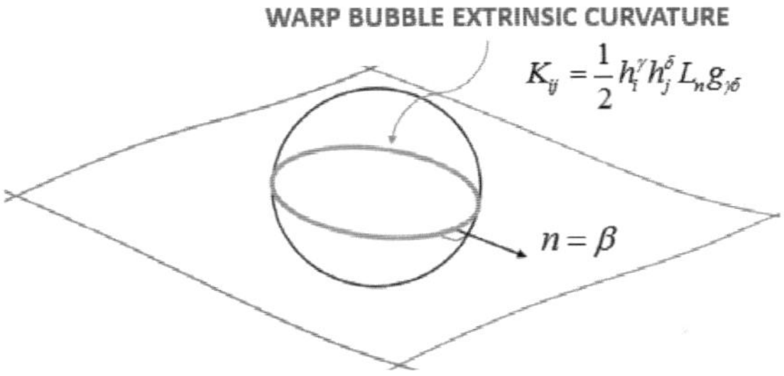

Figure 4. The extrinsic curvature between warp bubble and the surrounding spacetime

Beside flat spacetime base metric that represents locality the "surfing" concept is also about embedding a bubble of warped spacetime hypersurface onto the surrounding hyperspace. The transition region of warped spacetime properties between the bubble and the surrounding hyperspace can be elaborated further by the extrinsic curvature. Consider the definition of an extrinsic curvature that is the Lie derivative of the projection tensor $h_{ij} = h_i^\gamma h_j^\delta g_{\gamma\delta}$ which is the rate of change of the projection tensor along the normal vector field n_i [11]

$$K_{ij} = \frac{1}{2} h_i^\gamma h_j^\delta L_n g_{\gamma\delta} . \tag{28}$$

The normal vector of concern here is actually the vector of warp bubble direction of travel $n = \sqrt{n_i n^i} = \sqrt{\beta_i \beta^i} = \beta$ thus we may rewrite (28) as below

IOP Conf. Series: Materials Science and Engineering **785** (2020) 012040 doi:10.1088/1757-899X/785/1/012040

$$K_{ij} = \frac{1}{2} h_i^\gamma h_j^\delta L_\beta g_{\gamma\delta}, \tag{29}$$

which then can be expanded as the covariant derivatives of warp bubble direction of travel vector β_i and reduce toward the expression derived elaborately as the following

$$\begin{aligned}
K_{ij} &= \frac{1}{2} h_i^\gamma h_j^\delta \left(\nabla_\gamma \beta_\delta + \nabla_\delta \beta_\gamma\right) \\
&= \frac{1}{2} h_i^\gamma h_j^\delta \left(\partial_\gamma \beta_\delta - \Gamma_{\gamma\delta}^\lambda \beta_\lambda + \partial_\delta \beta_\gamma - \Gamma_{\delta\gamma}^\lambda \beta_\lambda\right) \\
&= \frac{1}{2}\left(h_i^\gamma h_j^\delta \partial_\gamma \beta_\delta + h_j^\delta h_i^\gamma \partial_\delta \beta_\gamma - 2 h_i^\gamma h_j^\delta \Gamma_{\gamma\delta}^\lambda \beta_\lambda\right) \\
&= \frac{1}{2}\left(\partial_i \beta_j + \partial_j \beta_i - 2\left(g^{\lambda\mu} \frac{1}{2}\left(\partial_i g_{\mu j} + \partial_j g_{\mu i} - \partial_\mu g_{ij}\right)\right)\beta_\lambda\right) \\
&= \frac{1}{2}\left(\partial_i \beta_j + \partial_j \beta_i - \beta_\lambda g^{\lambda\mu}\left(\partial_i g_{\mu j} + \partial_j g_{\mu i} - \partial_\mu g_{ij}\right)\right) \\
&= \frac{1}{2}\left(\partial_i \beta_j + \partial_j \beta_i - \beta_\lambda\left(\partial_i \delta_j^\lambda + \partial_j \delta_i^\lambda - g^{\lambda\mu} \partial_\mu g_{ij}\right)\right) \\
&= \frac{1}{2}\left(\partial_i \beta_j + \partial_j \beta_i + \beta_\lambda g^{\lambda\mu} \partial_\mu g_{ij}\right) \\
&= \frac{1}{2}\left(\partial_i \beta_j + \partial_j \beta_i + \beta^\mu \partial_\mu g_{ij}\right) \\
&= \frac{1}{2}\left(\partial_i \beta_j + \partial_j \beta_i + \beta^0 \partial_0 g_{ij} + \beta^1 \partial_1 g_{ij} + \beta^2 \partial_2 g_{ij} + \beta^3 \partial_3 g_{ij}\right) \\
&= \frac{1}{2}\left(\partial_i \beta_j + \partial_j \beta_i + \beta^1 \partial_1 g_{ij}\right) \\
K_{ij} &= \frac{1}{2}\left(\partial_i \beta_j + \partial_j \beta_i + \beta^1 \frac{\partial g_{ij}}{\partial x}\right) \quad g_{ij} = g_{xx} = g_{yy} = g_{zz} = 1
\end{aligned} \tag{30}$$

thus, reduced to

$$K_{ij} = \frac{1}{2}\left(\partial_i \beta_j + \partial_j \beta_i\right), \tag{31}$$

which is essentially an equation of motion of the warp bubble. The matrix of the extrinsic curvature is

$$K = \begin{pmatrix} K_{11} & K_{12} & K_{13} \\ K_{21} & K_{22} & K_{23} \\ K_{31} & K_{32} & K_{33} \end{pmatrix}, \tag{32}$$

where its trace describes the spacetime dynamic's characteristics θ indicating the warp bubble surrounding space of either expansion or contraction

$$\begin{aligned}
\theta &= -\alpha Tr K, \\
&= -\alpha\left(K_{11} + K_{22} + K_{33}\right), \\
&= -\alpha\left(\frac{1}{2}\left(\partial_1 \beta_1 + \partial_1 \beta_1\right) + \frac{1}{2}\left(\partial_2 \beta_2 + \partial_2 \beta_2\right) + \frac{1}{2}\left(\partial_3 \beta_3 + \partial_3 \beta_3\right)\right), \\
&= -\alpha\left(\partial_1 \beta_1 + \partial_2 \beta_2 + \partial_3 \beta_3\right),
\end{aligned} \tag{33}$$

$$\theta = -\alpha\left(\partial_x\beta_x + \partial_y\beta_y + \partial_z\beta_z\right), \quad \text{but since} \quad \beta_y = \beta_z = 0,$$

thus reduce to

$$\theta = -\alpha\partial_x\beta_x \,. \tag{34}$$

Expanding $\partial_x\beta_x$:

$$\begin{aligned}\partial_x\beta_x &= -\frac{\partial v f(r)}{\partial x} \\ &= -\left(v(t)\frac{\partial f(r)}{\partial x} + f(r)\frac{\partial v(t)}{\partial x}\right) = -v(t)\frac{\partial f(r)}{\partial x}\end{aligned} \tag{35}$$

Since,

$$v = v(t) = \frac{dx_b(t)}{dt}\text{, thus } \frac{\partial v(t)}{\partial x} = \frac{\partial\left(dx_b/dt\right)}{\partial x} = 0 \tag{36}$$

$$r = r(x,t) = \left[\left(x - x_b(t)\right)^2 + y_b^2 + z_b^2\right]^{\frac{1}{2}}\text{, } x_b = x_b(t) \text{ thus } dx_b = \frac{dx_b(t)}{dt}dt,\ \ x_b = \frac{dx_b(t)}{dt}t \tag{37}$$

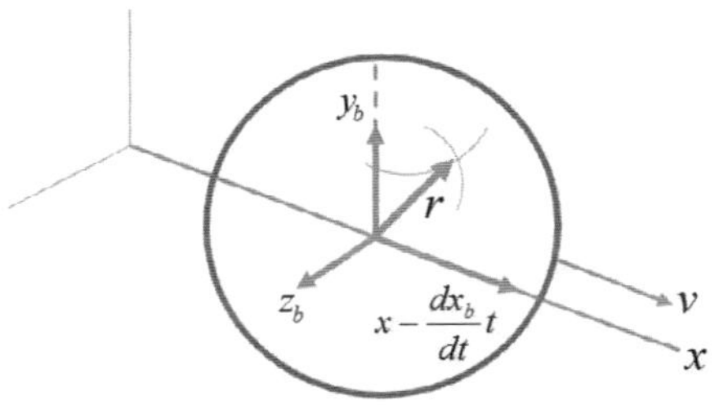

Figure 5. Warp bubble dynamics parameters

From Equation (35) and (37):

$$\begin{aligned}\partial_x\beta_x &= -v\frac{\partial f(r)}{\partial r}\frac{\partial r(x,t)}{\partial x} \\ &= -v\frac{\partial f(r)}{\partial r}\frac{\partial\left[\left(x - x_b(t)\right)^2 + y_b^2 + z_b^2\right]^{\frac{1}{2}}}{\partial\left(x - x_b(t)\right)}\frac{\partial\left(x - x_b(t)\right)}{\partial x}, \\ &= -v\frac{\partial f(r)}{\partial r}\frac{1}{2\left[\left(x - x_b(t)\right)^2 + y_b^2 + z_b^2\right]^{\frac{1}{2}}}2\left(x - x_b(t)\right), \\ &= -v\frac{\partial f(r)}{\partial r}\frac{1}{2r}2\left(x - x_b(t)\right), \\ \partial_x\beta_x &= -v\frac{\left(x - x_b(t)\right)}{r}\frac{\partial f(r)}{\partial r},\end{aligned}$$

$$\theta = -\alpha\partial_x\beta_x = v\frac{\left(x - x_b(t)\right)}{r}\frac{\partial f(r)}{\partial r}\,. \quad \text{since} \quad \alpha = 1 \tag{38}$$

Interior region of the warp bubble, $f=1$, while at the exterior region $f=0$ thus in these regions:

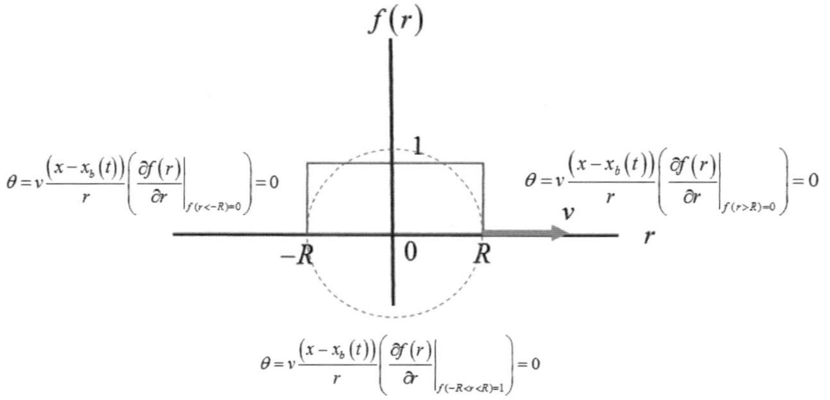

Figure 6. Spacetime dynamic's characteristics at the warp bubble interior and exterior regions.

$$\theta=v\frac{(x-x_b(t))}{r}\left(\left.\frac{\partial f(r)}{\partial r}\right|_{f(-R<r<R)=1}\right)=v\frac{(x-x_b(t))}{r}\left(\left.\frac{\partial f(r)}{\partial r}\right|_{f(|r|>R)=0}\right)=0. \tag{39}$$

At the *transition* where $f(r):0\Leftrightarrow 1$, $\theta\neq 0$, these regions constitute two thickening walls with thickness $\xi=1/\sigma$ along the direction of the warp bubble travel.

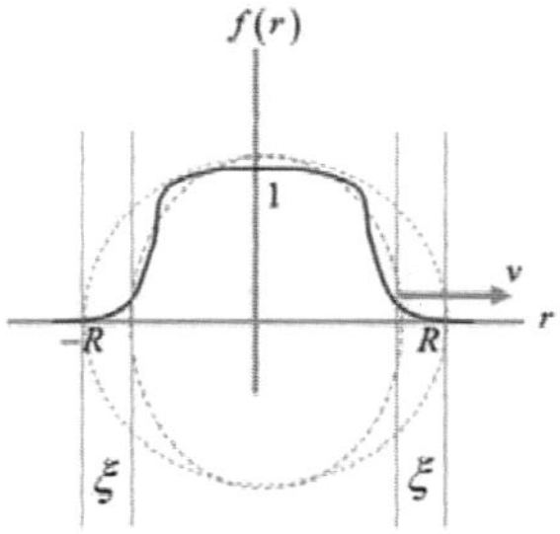

Figure 7. Spacetime dynamic's characteristics at the warp bubble's transition regions.

The transition region will show the characteristic of spacetime expansion and contraction.

At the front of the bubble in the direction of travel:

$$\theta=v\frac{(x-x_b(t))}{r}\left(\left.\frac{\partial f(r)}{\partial r}\right|_{(f=1)\to(f=0)}\right)<0, \tag{40}$$

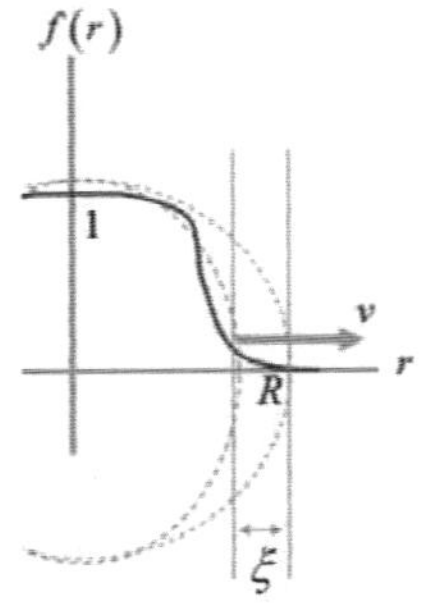

Figure 8. Transition region at the front of the bubble.

Since, $\left.\frac{\partial f(r)}{\partial r}\right|_{(f=1)\to(f=0)}<0$, thus $x-x_b(t)>0$, therefore $x_b(t)<x$, which represents space contraction. Thus, the transition region at the rear of the bubble, opposite to the direction of travel (as oppose to contraction condition ($\theta<0$)) , thus $\theta>0$ that is equation (38) is positive in value.

$$\theta=v\frac{(x-x_b(t))}{r}\left(\left.\frac{\partial f(r)}{\partial(r)}\right|_{(f=1)\to(f=0)}\right)>0 \quad , \tag{41}$$

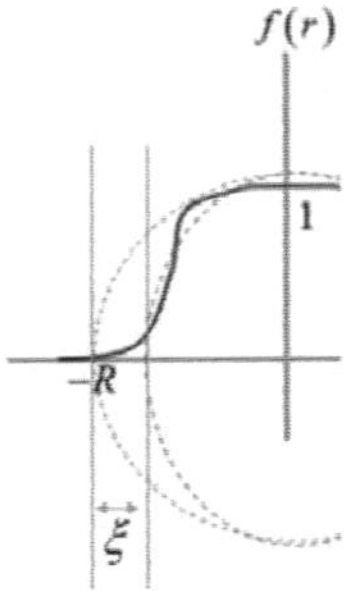

Figure 9. Transition region at the rear of the bubble.

Since also that $\left.\frac{\partial f(r)}{\partial(r)}=\frac{\partial f(-r)}{\partial(-r)}\right|_{(f=1)\to(f=0)}<0$, which implies that $\frac{\partial f}{\partial r}<0$ as always (of either during contraction or expansion), and $\theta>0$(as oppose to contraction condition), thus $x-x_b(t)<0$, therefore $x_b(t)>x$, which represents space expansion.

5. Velocity expression

By the Einstein field equation the relationship of between the warp bubble velocity and the energy density can be acquired. Expanding the Einstein curvature tensor $G_{\mu\nu}$ in term of Ricci tensor, rearrange in term of the energy momentum tensor and multiply both sides of the equation with timelike vector t^{μ} as below:

$$\frac{1}{8\pi}\left(R_{\mu\nu}-\frac{1}{2}Rg_{\mu\nu}\right)=T_{\mu\nu} \ , \qquad T_{\mu\nu}t^{\mu}t^{\nu}=\frac{1}{8\pi}\left(R_{\mu\nu}-\frac{1}{2}Rg_{\mu\nu}\right)t^{\mu}t^{\nu} \ . \tag{42}$$

$-T_{\mu\nu}=\rho_{warp}$ can be shown that the energy momentum tensor $T_{\mu\nu}$ in equation (42) will result to the following as proven by Lobo and Visser [13] but require a little modification since the warp bubble direction of travel is in the x direction (Fig. 5),

$$T_{\mu\nu}t^{\mu}t^{\nu}=-\frac{v^2}{32\pi}\left(\left(\frac{\partial f}{\partial y}\right)^2+\left(\frac{\partial f}{\partial z}\right)^2\right), \tag{43}$$

which is the energy density of the warp bubble that violates the energy condition $T_{\mu\nu}t^{\mu}t^{\nu}<0$ that is

$$T_{\mu\nu}t^{\mu}t^{\nu}=-\rho_{warp} \quad . \tag{44}$$

It is known that the shape function $f=f(r)$, thus $\partial f=\frac{\partial f(r)}{\partial r}\partial(r)$, therefore $\frac{\partial f}{\partial y}=\frac{\partial f(r)}{\partial r}\frac{\partial r}{\partial y}$ and $\frac{\partial f}{\partial z}=\frac{\partial f(r)}{\partial r}\frac{\partial r}{\partial z}$ from (37) as $y_b=y$ and $z_b=z$.

$$\frac{\partial f}{\partial y}=\frac{\partial f(r)}{\partial r}\frac{\partial\left[\left(x-x_b(t)\right)^2+y^2+z^2\right]^{\frac{1}{2}}}{\partial\left(\left(x-x_b(t)\right)^2+y^2+z^2\right)}\frac{\partial\left(x-x_b(t)\right)^2+y^2+z^2}{\partial y}=\frac{y}{r}\frac{\partial f(r)}{\partial r} \tag{45}$$

$$\frac{\partial f}{\partial z}=\frac{\partial f(r)}{\partial r}\frac{\partial\left[\left(x-x_b(t)\right)^2+y^2+z^2\right]^{\frac{1}{2}}}{\partial\left(\left(x-x_b(t)\right)^2+y^2+z^2\right)}\frac{\partial\left(x-x_b(t)\right)^2+y^2+z^2}{\partial z}=\frac{z}{r}\frac{\partial f(r)}{\partial r} \tag{46}$$

By Equations (43) to (46) the expression of the warp bubble energy density is:

$$\rho_{warp}=\frac{v^2}{32\pi}\left(\frac{y^2+z^2}{r^2}\right)\left(\frac{\partial f(r)}{\partial r}\right)^2 \quad . \tag{47}$$

The derivative $\partial f(r)/\partial r\approx\sigma$ which is actually the slope of the transition region in the condition where $f(r):0\Leftrightarrow 1$ and $\theta\neq 0$ which is also the thickness factor. The thickness of the bubble is $\xi=1/\sigma$ (Fig. 7) thus from (47) the warp bubble velocity expression can be written as:

$$v\approx\xi r\sqrt{\frac{32\pi\rho_{warp}}{y^2+z^2}} \quad . \tag{48}$$

This shows that the thickness of bubble and the energy density are the two most important factors to determine the warp bubble speed.

6. Conclusion

GR has shown that the superluminal travel is possible by not just through short cutting spacetime but also through the *surfing-like* dynamics of embedded hypersurface onto a hyperspace. Both have the shortcoming of requiring to violate energy condition, however being optimistic is always justifiable since more than 70% of the universe is still mysteriously encompassed by the dark energy that have the same characteristic of exotic matter for the energy required by both methods of superluminal travel. Superluminal capability of the surfing-like dynamics fundamentally based on the concept of separating inertial reference frames that is between the warp-bubble and its surrounding spacetime. The inertial reference frame is preserved in the respective region of either interior or exterior of the warp bubble. Spacetime condition in the bubble's interior is as normal as in a stationary non relativistic condition. Subject of travel could remain stationary inside the bubble whereas the warp bubble hypersurface would carry the subject via unlimited magnitude of velocity unconstrained by any relativistic restriction toward the subject using the hypersurface-hyperspace transition. The transition region of the warp bubble is fundamentally manipulated on how the shape-function characteristic of the warp bubble interact with the surrounding spacetime. This was done by using the extrinsic curvature of warp bubble. The trace of the extrinsic curvature represents the dynamics characteristic of warp bubble interaction with the surrounding spacetime of expansion and contraction. Deriving the dynamics characteristic expression θ from the trace of the extrinsic curvature provides its relationship with the warp bubble velocity and the slope of the transition region which is also representing the bubble thickness factor. These has shown how the warp bubble propelled through contraction and expansion of hyperspace that is the spacetime surrounding the warp bubble hypersurface. Finally using the Einstein field equation it has been shown that the energy density acquire exotic matter property of negative energy at the slope of the transition region which also indicates that beside the energy density ρ_{warp}, the thickness of the warp bubble boundary ξ also plays important factor in determining the speed of the warp bubble which by itself does not constrained by any relativistic limitations.

7. Reference

[1] David K 2017 Relativistic Light Sails (Department of Astronomy Columbia University) arXiv:1704.04310v3.
[2] Coffey and Jeffrey 2013 How big is the Milky Way? *Universe Today.*
[3] Alcubierre M 2000 The warp drive: hyperfast travel within general relativity *Classical and Quantum Gravity* **11**(5) (Department of Physics and Astronomy arXiv: gr-gc/0009013v1.
[4] Wheeler J A 1963 *Geometrodynamics* New York Academic Press.
[5] Guth Alan H 1997 The inflationary universe: the quest for a new theory of cosmic origins *Basic Books* p 233-234.
[6] Riess *et al.* 1998 Observational evidence from supernovae for an accelerating universe and a cosmological constant *The Astronom. J.* 116(3): 1009-1038.
[7] Morris M S and Thorne K S 1988 Wormholes in spacetime and their use for interstellar travel: A tool for teaching general relativity. *American J. of Phys.* **56**(5) p 395-412.
[8] Wang Y, Linde A *et al.* 2004 Current observational constraints on cosmic doomsday arXiv: astro-ph/0409264v2.
[9] Wheeler J A 1955 Geons *Phys. Rev.* **97**(2) p 511.
[10] Maldacena and Susskind L 2013 Cool horizons for entangled black holes *Fortsch. Phys.* **61**(9) arXiv:1306.0533 p 781-811.
[11] Carroll S M 2003 An Introduction to general relativity; spacetime and geometry (Pearson Addison Wesley).
[12] Shoshany B 2019 Lectures on faster-than-light travel and time travel (Perimeter Institute for Theoretical Physics) arXiv:1907.04178v1[gr-qc].
[13] Lobo F S N and Visser M 2004 Fundamental limitations on "warp drive" spacetimes *Class. and Quant. Grav.* **21**(24).

IOP Conf. Series: Materials Science and Engineering **785** (2020) 012041 doi:10.1088/1757-899X/785/1/012041

The role of nuclear science and technology (NST) in supporting the implementation of the sustainable development goals (SDGs) in Malaysia

F I A Rashid[1,a)], S S M Sali[1], A R A Kadir[1], M Z Zolkaffly[1], H Kassim[1], M A S Omar[1] and A Q A Bakar[1]

[1]Planning and International Relations Division, Malaysian Nuclear Agency, Bangi, 43000 Kajang, Selangor, Malaysia

[a]faisal_izwan@nuclearmalaysia.gov.my

Abstract. The United Nations 2030 Agenda for sustainable development (2030 Agenda) adopted in 2015 has underlined seventeen sustainable development goals (SDGs). The SDGs are used to chart the actions over the next fifteen years (2016-2030) in areas of critical importance for humanity and the planet towards sustainable development. Malaysia is among the countries that have put the SDGs in its development agenda and is on track to achieve the SDGs. Many countries including Malaysia use nuclear science and technology (NST) to meet their development objectives in various areas including food and agriculture, industry, human health, energy, water management, and environmental protection. The use of NST can help to achieve nine of the seventeen SDGs. This paper aims at highlighting the role of nuclear science and technology in supporting the implementation of the SDGs in Malaysia.

1. Introduction

On 25th September 2015, as part of the United Nations Resolution 70/1, the global leaders have adopted the sustainable development goals (SDGs) at the United Nations Summit [1]. SDGs are a collection of seventeen global goals with 169 targets, which designed to stimulate action over the next fifteen years (2016 -2030) in areas of critical importance for humanity and the planet. Countries including Malaysia have committed to undertake appropriate actions to integrate such goals and targets into their national development plans as well as to align policies and institutions behind them [2]. In this connection, the United Nations has highlighted the critical role of science, technology and innovation in achieving the SDGs [3].

In this connection, the International Atomic Energy Agency (IAEA) has emphasised the IAEA's role in transferring nuclear technology for peaceful uses to developing countries through its technical cooperation programme, to help them achieve their development objectives. The use of nuclear science and technology (NST) contributes directly to nine of the seventeen SDGs namely, SDG 2: Zero hunger, SDG 3: Good health and wellbeing, SDG 6: Clean water and sanitation, SDG 7: Affordable and clean energy, SDG 9: Industry, innovation and infrastructure, SDG 13: Climate action, SDG 14: Life below water, SDG 15: Life on land, and SDG 17: Partnerships for the goals [4]. Malaysia has long benefitted from the use of NST in various sectors including human health, food and

IOP Conf. Series: Materials Science and Engineering **785** (2020) 012041 doi:10.1088/1757-899X/785/1/012041

agriculture, industry, water management and environmental protection. The use NST contributes to national socio-economic development and therefore, could support the implementation of SDGs in the country. This paper aims at highlighting the role of NST in supporting the implementation of the SDGs in Malaysia.

2. SDGs in Malaysia

Malaysia has integrated the SDGs in its development agenda. SDGs were implemented in phases through the national development plans, particularly the Five Years Malaysia Plan. For period of 2016 to 2020, the Eleventh Malaysia Plan and its Mid-Term Review have incorporated the multi-dimensional nature of the SDGs, with the people at the centre of all development efforts. The Shared Prosperity Vision 2030 and the next two Malaysia Plans will further reinforce Malaysia's efforts to implement SDGs for period of 2021 to 2030. Figure 1 shows the SDGs implementation through the national development plans.

Figure 1. SDGs implementation through national development plans.

The first phase of the Malaysia Plan roadmap is being developed to provide guidance for the implementation of the seventeen SDGs. The roadmap takes into account the nation's capacities and capabilities in achieving the identified goals and targets of the SDGs. Mapping of SDGs with the Eleventh Malaysia Plan strategic thrusts was also performed [5].

3. Peaceful Uses of NST in Malaysia

Since the introduction in 1897, NST has steadily progressed and made direct contribution to national socio-economic growth [8]. The progression of NST is demonstrated by the increasing trend of license holders for medical and non-medical activities in Malaysia as shown in Figure 2 and 3, respectively. The NST is utilised in six technical sectors in the country, namely, medical and healthcare, food and agriculture, industry, water and environment, energy, and safety and security [6]. Consequently, the use of NST in those sectors has made positive contribution to national socio-economic growth. A study was carried out by the Malaysian Nuclear Agency (Nuklear Malaysia) in 2010 to assess the impact of nuclear technology to the national socio-economv for 2006 to 2008. The study found the percentage share of nuclear technology to the national Gross Domestic Product (GDP) increased from 0.024 percent in 2006 to 0.032 percent in 2008 [7].

The earliest application of nuclear technology in Malaysia was applied in the field of medicine with the first used of x-ray machine in Hospital Taiping in February 1897 [8]. Since then, the application of radiation in medical has expanded and has now become one of the most important techniques used for diagnosis, therapy and palliative of various medical conditions in Malaysia, particularly cancer. Besides that, NST is also applied in nutritional studies using stable isotope techniques to monitor situations and interventions for promoting infant, young child and adolescent nutrition.

In food and agriculture sector, the application of NST covers the complete plant life-cycle [9]. The acute and chronic gamma irradiations are used to produce beneficial new mutant plant varieties. Research activities in mutation techniques for crop improvement was also conducted, especially in producing new mutant varieties with desired characters such as high yield, resistance to abiotic and

IOP Conf. Series: Materials Science and Engineering **785** (2020) 012041 doi:10.1088/1757-899X/785/1/012041

biotic stress, and improved quality traits The use of nuclear and related techniques involving isotopic and elemental fingerprinting was also performed in determining the geographical origin of various food products and is currently being explored for traceability of Malaysian agricultural produce [10]. Other applications of NST include the food irradiation and food safety.

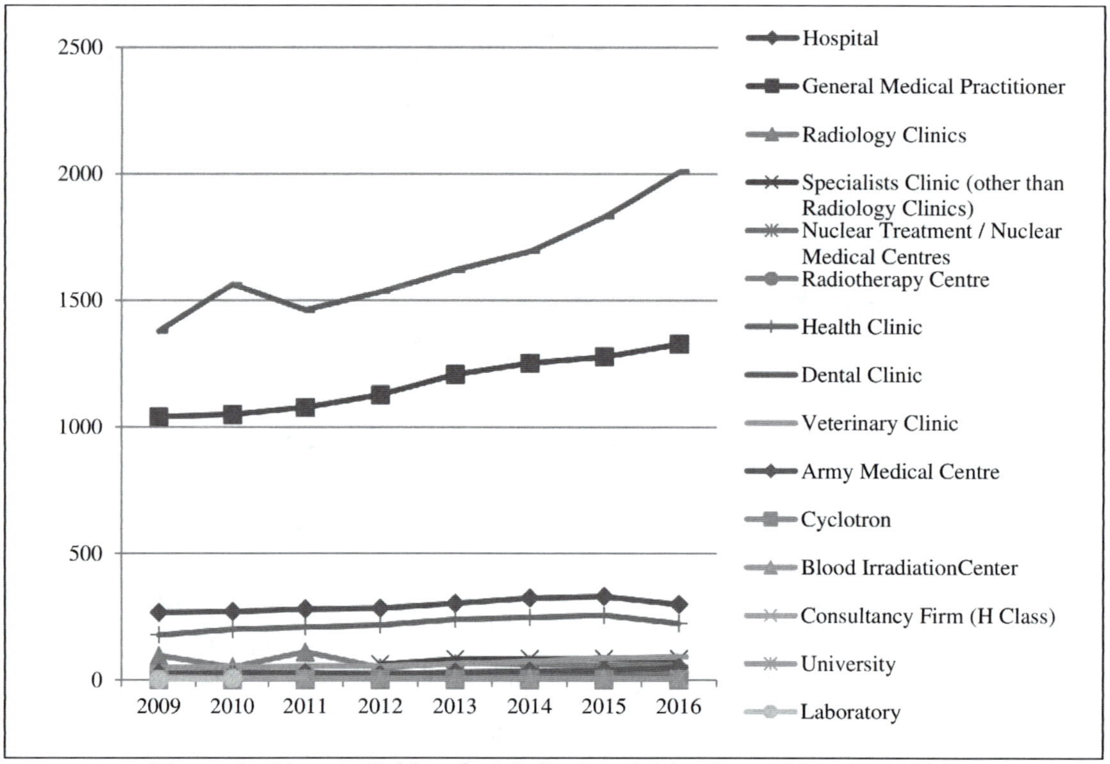

Figure 2: License Holders for Medical Activities from 2009 – 2016 [11]

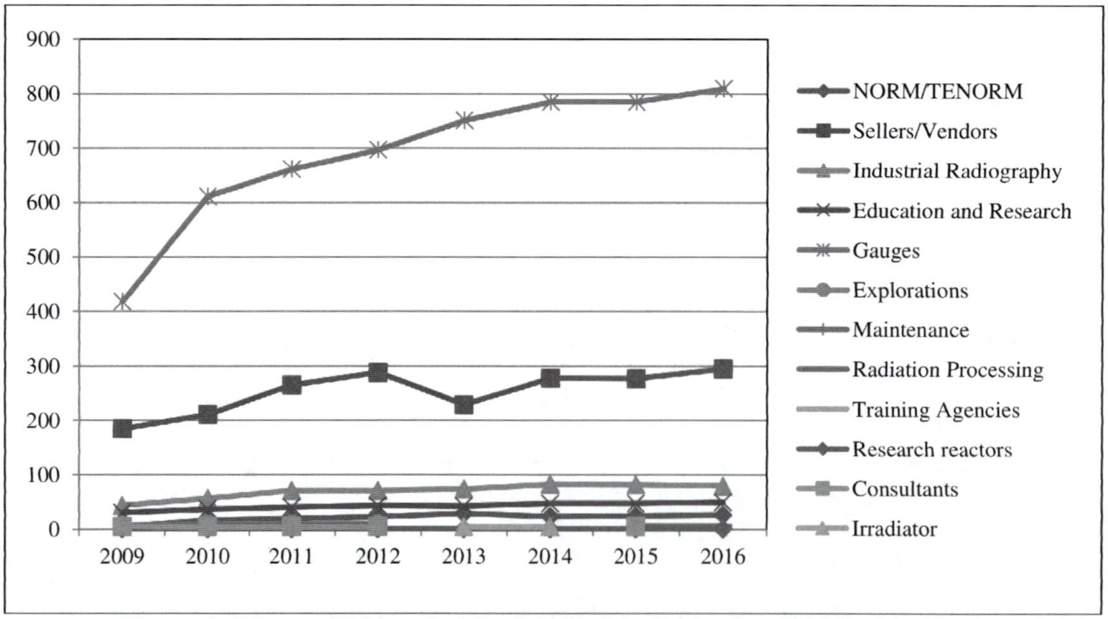

Figure 3. License Holders for Non-Medical Activities from 2009 – 2016 [11]

In industrial sector, NST is widely used for diagnostic and testing purposes in various industries such as oil and gas, conventional power generation, processing plants, transportation, aerospace and manufacturing. Such applications include non-destructive testing, nuclear and radiation gauges, radiotracer technology, plant assessment, and sealed sources technology. These applications have helped to optimise industrial productivity and operations as well as producing high quality goods. Radiation processing is also being used in many economic sectors in the country including for sterilisation, polymer cross-linking, rubber vulcanisation, and irradiation of certain food products for hygienisation and phytosanitation.

NST also plays an important role in water and environment sector, especially in the management of water resources. The use of environmental tracers and isotope hydrology techniques have helped to characterise ground and surface water resources to determine various factors and processes, including sources and history of water, past and present rainfall conditions, recharge of aquifers, mixing and interactions of water bodies, evaporation processes, geothermal resources and pollution processes. The nuclear and related techniques are also used for radiological monitoring, climate change studies, soil erosion and marine contamination studies.

4. The role of NST in supporting the implementation of the SDGs in Malaysia

The use of NST contributes directly to nine of the seventeen SDGs namely, SDG 2: Zero hunger, SDG 3: Good health and wellbeing, SDG 6: Clean water and sanitation, SDG 7: Affordable and clean energy, SDG 9: Industry, innovation and infrastructure, SDG 13: Climate action, SDG 14: Life below water, SDG 15: Life on land, and SDG 17: Partnerships for the goals. Based on our assessment using the selected publicly available publications related to the sectorial uses of NST in Malaysia, we summarised and mapped the utilisation of NST in related sectors in Malaysia with the SDGs as shown in Table 1.

4.1. Food and agriculture

The National Agrofood Policy developed in 2010 has put the main focus on improving the efficiency of the agro-food industry in the country [13]. Presently, Malaysia is still a net importer of rice with self-sufficiency level (SSL) hover between 60 to 70 percent [14]. Also, a research performed by the Khazanah Research Institute in 2019 to assess the status of paddy and rice industry in Malaysia has found that the nation recorded slow release of new rice varieties, which stand at less than 50 varieties for the past 50 years. The figure is lower than other countries in the South East Asia region such as the Philippines and Thailand which released more than 200 and 80 new rice varieties, respectively, over the same period [15]. Hence, Malaysia is targeting to achieve 80 percent SSL in rice by 2022 as well as to increase the release of new rice varieties [16]. On this note, the NST can play a role in helping the nation to achieve those targets. This argument is justified based on the development of NST in the country. For example, Nuklear Malaysia has started research on mutation breeding of rice since 2008 and has successfully developed two new rice mutants namely NMR 151 and NMR 152 [17]. These rice mutants were developed from gamma-irradiated seeds and have high tolerance to flooding and aerobic conditions as well as high yield. Apart from rice, the mutation breeding technology is also used to develop new varieties of fruits and vegetables including banana and mushroom.

Table 1. The summary and mapping of the utilisation of nuclear science and technology in related sector in Malaysia with the SDGs

SDG No.	IAEA TCP thematic areas[a]	Related fields	Utilisation of nuclear science and technology in Malaysia in respective areas
2	Food & agriculture	Food production, safety, security and sustainable agricultural practices	• Mostly used nuclear technique such as mutation breeding to produce new varieties of plants and crops with enhanced resilient to climate including rice, banana and mushroom • Food irradiation mostly for spices and herbs • Research in enhancing food safety including food traceability and authenticity
	Health & nutrition	Nutrition	• Used in nutritional studies including assessment of energy expenditure, assessment of body composition, breastfeeding study and bone health study
3	Health & nutrition	Disease management	• Applied in cancer management as well as production of medical radioisotopes • Sterilisation of healthcare products by irradiation • Research in controlling Aedes mosquito population using Sterile Insect Technique (SIT)
6	Water & environment	Water resource, quality and management	• Radiochemical and radioactivity contamination analysis including for food and drinking water • Used in management of water, local natural resources and environment conservation and climate change studies
7	Sustainable energy		• Education and training, and hands-on learning of the research reactor operation and nuclear reactor engineering
9	Industrial application & radiation technology	Industry productivity and efficiency	• Radiation processing for sterilisation, polymer cross-linking, and rubber vulcanisation • Used in determining and improving quality of material and structure such as non-destructive testing, radiotracer and sealed source technology • Nuclear gauge widely used in various industrial applications
13	Water & environment	Climate change monitoring, mitigation and adaptation	• Climate change studies using light stable isotope • Soil erosion and environmental studies using nuclear and radioisotopic techniques
14	Water & environment	Ocean, seas and marine protection	• Used in marine, ocean and coastal studies
15	Water & environment	Management of water and environment	• Soil erosion studies and analysis • Groundwater and natural resources studies • Environmental monitoring
17		International cooperation	• Numerous bilateral and multilateral platforms were established which enable sharing and building capacity in NST

[a] Thematic areas are classified based on the IAEA Technical Cooperation Programme (IAEA TCP) [12].

Malaysia as a trading nation involved in exports of food commodities to the international market. In this regard, the use of gamma irradiation for exported food commodities can be intensified to improve safety and extends the shelf life of foods by reducing or eliminating microorganisms and insects. The amount of irradiated food commodities from 1997 to 2014 in Malaysia is shown in Table 2 [18]. Research on the NST was also actively conducted in the country to study plant nutrition and soil fertility, manage insect pests, food safety and authenticity, and to promote soil and environmental preservation. The use of NST in this sector can help to support the implementation of SDG 2 in the country.

Table 2. The amount of irradiated food commodities from 1997 to 2014 in Malaysia [18].

Year	Metric Ton
1997 - 1999	100
2005	475
2006	800
2007	728
2008	691
2009	826
2010	785
2011	602.4
2012	567.8
2013	530.6
2014	984.1

4.2. Health and nutrition
Cancer constitutes an enormous burden on societies globally, including Malaysia. Cancer is the fourth most common cause of death in Malaysia, with approximately 37,000 cases reported every year [19]. A study conducted by the Malaysia Ministry of Health in 2015 found that a total of 103,507 new cancer cases were diagnosed in Malaysia from 2007 to 2011 [20]. On top of that, the National Health and Morbidity Survey 2015 unveiled an alarming concern regarding the number of Malaysians suffering from Non-Communicable Diseases (NCD) [21]. In this connection, the NST is used in cancer management areas such as diagnostic radiology, radiotherapy and nuclear medicine as well as prevention or control of NCD. Production of medical radioisotopes is also conducted at a few medical facilities that operate cyclotron. Medical facilities equipped with nuclear technology-related equipment were established throughout the country to provide cancer-related services. Besides, Malaysia also participated in the IAEA Programme of Action for Cancer Therapy (PACT) which has helped Malaysia to develop its national cancer control plan. In the area of nutrition, NST has played a role in nutritional studies including assessment of energy expenditure, assessment of body composition, breastfeeding study and bone health study. These studies have contributed to the prevention and control of nutritional deficiency among the people. The use of NST in this sector can help to support the implementation of SDG 2 and 3 in Malaysia.

4.3. Water and environment
Malaysia has always put environmental and climate change issues in its development agenda. The alarming trend of climate change and environment has made the Government to take necessary measures to tackle the issues. In addition, Malaysia also faced challenges in surface and groundwater resources, in terms of deteriorating quality and quantity due to several factors, including rapidly growing populations, increasing agricultural demands, and threat of pollution in the surrounding areas. In this regard, NST can play a significant role in climate change monitoring, mitigation and adaptation as well as management of water, local natural resources and environmental conservation. Such applications include the mitigation of sedimentation, determination of atmospheric and marine pollutants, and groundwater studies. The environmental tracers and isotope hydrology techniques are

widely used to characterise ground and surface water resources in Malaysia. Consequently, it helped to improve the management of water resources across the country. The use of NST in this sector can help to support the implementation of SDG 6, 13, 14 and 15 in the country.

4.4. Sustainable energy
Presently, the nuclear power industry is non-existent in Malaysia. Thus, in this sector, focus a more towards developing national capacity and capability in nuclear science and nuclear reactor engineering. For this purpose, the only research reactor in the country, the TRIGA PUSPATI Reactor, is used for education and training on the operation of research reactor as well as nuclear physics and nuclear reactor engineering. Since the nuclear power is non-existent in the country, the use of NST is centred at education and training and therefore, will have minor role in supporting the implementation of SDG 7 in the country.

4.5. Industrial application and radiation technology
Manufacturing and services remain as the main driver of growth in Malaysia, recorded 78.3 percent share of national GDP in 2018 [22]. This sector has a long history in utilising NST. In manufacturing sector, nuclear and radiation gauges are widely used for measuring and monitoring of product and process parameters such as thickness, density, moisture or fill level.

Radiation processing is a technology that offers clean and additive-free method for preparation of value-added novel materials based on renewable, non-toxic and biodegradable natural polymers, and natural polymer waste. Research on this technology is extensively conducted in the country focusing on polymer and natural polymer [23]. Commercial processing technology such as radiation cross-linking is also maturely developed in Malaysia for the production of cross-linked cables and wires. The cables and wires exhibit enhanced properties and are used in automotive and machinery industries. In addition, radiation technology is also used for sterilisation of health care products in the country. All of these usages have made direct contribution to national socio-economic development.

Malaysia is among the largest producer of natural rubber globally. Rubber industry contributed RM 40.36 billion to national exports, with sixteen percent coming from natural rubber [24]. In Malaysia, nuclear technology such as radiation vulcanisation of natural rubber latex, or RVNRL is used to initiate vulcanisation, a process that chemically bonds molecules to produce rubber elasticity and strength. This technology has replaced the use of a sulphur-based process and produces a material with remarkable qualities such as low cytotoxicity, absence of sulphur and zinc oxide, and high transparency and softness. This technology has proved that RVNRL can provide a technically and economically viable alternative to the existing vulcanisation process, thus benefit the rubber industry [25]. The use of NST in this sector can help to support the implementation of SDG 9 in the country.

5. Conclusion
Since its introduction, NST has long benefitted various key sectors in the country and has been an integral part of the country's economic development. In tandem with the Government's commitment to achieve the targets set in the SDGs, NST can play a role in achieving nine SDGs. Most of the NST used in the country has reached a mature phase and is commercially available. As NST undergoes advancement, it will continue to support national socio-economic growth and ultimately, towards sustainability.

6. References

[1] United Nations 2015 Resolution adopted by the general assembly on 25 September 2015 no A/RES/70/1.
[2] United Nations 2019 *The Sustainable Development Goals Report 2019* p 3.
[3] United Nations 2017 UN highlights critical role of science, technology and innovation in achieving the SDGs.

[4] IAEA 2016 Atoms for peace and development: How the IAEA supports the sustainable development goals pp 1-11.
[5] Ministry of Economic Affairs 2018 *Mid-Term Review of Eleventh Malaysia Plan 2016-2020 New Priorities and Emphases.*
[6] Jamal N, Sali S S M, Rashid F I A, Hedar R J R and Baharuddin B 2018 An overview of development of nuclear technology in Malaysia after 2008 *J. of Nucl. and Related Tech.* **15** 1 p 17-24.
[7] Malaysian Nuclear Agency 2010 *The impact of nuclear technology to the national socio-economy in Malaysia* p 198.
[8] Wastie M L 1994 Radiology in Malaya in the 1920s *Med. J. Malaysia* **49**(3) p 199-204.
[9] Kasim H, Daud A H, Ibrahim J K and Musa A 2011 Impact of nuclear technology to the national socio-economy: technical support by Nuclear Malaysia *Proc. Conf. Nucl. Tech. Conv. 2011* (Kajang: Malaysian Nuclear Agency).
[10] Othman Z, Abdullah S N A, Moosa S and Mostapha R 2015 Establishing food traceability system using nuclear and related techniques in Malaysia *Proc. Conf. Nucl. Tech. Conv. 2015* (Kajang: Malaysian Nuclear Agency).
[11] Jamal N, Rashid F I A, Baharuddin B, Omar M A S, Hedar R J R, Bakar A Q A, Zolkaffly M Z, Sali S S M and Kadir A R A 2018 Overview on the development of utilisation of nuclear technology in Malaysia *Proc. Conf. Research and Dev. Sem. 2018* (Kajang: Malaysian Nuclear Agency).
[12] Arias G and Nebel S 2017 The international development context and the IAEA technical cooperation programme.
[13] Dardak R A 2019 Malaysia's Agrofood Policy (NAP 2011-2020): performance and new direction *FFTC agricultural policy platform* p 1-13.
[14] Arshad F M 2018 Agriculture and food sector development in Malaysia: the alternative path
[15] Omar S C, Shaharudin A and Tumin S A 2019 *The status of the paddy and rice industry in Malaysia* p 78.
[16] FMT 2018 Malaysia targets 80% rice self-sufficiency by 2022.
[17] Salleh S, Ahmad F and Hussein S 2016 Study on the effects of oligochitosan and bioliquifert on two rice mutants NMR 151 and NMR 152 *Proc. Conf. Research and Dev. Sem. 2016* (Kajang: Malaysian Nuclear Agency).
[18] Kume T and Todoriki S 2013 Food irradiation in Asia, the European Union, and the United States: A status update *Radioisotopes* **62**(5) p 291-99.
[19] National Cancer Institute 2018 Malaysian study on cancer survival (MySCan) p 3.
[20] Manan A A, Tamin N S I, Abdullah N H, Abidin A Z and Wahab M 2015 *Malaysian Nat. Cancer Registry Report 2007-2011* p 1.
[21] Institute for Public Health 2015 *The National Health And Morbidity Survey 2015 Non-Communicable Diseases, Risk Factors & Other Health Problems* **2** p 13.
[22] Ministry of Economic Affairs 2018 The Malaysian economy in figures 2019.
[23] Saeid M H, Safrany A, Sampa M H and Ramamoorthy N 2009 *Rad. Phys. and Chem.* **79** p 255–260.
[24] Malaysian Rubber Board 2018 *Natural Rubber Stat. 2018*
[25] Makuuchi K and Markovic V 1991 Radiation processing of natural rubber latex IAEA Bulletin **1**(1999) p 25-8.

IOP Conf. Series: Materials Science and Engineering **785** (2020) 012042 doi:10.1088/1757-899X/785/1/012042

The effect of photomultiplier tube glass entrance window on plastic scintillator cast sheet photon dosimetry: A Monte Carlo study

N S Mohamed[1] , A Aydın[2] S M Tajudin[1,3,a)]

[1]Faculty of Health Sciences, Universiti Sultan Zainal Abidin (UniSZA), Terengganu, Malaysia.
[2]Faculty of Arts and Sciences, Balikesir University, 10145 Balikesir, Turkey.

[a] suffian@unisza.edu.my

Abstract. Plastic scintillators are commonly used for medical dosimetry due to the density and effective atomic numbers that are closer to human soft tissue. When a photomultiplier tube (PMT) is used with a thin scintillator, reflected photon and electrons as a result of Compton scattering either inside the scintillator or PMT entrance window, might contribute to a significant source of additional absorbed dose. Monte Carlo simulation was used to study the effect of different PMT window materials on the absorbed dose of a 0.5 mm plastic scintillator cast sheet for parallel photon beam energy up to 1 MeV. The additional dose in the plastic scintillator from 400 keV to 1 MeV due to sapphire and lithium fluoride (LiF) glass are increases from 11 % to 47 % and 8 % to 31 %, respectively. Despite of the lower density of Quartz among other materials, Quartz and magnesium fluoride (MgF_2) demonstrate almost similar trends of additional dose throughout the energies, which is closer to the sapphire. To reduce the unexpected additional dose, a plastic scintillator with a considerable thickness of a PMT window should be adopted for the 'soft-tissue' dose response.

1. Introduction

There are significant advances in new detector materials for a range of applications due to the use of radiation in space exploration, nuclear environments, industrial and medical applications. Inadequate number of detectors due to their cost, size and complexity has been a strong driving force for the development of accurate, vigorous and reliable instruments [1]. Commercially available detectors are air ionization chamber, Geiger Muller counter, NaI(Tl), CsI(Tl) and semiconductor [2][3]. To be more reliable as a photon dosimeter, several studies have been conducted to understand their basic properties, which are include the calibration of detectors, energy dependence and detection efficiency. For medical applications such as interventional radiology (IVR) or depth dose measurements in phantom, plastic scintillator is a good candidate of photon dosimeter since its effective atomic number approximately equals to that of human soft tissue. The dose measurements for a small photon fields and area drives the needs of smaller size dosimeter with high accuracy. The use of detector system with a small size plastic scintillator had also been studied by researchers in radiotherapy field [4, 5]. The use of other dosimeter such as NaI(Tl) is available with high efficiency, however, it is known to exhibit a strong energy dependence [6].

The scintillation process is the most applied methods in the detection and spectroscopic measurement of a different type of radiations. A plastic scintillator requires a device that collects scintillation light and converts it to an electronic signal as shown in Figure 1. When a small plastic scintillator coupled with a photomultiplier tube (PMT) to be employed as a dosimeter, a PMT window of considerable size and weight exists on the side of the plastic scintillator. There is a possibility that the PMT window glasses will influence the absorbed dose when thin plastic scintillators are used, which contribute to a source of unwanted scintillations. In this paper, the contribution of backscattered photon and electron from the window were evaluated by using Monte Carlo simulation. The insight would stimulates important questions concerning soft-tissue dosimetry by using a thin plastic scintillator.

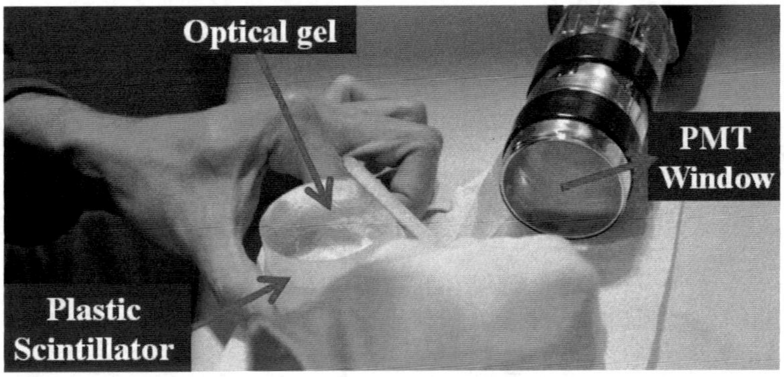

Figure 1. The exit face of the scintillator (non-wrapped part) is optically coupled to the PMT of the same diameter with optical silicon grease.

2. Monte Carlo simulation

Monte Carlo code for electron and photon transport (EGS5 code [7]) was used to calculate the absorbed energy and dose in scintillator for total effect with PMT window for 0.1 to 1 MeV photon energy. The plastic scintillator and PMT window material data and density for the calculations correspond to the manufacturer's data [8] and National Institute of Standard and Technology, NIST database [9], respectively as per our previous study [10]. The plastic scintillator material used in this study is EJ200 (Eljen Technology, Sweetwater, TX, USA) [8], while the readout device used is a photomultiplier tube Model R375 from Hamamatsu [11] with a diameter of 5.0 cm. To date, the thickness of standard plastic scintillator cast sheet available in the industry is within 0.5 mm to 150 mm, while for special large cast sheet, the thickness ranged from 1 to 5 cm with maximum diameter and length of 30 cm and 450 cm, respectively [12]. In this study, the calculations were performed for a 0.5 mm plastic scintillator for total effect with five different types of PMT window materials; quartz, sapphire, borosilicate, lithium fluoride (LiF) and magnesium fluoride (MgF_2).

In the simulation process, photons interaction such as photoelectric absorption, Compton scattering and pair productions were considered. The parallel photons beam of the same radius with the scintillator was set to incident on the front face of the scintillator. In EGS5 calculation, the calculated absorbed doses in the scintillator were based on calculation of absorbed photon energy in some volume of material. When material is exposed to radiation, energy is absorbed in material. Absorbed dose is then calculated as the average energy transferred to small volume mass of plastic scintillator material by photon. Following conversion were used to convert the unit of MeV/g to Gy as outlined in equation 1 and 2.

$$1 \text{ MeV} = 1.602 \times 10^{-13} \text{ J} \tag{1}$$

$$1 \text{ MeV/g} = 1.602 \times 10^{-13} \text{ (J/MeV)} \times 1000\text{(g/kg)} = 1.602 \times 10^{-10} \text{ Gy} \tag{2}$$

IOP Conf. Series: Materials Science and Engineering **785** (2020) 012042 doi:10.1088/1757-899X/785/1/012042

The electrons or photons generated in the PMT window that were backscattered to the scintillator were identified. The absorbed energy in the scintillator was scored separately by considering the contribution of reflected photons or electrons from the window. For comparison, the absorbed energy was also scored without the window. Consequently, the absorbed energy and dose in scintillator were scored with the total effect of PMT window as follows; (a) with PMT window, (b) with reflected photon only, (c) with reflected electron only and (d) without PMT window.

3. Result and discussion

3.1. Calculated transmitted-reflected electrons of the aluminium for 159 keV electron

EGS5 code had been used widely to transport or simulate radiation like photons [13] in materials. For electrons, as a benchmark of our study, we have calculated the fraction of electrons that reflected and transmitted after interaction with aluminium for comparison with literature data. Such comparison is very important to ensure that the electrons interactions within the material are properly handled by the code. Figure 2(a) and (b) shows the calculation of transmitted and reflected electron as a function of aluminium thickness for electron energy of 159 keV. The calculated values have a considerably good agreement with the literature [14], to validate the subsequent calculations result.

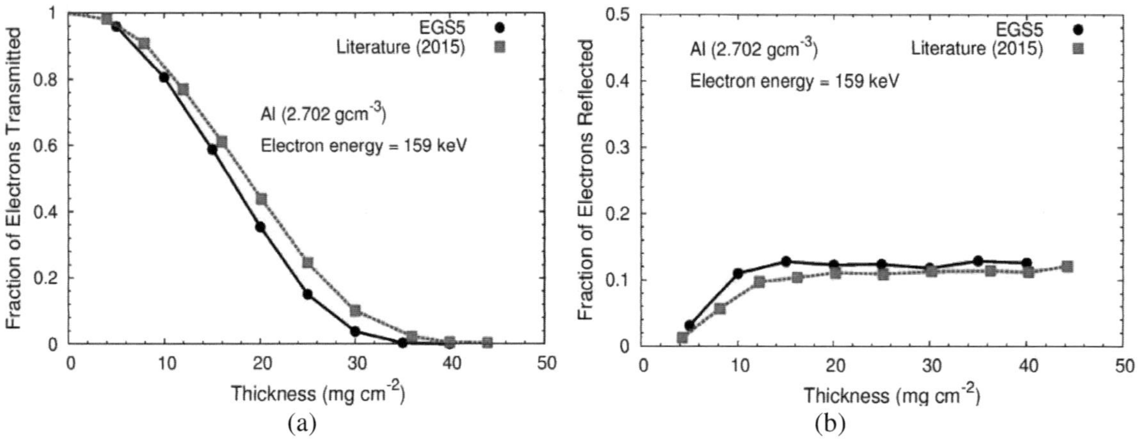

Figure 2. Comparison for the fractions of electrons that (a) transmitted and (b) reflected as a function of aluminum thickness.

3.2. Absorbed dose of plastic scintillator without and with PMT window materials; reflected electron and photons

Figure 3 (a), (b), (c), (d) and (e) demonstrate the absorbed dose of plastic scintillator without and with (reflected photon and electron) PMT for lithium fluoride (LiF), quartz, magnesium fluoride (MgF_2), borosilicate and sapphire, respectively. It can be observed that the absorbed dose linearly increases from 0.1 MeV up to 0.4 MeV for both reflected photon and electron. However, absorbed dose started to slightly reduced after 0.4 MeV up to 1 MeV. A prerequisite to understand the response of a specific type of detector is the knowledge of the dominant mechanism by which radiation interacts and hence loses energy in matter [6]. This can be due to the dominant interaction mechanism for photon energies below 0.4 MeV is the photoelectric effect, whereas the Compton scattering begins to be relevant above 0.4 MeV. Moreover, the different energies of the photon beam resulted in different penetrating depths and absorption mechanisms of the beam inside the plastic scintillator. The penetration depth is governed by the energy of the individual photons, the atomic number, density and thickness of the object, which result from the different absorption mechanisms that occur in the plastic scintillator [15]. Increasing the photon energy (0.4 MeV up to 1 MeV) causes increment in the penetrating ability and decrement in the probability of interaction, and hence resulted in less absorbed photon energies in the plastic scintillator. Whilst, the low energy photon beam (<0.4 MeV) resulted in most radiation attributed from the photoelectric effect, which is the dominant mechanism that contribute to the higher absorption and hence excitation.

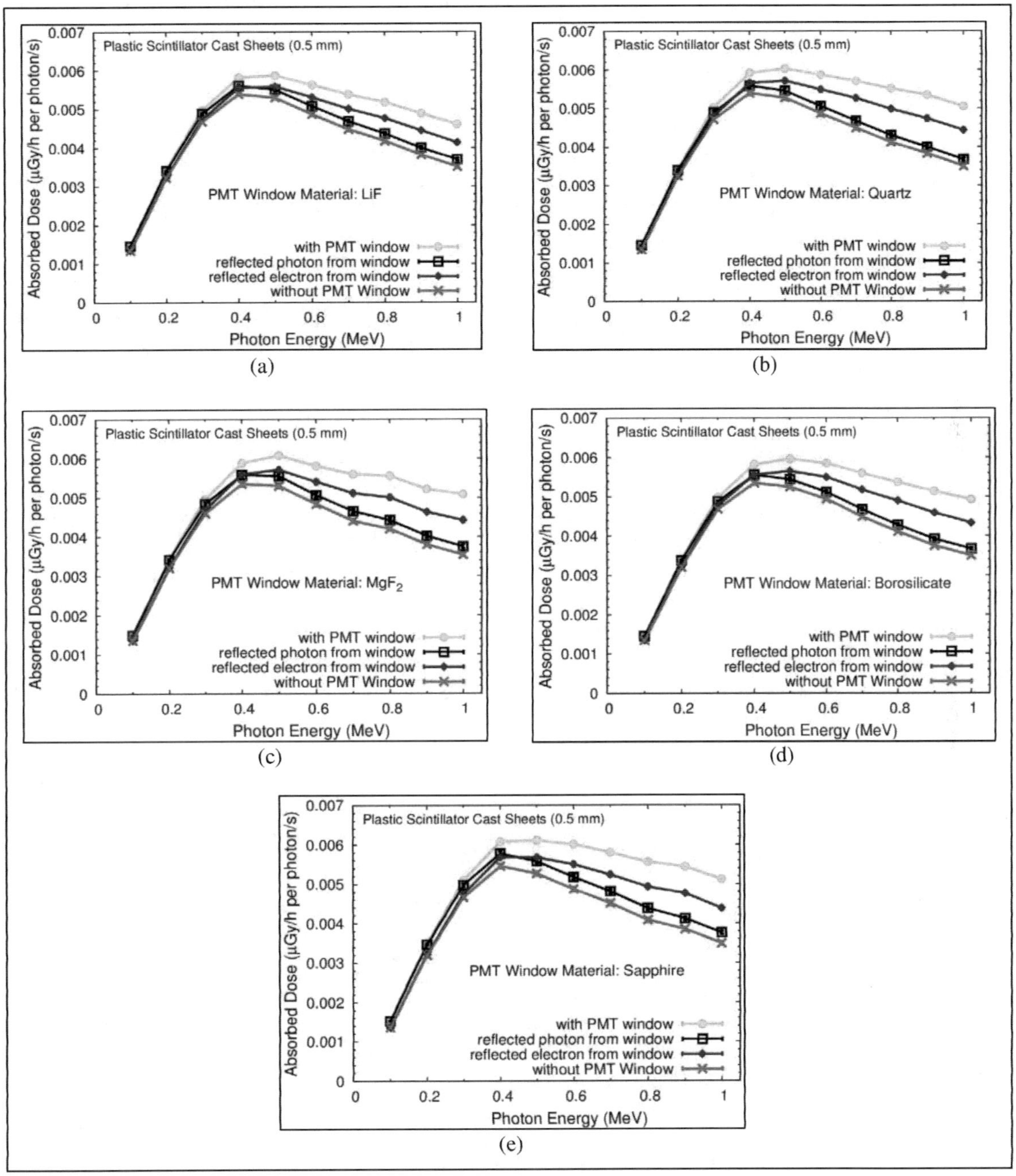

Figure 3. Absorbed dose of plastic scintillator without and with (reflected photons and electrons) for PMT window materials; **(a)** lithium fluoride (LiF), **(b)** quartz, **(c)** magnesium fluoride (MgF_2), **(d)** borosilicate, and **(e)** sapphire.

Generally, the reflected photon and reflected electron exhibited almost similar trend of absorbed dose for photon energies below 0.4 MeV for all PMT window materials. Whilst, reflected electrons contribute significantly higher to the additional absorbed dose in comparison to the reflected photon for photon energies 0.4 MeV up to 1 MeV. Within that range also, the absorbed dose without PMT window can be observed to be significantly lowered than absorbed dose with the PMT window materials. This can be ascertained from the significant additional dose of a PMT quartz window for 0.4 to 1 MeV photons (20 % to 43 %) in comparison to one without quartz as shown in Figure 3(b). These circumstances can be explained from the dominant interactions occur inside the plastic scintillator below 0.4 MeV is the photoelectric effect, whereas the Compton scattering begins to be relevant above 0.4 MeV, as discussed previously. Compton interaction is the scattering of incoming photons, which is deflected through an angle, θ with respect to its original direction [6]. During interaction, the incident photons will transfer part of its energy to the scattered electron, which results in a decrease in energy of the incident photon and contribute to an unwanted scintillation being produced in the plastic scintillator. The probability of Compton scattering per atom of the absorber depends on the number of available electrons as scattering targets and therefore increases linearly with Z. Therefore, the effect of the reflected electron and reflected photon on the absorbed dose can be significantly differentiate at 1 MeV in the high-Z material; sapphire (Z_{eff} = 10.454) [16] in Figure 3(e), which increase up to 19% in comparison to low-Z material; LiF (Z_{eff} = 8.14) [17] in Figure 3(a), which increases up to 13.5%.

3.3. Total absorbed dose of plastic scintillator with PMT window materials; quartz, sapphire, borosilicate, lithium fluoride (LiF) and magnesium fluoride (MgF_2)

The data in Figure 4 indicates that sapphire glass contributes the highest additional dose to the plastic scintillator, while the lowest dose given by LiF material. The additional dose in the plastic scintillator for energy ranged from 0.4 MeV to 1 MeV due to sapphire and LiF glass are increases from 11 % to 47 % and 8 % to 31 %, respectively. The second lowest material that contributes to the additional dose is borosilicate. This can be explained from the probability of the different interactions occurring within the material, which depends on the density of the material and photon energy. Different materials will have different threshold energy and regions of high cross-sections, which results in different mechanism (either Photoelectric or Comptom) dominating the characteristics as explained in Section 3.2. Both processes lead to the partial or complete transfer of the photon energy to the electron, which results in the incident photon undergoing scatter over a range of angles or being completely absorbed.

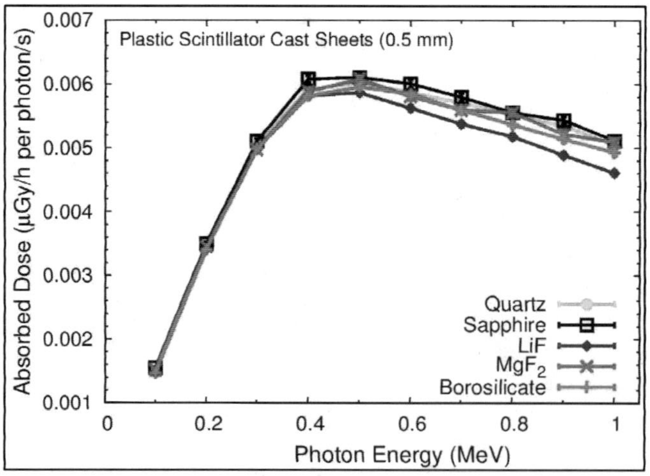

Figure 4. Total absorbed dose of plastic scintillator for five PMT window materials

The rate of the photon attenuation by the material depends on the atomic number, atomic mass and density of the material [15]. Due to the change in the photon direction after the interaction with the electron, Compton scattering results in a portion of the incident radiation 'bouncing off' or being fully scattered by the material [6]. Therefore, material within the primary photon beam, which includes the detector itself becomes the actual source of unwanted scintillation. The low–Z number (Z_{eff} = 8.14) [17] and density (ρ = 2.63 g/cm^3) of LiF result in a lower photon absorption coefficient, where the dominant mechanism is through photoelectric absorption. Otherwise, sapphire glass consists a relatively high percent of aluminum (Al) element in its mixture, which contribute to the high Z_{eff} (Z_{eff} = 10.454) [16] and density (ρ = 3.987 g/cm^3), and hence increasing the unwanted scintillations in the plastic scintillator.

4. Conclusion

The use of plastic scintillator along with the PMT has contributed to the significant additional absorbed dose. Contribution of the additional absorbed dose in plastic scintillator mainly depends on the thickness and types of PMT window material, which consist of different density and Z number, and therefore different absorption and scattering mechanism. In conclusion, a plastic scintillator with considerable thickness along with a good choice of PMT window thickness and material should be adopted for the 'soft-tissue' dose response, in order to reduce the unwanted additional dose.

5. References

[1] Braccini S, Ereditato A, Giacoppo F, Kreslo I, Nesteruk K P, Nirkko M, Weber M, Scampoli P, Neff M, Pilz S and Romano V 2012 *J. of Instr.* **7**(02,) p T02001– T02001.

[2] Mohamed N S, Wright N G, and Horsfall A B 2017) *IEEE Trans. Nucl. Sci.* **64**(7) p 1912-1919.

[3] Tajudin S M, Namito Y, Sanami T, & Hirayama H (2019) *Malaysian J. of Fund. and Appl. Sci.* **15**(4) p 580–584.

[4] Beddar A S, Mackie T R and Attix F H 1992 *Phys. Med. Biol.* **37** p 1883.

[5] Beddar A S 2007 *Radiat. Meas.* **41** p 124–133.

[6] Knoll G F 2010 *Radiat. Detection and Meas. 4th ed.* (John Wiley & Sons) p 229-232.

[7] Hirayama H, Namito Y, Bielajew A F, Wilderman S J and Nelson W R 2010 EGS5 Code System *SLAC Report* SLAC-R-730 *and KEK Report 2005-8.*

[8] http://www.eljentechnology.com/index.php/products/plastic-scintillators

[9] https://physics.nist.gov/cgi-bin/Star/compos.pl?matno=185

[10] Tajudin S M, Namito Y, Sanami T and Hirayama H 2014 *In 18th Topical Meet. of the Radiat. Protect. and Shielding Div. of ANS.*

[11] http://www.hamamatsu.com/jp/en/R375.html

[12] https://www.crystals.saint-gobain.com/products/cast-sheet-assemblies

[13] Tajudin S M and Hanim A S 2019 *The J. of Polish Soc. of Med. Phys.* **25**(3) p 189–192.

[14] Aydin A and Peker A 2015 *Nukleonika,* **60**(2), 361–366.

[15] Sprawls P 1995 *Physical Principles of Med. Imaging 2nd ed.* (Medical Physics Publishing Corporation).

[16] Yalcın Z, Icelli O, Okutan M, Boncukcuoglu R, Artun O, Orak S 2012 *Nucl. Instr. and Meth. in Phys. Res.* A **686** p 43–47

[17] Furetta C 2003 *Handbook of Thermoluminescence* (World Scientific).

IOP Conf. Series: Materials Science and Engineering **785** (2020) 012043 doi:10.1088/1757-899X/785/1/012043

Polyvinyl alcohol (PVA) based bolus material for high energy photons and electrons

Franca Ujah Okoh[1], Mohd Fahmi Mohd Yusof[2,a)], Reduan Abdullah[2], Norlaili Ahmad Kabir[1]

[1]School of Physics, Universiti Sains Malaysia, 11800 Penang, Malaysia
[2]School of Health Sciences, Universiti Sains Malaysia, 16150 Kelantan, Malaysia

[a]mfahmi@usm.my

Abstract. The study focused on the use of fabricated bolus materials based on the polyvinyl alcohol (PVA) to increase the surface dose on high energy photons and electrons. The PVA based bolus materials were fabricated at 2.5, 5, 7.5 and 10% PVA in water at target density of 1.0 g/cm^3. The PVA based boluses were scanned by using the computed tomography (CT) scanner along with the solid water phantoms. The percentage depth (PDD) dose of the solid water phantoms and the PVA based bolus materials were measured by using the treatment planning software (TPS) at 6 and 10 MV photons and 6 and 9 MeV electrons. The PDD curved obtained were compared to that of the standard bolus material and the surface dose were measured. The results showed that the PDD curves of solid water phantoms with PVA based bolus materials were consistent with that in standard bolus material within 1.4% percentage of difference for both 6 and 10 MV photons and within 2.9 and 1.4% percentage of differences in 6 and 9 MeV electrons respectively. The increment of surface dose in PVA based bolus materials were in excellent agreement to the standard bolus material and the percentage of PVA did not significantly affect the surface dose. The overall results showed the potential use of the PVA to be used as bolus material for clinical photons and electrons.

1. Introduction

Bolus has been used in as tissue compensator to compensate the deficit of the anatomical surface of patients undergoing radiotherapy [1][2]. Bolus materials usually have flabby characteristics and often constructed in the form of semi-solid gel to provide better contact with the skin surface. It is constructed with density close to water and human soft tissue (1.0 g/cm^3) to provide similar attenuation to the human soft tissue towards the photons and electrons. In other application, bolus materials are used to increase the surface dose especially for the treatment of near surface [3][4]. This is achieved by placing the bolus material on the surface while maintaining the source to surface distance (SSD) between the gantry and the surface of the medium. This would reduce the skin sparing effect by bringing the build-up region and the depth of maximum dose, d_{max} closer to the surface. The semi-solid characteristic of the bolus would reduce the air gap between the bolus and the medium that would cause dose perturbation beyond the air gap [5].

Typical bolus materials used in the radiotherapy are made of various carbohydrate based substances to simulate the human soft tissues such as rice, sugar pellets and sodium bicarbonate [1]. Several studies had been carried out by researchers to fabricate bolus materials with various non-carbohydrate substances [6-10]. This study focused on the use of polyvinyl alcohol (PVA) as bolus material. PVA is a water-soluble synthetic polymer that would form gel-structure materials through the process of freezing and defrosting [11]. It has approximate mass density of 1.19 g/cm^3 which is near to the value of water made it suitable to be developed as water and tissue equivalent bolus materials for photons and electrons. A previous work suggested the suitability of PVA gel to be used as phantom for the magnetic resonance imaging due to its close MRI parameters to the human soft [12]. This study fabricated the bolus material from the PVA-water solution and measured the percentage depth dose (PDD) and surface doses in high energy photons and electrons.

2. Methodology

2.1. Preparation of the Polyvinyl Alcohol Based Bolus Materials

The polyvinyl alcohol (PVA) was purchased from the Sigma-Aldrich Inc. (Sigma-Aldrich, Darmstadt, Germany). It has the monomer unit of $(CH_2CHOH)_n$ and was prepared in a crystalline powder form by the manufacturer. It's close density to water made it suitable to be considered as water equivalent phantom material. The PVA based bolus materials were achieved by mixing the PVA crystals into the water. An amount of water was heated during the process to enhance the solubility of PVA crystals into water. The PVA-water solutions were prepared at 4 different percentages of 2.5, 5.0, 7.5 and 10% based o the weight of the water. The PVA-water solutions were then cooled down to the room temperature to eliminate the air bubbles from the solutions. The solutions were then were filled in a square-shaped containers with external dimensions of 10 × 10 cm^2. Each percentage of the PVA-water solutions was prepared with two different thicknesses of 0.5 and 1.0 cm based on the readily available bolus materials for radiotherapy. The PVA-water solutions were then was cooled and being freeze by using the refrigerator for approximately 12 hrs to from the solid PVA-water materials. The solid PVA-water materials were then being defrosted at room temperature until the translucent gel PVA bolus materials were achieved as shown in Figure 1.

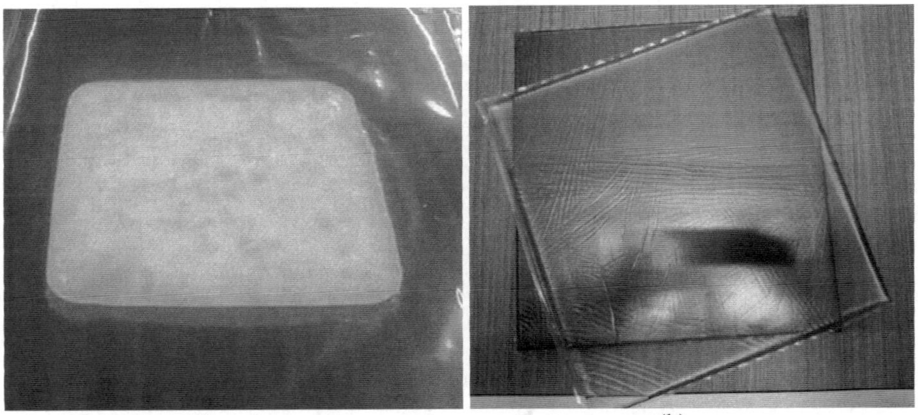

(a) (b)

Figure 1. (a) The fabricated PVA bolus material and (b) the standard carbohydrate based bolus material commonly used in radiotherapy.

2.2. Measurement of Percentage Depth Dose in High Energy Photons and Electrons

The PVA bolus materials were placed onto a set of solid water phantoms with approximate thickness of 25 cm and being scanned by using a computed tomography (CT) scanner. The PVA bolus materials and the solid water phantoms were scanned at abdominal scanning protocols with exposure factors of 120 kVp and 250 mAs and slice thickness of 0.1 cm. The CT images of the PVA bolus materials and the solid water phantoms were transferred into the treatment planning software (TPS) commonly used in radiotherapy. The photon and electron beams were simulated from the TPS on the solid water

IOP Conf. Series: Materials Science and Engineering **785** (2020) 012043 doi:10.1088/1757-899X/785/1/012043

phantoms and the PVA bolus materials. The simulations were based on the IAEA TRS 398: 2000 codes of practice (100 cm source to surface distance and 10 × 10 cm^2 field size). Two energies of photons of 6 and 10 MV and two energies of electrons of 6 and 9 MeV were simulated as shown in Figure 2. The percentage depth dose (PDD) in the solid water phantoms were measured based on the equation:

$$Percentage\ Depth\ Dose = \frac{D}{D_{max}} \times 100\% \tag{1}$$

with D and D_{max} is the dose at depth and maximum dose respectively.

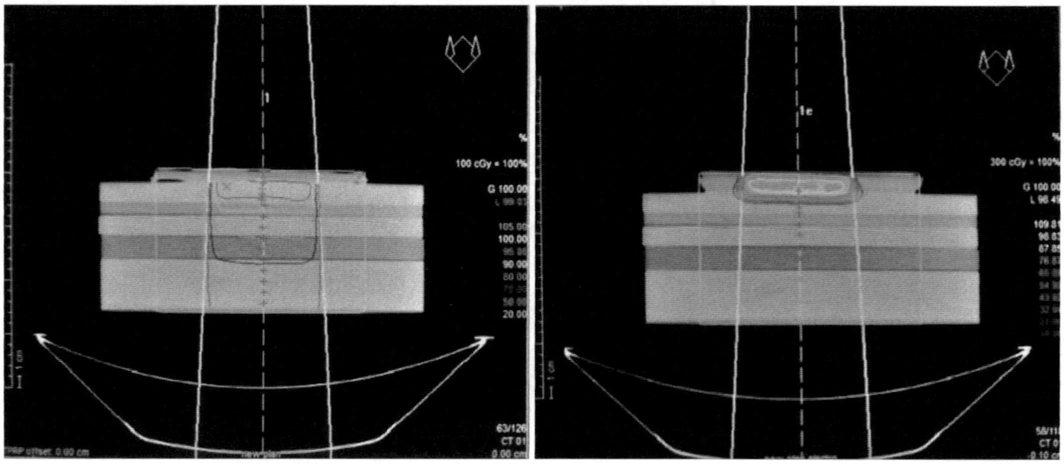

(a) (b)

Figure 2. The simulation of (a) photon and, (b) electron beams on the solid water phantoms by using the treatment planning system (TPS) software.

The PDD curves were plotted for all photon and electron energies. The PDD curves were compared to that in the available bolus material and the open field beams (without bolus). The surface doses at the presence of bolus and without the use of bolus were determined and the percentage differences of the surface dose were calculated by using the equation:

$$Percentage\ difference = \frac{dose\ with\ bolus - dose\ without\ bolus}{dose\ without\ bolus} \times 100\% \tag{2}$$

3. Results and Discussion

The PDD curves were plotted from all the photons of 6 and 10 MV and electrons of 6 and 9 MeV based on the calculation in Equation 1. Figure 3 shows the PDD curves in the 6 and 10 MV photons measured with the use of PVA bolus materials in comparison to the available bolus material in the radiotherapy department. The results showed that the PDD curves obtained by using the PVA bolus materials were in excellent agreement to the standard bolus material. The different percentage of PVA also did not significantly change the measured PDD. The comparison to the PDD curve measured in open field (without bolus) showed a slight shift of the PDD measured in the PVA bolus materials and standard bolus material to the left. The use of bolus significantly increased the surface dose on the solid water phantom as shown in Table 1. The use of bolus increased the surface dose in comparison to the open field to 95.44 and 95.66% to the D_{max} [4,8,9]. This satisfied with the primary function of the standard bolus materials used in radiotherapy. The surface dose was increased when bolus is applied due to the shortening of the source to surface distance (SSD) as well as the additional attenuation volume by the bolus materials before the beams interacts with the phantom or human soft tissues. The surface dose measured by using the PVA bolus showed an agreement to the standard bolus material within maximum percentage of discrepancy of 1.4% for both 6 and 10 MV photons. The percentage of the PVA used in the bolus did not significantly change the percentage of surface dose.

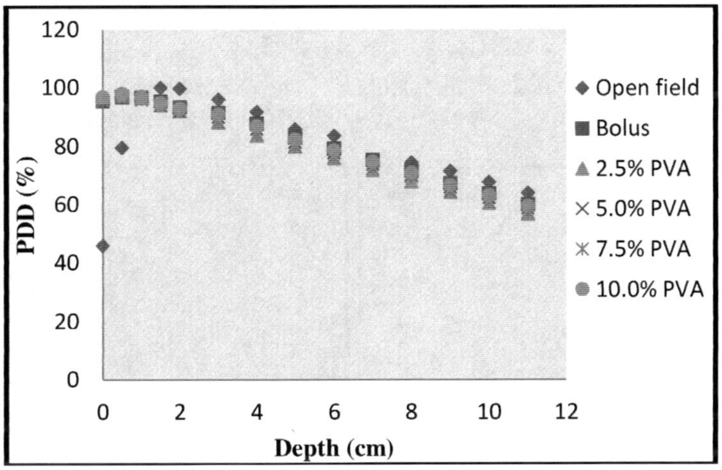

(a)

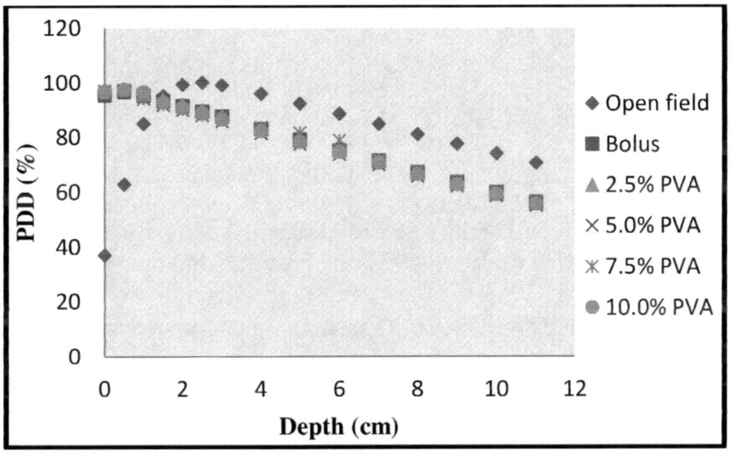

(b)

Figure 3. The PDD curves in high energy photons of (a) 6 MV and (b) 10 MV for all PVA bolus materials in comparison to the available bolus material and open field.

Table 1. The percentage of surface dose in comparison to the maximum dose, D_{max} measured with the presence of PVA bolus materials in comparison to the standard bolus and open field (without bolus) measured in 6 and 10 MV photons.

Energy (MV)	**Percentage of surface dose to D_{max} (%)**					
	Open field	**Standard bolus**	**2.5% PVA**	**5.0% PVA**	**7.5% PVA**	**10.0% PVA**
6	45.91	95.44	96.77	95.77	95.56	96.78
10	36.95	95.66	96.71	96.69	97.00	96.78

The PDD curves in electrons of 6 and 9 MeV energies are illustrated in Figure 4. Similarly to that in high energy photons, the results showed that the use of bolus shifted the PDD curves to the left in comparison to the open field. The displacement of PDD curve was more significant at higher electron energy (9 MeV) compared to that in 6 MeV. The surface dose was also increased between 94 and 97% to the D_{max} compared to 76.34 and 83.85% in the open field of 6 and 9 MeV electrons respectively [3,10]. The PDD curves measured by using the PVA bolus materials showed excellent agreement to the standard bolus material in both 6 and 9 MeV electrons. The measured surface dose in the PVA bolus materials in comparison to the standard bolus and open field is shown in Table 2. The results showed that the use of different percentage of PVA in the fabricated bolus materials did not significantly influence the surface dose in the electrons. The overall results suggested that the PVA gel materials can be fabricated as bolus at any percentage of PVA between 2.5% and 10%.

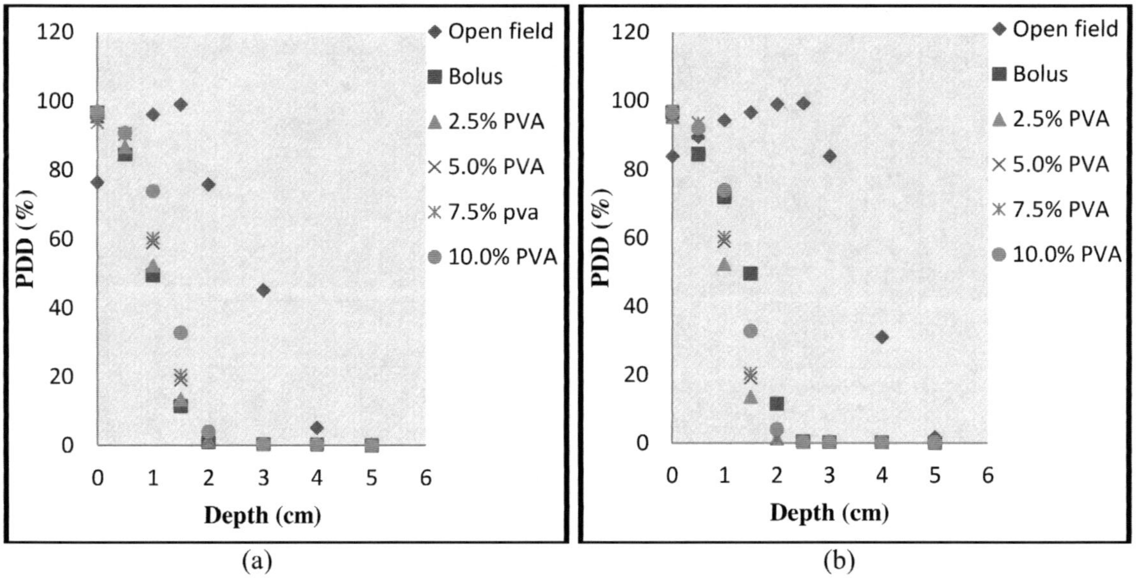

Figure 4. The PDD curves in electrons of (a) 6 MeV and (b) 9 MeV for all PVA bolus materials in comparison to the available bolus material and open field.

Table 2. The percentage of surface dose in comparison to the maximum dose, D_{max} measured with the presence of PVA bolus materials in comparison to the standard bolus and open field (without bolus) measured in 6 and 9 MeV electrons.

Energy (MeV)	Percentage of surface dose (%)					
	Open field	**Standard bolus**	**2.5% PVA**	**5.0% PVA**	**7.5% PVA**	**10.0% PVA**
6	76.34	96.68	95.65	95.48	93.8	96.91
9	83.85	96.68	95.28	95.48	95.8	96.7

4. Conclusion

The measurement of PDD showed that the PVA bolus materials increased the surface dose close to the values measured in the standard bolus material at all measured photon and electron energies. The percentage of the PVA used did not significantly change the measured surface dose. The overall results indicated the potential use of the fabricated PVA based bolus materials to be used as bolus in for high energy photons and electrons.

5. References

[1] Mano I, Goshima H, Nambu M and Lio M. 1986 New polivynil alcohol gel for MRI phantoms. *Magn. Reson. Med.* **3**(6) p 921-6.

[2] Butson M J, Cheung T, Yu P and Metcalfe P 2000 Effects on skin dose from unwanted air gaps under bolus in photon beam radiotherapy. *Radiat. Maes.* **32**(3) p 201-204.

[3] Hsu S H, Roberson P L, Chen Y, Marsh R B, Pierce L J and Moran J M 2008 Assessment of skin dose for breast chest wall radiotherapy as a function of bolus material *Phys. Med. Biol.* **53**(2008) p 2593-2606.

[4] Myer M N, Yoshikawa H, Moriarity L and Sidhu N 2009 Use of a petroleum-based bolus for photon radiation therapy of distal distal extremities in dogs. *Vet. Radiol. Ultrasound* **50**(2), p 235-238.

[5] Vyas V, Palmer L, Mudge R, Jiang R, Fleck A, Schaly B, Osei E and Charland P 2013 On bolus for megavoltage photon and electron radiation therapy *Med. Dosim.* **38**(3) p 268-273.

[6] Visscher S and Barnett E 2017 Comparison of bolus materials to highly absorbent polypropylene and rayon cloth. *J. of Med. Imaging and Radiat. Sci.* **48**(1) p 55-60.

[7] Sivarman A, Ganti S S, Nguyen H X, Birk G, Wieber A, Lubda D and Banga A K 2017 Development and evaluation of polivynil alcohol based topical gel.

[8] Supratman A S, Sutanto H, Hidayanto E and Jaya G W 2018 Characteristic of natural rubber as bolus material for radiotherapy *Mater. Res Express* **5**(9) 095302.

[9] Dipasquale G, Poirier A, Sprunger Y, Uiterwijk J W E and Mirabell R. 2018 Improving 3D-printing of mgavoltage x-rays radiotherapy bolus with surface scanner *J. Drug Deliv. Sci. Technol.* **39 p** 210-216.

[10] Rancangkapti N, Hariyanto A P, Mariyam F U, Almira L, Endarko and Haris B 2019 Dosimetry analysis of homemade bolus using propylene glycol for photon mega voltage and electron radiation therapy *J. of Phys: Conf. Ser.* **1248** p 012051.

[11] Khan F M 2010 *The Phys. of Radiat. Therapy* 4th ed. (Philadelphia: Lippincott Williams and Wilkins).

[12] Stanton R and Stinson D 2009 *Appl. Phys. for Radiat. Oncol.* (Wisconsin :Medical Physics Pub.)

Acknowledgement
The authors would like to acknowledge the financial support from the USM Bridging Grant scheme no 304.PPSK.6316324. The authors also would like to acknowledge the Nuclear Medicine, Raiotherapy and Oncology Department, Hospital Universiti Sains Malaysia (HUSM) for the use of CT scanner and TPS software in this study.

IOP Conf. Series: Materials Science and Engineering **785** (2020) 012044 doi:10.1088/1757-899X/785/1/012044

X-ray diffraction method to identify epithelial to mesenchymal transition in breast cancer tissue

A M Alsharif[1,2], S F Abdul Sani[1,a)], F Moradi[1]

1 Department of Physics, University of Malaya, Kuala Lumpur, Malaysia
2 Department of Physics, University of Sebha, Sebha, Libya

[a]s.fairus@um.edu.my

Abstract. Breast Stromal tissue is significantly involved in the spread of cancer because of molecular variations. The main component, collagen, has a spatial arrangement allowing the investigation of its structural patterns in tissues by X-ray diffraction (XRD). Since detection and classification of tumor type and stage from the pathological examinations is a laborious task that additionally does not allow cancer detection in the early stages, development of new tools to improve the diagnostic power of pathologists is still needed. In this study, the potential used of XRD tecniques as one of the method for early detection of breast cancer was done. Four pathological cancerous breast tissue samples were taken from four different patients; where two samples are with epithelial-to-mesenchymal transition (P-EMT) and two samples are without epithelial to mesenchymal transition (N-EMT). The latter was incorporated at an angle interval of 10° to 80° to obtain interference diffraction pattern of XRD of human malignant tissues. This was done to determine molecular structure changes of collagen fiber within the tissue and their potential relation to the changes in cancerous tissue toward metastasis. The results showed that the momentum transfer values for the first region of N-EMT,P-EMT cancerous tissue are not significantly different for both cancerous breast tissue types, being at 1.61 ± 1.74 nm^{-1}. The average peaks of the second region were determined for N-EMT at 3.4 ± 3.5 nm^{-1} and P-EMT at 3.6 ± 3.7 nm^{-} which is attributed to water content of the tissues, due to the greater intensity of the P-EMT compared to that of N-EMT. Principal component analysis (PCA) used to confirm statistical appropriateness of the results, showed a normal distribution within 95% confidence level. P-EMT clusters have a larger number of scatter plots compared to N-EMT, which indicates a higher similarity between P-EMT samples than between N-EMT samples, thus confirming the difference between positive and negative-EMT clusters. The curent analysis can differentiate XRD features and thus confirms its effectiveness in detecting the molecular correlation of abnormal collagen fiber structure within epithelial change, and is useful for early diagnosis even case of dense breasts.

1. Introduction

Breast cancer is the most frequent malignancy in females with over 1.6 million newly-diagnosed cases yearly [1]. The early stage of breast cancer starts with uncontrolled growth of aggressive cells. When breast epithelial cells undergo regular periods of proliferation and apoptosis, their mutations may cause cancer. These tumor cells are usually diagnosable in a physical checkup or X-ray graph. However, the pathology can determine the grade of malignancy, which is extremely important as a clinical guide to the treatment options [2]. It is widely accepted that early-stage detection of breast cancer before the metastasis stage gives the patient a higher chance of cure [3]. The regular procedure of breast cancer diagnosis after a physical check-up is an ultrasound, mammography and magnetic resonance imaging

(MRI). Apart from methods to detect breast cancer, breast biopsy techniques are also used to differentiate between cancer types, but these are costly, and time-consuming with inherent limitations. for example, ultrasound may fail to detect some tumors due to the similarity in the acoustic properties of normal and abnormal tissues [4], while in the case of MRI, in addition to being relatively costly and time-consuming, it is also not appropriate for patients with claustrophobia or metallic implants [5]. Typically, in mammography, contrast between abnormal and fibroglandular tissues is low, which makes it difficult to detect small lesions, causing up to 50% of breast carcinomas not being diagnosed in the first round of mammography tests. Mammography is known to be generally less efficient with patients under the age of 40 years and in detecting tumor sizes of below millimeter dimension or less than 105 cells [5]. Statistically, most of the tumors detected by mammography have a size ranging from 1 to 2 cm, while most of the MRI detected lesions are smaller than 1 cm.

The use of XRD to analyze the molecular structure of crystalline materials has been the traditional practice. It has been shown that coherent scattering can be used to determine the average inter-molecular spacing in biological samples [6]. Many XRD measurements on biological materials have highlighted the changes in unhealthy tissues which can also be seen as cancer fingerprint [7]. These changes can be related to an improper structure of collagen, caused by a breast malignant lesion [8]. Many studies have indicated the potential of XRD to discriminate tissue types due to the alteration of collagen fiber within tissues [6][9][10]. The early stages of metastasis are typified by the tumor cells migrating and advancing by way of the stroma into the circulation via epithelial-to-mesenchymal transition (EMT) [11]. The EMT is a well-defined developmental progress frequently adopted by tumor cells during the metastatic process. Cells acquire a more spindle-like fibroblastic morphology and reach the distant organs, losing epithelial cell adhesion and cell-cell contacts increase their stroma interactions and invasive behavior [11][12]. But, the biological-effects of EMT are assorted and complicated. Linked features of a tumor-cell may be different depending on the tissue type [13]. In the specific case of breast cancer, many studies have proposed that cancer cells can hijack EMT to invade, showing the EMT's importance in breast cancer's invasion, growth, and spread. Elevation of cell adhesion and collagen, as well as high motility of cells, are signs of the tumor invasion [14].

In this study, the molecular structure of the breast cancer was studied in terms of collagen fiber alteration and its association with EMT, to provide improvement for early diagnosis of breast cancer. To achieve this, two different breast cancerous tissue types were used to examine the ability of the XRD technique to discriminate the histopathological behavior between malignant tissues based on their stage.

2. Materials and methods

2.1 Tissue samples

Four cancerous breast tissue samples were taken from histopathology examination of four patients; two samples P-EMT and two samples N-EMT. These patients underwent a breast-conserving mastectomy where samples were between 3 to 5 μm thickness and not more than 2×2.5 cm^2 dimensions. To allow tissue fixation and preservation, samples were immersed into 10 % neutral buffered formalin and then impregnated with paraffin.

2.2 X-ray diffraction

XRD is a non-destructive test to analyse crystalline material and can be used to provide detailed information on molecular structure of the material. It works based on the elastic or coherent scattering of x-ray photons by atoms in an intermittent lattice. When X-rays incidence on material, they may interact with the atomic electrons coherently or incoherently (Compton scattering). Diffraction occurs through coherent scattering from the space between atomic planes (d) resulting in constructive interfere with the scattering angle (2θ). Equation 1,

$$n\lambda = 2\mathrm{d} \sin\theta \tag{1}$$

where n is the order of reflection and is an integer, while λ is the X-ray wavelength. XRD patterns data were converted into values of momentum transfer by Equation 2, as follows:

$$X\,(nm^{-1}) = ({}^{E}/_{hc})\sin{}^{\theta}/_{2} \tag{2}$$

where: E, h, c, and θ are incident photon's energy, the Planck constant, the light's speed, and scattering angle, respectively. The XRD scan was performed with a BRUKER-binary V3 (RAW) diffractometer at room temperature with copper as the anode target material and value of $K\alpha 1$, $K\alpha 2$ and $K\beta$ being 1.540, 1.544 and 1.390 keV respectively. The accelerating potential of electrons to produce X rays was 40 kV while the current was 40 mA. The range of scattering angles measured varied from 10° to 80° of the momentum transfer between 1.40 nm^{-1} and 11.03 nm^{-1}.

The XRD scan was performed with a BRUKER-binary V3 (RAW) diffractometer at room temperature in a setup shown in Figure 1 with copper as the anode target material. The accelerating potential of electrons to produce X rays was 40 kV while the current was 40 mA. The range scattering angles measured varied from 10° to 80° corresponding to the momentum transfer interval of between 1.40 nm^{-1} and 11.03 nm^{-1}).

2.3 Statistical analysis

The XRD data were analysed statistically by IBM SPSS Statistics 22.0 software to distinguish between tissue types and to classify the XRD pattern. The graphs were plotted using Minitab.

3. Result and discussions

3.1 XRD results

The momentum transfer values were calculated from XRD scattering results of P-EMT and N-EMT cancerous samples of four different patients as shown in Figure 1. Figure 1(a) shows XRD spectrum for N-EMT samples and Figure 1(b) shows XRD spectrum for P-EMT samples. In order to compare the results, XRD patterns were divided into two regions of scattering angle between 5° to 9° (region 1) and between 9° to 17° (region 2) that corresponded with the momentum transfer values of 1.4 to 2.4 nm^{-1} and 2.4 to 5 nm^{-1}. In region 1 of both P-EMT1, P-EMT2, and N-EMT1, N-EMT2 specimen types (a and b), sharp peaks were observed at 1.61 to 1.74 nm^{-1} at scattering angle around 6° which is attributed to the prevalence of fibrous tissue characterized by water components, and in agreement with previous studies [9, 15-20] in peak position for initial photon energy of 40 keV, malignant tissue and, abnormal fibre portion around 47.8%. The number of uncommon structures of collagen fibres for different pathological types of tissue has been reported around 52.2 % [21].

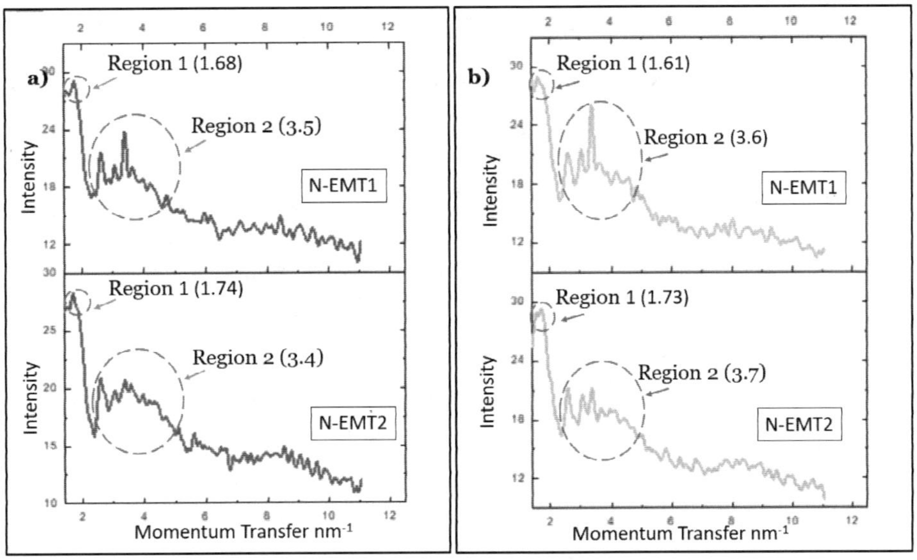

Figure 1. XRD spectra of all of the cancerous breast tissues a) (N-EMT) b) (P-EMT).

Momentum transfer values are characteristics of the material and can reveal information on the spacing between atomic planes in its crystalline or amorphous structure [15]. Due to the molecular space the first two neighbours in these cells being larger and interconnected on a large spatial range, it results in new order with a higher degree [22]. The region of momentum transfer below 2.5 nm^{-1} represents specimen interference found in most of biological samples [23]. Oliveira, *et al.* (2008) reported the signature peak of the connective tissue and the tumor to be at 1.5 ± 0.7 nm^{-1}[10].

Table 1. The values of the diffraction pattern parameters obtained from region 1.

Tissue type	Peak position (Θ)	FWHM (nm-1)	d-spacing
N-EMT1	5.9	0.49	0.33
N-EMT2	6.2	0.42	0.34
P-EMT1	5.7	0.53	0.31
P-EMT2	6.1	0.31	0.77

From Table 2, the differences in the peak height noticed in Region 1 between N-EMT1 and N-EMT2 were found to be 0.116 to 0.119 while P-EMT1 and P-EMT2 were higher at around 0.121 to 0.123 respectively. Poletti *et al.* [9] stated that the values of the scattering intensity of tumor parts are greater than the health tissue region. Due to the existence of the collagen fibre [24], there is ample evidence indicating that of the extracellular matrix of breast tissue carcinoma undergoes change on its structure and composition [25].The supra-molecular structure of collagen fibrils' alteration has been linked as a major component of the extracellular matrix because of the tumor existence [25][26]. Weaver [27] found that the healthy or malignant epithelial cells start migrating when they come into direct contact with collagen. Where interference effects are big and the angular distribution peak is at a certain angle, this angle depends on the photon's energy, type, and structure of the tissue. This lends support to previous findings [26]. Also, the peak heights among tissues can be related to the interaction between water molecules, which are a major component of glandular tissue. For instance, it has been found that a full width half maximum (FWHM) is more suitable than peak location for differentiation of water content between tissues [28]. From Table 2, the value of FWHM for all cancerous tissues was found higher in P-EMT from 0.31 nm^{-1} to 0.53 nm^{-1} than N-EMT with 0.42 to 0.49 nm^{-1}. Average peaks shown in Table

2 refer to the second region, showing the effects of the molecular interference for the case of all specimens. These sharp peaks started at position θ = 9° to17°. On average, it is concluded that momentum transfer of between 2.5 and 4.4 nm^{-1}, refers to the water content, which is in agreement with previous reports [18][19]. Also, [10] [9] reported breast tissue momentum transfers of between 0.5 and 4 nm^{-1} in agreement with the present findings.

Table 2. The values of the diffraction pattern parameters averagely obtained from region 2.

Tissue type	Peak position (Θ)	FWHM (nm-1)	d-spacing
N-EMT1	3.5	0.32	0.0119
N-EMT2	3.4	0.25	0.0105
P-EMT1	3.6	0.28	0.0111
P-EMT2	3.7	1.89	0.0101

3.2 Principal component analysis result

PCA score plots were performed for all samples and two PCA models were built. From Fig. 2, score plots of the first component (PC1) for N-EMT samples and those of the second component (PC2) for P-EMT samples were categorized. Randomly, 30% of the two subgroups were picked out for all 4 samples. The score plots of the same tissue type samples have similar perpendicular alignment [29]. From the result, the level of confidence is 95%; cluster 1 has less similarity than cluster 2, which confirms the statistical difference between two clusters. The smaller number of scatter plots in cluster 1 indicates a big difference between N-EMT1 and N-EMT2. On the other hand, cluster 2 has a large number of score plots that indicates the big similarity between P-EMT1 and P-EMT2 tissues. As a result, this analysis has the potential to differentiate between cancer stages.

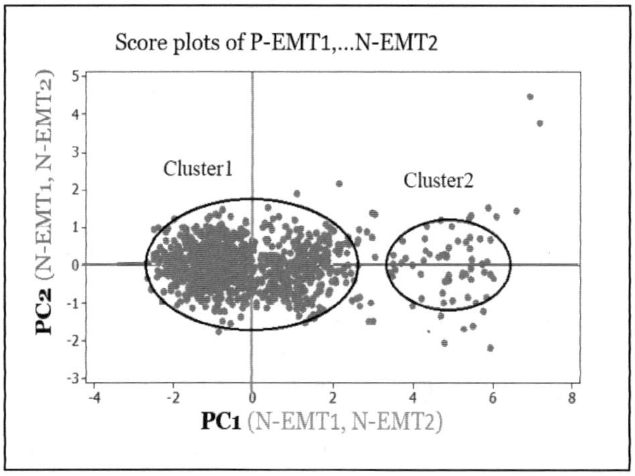

Figure 2. Principal component analysis of tissue type

Thus, the molecular structure of collagen fibre from a sample tissue type can help in identification of composition change in tissue. Furthermore, usefulness of PCA to detect a correlation between breast tissue types was shown. Therefore, this result encourages and suggests the use of XRD for bulk tissues classification and early detection of breast cancer [30].

4. Conclusion

In this work, a comparison was made between the malignant breast tissue types by XRD method considerable change of scattering angle was found between tissue types, and the intensity of the diffraction pattern was the only parameter changed by difference in the molecular structure of collagen. This was found useful to distinguish between positive and negative EMT pathological tissue types. The peaks correspond to water and fibrous content within the tissue, from the P-EMT was found to have greater intensity compared to that of N-EMT.

The results of statistical analysis demonstrated a significant difference between P-EMT and N-EMT samples. PCA data provided support in differentiating various types of tissue using XRD pattern. The current work suggests that XRD has the potential to be used in early screening tests for detection of breast cancer patients, and thus can assist to classify the cancer tissue histopathology

5. References

[1] Ferlay J, Soerjomataram I, Dikshit R, Eser S, Mathers C, Rebelo M, Parkin D M, Forman D and Bray F 2015 Cancer incidence and mortality worldwide: sources, methods and major patterns in GLOBOCAN 2012 *Int. J. of Cancer* **136** p E359-E386.

[2] Samah A A, Fauzi M F A and Mansor S 2017 Classification of benign and malignant tumors in histopathology images *IEEE Int. Conf. on Signal and Image Processing App. (ICSIPA)* p 102-106.

[3] Laronga C, Chagpar A B and Vora S R 2016 Patient education: breast cancer guide to diagnosis and treatment (beyond the basics) *UpToDate* (Waltham).

[4] Kidane G, Speller R D, Royle G J and Hanby A M 1999 X-ray scatter signatures for normal and neoplastic breast tissues *Phys. in Med.& Bio.* **44** p 1791.

[5] Darambara D G, Sellin P J and Maehlum G 2004 A novel detection system consisting of a large-area sensor and a multicell Si-pad array operated in spectroscopic mode for X-ray breast imaging *IEEE Trans. on Nucl. Sci.*, **51**, p1641-1647.

[6] Costa M, Benseny-Cases N, Cócera M, Teixeira C V, Alsina M, Cladera J, López O, Fernández M and Sabés M 2009 Diagnosis applications of non-crystalline diffraction of collagen fibres: breast cancer and skin diseases *App. of Sync. Light to Scattering and Diffraction in Mat. and Life Sci.*. **776** p 265-280 (Berlin, Heidelberg: Springer).

[7] Torzilli P A, Bourne J W, Cigler T and Vincent C T 2012 A new paradigm for mechanobiological mechanisms in tumor metastasis *Sem. in Cancer Bio.* **22** p 385-395.

[8] Krisanaprakornkit S and Iamaroon A 2012 Epithelial-mesenchymal transition in oral squamous cell carcinoma *ISRN oncology* p 1-10.

[9] Lo H C and Zhang X H F 2018 EMT in metastasis: Finding the right balance *Dev. Cell* **45** p 663-665.

[10] Liu F, Gu LN, Shan BE, Geng C Z and Sang M X 2016 Biomarkers for EMT and MET in breast cancer: An update *Onco. Lett.* **12**(6), p 4869-4876.

[11] Marticke F, Montémont G, Paulus C, Michel O, Mars J I and Verger L 2017 Simulation study of an X-ray diffraction system for breast tumordetection *Nucl. Inst. and Meth. in Phys. Research Sec, A: Accelerators, Spectrometers, Det.s and Assoc. Equip.*, **867**, pp.20-31.

[12] Lewis R A, Rogers K D, Hall C J, Towns-Andrews E, Slawson S, Evans A, Pinder S E, Ellis I O, Boggi, C R M, Hufton A P and Dance D R 2000 Breast cancer diagnosis using scattered X-rays *J. of Synch. Rad.* **7**, p 348-352.

[13] Poletti, M.E., Gonçalves, O.D. and Mazzaro, I., 2002. Coherent and incoherent scattering of 17.44 and 6.93 keV x-ray photons scattered from biological and biological-equivalent samples: characterization of tissues. *X-Ray Spectrometry: An International Journal*, **31**, pp.57-61.

[14] Farquharson M J, Al-Ebraheem A, Cornacchi S, Gohla G and Lovrics P 2013 The use of X-ray interaction data to differentiate malignant from normal breast tissue at surgical margins and biopsy analysis *X-Ray Spectr.* **42** p 349-358.

[15] Johnston E M, Dao E and Farquharson M J 2019 Advances in the histopathological characterization of breast tissue using combined X-ray fluorescence and X-ray diffraction data in a multivariate analysis approach. *X-Ray Spectr. p* 1–6

[16] Alaroui D F, Johnston E M and Farquharson M J 2019 Comparison of two X-ray detection systems used to investigate properties of normal and malignant breast tissues *Nucl. Inst. and*

IOP Conf. Series: Materials Science and Engineering **785** (2020) 012044 doi:10.1088/1757-899X/785/1/012044

Meth. in Phys. Research A.

[17] Lakshmanan M N, Greenberg J A, Samei E and Kapadia A J 2017 Accuracy assessment and characterization of x-ray coded aperture coherent scatter spectral imaging for breast cancer classification *J. of Med. Imaging* **4** p 013505-013505.

[18] Ambekar R, Lau T Y, Walsh M, Bhargava R and Toussaint K C 2012 Quantifying collagen structure in breast biopsies using second-harmonic generation imaging *Biomed. optics express* **3** p 2021-2035.

[19] Kosanetzky J, Knoerr B, Harding G and Neitzel U 1987 X-ray diffraction measurements of some plastic materials and body tissues *Med. Phys.* **14** p526-532.

[20] Sosa C, Malezan A, Poletti M E and Perez R D 2017 Compact energy dispersive X-ray microdiffractometer for diagnosis of neoplastic tissues *Rad. Phys. and Chem.* **137** p 125-129.

[21] Oliveira O R, Conceição A L, Cunha D M, Poletti M E and Pelá C A 2008 Identification of neoplasias of breast tissues using a powder diffractometer *J. of Rad. Research* **49** p 527-32.

[22] Sidhu S, Siu K K, Falzon G, Nazaretian S, Hart S A, Fox J G, Susil B J and Lewis R A 2008 X-ray scattering for classifying tissue types associated with breast disease *Med. Phys.* **35** p 4660-4670.

[23] Burke K, Tang P and Brown E B 2012 Second harmonic generation reveals matrix alterations during breast tumor progression *J. of Biomed. Optics* **18** p 031106.

[24] Conceição A L C and Poletti M E 2010 Identification of molecular structures of normal and pathological human breast tissue using synchrotron radiation *AIP Conf. Proc.* **1266** p 72-77.

[25] Weaver V M, Bissell M J, Fischer A H and Peterson O W 1996 The importance of the microenvironment in breast cancer progression: recapitulation of mammary tumorigenesis using a unique human mammary epithelial cell model and a three-dimensional culture assay *Biochem. and Cell Bio.* **74** p 833-851.

[26] Theodorakou C and Farquharson M J 2008 Human soft tissue analysis using x-ray or gamma-ray techniques *Phys. in Med. and Bio.* **53** p R111-R149.

[27] Pani S, Cook E J, Horrocks J A, Jones J L and Speller R D 2010 Characterization of breast tissue using energy-dispersive X-ray diffraction computed tomography *Appl. Rad. and Iso.* **68** p 1980-1987.

[28] Moss R M, Amin A S, Crews C, Purdie C A, Jordan L B, Iacoviello F, Vinnicombe S J 2017 Correlation of X-ray diffraction signatures of breast tissue and their histopathological classification *Sci. Rep.* **7** p 12.

Acknowledgment

The first author would like to thank the Libyan Ministry of Education and the Libyan Embassy in Malaysia for financial support. This work is supported by the University of Malaya (UM), providing a grant under the Bantuan Kecil Penyelidikan (BK093-2016), and the Ministry of Higher Education, Malaysia for support under the Fundamental Research Grant Scheme (FRGS) FP032-2017A.

IOP Conf. Series: Materials Science and Engineering **785** (2020) 012045 doi:10.1088/1757-899X/785/1/012045

Validation of in-house algorithm for SPECT-CT 3D segmentation and high-accuracy cancer internal dosimetry protocol

M Z Ngali[1,a)], K Y H Lawrence[1], M A Masud[2], M S Aminudin[2], M S Salihatun[1], T Ishkrizat[1]

[1]Sports Engineering Advanced Research Group, Faculty of Mechanical and Manufacturing, Universiti Tun Hussein Onn Malaysia
[2]National Cancer Institute, Ministry of Health Malaysia

[a]zamani@uthm.edu.my

Abstract. Tumour volume measurement requires physicist to perform manual delineation for every slice of 2D image produced by SPECT-CT to obtain the total number of pixel of abnormal activities. The total number of pixel is multiplied with spatial resolution of each pixel and thickness between each slice. Physicist delineate abnormal activities of tumour subjectively which lead to inconsistent accuracy of the measured volume. Manual delineation process requires high consumption of time and effort. This work is aimed to verify and validate an in-house developed algorithm that capable of numerically quantify the tumour volume. The results were compared with the actual and manually calculated volume of Phantom Body, scanned with SPECT-CT apparatus. The algorithm verification is performed by comparing the algorithm structure and preliminary results with previous works. The data for validation were obtained by conducting experiment with NEMA-IEC body phantom and Iodine-131. The algorithm structure and outcome were comparable with previous works with slight modification. The algorithm measured volume is 26.9579 cm^3 which is only 1.6% error as compared to the actual tumour volume of 26.521 cm^3. The algorithm is expected to be used for actual patient data in the near future once sufficient verification and validation procedures are achieved.

1. Introduction

Cancer is one of the biggest death cause of Malaysian. According to Deputy Health Minister Malaysia, 100000 Malaysians diagnose with cancer every year [1]. Unhealthy lifestyle such as smoking, unhealthy diet increase risk of Malaysian suffers from cancer. Malaysia has acquired with high technology and devices in medical which improve the health level of Malaysian and medical fee in Malaysia is affordable for most of the Malaysian. However, mortality rate of cancer patient in Malaysia is relatively high which is 61% [2]. According to Malaysia Health Minister, Datuk Seri Dr Dzulkefly Ahmad, most of the cancer cases of Malaysian are detected at late stage due to Malaysian low health awareness and medical check-up constancy [3]. The report also insisted that the patients missed the treatment in early stage had been the main cause of decreasing survival rate among cancer patients.

The manual process of delineation of 2D image produced by CT scan or SPECT-CT are time consuming and requires skilled workers [4][5]. Physicists are required to circle the abnormal activities from 2D image to identify the location of the tumour and the size of tumour. The overall process for manual delineation are time consuming, training for skilled workers required effort and time. Besides that, potential error occurs due to contrast of the 2D imaging has low distinguish between normal and cancer cells which lead to physicist include or exclude the abnormal activities pixel during manual delineation. This scenario led to inaccuracy of the internal dosimetry procedure. Besides that, diagnosis results for patient took longer time to process and caused patients to miss the best treatment time.

The aims of this work are to validate result of algorithm measured volume of tumour with NEMA-IEC phantom and to compare the result of algorithms calculated for NEMA-IEC phantom and manual delineation results.

This research enables physicist to implement personalised radionuclide treatment for patients. Based on the volume of interest, the accurate amount of radiopharmaceutical calculated for the specific patient. The suitable amount of radiopharmaceutical injected to patient which minimise the harm to patient and achieve cancer treatment with highest efficiency without lack of dosage or overdose.

2. Literature Review

2.1. Targeted radionuclide therapy

Targeted radionuclide therapy used radiopharmaceutical which capable to localise the position of the tumour and released radiation energy through decay to kill cancer cell. However, targeted radionuclide therapy is inefficient due to fixed dosage of dosimetry for every patient which create two scenarios. First, lack of dosage of radiopharmaceutical to kill larger volume of cancer cell. Second, overdose of radiopharmaceutical lead to healthy cell nearby cancer cell affected. This indicated that dosage of radiopharmaceutical are crucial to perform high accuracy therapy. Targeted radionuclide therapy efficiency and survival rate increased with individualised targeted radionuclide therapy [6]. Individualised targeted radionuclide therapy with the data of volume of tumour are able to calculate dosimetry for each patient.

2.2. Internal dosimetry

Internal dosimetry deals with the absorption dosage rate of radiation energy emitted by radionuclide in targeted organ [7]. Internal dosimetry is crucial in both diagnosis and therapy process to maximise the efficiency for both process. For diagnosis process, dosage of radionuclide is essential to maximise quality of nuclear medicine imaging and minimise the effect of radionuclide to patients. For therapy process, dosage of radionuclide is crucial to maximise absorbed dose for tumour and minimise the toxicities to surrounding normal cells [7]. The formula of calculate the dosage absorb rate is defined as below:

$$\dot{D} = \frac{k \sum_s A_s \sum_i n_i E_i \phi_i (T \leftarrow S)}{m_T} \tag{2.1}$$

The formula can be simplified as below:

$$\dot{D} = \frac{\varepsilon}{m} \tag{2.2}$$

where ε is the activity and energy radiation from radionuclide sources absorbed by targeted tumour and m is the mass of tumour. Each patient has different size of organ and different size of tumour which lead to the dosage required by each patient is vary based on their situation. From Equation 1, mass of tumour is the parameter to determine dosage needed. Mass of tumour obtained through nuclear imaging for better accuracy of dosage calculation.

2.3. Review on comparable algorithms

Visualisation of 3D image based on planar image is interested by physicist as the information in planar image is limited compare to 3D image. Research with regards to 3D image reconstruction using planar image of image modality is conducted. Various type of methods of 3D reconstruction algorithms is used.

3D image simplified the workload and reduced time consumption to calculate the volume of tumour compare to manual delineation. Measurement of volume for tumour have been research by researcher to perform high accuracy of result. Method for measurement of volume consist of various method used in algorithms. The researches data is tabulated as shown in Table 1 and Table 2.

Table 1. 3D reconstruction study.

Study	Modalities	Data	Radionuclide	3D reconstruction method and process
Synefia et al., 2014	SPECT	Stress and rest myocardium DICOM image	Tc-99m	• Isosurface – Extract DICOM image – Determine isocontour surface for isovalue – extract pixel value from DICOM data
Kyriacou, Christofides, & Pattichis, 2016	SPECT	Kidney image	Tc-99m	• Isosurface – Used Otsu threshold to determine optimal isovalue – extract pixel value from DICOM data
Tory et al., 2005	SPECT	4D kidney SPECT data	-	• Isosurface – Isovalue value determine by user • Direct volume rendering – High intensity region is highlighted

2.4. Measurement of volume

Method to measure volume of tumour is an interested topic for researcher to find a quantitative method with high accuracy. Accuracy of volume of tumour enable physicist to perform more accurate diagnosis and treatment. The data used for initial experiment in imaging are mostly Phantom. Phantom is an object that used for imaging in validation of imaging machine. Phantom data from imaging is used to analyse, evaluate the performance of machine. Phantom used to represent as the body of patient send to imaging. There are many type and shape of phantom even phantom that can mimic lung, liver of human. Each phantom type help in physicist for research and validation.

From Table 2, there are two main method used in measure volume of interest which is used of 3D image to obtain pixel value and used of planar image to obtain pixel value. The pre-process for used of 3D image to measure volume is similar to 3D reconstruction which discussed in section 2.6.1. The maximum pixel value for threshold is determined using grey-level histogram in region of interest. The threshold level is either fixed or change based on the user to determine the suitable pixel value in calculation for volume of tumour. If the pixel value overestimated, the volume of the tumour in calculation is smaller than the exact size and vice versa. Thus, pixel value range threshold for tumour calculation must be correct to ensure that the accurate calculation of affected area is achieved. The number of pixel with pixel value within the threshold level is calculated. Volume of phantom is obtained by multiply total number of pixel value with pixel area and slice thickness.

Table 2. Measurement of volume study.

Study	Modalities	Data	Reconstruct 3D image	Process for measurement of volume	Formula
Erdi, Wessels, Loew, & Erdi, 1995	SPECT	6 Lucite sphere	Yes	– Find highest counts for pixel based on grey-level in ROI – Change threshold level based on maximum counts – Count number of pixel with pixel value	Total number of pixel x pixel area x slice thickness
Zingerman, Golan, & Moalem, 2009	SPECT-CT	Cube and sphere phantom	No	– Attenuation correction for SPECT and CT data – Draw FWHM intensity profile across phantom for each row and column for all slices – Count total number of pixel value for each slice	Overlapping pixel are used for VOI boundary
Mortelmans, Pamel, Maegdenbergh, Roo, & Suetens, 1986	SPECT	Cylinder phantom with 20cm diameter	No	– Fixed threshold and grey level histogram determine pixel value	Total number of pixel counted x voxel size
Synefia et al., 2014	SPECT	Stress and rest myocardium DICOM image	Yes	– Determine isocontour surface for threshold value – Based on threshold level to determine volume of myocardium	

In order to obtain planar image, the result of SPECT and CT is attenuated to merge both data. The merged planar image is drawn with Full Width Half Maximum (FWHM) intensity profile across the phantom for each row and column. Every slices are drawn with FWHM intensity profile. FHWM profile is drawn column by column and row by row for all slices. The profile for each row and slice and analyse individually on the original slices. The overlapping pixel is used as border of volume of interest. Figure 1 shows the FWHM intensity profile across a row of sphere phantom.

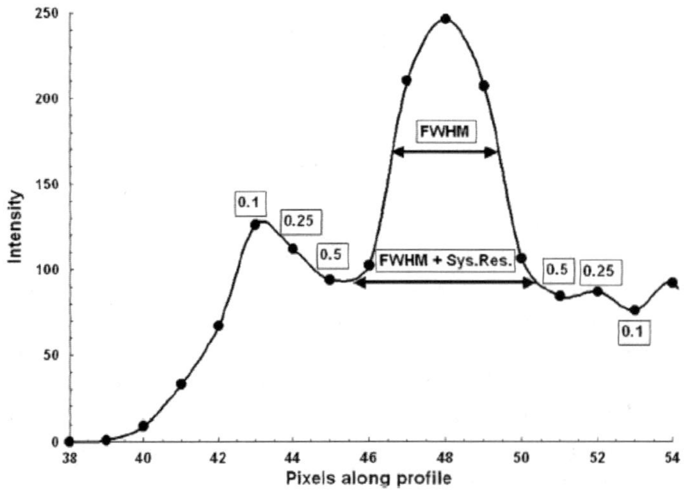

Figure 1. FWHM intensity profile across a row of sphere phantom [8].

3. Methodology

Actual Phantom volume for each diameter of sphere is recorded. The Phantom then referred to nuclear imaging, SPECT-CT machine to produce SPECT and CT DICOM data. The DICOM data used as input for 3D reconstruction algorithms to produce 3D image. Measurement of volume algorithm is used to measure pixel value of reconstructed phantom in 3D structure. Volume of phantom is calculated pixel size multiplied spatial resolution of one pixel and thickness between each slice. The volume of phantom calculated by algorithms is compared with the actual volume of the sphere. The result is validated as the result of algorithms is matched with actual volume. If the result of algorithm has high percentage of error, reconstruction of 3D image process is repeated to ensure the image of phantom reconstructed is correct.

The pixel value obtained from the experiment is measured manually by experienced physician. The pixel value is multiplied with spatial resolution of SPECT to calculate the diameter of the phantom sphere. The diameter is used in formula of volume of sphere. The calculated volume of phantom is compare with the result of volume of phantom calculated by using algorithms. Minimum pixel value and maximum pixel value is set. Minimum pixel value is 395 which is calculated based on phantom sphere experiment. Specific sphere phantom image is reconstructed by using isosurface volume rendering. Phantom image is reconstructed in 3D image and allows physicist to rotate the image for more accurate diagnosis. The pixel size of the specific phantom is calculated by summation of pixel value in between maximum and minimum value pixel value. Volume of phantom is calculated by multiplying spatial resolution, thickness of slices and pixel size.

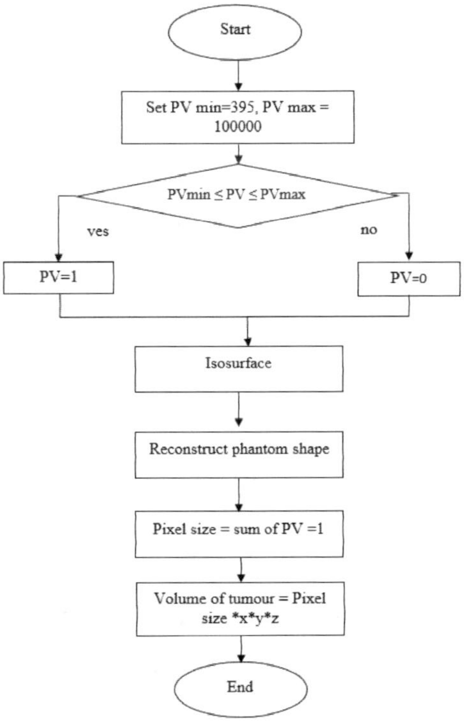

Figure 2. Algorithms flowchart for measurement for volume.

4. Results and discussion

To validate the result of algorithms measured volume, Table 3 shows result of comparison between algorithms calculated volume with actual volume of phantom and manual measured volume of phantom. The actual phantom volume result for 37mm sphere is 26.5218 cm^3. The error between algorithms measured volume and actual volume is 1.6%. Manual measurement of volume using pixel value obtain is 27.3913 cm^3. Compare with the actual algorithms measured volume of phantom, the error is 1.58%. Error of manual calculated volume compare to actual volume is 3.28%.

Based on the result, the algorithms measured volume result is validated with small error of deviation. The algorithms calculated volume have smaller error to the manual calculated volume. This shows that algorithms calculated volume can reduce the error in measurement of volume. The result obtained is a numeric result which imply the algorithms calculated volume for the result is correct for future set of data as the result in the research study is correct. Hence, the validation of algorithms is achieved.

Table 3. Error of volume of phantom compare with algorithms calculate and actual calculated.

Method of Measurement	Volume (cm^3)	Error of volume compare with algorithms measured	Error of volume compare with actual volume
Algorithms measured	26.9579cm^3	-	1.6%
Actual volume	26.5218cm^3	1.6%	-
Manual calculate	27.3913cm^3	1.58%	3.28%

5. Conclusion

The combination of 3D reconstruction and measurement of volume algorithms are able to reduce the time consumption in diagnosis process. The location of the 3D image is based on the CT topographic image provided location of phantom. The numerical measurement of volume through the algorithm is validated with the actual phantom volume. The error between algorithms measured volume and actual phantom volume is less than 2%. Algorithms measured volume is compared with the manual delineation and calculation based on the planar image. The error percentage is 1.6%. The objective of the research is achieved.

6. References

[1] The Star 2016 *About 100,000 Malaysians suffer from cancer each year.* https://www.thestar.com.my/news/nation/2016/04/03/about-100000-malaysians-suffer-from-cancer-each-year/.

[2] Express D 2018 *Malaysia has high cancer mortality rate | Daily Express Online, Sabah, Malaysia.* http://www.dailyexpress.com.my/news.cfm?NewsID=129495

[3] New Straits Times 2019 Majority of cancer cases detected at later stages in Malaysia (New Straits Times | Malaysia General Business Sports and Lifestyle News) https://www.nst.com.my/lifestyle/heal/2019/02/457189/majority-cancer-cases-detected-later-stages-malaysia.

[4] Makris N E, Van Velden, F H P, Huisman M C, Menke C W, Lammertsma A A and Boellaard R 2014 Validation of simplified dosimetry approaches in 89Zr-PET/CT: The use of manual versus semi-automatic delineation methods to estimate organ absorbed doses *Med. Phys.* **41** p 10.

[5] Mittal S, Kaur H, Gautam N and Mantha A K 2017 Biosensors for breast cancer diagnosis: A review of bioreceptors, biotransducers and signal amplification strategies *Biosensors and Bioelectronics* **88** p 217–231.

[6] Strigari L, Konijnenberg M and Chiesa C 2014 The evidence base for the use of internal dosimetry in the clinical practice of molecular radiotherapy https://doi.org/10.1007/s00259-014-2824-5.

[7] Ljungberg M and Sjögreen G K 2016 Personalized Dosimetry for Radionuclide Therapy Using Molecular Imaging Tools *Biomed.* **4**(4) p 25.

[8] Zingerman Y, Golan H and Moalem A 2009 Spatial linear recovery coefficients for quantitative evaluations in SPECT *Nucl. Instr. and Meth. in Physics Research A: Accel., Spectro., Det. and Assoc. Equip.*, **602**(2) 607–613.

Acknowledgments
This research is supported by the Fundamental Research Grant Scheme (FRGS), Grant No: FRGS/1/2019 Vot K200 by Ministry of Education, Malaysia.

IOP Conf. Series: Materials Science and Engineering **785** (2020) 012046 doi:10.1088/1757-899X/785/1/012046

Evaluation of contrast and recovery coefficients as performance parameters in planar and SPECT imaging

Mohd Fahmi Mohd Yusof[1,a)], Ummi Solehah Ab Ghani[1], Nor Amalyna Ghazali[1], Ahmad Thaifur Khaizul[1], Abdullah Waidi Idris[1]

[1]School of Health Sciences, Universiti Sains Malaysia, 16150 Kelantan, Malaysia

[a]mfahmi@usm.my

Abstract. The study focused on the performance evaluation of the gamma camera on the planar and SPECT imaging mode in nuclear medicine imaging. Several parameters were studied including the partial volume effect, contrast evaluation at different radiopharmaceutical size of localizations and the recovery coefficients. The localizations of Technetium-99m (^{99m}Tc) radionuclide were achieved by using Perspex® vials with different diameter sizes between 1.3 and 3.5 cm in a Jaszczak phantoms filled with water. The images were obtained in both planar and SPECT modes. The contrast in planar images showed no significant change when different activity of radionuclide was used. The contrast in SPECT images showed that the higher contrasts were achieved when higher time of projection used when lower activity of ^{99m}Tc were used. The contrast at higher ^{99m}Tc activity however was the highest when lower time of projection was used. The recovery coefficient in planar images showed a linear relationship to the size of radionuclide localizations and did not significantly affect the activity of the radionuclide. The recovery coefficient of SPECT images however were significantly affected by activity of the radionuclide. The recovery coefficient in SPECT increased dramatically when higher activity of the radionuclide was used. The overall results showed that the measurement of contrast and recovery coefficient can be used as performance parameters for both the planar and SPECT imaging in nuclear medicine.

1. Introduction

The performances characteristic of the gamma camera system is related to the partial volume effect (PVE) and recovery coefficient (RC) [1]. The partial volume effect is important in qualitative and quantitative of the image [2][3]. The small sized image that is near to the resolution limits of the imaging device are often appears to show smaller activity and concentration in comparison to the actual value. The ratio between the apparent activity concentrations to the true activity concentration is known as recovery coefficient (RC). Image contrast also plays an essential role in both qualitative and quantitative aspects of the image quality in nuclear medicine imaging. Contrast in radiography is defined as the ratio of signal change of an object interest to the signal level in surrounding part of the images [4]. It is an important indicator to the imaging system's performance with respect to detection of small lesions.

Single photon emission computed tomography (SPECT) is an imaging technique used in nuclear medicine imaging by producing cross-sectional or three-dimensional nuclear medicine image with multiple slices [5]. This imaging technique is in-vivo imaging and conducted by using the unsealed sources commonly the Technetium-99m (^{99m}Tc). The SPECT image has advantages to give lesion detection image compared with planar imaging by removing the out of plane information for increase the contrast of the image [6]. The SPECT system has several parameters that would affect performance and image produces by that machine. The common problem occurs to SPECT system including the limitation of spatial resolution and contrast. As general, spatial resolution is the ability of system to differentiate two or more size of tumour according to the radionuclide uptake. The limitation of spatial resolution leads to the PVE [7]. PVE occurs when the activity at region of interest increase or decrease from the actual activity. However, the PVE can be corrected by using partial volume correction (PVC) or by using RC correction.

This study is focused on determination of the recovery coefficient and image contrast for varying cylindrical diameters phantom with known activity concentration in planar and SPECT images. The ratio of the apparent concentration to the true activity concentration and the ratio of signal change in object interest to the signal level in surrounding were calculated.

2. Methodology

2.1. Phantom Preparation
The localization of radiopharmaceutical of different sizes were simulated by using an cylindrical vials of different diameters between 1.3 and 3.5 cm based from the study by [8] as shown in Figure 1. The vials were made up from Perspex® with approximate length of 6 cm made compatible for Jaszczak phantom commonly used in quality control of SPECT imaging. All vials were attached on the circular Perspex® plate having the same diameter with the internal diameter of Jaszczak phantom. The details of the cylindrical vials are presented in Table 1. The cylindrical vials were filled with ^{99m}Tc of three different activity levels of 3, 10 and 18 μCi. The Jaszczak phantom was filled with water and an amount of ^{99m}Tc with activity ratio of 1:12 to the activity used in the vials.

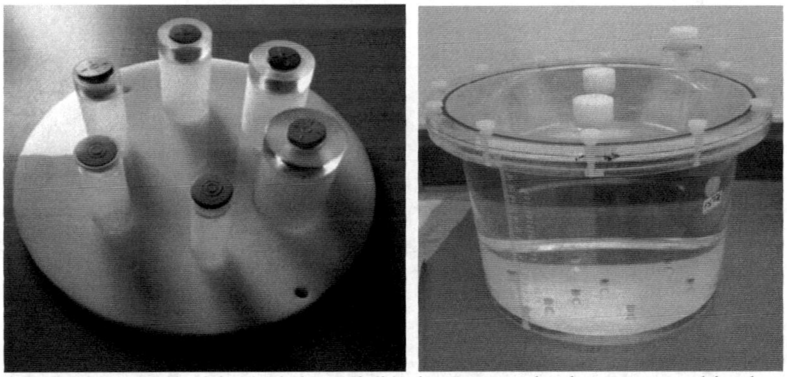

Figure 1. (a) The cylindrical vials and (b) the Jaszczak phantom used in the study.

2.2. Contrast and Recovery Coefficient Analysis
The phantom was scanned by using gamma camera machine model NM/CT 670 Pro at planar and SPECT mode. The planar mode was performed by using the matrix size of 128 x 128. The SPECT mode was performed the incremental view 3 degree per rotation and matrix size of 128 x 128 [9,10]. The SPECT images were reconstructed by using Ordered-Subset Expectation Maximization (OSEM) algorithm and Butterworth filter at 0.50 cut-off frequencies and 10.0 order value at all experimented time of acquisition per projection set up based on the standard scanning protocol in the Department of Nuclear Medicine HUSM.

The contrast of SPECT/CT image was evaluated based on the volume of interest (VOI) on each of vials and the around the background region. The contrast of SPECT/CT image was calculated based on the equation:

IOP Conf. Series: Materials Science and Engineering **785** (2020) 012046 doi:10.1088/1757-899X/785/1/012046

$$contrast = \frac{c_{vial} - c_{background}}{c_{background}} \qquad (1)$$

with c_{vial} is count per pixel of vial and $c_{background}$ is the count per pixel in background region. The graph diameter of vial vs contrast was plotted [3].

The recovery coefficient was calculated based on the equation:

$$Recovery\ coefficient = \frac{A_a}{A_t} \qquad (2)$$

with A_a is apparent activity of radionuclide concentration radionuclide and A_t is actual activity of concentration radionuclide [3]. The apparent activity concentration of radionuclide was calculated by using equation:

$$Apparent\ activity, A_a = \frac{R_{voi}}{S_{vol}} \times \frac{VS}{T_{acq}} \qquad (3)$$

with, R_{voi} is count per pixel of volume of interest vial, S_{vol} is volume sensitivity, VS is voxel size and T_{acq} is total time of scanning [9,10]. The count per pixel for each vial was measured by manually drawn the VOI at each vial based on the CT image using the Volumetric MI Evaluation software on the Xeleris workstation. The value of volume sensitivity is 2.667 cps/μCi. While the value voxel size for matrix size 128 x 128 is 4.42 x 4.42 x 4.42 mm^3 was based from the department's recent quality assurance for the gamma camera. The actual activity concentration of radionuclide was calculated by using the equation:

$$Actual\ activity, A_t = A_o e^{-\lambda t} \qquad (4)$$

with A_t is actual activity concentration of radionuclide, A_O is initial activity concentration of radionuclide source (^{99m}Tc), λ is decay constant of radionuclide and the t is time elapsed from the initial activity. The value of the decay constant of radionuclide source was 2.5 x 10^{-5} s^{-1}.

3. Results and Discussion

The planar and SPECT images of radionuclides in the cylindrical vials of different diameter sizes are illustrated in Figure 2 and 3 respectively. The planar and SPECT images obtained showed good visualisation of radionuclide of different diameters as well as the regions of high and low concentration of the radionuclide. The background image in planar images has better visualization when lower activity level was used compared to that in high activity levels. The background in SPECT images on the other hand did not show any significant different when different activity levels of radionuclide were used.

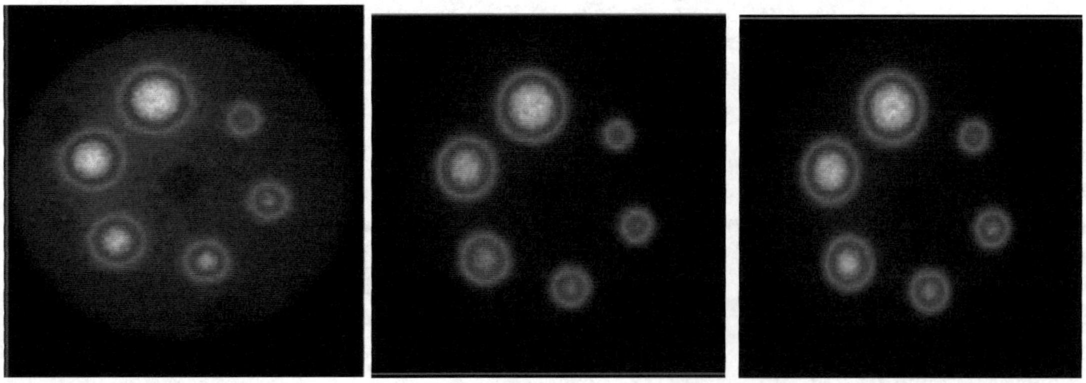

(a) (b) (c)

Figure 2. The planar images of radionuclide vials of different diameter of (a) 3 μCi (b) 10 μCi and, (c) 18 μCi.

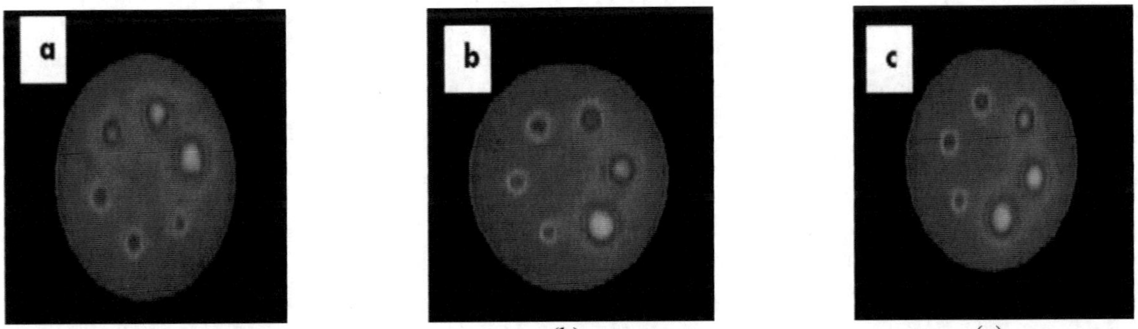

(a) (b) (c)

Figure 3. The SPECT images of radionuclide vials of different diameter of (a) 3 μCi (b) 10 μCi and, (c) 18 μCi.

The contrast values at different diameter sizes of and activity levels of the radionuclide in planar and SPECT images are presented in Figure 4. The results showed that the contrast values in planar and SPECT increased when the diameter of the radionuclide increased [10]. The increment of contrast values in planar images was more significant when higher activity level of radionuclide was used compared to that in SPECT images. This can be observed in the exponential pattern of the contrast curve in planar from low to high activity levels of the radionuclide. The contrast curve in SPECT images was observed to be more linear and shifted upward when higher activity levels of radionuclide were used.

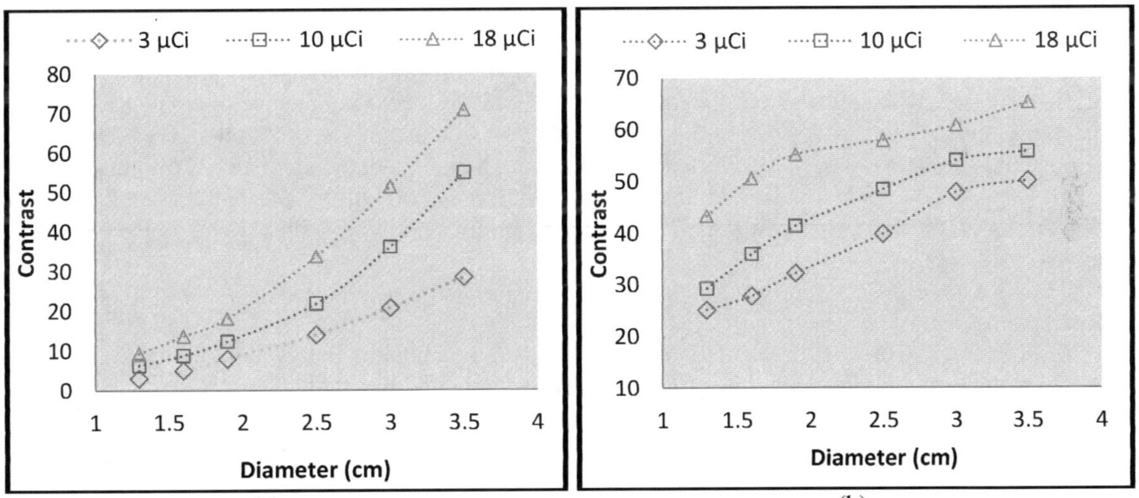

(a) (b)

Figure 4. The contrast values at different radionuclide diameter sizes in (a) planar and (b) SPECT images.

The recovery coefficients, RC of planar and SPECT images at different radionuclide diameter sizes are illustrated in Figure 5. The results showed that the RC increased when the size of radionuclide increased for both planar and SPECT images [10]. The values of the RC were also found to decrease when higher activity levels of radionuclide were used. The ideal value of RC in a nuclear medicine image is 1.0 indicating equal ratio between the apparent and actual activity of the radionuclide [11]. The RC at larger sizes of radionuclides had the tendency to exceed the value of 1.0 [8]. Furthermore, the RC values were found to be higher when lower activity levels of radionuclide were used. This condition is called the 'spill over' effect occurred due to contamination of activity from the neighbouring tissue or spot to these 'hot' areas or the areas of higher concentration of radionuclide [3]. The results also showed that the 'spill over' in planar images were not significant compared to those in the SPECT images. The overall results showed that the planar and SPECT images showed comparable results in the evaluations of contrast and recovery coefficient measurements.

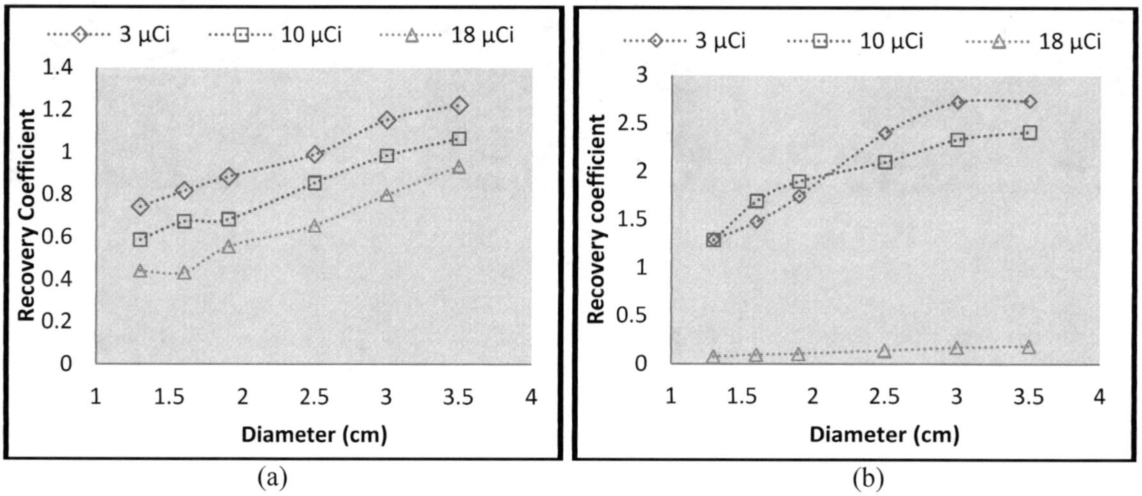

Figure 5. The recovery coefficient values at different radionuclide diameter sizes in (a) planar and, (b) SPECT images.

4. Conclusion

The contrast of planar and SPECT images increased when the diameter of the radionuclide increased and the contrast values increased when higher activity levels of radionuclide were used. The recovery coefficient of planar images was consistent to that in the SPECT image. The 'spill over' effect in the planar images was lower than those in the SPECT images. The use of lower activity levels in SPECT imaging would increase the 'spill over' effect that would disrupt the diagnostic information of the images. The overall results indicated the suitability of the evaluations of contrast and recovery coefficient based on the partial volume effect as image parameters in both planar and SPECT imaging in nuclear medicine.

5. References

[1] Erlandsson K, Buvat I, Pretorius P H, Thomas B A and Hutton B F. 2012 A review of partial volume correction technique for emission tomography and their application in neurology, cardiology and oncology. *Phys. Med. Biol.* **57** p 119-159.

[2] Sjogreen K, Ljungberg M, and Strand S 2009 Parameter influencing volume and activity quantitation in SPECT *Acta Oncol.* **35**(3) p 323-330.

[3] Cherry S R, Sorenson J A and Phelps M E 2012 *Phys. in Nucl. Med. 4th ed.* (Philadelphia: Saunders Elsevier).

[4] Saad I E, Helal N, El-din H M and Moneam R A 2012 Evaluation of varying acquisition parameter on the image contrast in SPECT studies *IJRRAS* **13**(2) p 485-491.

[5] Groch M W, and Erwin W D 2000 SPECT in the year 2000: Basic principles *J. Nucl. Med. Technol.* **28(4)** 233–44.

[6] Jacene H A, Goetzel S, Patel H, Wahl R L and Ziessman H A 2008 Advantage of hybrid SPECT/CT vs SPECT alone. *TOMIJ* **2** p 67-79.

[7] Rit P, Vija H, and Hornegger J 2011 Absolution quantification in SPECT *J. Nucl. Med.* **38** p 69-77.

[8] Hamid P N K A, Yusof M F M, Tajuddin A A, Hashim R and Zainon R 2018 Design and evaluation of corn starch-bonded *Rhizophora* spp. particleboard phantoms for SPECT/CT imaging. *IOP Conf. Ser. Mater. Sci. and Eng.* **298** p 012041.

[9] IAEA Human Health Reports No. 9 2014 Quantitative Nuclear Medicine Imaging Concepts, Requirements and Methods (Vienna: International Atomic Energy Agency).

[10] Hamid P N K A, Yusof M F M, Tajuddin A A, Hashim R and Zainon R 2018 Evaluation of image quality and recovery coefficient of corn starch-bonded *Rhizophora* spp. particleboards as phantom for SPECT/CT imaging. *IOP Conf. Ser.: Mater. Sci. and Eng.* **555** p 012031.

[11] Gopal B S 2010 *Basics of PET Imaging 2nd ed.* (New York, London.:Springer).

Acknowledgement

The author would like to acknowledge the financial support of USM Bridging Grant scheme no. 304.PPSK.6316324. The author would also like to acknowledge the Department of Nuclear Medicine, Radiotherapy and Oncology, Hospital University Sains Malaysia (HUSM) for the use of the linear accelerator facility in this study.

IOP Conf. Series: Materials Science and Engineering **785** (2020) 012047 doi:10.1088/1757-899X/785/1/012047

Evaluation of default mode network in temporal lobe epilepsy patients and healthy subjects: a preliminary result

Shuhada JM[1,a)], Husbani MAR[2], A I A Hamid[3] and Muhammad AR[1]

[1]School of Medical Imaging, Faculty of Health Sciences, Universiti Sultan Zainal Abidin, 21300 Kuala Nerus, Malaysia.
[2]Faculty of Medicine, Universiti Sultan Zainal Abidin, 21300 Kuala Nerus, Malaysia.
[3]Department of Neurosciences, School of Medical Sciences, Universiti Sains Malaysia, 16150 Kubang Kerian, Malaysia

[a]shuhada@unisza.edu.my

Abstract. The default mode network (DMN) is thought to be impaired in epileptic patients, where the functional integrity of its core regions compromised. Fourteen healthy participants and fourteen aged and gender-matched patients with temporal lobe epilepsy (TLE) took part in this study. The resting-state fMRI (rsfMRI) imaging protocol was executed using a 3-T Phillips Achieva MRI scanner at the Department of Radiology, HUSM. The TLE patient displays decreased activation clusters in precuneus and supramarginal gyrus compared to the healthy group. However increased activation clusters were found on middle frontal gyrus, superior temporal gyrus, and angular gyrus compared to healthy subjects. The result from random effects (RFX) on healthy TLE revealed that the left middle frontal gyrus, bilateral superior temporal gyrus, bilateral precuneus, and right supramarginal gyrus were significantly activated but the NOV survive was so small. The findings suggested that TLE patients suffer from impairment in some DMN region, which may cause certain neuropsychological and cognitive degradation. Keywords: TLE=Temporal lobe epilepsy; DMN=default mode network.

1. Introduction

1.1. Epilepsy

Epilepsy is characterized by the occurrence of abnormal neuro-electrical activity in the brain. In focal seizures and epilepsy, the epileptic activity arises primarily within networks limited to one hemisphere. The origin of focal seizures may be associated with the localization of radiologically visible brain lesions or electrophysiological abnormalities. However, there is growing evidence from neuroimaging studies that focal epilepsies involve abnormal functional networks rather than a single epileptogenic region [1].

In resection treatment for epilepsy, vascular lesions, and tumors, it is important to preserve healthy functional tissue. Pre-surgical mapping involves identifying healthy functional tissue that should be preserved during resection. For example, damage to the default-mode network may affect cognitive functions such as planning for the future, navigation of social interactions and memory retrieval [2]. Furthermore, any damage to the

IOP Conf. Series: Materials Science and Engineering **785** (2020) 012047 doi:10.1088/1757-899X/785/1/012047

motor network may lead to deficits in motor function [3].

1.2. Default Mode Network (DMN)
The DMN is a brain network of interacting brain regions known to have activity highly correlated with each other [2] with major hubs being located in the posterior cingulate cortex and precuneus, the medial prefrontal cortex, and the angular gyrus [4]. In neuroscience, the default network, also known as the default mode network (DMN), is a large scale brain network of interacting brain regions known to have activities that are highly correlated with each other and distinct from other networks in the brain [5].

1.3. Functional Magnetic Resonance Imaging (fMRI)
fMRI has the potential to predict possible deficits in language, visual, motor and sensory functions that would arise from surgical intervention or post-operation [6]. This would be of great helps to the treating physician or the surgeon in explaining the relative risk of intervention against non-intervention so that a decision about treatment options can be made after considering their costs and benefits. In many cases, the decision making requires a judgment to be made regarding the balance between reduction of seizures and maintenance of function, to ensure minimal effects on the quality of life (QOL).

It is well known that fMRI is a noninvasive technique that has demonstrated great promises for pre-surgical planning [7] especially when used preoperatively to localize areas of the eloquent cortex [8]. Connectivity study using resting-state fMRI in neurosurgery has also become a promising tool that can be used in diagnosing functional abilities and abnormalities [9]. It can also make functional mapping more interactive by allowing ongoing paradigms to be adjusted if the need arises, making fMRI a more flexible tool for neurological investigations [10].

The aim of the study was to determine whether there was a significant activation present in the resting brain. Thus, in this study, we explore the location of the brain activation area in patients with temporal lobe epilepsy (TLE) and healthy subject. It might be useful for providing complementary information to other clinical diagnostic measures.

2. Methodology

2.1. Subjects
Fourteen healthy participants (3 males,11 females) (average age ± standard deviation = 36.93 ± 8.75 years) and fourteen patients (3 males,11 females) with TLE (average age ± standard deviation = 37.00 ± 8.79 years) took part in this study. The participants from the two groups were matched by age and gender. The healthy participants were purposely recruited from those who accompanied their relatives at the polyclinic and also among staff from Hospital Universiti Sains Malaysia (HUSM). The TLE patients were screened by a neurologist based on reported clinical semiology of the seizures and were confirmed by video-EEG monitoring. The TLE patients were reported to be on drug treatment. The participants agreed to participate in the study by signing the informed consent form after being given full explanation and understand the nature and risks of the research. All participants were screened for MRI compatibility. This study has been approved by the institutional ethics committee (IEC) USM/JEPeM/16050175. All participants with major brain abnormalities caused by surgery or injury, tumors, significant malformation of cortical development leading to distortion of brain anatomy, had a previous history of brain injury and psychiatric disorders were excluded.

2.2. Data acquisition for Resting-state fMRI.
The rsfMRI blood oxygen level-dependent (BOLD) imaging protocol was executed using a

3-T Phillips Achieva MRI scanner equipped with a 32-channel head and neck system at the Department of Radiology, HUSM. The first three scans were discarded by the BOLD imaging protocol to eliminate the magnetic saturation effect. The echo-planar imaging (EPI) parameters for acquiring functional T2* weighted images are echo time (TE) = 33 ms, repetition time (TR) = 1.7 s, flip angle = 75°, slice thickness = 4 mm, slice gap = 0 mm, field of view (FOV) = 192 x 192 mm, matrix size = 64 x 64, voxel size = 2 x 2 x 4 mm, number of scans = 250 and total imaging time = 425 s.

The participants were instructed not to move their head during the scan. They were required to be in a state of relaxation, empty their mind and passively focus on a fixation point "×" symbol on the screen throughout the session. Participants must not fall asleep during the scan because "sleeping brain" is very much different from "resting brain"[11]. A demographic and clinical data from both patients and control subjects were obtained prior to the study. We studied a continuous response variable from both groups using t-test statistical analysis.

2.3. Pre-processing and data analysis

A total of 247 functional images were analyzed using Statistical Parametric Mapping (SPM12) obtained from the Functional Imaging Laboratory, Wellcome Trust Centre for Neuroimaging, in the Institute of Neurology at the University College London (www.fil.ion.ucl.ac.uk/spm/) which runs on MATLAB R2016b (The Mathworks, Inc. USA) platform. The functional images in each measurement were first entered into a slice timing module for time correction. They were then realigned using the 6- parameter affine transformation in translational (x, y, and z) and rotational (pitch, roll, and yaw) directions to reduce the effects on the overall signal intensity from participant's movements. After realigning the data, a mean image of the series was used to estimate some warping parameters that mapped it onto a template that already conforms to standard anatomical space (EPI template provided by the Montreal Neurological Institute – MNI). The normalization procedure used a 12-parameter affine transformation. The images were then smoothed using a 6-mm full-width-at-half-maximum (FWHM) Gaussian kernel. Low-frequency responses caused by aliased biorhythms, cardiac effects, and other oscillatory signal variations were removed using a high-passed filter.

The human brain has been found to exhibit low-frequency fluctuations signal (LFF) when it is at rest. This LFF ranges from 0.01 to 0.08 Hz [12]. The LFF detected during resting state is associated with internally and externally oriented consciousness [13]. Thus, to model the LFF, a combination of mathematical functions such as sine and cosine functions of different frequencies that show a resemblance of LFF, can be used. This Fourier basis set is thought to be suitable to model brain responses during the resting state which is assumed to oscillate between 0.01 – 0.08 Hz [14].

The functional T2* weighted images were also separately entered into a group analysis design matrix (GLM) to search for voxels that survived the uncorrected significant level of 0.001. This was done for both groups by means of fixed effects analysis (FFX) due to the constraint in the number of subjects for each group. Random (RFX) effects analyses and inferences based on the group responses were made onto the whole subject.

3. Results and discussion

3.1. Study sample

We collected data from 14 healthy participants (mean age: 36.93) and 14 TLE patients (mean age 37). The healthy participants were matched to the patients on their gender, age, race, and education level. Table 1 shows the mean age of the fourteen healthy subjects (average age ± standard deviation = 36.93 ± 8.75 years) and fourteen TLE patients (average

IOP Conf. Series: Materials Science and Engineering **785** (2020) 012047 doi:10.1088/1757-899X/785/1/012047

age ± standard deviation = 37.00 ± 8.79 years).

Table 1. Mean Age and gender

GROUP		Number of subjects (n)	Mean	Std. Deviation
AGE	HEALTHY	14	36.93	8.748
	TLE	14	37.00	8.788

3.2 Activations

There was a difference between cluster size of the activation in bilateral middle frontal gyrus (MFG), superior temporal gyrus (STG), precuneus (PRE) and supramarginal gyrus (SMG) between healthy and TLE, where healthy subjects display higher activation in PRE and SMG compared to TLE subjects. The TLE patient displays higher activation clusters in MFG, STG, and angular gyrus compared to healthy subjects. Table 2 showed the clusters of activation between subjects in the same group using a one-sample t-test in SPM12 (thresholded at p= 0.001), while Figure 1 shows the activations overlaid onto standard brain space.

The present study found that there was a significant difference between cluster size of the activation in bilateral MFG, STG, PRE and SMG between healthy and TLE, where healthy subjects display higher activation in PRE and SMG compared to TLE subjects. Similar results were found in the previous study on TLE with bilateral hippocampal sclerosis shows decreased activation in posterior cingulate cortex/ PRE [15]. However, in this study the TLE patient displays higher activation clusters in MFG, STG and angular gyrus (ANG) compared to healthy subjects but contrary with the findings from Zhang (2008) which the TLE subjects display decreased activation in MFG and STG as well as other structures such as dorsal lateral prefrontal cortices, superior temporal gyri, caudate heads, dorsal brain stem and the posterior cerebellum [15]. By comparison, rsfMRI study of the brain's network of TLE and TLE with depression, they found more strong active brain areas including thalamus and the default-mode network which involved in prefrontal cortex, PRE, ventral anterior cingulate and hippocampus. They found that the extra-temporal activation in DMN is increased in TLE with depression compared to non-depression group [16]. No SMG activation mentioned in these two journals. This is probably due to the different conditions of the disease or types of TLE.

Table 2. The number of voxels (NOV) of the region of interest (ROIs) on FFX in healthy and TLE.

REGION	HEALTHY		TLE	
	No of voxel	Coordinates	No of voxel	Coordinates
Left Middle Frontal Gyrus (L MFG)	5	-30 58 22	50	44 32 46
Right Middle Frontal Gyrus (R MFG)	63	30 60 6	97	-28 38 34
Left Superior Temporal Gyrus (L STG)	0	0	187	-62 -30 14
Right Superior Temporal Gyrus (R STG)	2	56 -14 6	53	56 -20 6
Left Angular (L ANG)	0	0	35	-46 -66 48
Right Angular (R ANG)	50	30 -68 46	57	46 -62 38
Left Precuneus (L PRE)	145	-6 -62 14	98	-2 -52 54
Right Precuneus (R PRE)	208	8 -66 62	33	2 -38 54
Left Supramarginal Gyrus (L SMG)	82	-50 -46 30	33	-64 -28 30
Right Supramarginal Gyrus (R SMG)	157	62 -34 42	34	58 -24 30

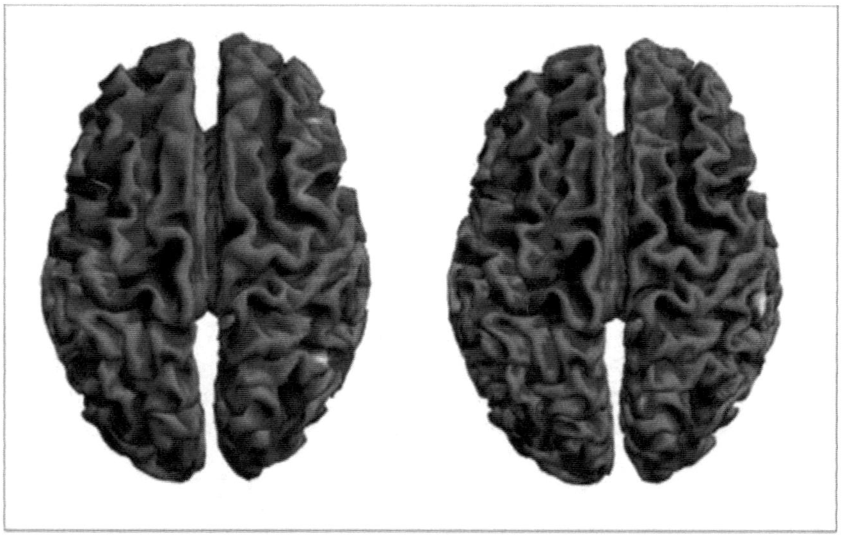

(a) **(b)**

Figure 1. (a) ROI on FFX in TLE and (b) healthy brain.

The result from random effects (RFX) analysis (2-sample t-test thresholded at $p = 0.001$) on healthy < TLE revealed that the left MFG, bilateral STG, bilateral precuneus, and right SMG were significantly activated (shown in Table 3) but the NOV survive was so small. This may be attributed to small activation clusters or may be influenced by the fact that all the TLE patients were already stabilized with epileptic medication. These results were compared with the activation from both groups across within and between subjects. The fixed effect analysis is a more sensitive analysis than the random effect method [17]. A robust random-effect analysis needs a large number of subjects to achieve high confidence in parameter estimation. A small number of subjects lead to low degrees of freedom in variance estimation.

Table 3. NOV of ROIs on RFX analysis in healthy and TLE at $p = 0.001$ uncorrected.

Region	*NOV- H<TLE*	*H>TLE*
Left Middle Frontal Gyrus (L MFG)	7	0
Right Middle Frontal Gyrus (R MFG)	0	0
Left Superior Temporal Gyrus (L STG)	3	0
Right Superior Temporal Gyrus (R STG)	2	0
Left Angular (L ANG)	0	0
Right Angular (R ANG)	0	0
Left Precuneus (L PRE)	2	0
Right Precuneus (R PRE)	1	0
Left Supramarginal Gyrus (L SMG)	0	0
Right Supramarginal Gyrus (R SMG)	9	0

4. Conclusion

These findings suggested that TLE patients suffers from impairment in some DMN regions, that may cause certain neuropsychological and cognitive degradation. Further studies should be conducted to investigate the impact of these impairments to the integrity of brain function in TLE patients. In spite of its limitations, the study certainly adds to our understanding of the activation in TLE patient is beyond the temporal area. This understanding is important to help the clinicians in order to support their decision making.

There are some limitations to this study. First, the number of TLE patients is small, which permitted meaningful results but shall be treated as preliminary at best. Second, the structural integrity of the DMN may be useful in analyzing any functional degradation in the network, to support the findings and further clarify its impairments.

5. References

[1] Kramer M A and Cash S S 2012 Epilepsy as a disorder of cortical network organization *Neuroscientist*.

[2] Buckner R L, Andrews-Hanna J R and Schacter D L 2008 The brain's default network: Anatomy, function, and relevance to disease *Ann. N. Y. Acad. Sci.* **1124** p 1–38.

[3] Vergun S *et al.* 2016 Classification and extraction of resting state networks using healthy and epilepsy fMRI data *Front. Neurosci.* **10** p 1–13.

[4] Andrews-Hanna N, Jessica R; Smallwood, Jonathan and Spreng 2014 NIH Public access **2016**(1) p 29–52,.

[5] Raichle M E 2015 The brain's default mode network *Annu. Rev. Neurosci.* **38**(1) p 433–447,.

[6] Nadkarni T N *et al.* 2015 Usage of fMRI for pre-surgical planning in brain tumor and vascular lesion patients: Task and statistical threshold effects on language lateralization *NeuroImage Clin.* **7** p 415–423.

[7] Tang Y *et al.* 2014 Short-term cognitive changes after surgery in patients with unilateral mesial temporal lobe epilepsy associated with hippocampal sclerosis *J. Clin. Neurosci.*

[8] Lee M H, Miller-thomas M M, and Benzinger T L 2017 Clinical Resting-state fMRI in the Preoperative Setting **25**(1) p 11–18,.

[9] Böttger J, Margulies D S, Horn P and Thomale U W 2011 A software tool for interactive exploration of intrinsic functional connectivity opens new perspectives for brain surgery p. 1561–1572.

[10] Chen J E and Glover G H 2016 HHS Public Access **25**(3) p 289–313.

[11] Yeo T B T, Tandi J, and Chee M W L 2015 Functional connectivity during rested wakefulness predicts vulnerability to sleep deprivation *Neuroimage* **2014**.
[12] Cordes D *et al.* 2001 Frequencies contributing to functional connectivity in the cerebral cortex in 'Resting-state' *Data*," p 1326–1333.
[13] Vanhaudenhuyse A *et al.* 2010 Two distinct neuronal networks mediate the awareness of environment and of self p 570–578,.
[14] Di X and Biswal B B 2014 Identifying the default mode network structure using DynamicCausal modeling on resting-state functional magneticresonance imaging, *Nih*, p 53–59,.
[15] Zhang Z, Lu G M, Zhong Y, Tan Q F, Zhu J G, Jiang L, Chen Z L, Wang Z Q, Shi J X, Zang Y F, Liu Y J 2008 Application of amplitude of low-frequency fluctuation to the temporal lobe epilepsy with bilateral hippocampal sclerosis: an fMRI study **88**(23):1594.
[16]. Yi Z, Zhi X Z "Resting-state functional magnetic resonance study of the brain's network of the temporal lobe epilepsy patients with depression **96**(21) p1696-8. doi: 10.3760/cma.j.issn.0376-2491.2016.
[17] Friston K J, Holmes A P Price C J, Büchel C and Worsley K J 1999 Multisubject fMRI studies and conjunction analyses *Neuroimage* **10**(4) p 385–396.

Acknowledgment
This project was funded by FRGS grant (FRGS/1/2015/SKK02/UNISZA/02/1).

IOP Conf. Series: Materials Science and Engineering **785** (2020) 012048 doi:10.1088/1757-899X/785/1/012048

Research reactor operator performance based on the human error assessment and reduction technique (HEART)

Ahmad Hassan[1,a)], Faizal Mohamed[2], Mazleha Maskin[3]

[1]Physics and Chemistry Department, Faculty of Applied Sciences and Technology, Universiti Tun Hussein Onn Malaysia,86400 Muar, Johor, Malaysia.
[2]School of Applied Physics, Faculty of Science and Technology, National University of Malaysia, 43600 Bangi, Selangor, Malaysia.
[3]Malaysian Nuclear Agency (Nuclear Malaysia), Bangi, 43000 Kajang, Selangor, Malaysia.

[a]hassan@uthm.edu.my

Abstract. This study is an attempt to apply the Human Error Assessment and Reduction Technique (HEART) method in order to quantify the human failure event (HFE) for probabilistic safety assessment (PSA) of a TRIGA Mark II reactor, situated at the Malaysia Nuclear Agency, Malaysia. This serves the qualitative part of the human reliability analysis (HRA) framework and HEART method is considered due to its potential applicability in quantifying human error probability (HEP) for other industries, beside nuclear power. This study addresses the application of HEART method in details: classification of selected research reactor human task into the Generic Task Type, the assignment of the nominal HEP, evaluation of Error Producing Condition and the resultant Assessed Proportion of Affect. The results of this study will be applied in the quantification of PSA's event sequence end state frequencies.

1. Introduction

Probabilistic Risk Assessment (PRA), or Probabilistic Safety Assessment (PSA) is a tool used to determine risk, and have been benefited much in the nuclear power industry[1], [2]. The risk can be defined as the product of a prescribed undesirable occurrences, i.e. consequences and their likelihood, i.e. frequency. Simply put, risk can be decreased by making the likelihood of a consequence lower, and/or by making the occurrences less severe. PSA uses event tree (ET) and fault tree (FT) in modelling postulated accident sequences that lead to the undesirable consequences and reliability of mitigations systems, respectively. The quantification of a consequence frequency in an accident sequence is eventually involve the conversion of the ET to a large FT[3].

Human contribution to the risk is incorporated in a PSA through the modelling of human action as the basic events in the mitigation systems' FTs, similarly as to the modelling of hardware. The identification and evaluation of the human performance is accomplished in Human Reliability Analysis (HRA) which is one of the major tasks in PSA. The quality of a PSA relies heavily on many factors, including the input of the reliability data[1]. Contrary to hardware, the statistical human reliability data that could match a particular accident scenario/context of interest is not much available and alternative methods have been used to produce these data[4]. Although the general framework of HRA for PSA has been long established in the nuclear power industry domain, there are various human error rate, i.e. human error probability (HEP) quantification methods which are available and yet, they are still continuously evolving[5–8].

Besides for the purpose of civil power generation, nuclear fuel is also utilized in other nuclear installation such as a research reactor (RR). A PSA is also applicable in assessing the possible radiological risk posed by RR due to the existence and the use of nuclear fuel in its operation. PSA for RR is unique owing to the inherent flexibility in the RR operation, in which a wide range of human interactions can be expected during either normal or abnormal condition[9], [10]. These various reasons have led to a numbers of PSA study on RR[11–16]. In the previous studies of HRA for RR PSA, HEP quantification methods that are being applied are: Technique for Human Error Rate Prediction (THERP)[15][17][11][18], Standardized Plant Analysis Risk-Human Reliability Analysis (SPAR-H)[19], A Technique for Human Event Analysis (ATHEANA)[20] and Empirical Technique to Estimate Operator Errors (TESEO)[18].

This study is an attempt to employ the Human Error Assessment and Reduction Technique (HEART)[21] method in order to quantify the human failure events (HFEs) for the PSA of the TRIGA Mark II research reactor, situated at the Malaysian Nuclear Agency, Malaysia. This would serve the qualitative part of the HRA framework and HEART method is considered due to its generic approaches that can be applied in quantifying HEP for other industries, beside the nuclear power[5]. The details on the methodology and results for quantitative part of the HRA can be found in our previous study[22]. The quantification of HFEs by HEART method in this study is limited only for those modelled in accident sequences postulated for reactivity insertion accident (RIA) initiating event[23]. Eventually, the HEART outputs in this study is verified with the SPAR-H method[24].

The next section of this paper describes the HEART methodology in details: classification of selected research reactor human task into the Generic Task Type, the assignment of the nominal HEP, evaluation of Error Producing Condition and the resulting Assessed Proportion of Affect. In section Result and Discussion, the quantified HEPs are presented and verified with those of from SPAR-H method, and insights obtained through the application of the HEART method for RR PSA are addressed. Finally, the summary of the methodology and results of the study are given in Conclusion.

2. Methodology

The approach of HEART[21] in deriving HEP of a task begins with selection of the generic task type (GTT) that match the description of the analysed task. There are nine GTTs (Table 1) defined within HEART:

The given nominal human unreliability values are the basic HEP values (mean) that are tend to be achieved by human when performing the prescribed generic tasks.

The value and the probabilistic limits are applied when the context within which the task is performed is in perfect condition. When there are circumstances that degrade the human performance, the perfect condition does no longer exist and the higher HEP value has to be applied. There are 38 circumstances that are defined as Error Producing Condition (EPC), each of which with maximum weighting factor that degrade the human performance (Table 3)[21]. The weighting factor of the EPCs can be adjusted by multiplying them with value ranging between (best, positive) to 1 (worst, negative). This step is termed as Assess Proportion of Affect (APAO). The final HEP of a task is calculated using Equation (1).

$$\mathbf{Final\ HEP} = \mathbf{GTT} \times \prod_{i=1}^{n}[(\mathbf{EPC_i} - \mathbf{1}) \times \mathbf{APAO_i} - \mathbf{1}\,] \qquad (1)$$

IOP Conf. Series: Materials Science and Engineering **785** (2020) 012048 doi:10.1088/1757-899X/785/1/012048

Table 1. HEART GTTs and the corresponding mean HEP value.

Item	Generic Task	Proposed Nominal Human Unreliability (5th – 95th Percentile Bounds)
A	Totally unfamiliar, performed at speed with no real idea of likely consequences	0.55 (0.35 - 0.97)
B	Shift or restore system to a new or original state on a single attempt without supervision or procedures	0.26 (0.14 - 0.42)
C	Complex task requiring high level of comprehension and skill	0.16 (0.12 - 0.28)
D	Fairly simple task performed rapidly or given scant attention	0.09 (0.06 - 0.13)
E	Routine, highly-practiced, rapid task involving relatively low level of skill	0.02 (7E-3 - 4.5E-2)
F	Restore or shift a system to original or new state following procedures, with some checking	3E-3 (8E-4 - 7E-3)
G	Completely familiar, well-designed, highly practiced, routine task occurring several times per hour, performed to highest possible standards, by highly-motivated, highly-trained and experienced person, totally aware of implications of failure, with time to correct potential error, but without the benefit of significant job aids	4E-4 (8E-5 - 9E-3)
H	Respond correctly to system command even when there is an augmented or automated supervisory system providing accurate interpretation of system state	2E-5 (6E-6 - 9E-4)
I	Miscellaneous task for which no description can be found	0.03 (8E-3 - 0.11)

In general, the HEART quantification process for a HFE can be illustrated as in Figure 1.

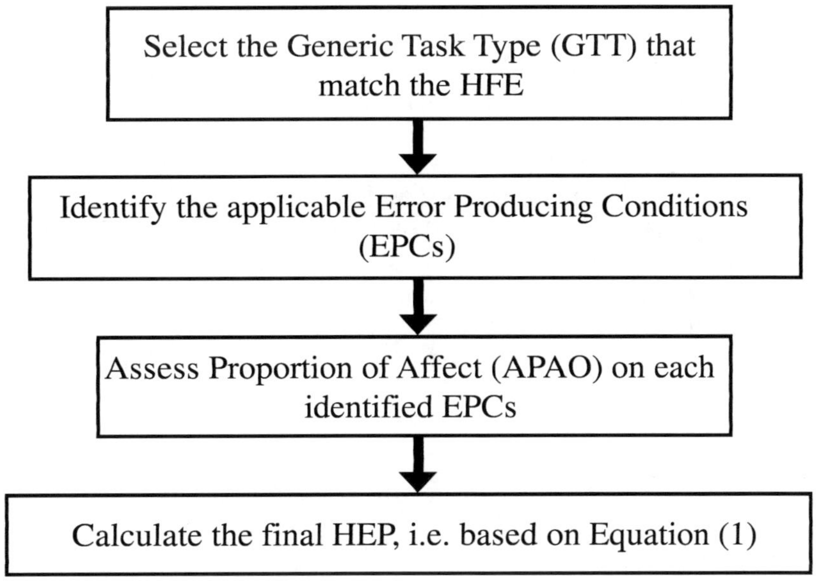

Figure 1. HEP quantification process based on HEART.

3. Results and discussion

The explanation in the next two paragraphs is regarding two HFEs postulated in RIA accident sequence. RIA is postulated to occur due to the existence of sample with positive reactivity which leads to excess reactivity of the core.

During RIA event, reactor safety channel will detect the excess reactivity and subsequently generates trip signal. Operators would have been attempting scram the reactor when realizing no reduction on the reactor power. The operators have to trip the reactor by improvisation. The operators would have to directly de-energize the control rods withdrawal mechanism by disconnecting its power supply: this task is identified as RIA-LOIC-H. The manual operators action within this HFE comprise of a simple turning a switch-off, which is able to be completed in a single second, once the decision is made. The operators then would attempt to push scram button a few times followed by physical field inspection to verify the control rods response. The diagnosis can be assumed to require maximal time of 20 minutes. The available time can be regarded as positive performance driver for the operators because it is estimated that the time until the complete depletion of pool water due to heat from fission reaction in the core is approximately 11 hours.

The reactor pool water has to be manually added when considering that all means to terminate the chain reaction have failed in order to prevent core uncovery due to the water depletion. This human task is identified as PW-RIA-H. Again, with regard to the fact that pool water depletion up to the beginning of core uncovery would take approximately 11 hours long, the available time can be regarded as positive performance driver for the operators.

This study identifies that the GTT that match RIA-LOIC-H task is Item F. The consideration of others GTT on this task is as addressed in Table 3. The task physically involves the changing of the instrument and control (I&C) system to a new state, and the checking by human redundancy. Although there is no formal procedure for this task specific for this abnormal condition, the operators are experience with the system since they perform the maintenance annually. This complements the unavailability of formal procedure. With regards to EPCs, the consideration is denoted in Table 3. As a result, the EPCs that are taken into account for the final HEP calculation are Item 1 and 34. A conservative approach is taken for APOA, by using the weighting factor 1. Putting these numerical values into Equation (1), the final HEP = 5.61E-2.

In classifying GTTs for a HFE, the one that is closely matching the task description is selected. The rest GTT which are not selected are provided with specific justifications. Intuitively, the result would be heavily depending on the task description, and therefore the familiarization of an analyst with the HFE should be in according to the HRA framework for PSA[25]. Likewise, the same explanation applied for the identification of EPCs. This study is unable to systematically perform the APOA, the part which is considerably relying on expert judgment[4]. Instead a conservative approach is used and yields HEP which is conservative. For the second human task postulated in RIA, i.e. PW-RIA-H, the context is similar as to the RIA-LOIC-H task and thus the same final HEP value is deduced.

The study using SPAR-H gave HEPs of 1.00E-2 for I&C-RIA-H and 2.55E-3 for PW-RIA-H [26]. The value of the former is considerably close to that calculated by HEART while the latter is an order of magnitude lower than that calculated by the HEART methodology in this study, due to the conservative approach taken with regards to its APAO.

IOP Conf. Series: Materials Science and Engineering **785** (2020) 012048 doi:10.1088/1757-899X/785/1/012048

Table 2. HEART GTTs and the consideration for RIA-LOIC-H task.

Item	Generic Task	Consideration for RIA-LOIC-H task
A	Totally unfamiliar, performed at speed with no real idea of likely consequences	The operators are familiar with the control system, and they know the consequence of the action
B	Shift or restore system to a new or original state on a single attempt without supervision or procedures	Although there is no formal procedure for this task which is specific for this abnormal conditions, the operators are experience with the system: they perform the maintenance
C	Complex task requiring high level of comprehension and skill	The action is not complex. The displays and annunciators provide the information for the operators to do diagnosis
D	Fairly simple task performed rapidly or given scant attention	This is not directly a simple task such as the skilled-based action. It requires analysis on the reactor parameters
E	Routine, highly-practiced, rapid task involving relatively low level of skill	This is not a routine task
F	Restore or shift a system to original or new state following procedures, with some checking	-
G	Completely familiar, well-designed, highly practiced, routine task occurring several times per hour, performed to highest possible standards, by highly-motivated, highly-trained and experienced person, totally aware of implications of failure, with time to correct potential error, but without the benefit of significant job aids	This task only demanded during abnormal condition (not hourly), which is never been occurred
H	Respond correctly to system command even when there is an augmented or automated supervisory system providing accurate interpretation of system state	The augmented or automated supervisory system specific for this case does not available
I	Miscellaneous task for which no description can be found	The task is identified with respect to the postulated accident sequence

Table 3.1 HEART EPCs and the corresponding maximum weighting factors. For task RIA-LOIC-H's GTT Type F, applicable EPCs are denoted as: **A**, not applicable (out of the context) : **NA**, and opposite to the context in which the task is performed: **O**.

Item	Error Producing Conditions (EPCs)	Maximum predicted nominal amount by which unreliability might change going from 'good' condition to 'bad'	Consideration for RIA-LOIC-H's GTT Type F
1	Unfamiliarity with a situation which is potentially important, but which only occurs infrequently, or which is novel	17	A
2	Time shortage	11	O
3	Low Signal/ Noise ratio	10	NA
4	Features override allowed	9	NA
5	Spatial and functional incompatibility	8	NA
6	A mismatch between an operator's model of the world and that imagined by a designer	8	NA
7	No obvious means of reversing an unintended action	8	O
8	A channel capacity overload, particularly one caused by simultaneous presentation of non-redundant information	6	NA
9	A need to unlearn a technique and apply one which requires the application of an opposing philosophy	6	NA
10	The need to transfer specific knowledge from task to task without loss	5.5	NA
11	Ambiguity in the required performance standards	5	NA
12	A mismatch between perceived and real risk	4	NA
13	Poor, ambiguous or ill-matched system feedback	4	O
14	No clear direct and timely confirmation of an intended action from the portion of the system over which control is to be exerted	4	O
15	Operator inexperience (e.g. a newly-qualified tradesman, but not an " expert")	3	O
16	An impoverished quality of information conveyed by procedures and person/person interaction	3	O
17	Little or no independent checking or testing of output	3	O
18	A conflict between immediate and long-term objectives	2.5	NA
19	No diversity of information input for veracity checks	2.5	O
20	A mismatch between the educational achievement level of an individual and the requirements of the task	2	NA

IOP Conf. Series: Materials Science and Engineering **785** (2020) 012048 doi:10.1088/1757-899X/785/1/012048

Table 3.2 HEART EPCs and the corresponding maximum weighting factors. For task RIA-LOIC-H's GTT Type F, applicable EPCs are denoted as: **A**, not applicable (out of the context) : **NA**, and opposite to the context in which the task is performed: **O** (continue).

	Error Producing Conditions (EPCs)	**Maximum predicted nominal amount by which unreliability might change going from 'good' condition to 'bad'**	**Consideration for RIA-LOIC-H's GTT Type F**
21	An incentive to use other more dangerous procedures	2	NA
22	Little opportunity to exercise mind and body outside the immediate confines of a job	1.8	NA
23	Unreliable instrumentation (enough that it is noticed)	1.6	NA
24	A need for absolute judgments which are beyond the capabilities or experience of an operator	1.6	O
25	Unclear allocation of function and responsibility	1.6	NA
26	No obvious way to keep track of progress during an activity	1.4	NA
27	A danger that finite physical capabilities will be exceeded	1.4	NA
28	Little or no intrinsic meaning in a task	1.4	O
29	High-level emotional stress	1.3	NA
30	Evidence of ill-health amongst operatives, especially fever	1.2	NA
31	Low workforce morale	1.2	NA
32	Inconsistency of meaning of displays and procedures	1.2	NA
33	A poor or hostile environment (below 75% of health or life-threatening severity)	1.15	NA
34	Prolonged inactivity or highly repetitious cycling of low mental workload tasks	1.1	A
35	Disruption of normal work-sleep cycles	1.1	NA
36	Task pacing caused by the intervention of others	1.06	NA
37	Additional team members over and above those necessary to perform task normally and satisfactorily	1.03	NA
38	Age of personnel performing perceptual tasks	1.02	NA

4. Conclusion

In this paper, we demonstrate the application of HEART method for the quantification of HEP for two selected HFEs modelled in a TRIGA research reactor. The process involves the classification of GTTs, identification of EPCS and APAO. A conservative approach is taken in dealing with APOA so that there would be no unnecessary optimism when interpreting the overall risk at this stage. HEPs calculated in this study are verified by the results from the previous study that has employed SPAR-H method, and therefore they would be used as input parameter for human basic events of the PSA logic models. In addition, remedial measures related to each EPCs can be proposed into the reactor operation and management so that a better operator performance and hence the more risk reduction can be achieved. The future work of this study requires the development of a systematic scheme for rating the APAO specifically for the context of the RR PSA.

5. References

[1] Bier V M 1999 Challenges to the acceptance of probabilistic risk analysis *Risk Anal.* **19**(4) p 703–710,.
[2] Apostolakis G and Michal R 2000 Apostolakis: On PRA *Nucl.News* **43**(3) p 27–31.
[3] Rasmuson D M A 1992 Comparison of the small and large event tree approaches used in PRAs *Reliab. Eng. Syst. Saf.* **37**(1) p 79–90.
[4] Spurgin A J 2010 *Human Reliability Assessment Theory and Practice* (CRC Press).
[5] Chander F T, Chang Y H, Mosleh A, Marble J L, Boring R L, and Gertman D I Human 2006 Reliability Anal. Meth. Selection Guidance for NASA.
[6] Bell J and Holroyd J 2009 Review of Human Reliability Assessment Methods.
[7] Kolaczkowski A, Forester J, Lois E, and Cooper S 2005 *Good Practices for Implementing Human Reliability Anal. (HRA)* (Washington, DC: U.S Nuclear Regulatory Commission).
[8] Whaley A M *et al.* 2016 *Cognitive Basis for Human Reliability Anal. (NUREG - 2114).*
[9] IAEA 1986 *Probabilistic Safety Assessment for Research Reactors* (Vienna: International Atomic Energy Agency).
[10] IAEA 1989 *Application of Probabilistic Safety Assessment to Research Reactors* (Vienna: International Atomic Energy Agency).
[11] Aneziris O N, Housiadas C, Stakakis M, and Papazoglou I A 2004 Probabilistic safety analysis of a Greek Research Reactor *Ann. Nucl. Energy* **31**(5) `p 481–516.
[12] Barati R and Setayeshi S 2014Probabilistic Safety Assessment of Tehran research reactor based on a synergy between plant topology and hierarchical evolutions *Prog. Nucl. Energy* **70** p. 199–208.
[13] Baron J H 2010 PSA applications for research reactors towards a very low risk *Reliability, Safety and Hazard (ICRESH) 2010- 2nd Int. Conf. on Reliability, Safety and Hazard - Risk-Based Technol. and Phys. of Failure Meth. (ICRESH),.*
[14] Ghofrani M Band Damghani S A 2002 Determination of the safety importance of systems of the Tehran research reactor using a PSA method *Ann. Nucl. Energy* **29**(16) p 1989–2000.
[15] Jordan-Cizelj R and Mavko B 1996 Influence of specific data on a research reactor probabilistic model *Reliab. Eng. Syst. Saf.* **52**(1) p 55–63.
[16] Kamyab S. and Nematollahi M 2011 Evaluating the core damage frequency of a TRIGA research reactor using risk assessment tool software *Nucl. Eng. Des.* **241**(8) p 2942–2947.
[17] Barati R and Setayeshi S 2012 Human reliability analysis of the Tehran research reactor using the SPAR-H method *Nucl. Technol. Radiat. Prot.* **27**(3) p 319–332.
[18] Mladin D and Prisecaru I 2014 Aspects of probabilistic safety assessment for TRIGA research reactor *Univ. Politech. Bucharest Sci. Bull. Ser. C* **76**(1) p 261–270.
[19] Barati R and Setayeshi S 2014 On the operator action analysis to reduce operational risk in research reactors," *Process Saf. Environ. Prot.* **92**(6) p 789–795.
[20] Baron J H, Leod J E N M and Rivera S S *2003* A prospective application of ATHEANA for a research reactor PSA *Safety and Reliability: Proc. of the ESREL Conf., Maastricht, the Netherlands, 15-18 June 2003* (Maastricht, the Netherlands: CRC Press) p 91.
[21] Williams J C 1988 A data-based method for assessing and reducing human error to improve operational performance *Proc. IEEE Fourth Conf. Hum. Factors Power Plants, Monterey, California* p436–450.
[22] Mohamed F *et al.* 2015 Operator reliability study for probabilistic safety analysis of an

operating research reactor *Ann. Nucl. Energy* **80** p 409–415.

[23] Maskin M, Charlie F, Hassan A, Prak Tom P, Ramli Z and Mohamed F 2016 Selection of important initiating events for Level 1 probabilistic safety assessment study at Puspati TRIGA Reactor *Ann. Nucl. Energy* **92** p. 198–210.

[24] Gertman D I, Blackman H S, Marble J, Byers J, and Smith C 2005 *The SPAR-H Human Reliability Analysis Method* (Washington, DC: U.S Nuclear Regulatory Commission).

[25] IAEA 2010 *Development and Application of Level 1 Probabilistic Safety Assesment for Nuclear Power Plants* (Vienna: International Atomic Energy Agency).

[26] Hassan A, Maskin M, Tom P P, Brayon F, Hlavac P andMohamed F 2017 Operator response modeling and human error probability in TRIGA Mark II research reactor probabilistic safety assessment *Ann. Nucl. Energy*, **102** p 179–189.

Acknowledgement
The authors would like to acknowledge the financial support from the IAEA – Norwegian Extra-Budgetary Program (NOKEBP) Fund and the University Tun Hussein Onn Malaysia through the TIER 1 Grant Scheme (H067). We also thank the Reactor TRIGA Puspati operators for providing the operational procedures and details pertaining to their response during reactor abnormal conditions.

IOP Conf. Series: Materials Science and Engineering **785** (2020) 012049 doi:10.1088/1757-899X/785/1/012049

Internal fire probabilistic safety assessment preliminary work at research reactor TRIGA PUSPATI

Zaredah Hashim[1], Devintharaaj Kumar[2], Mohd Fazli Zakaria[1], Mohd Huzair Hussain[1], Ahmad Hassan Sallehudin Mohd Sarif[3] and Mazleha Maskin[1,a)]

[1]Reactor Technology Center, Technical Services Division, Malaysian Nuclear Agency, MESTECC, 43000 Bangi, Kajang, Selangor, Malaysia
[2]School of Chemical and Energy Engineering, Faculty of Engineering, Universiti Teknologi Malaysia, 81310 Skudai, Johor
[3]Faculty of Applied Sciences and Technology, Universiti Tun Hussein Onn Malaysia (UTHM), Hab Pendidikan Tinggi Pagoh, KM 1, Jalan Panchor, 84600 Panchor, Johor

[a]mazleha@nuclearmalaysia.gov.my

Abstract. An internal fire analysis Level-I probabilistic safety assessment (PSA) for TRIGA PUSPATI (RTP) nuclear research reactor was initiated. Previous study on half-scope mode Level-I PSA, focusing only on internal initiating event, were developed for both full power and shutdown operational mode. Hence, in ensuring PSA remains relevant, a subsequent study to kept PSA up to date is a compulsory. The aim of this study is to present the step-by-step work done and the insights gained in identifying and screening significant compartments that are potential to had the worst consequences as an ignition sources. This preliminary study will only focus on RTP's basement which consist of 3 compartments: switch room, battery room and basement hall. Procedure in data collection were based on IAEA document, Specific Safety Guide No. 3: Development and Application of Level 1 Probabilistic Safety Assessment for Nuclear Power Plants (SSG-3). Microsoft Excel was used as data collection media and internal fire hazard database development. A brief qualitative analysis using failure mode and effect analysis is also presented.

1. Introduction

Probabilistic safety assessment (PSA) is a comprehensive structured approach to identify failure scenarios and to quantify estimation of risk and/or safety which describes accident sequences with the help of fault tree and event tree analyses [1], [2]. Figure 1 portrays three types of data needed in conducting a full scope PSA study, which are: (1) Internal events: data consists of random failure of systems or components or human error; (2) Internal hazards: data consists of internal fire or internal flood; and (3) External hazards: data consists of natural phenomenon or man-induced. Each of these data is further divided according to three reactor's operation modes: (1) Shutdown state; (2) Low power; and (3) Nominal power. The result of Level I PSA is core damage frequency (CDF), Level II, large early release frequency (LERF), and Level III, is radiological consequences.

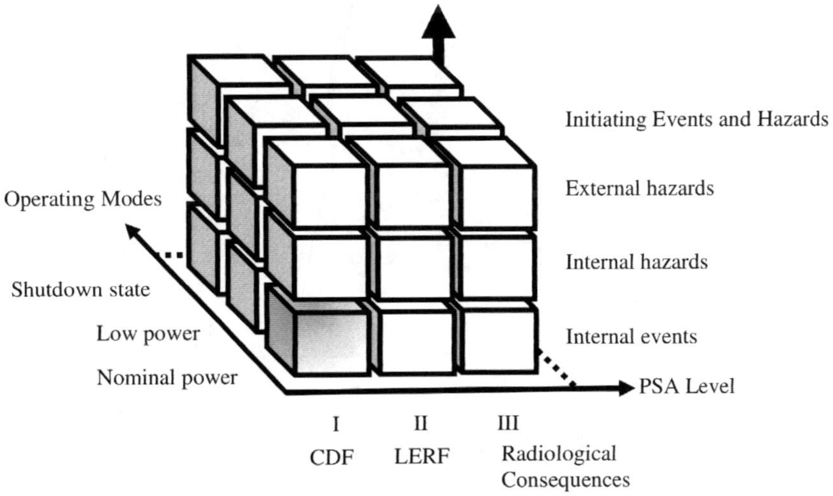

Figure 1. Full scope PSA illustration.
Source: IAEA [2].

PSA was introduced in the sixties as an ultimate safety assessment approach for space shuttle by National Aeronautics and Space Administration (NASA), when three astronauts were killed from the Apollo test AS-204 on January 27, 1967 [3]. Early applications of PSA on conventional nuclear power plant (NPP) was traced back in the United Kingdom (UK) [4] and the United Stated (US) [5] for their gas-cooled reactors (GCR's) and light water reactors (LWR's) respectively. Meanwhile, the application of PSA on nuclear research reactor was documented by the IAEA [6] as early as 1980.

In Malaysia, at the only nuclear research reactor been built, reactor TRIGA PUSPATI (RTP), an initial work in attempting to incorporate internal fire probabilistic safety assessment (PSA) into Level-I PSA model was implemented in early June 2019. Recently, the Malaysian PSA team developed a level 1 PSA focusing on random failure of systems, components and human error [7–11]. As per mention in Maskin *et al.* [9], Level-I PSA for RTP is a new adapted project which is develop without any official database.

The main purpose of this paper is to prepare necessary data in assessing internal fire hazard Level-I PSA. In particular, this paper aims at introducing a systematic approach in data collection and qualitative assessment throughout the process in preparing the necessary data. It is also intended to serve as a guide in a step-by-step guidelines for the development of fire PSA studies for a nuclear research reactor, as per recommendation by the IAEA [12], requirement 20, which is: "…The Level-I PSA for internal fire should be documented in a manner that facilitates review, applications and updating of the Level 1 PSA". Finally, a success in this initial work, will be a result to expand the current study from only basement to the entire RTP building.

2. Methodology

An illustration of the adapted methodology is illustrated in Figure 2, showing two major process, which are: (1) Data collection; and (2) Qualitative analysis.

The scope of this study covers Level-I internal fire PSA, perform at full power and daily shutdown operational conditions (the shutdown for annual maintenance is not included in this study). Meanwhile, the location considered is focusing only at the reactor basement.

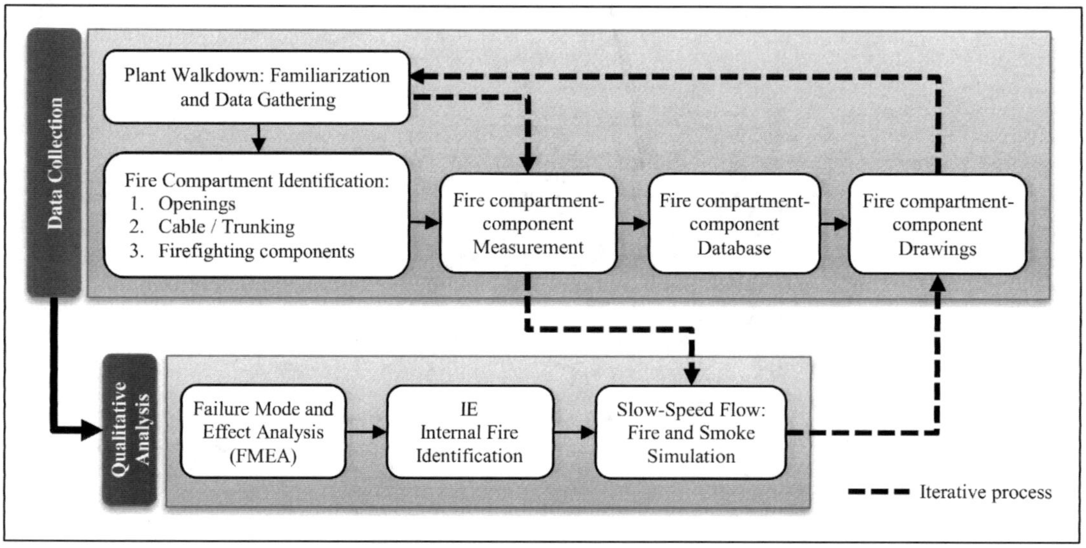

Figure 2. Data collection and qualitative analysis approach for fire PSA.

2.1. Data Collection

2.1.1. Plant walkdown. On preparation for Level-I fire PSA, a thorough plant walkdown is performed to verify information on fire hazard (potential ignitor) sources and plant features susceptible to damage due to fire hazard.

Familiarization of the plant, especially on the identification and how measurement need to be taken for compartment physical characteristics such as, boundaries (walls, doors, openings, etc.), cables route (conduits, trays, trunkings, etc.) is the key element in a fire PSA study. Fail to record precise information is a failure in simulating fire and smoke flow correctly. A walkdown sheet used in this study is partially depicted in Table 1.

Table 1. Walkdown sheet (partial)

INTERNAL FIRE WALKDOWN SHEET

Date:	**Completed by:**
Building:	**Elevation /floor:**
Room#:	**Room name:**

1. Compartment attributes

Attribute	Comments / Quantity
Area / Height (m^2, approximate or L x W)	
Wall material / covering	
Floor material / covering	
Ceiling material / covering	
HVAC system ☐ Present	

2. Compartments openings (doors, access, HVAC, etc.)

Opening	Location (X-Y or degree)	Height of Floor (m)	Opening Height (m)	Opening Width (m)	Open / Closed / Changes	Comments (note of doorway)

2.1.2. Fire compartment identification. According to IAEA [13], in the context of a PSA for internal fires: "…a fire compartment could be a well-enclosed room that is not necessarily surrounded by fire resistant barriers". Therefore, as illustrated in Figure 3, RTP basement is identified to have three compartments, consist of: (1) Switch room; (2) Battery room; and (3) Basement hall.

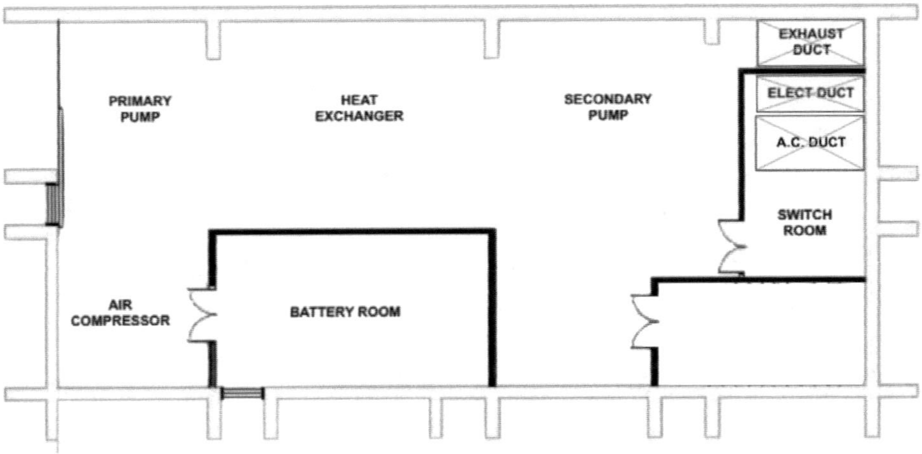

Figure 3. Basement layout diagram (landscape)

2.1.3. Fire compartment-components database. All information (recorded manually in Table 1), including the measurement of each physical characteristic (width, height, location), both boundaries and cables route were recorded systematically into an Excel database. This database is a living document and can be altered accordingly, based on additional important information.

The objective in developing such database is to quickly and accurately assess all relevant information related to the potential equipment failures for fire scenarios of interest, e.g. total failure of all circuits in a compartment, failure of cables within a specific raceway, impact of failures based on *specific equipment failure modes etc.*

2.1.4. Fire compartment drawings. Autodesk Inventor, a 2D and 3D mechanical design software is used (student's licensed trial version). Four templates 3D view layout drawings for each compartment were developed. These layouts are consisting of: (1) Compartment openings; (2) Cables and trunkings, (3) Electrical cabinets; and (4) Fire suppression components, with exact measurement applied to all layouts. Later, a 2D drawings for the same four layouts in the 3D drawings mentioned above were developed, which mainly to determine the coordination location of each component.

2.2. Qualitative Analysis and Screening

2.2.1. Qualitative Analysis: Failure Mode and Effect Analysis (FMEA). An inductive analysis, FMEA to examine each system potential failure mode due to occurrence of fire. It may be loss of function, unwanted function, out-of-tolerance condition, or any related failure. The significance of a failure mode depends on how the system responses to the failure [14], [15]. Table 2 indicates designator failure mode for each component in the basement, together with an identification consequence on the system for each failure mode, if the component fails.

2.2.2. Qualitative Screening. Uniform approach is perform in qualitative screening to assign appropriate initiating event (IE) as depicted in Figure 4 [16].

Table 2. An independent failure mode and effect designator for component type

Failure Mode	Consequences / Effects
1. Fail to start [FS]	1. Lead to an initiating event (IE)
2. No power supply [NP]	2. Affect the ability of safety functions to mitigate an IE
3. Operator action [OA]	3. Affect operator actions after the occurrence of an IE
4. No signal [NS]	4. Lead to spurious actuation of functions that could induce other unsafe effects
5. Fire spark [FiS]	

Source: IAEA, pg. 72 [17].

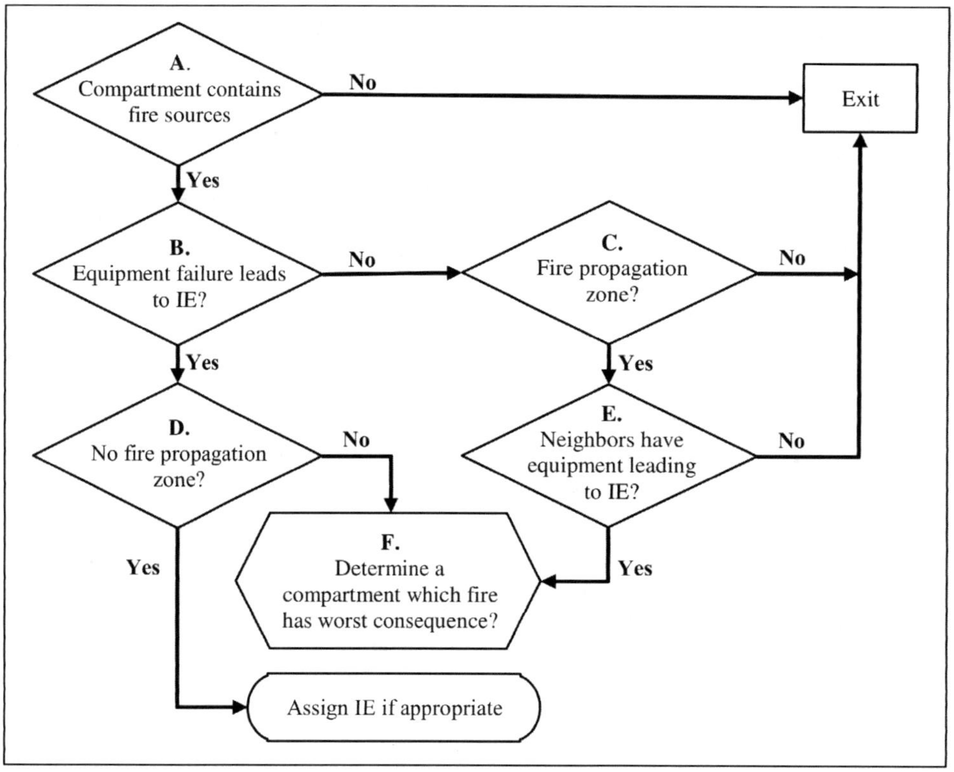

Figure 4. Flowchart for qualitative screening

Source: Šimurka [16].

3. Results and Discussion

3.1. Data Collection and Drawings

Inaccurate data collection typically in measurement, may lead to invalid result and affect the outcome of a study. This is a valid lesson learnt whilst trying to merge inventor template files to form one complete compartment consist of 4 templates: (1) compartment openings; (2) cables and trunkings, (3) electrical cabinets; and (4) fire suppression components. Failing in merging the templates, stemmed in measuring the same compartment repeatedly. Figure 4a and 4b illustrated 2D drawing for switch room compartment openings and fire suppression systems respectively. Meanwhile, Figure 5 depicted a successful merged 3D drawing for switch room.

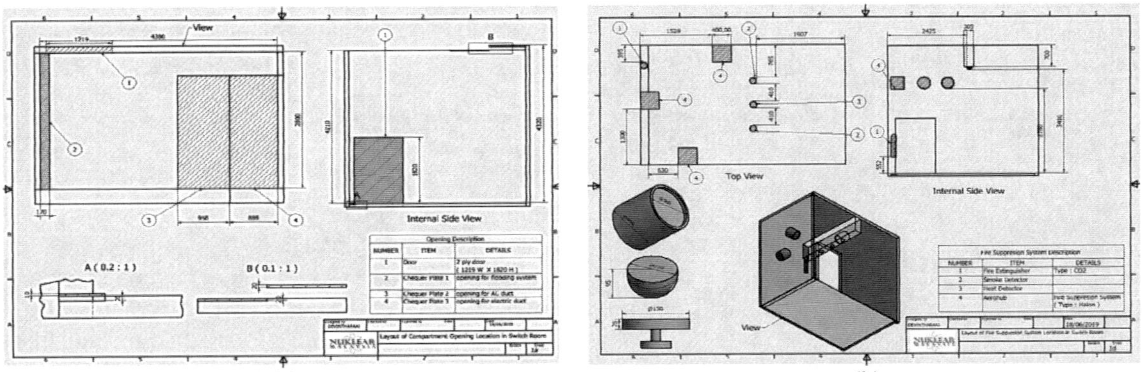

(a) (b)

Figure 4. Switch room: (a) Compartment openings layout; and b) Fire suppression systems.

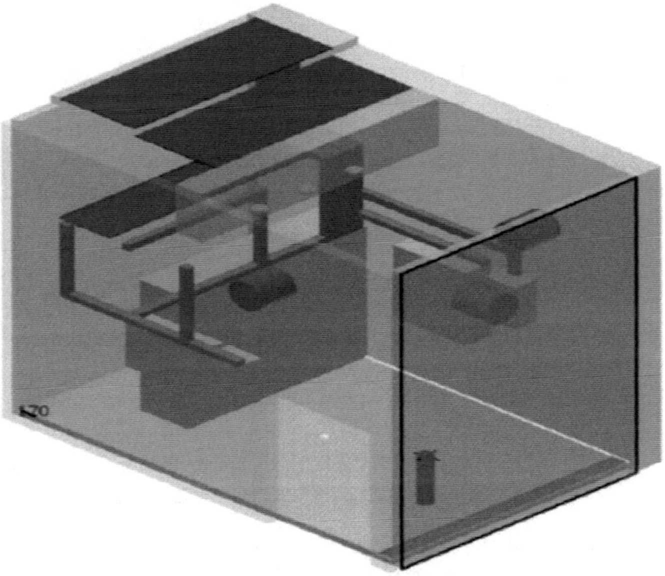

Figure 5. Switch room: 3D view.

Drawings for RTP basement resulted with 51 template files for switch room, 23 template files for battery room and 212 template files for basement hall.

3.2. RTP Internal Fire Database Development

Figure 6 illustrate a screenshot of the developed internal fire database. As per stated earlier, this database is a living document which will be updated from time to time to reflects any current modification and component installation.

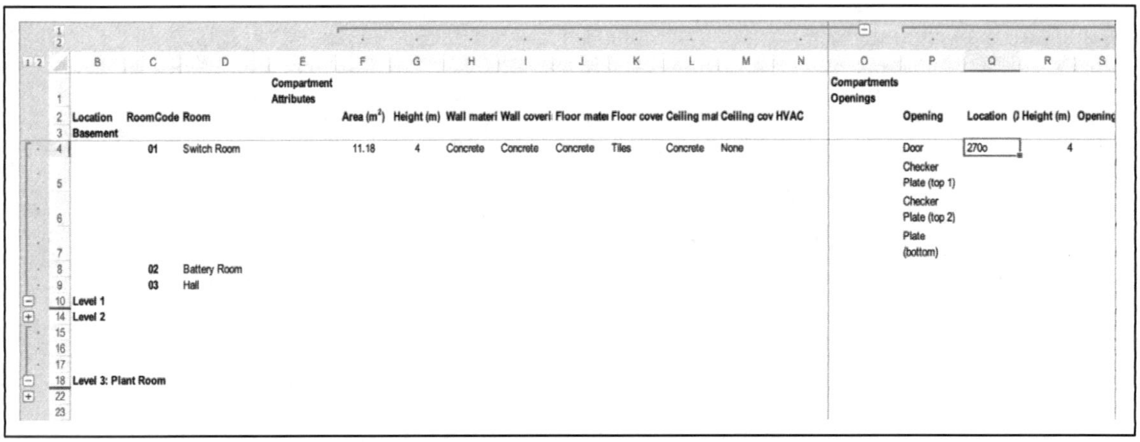

Figure 6. RTP internal fire database (screenshot).

3.3. Failure Mode and Effect Analysis

Table 3 presents the lists of components with its failure mode and the consequences or effects to RTP if it fails.

Table 3. List of FMEA internal fire for RTP basement for switch and battery room

No.	Compartment	Failure Mode	Consequences or Effects	IE
A.	**Switch room**			
	1. Fire suppression: Heat detector	FS	Operator missing compelling signal/ annunciator after the occurrence of fire	No
	2. Fire suppression: Smoke detector	FS	Operator missing compelling signal/ annunciator after the occurrence of fire	No
	3. DB 'RE'	FiS	Lead to an initiating event (IE)	Yes
	4. DB 'REB'	FiS	Lead to an initiating event (IE)	Yes
	5. Main switchboard (MDB)	FiS	Lead to an initiating event (IE)	Yes
	6. DB 'RNB'	FiS	Lead to an initiating event (IE)	Yes
	7. TNB source	NP	Reactor blackout	
B.	**Battery room**			
	8. Battery	NP	Reactor blackout	No
	9. Uninterruptable power supply (UPS)	NP	Reactor blackout	No
	10. Switchboard	FiS	Lead to an initiating event (IE)	Yes
	11. Fire suppression: Heat detector	FS	Operator missing compelling signal/ annunciator after the occurrence of fire	No

Based on the initial FMEA screening analysis, the 'Yes' value indicated in Table 3 above, only shows one initiating event for both compartments, which is fire spark. However, this result will be re-evaluated based on which compartment has the worst fire consequences.

4. Conclusion

To have a typical and comprehensive PSA plant model, a complete list of representative IEs is essentially non-negligible. Incomplete consideration of IEs adversely affects the quality of a PSA, thus leading to results that underestimate the level of risk. Therefore, this preliminary study on the development of IEs for RTP internal fire is part of the effort towards fulfilling the completeness of a full scope Level-I PSA. This study has addressed both the systematic approach in data collection and qualitative assessment in preparing the necessary data for Fire PSA input. Finally, success in this initial work will be a result to expand the current study from just basement to the entire RTP building.

Data collection is the most crucial part of any study — failure to collect accurate data resulted in an incorrect result. Lesson learnt in a wrongly measured dimension stemmed in measuring the same compartment repeatedly. However, despite the repetitive measurement, all 3D drawings for three compartments were successfully drawn and merged. At the same time, a compartment-component database that records all the plant walk-down sheet which reflects any current modification and component installation was developed as soon as the project starts. Meanwhile, from a qualitative assessment using inductive analysis, for all three compartments, only one IE was identified: fire spark.

Hence, our future works will be focused for the rest of the RTP's building and proceed on the accident sequence modelling, and event tree development for the internal fire.

5. References

[1] British Electricity International Nuclear safety 1992 *Modern power station practice: nuclear power generation*, 3rd ed British Electricity International, (Pergamon) chapter 3 p. 484–570.

[2] IAEA 2015 *BPTC module VII: probabilistic safety assessment* (Vienna: Austria: International Atomic Energy Agency).

[3] Bedford T and Cooke R 2001 *Probabilistic risk analysis: foundations and methods* (United Kingdom: Cambridge University Press).

[4] Cave L and Holmes R E 1967 Suitability of the advance gas-cooled reactor for urban siting *IAEA Symp. on Containment and Siting of Nucl. Power Plants* p 273–299.

[5] Garrick B J, Gekler W C, Goldfisher L, Karcher R H, Shimizu B and Wilson J H 1967 *Reliability analysis of nucl. power plant protective systems* (Los Angeles, CA: Nuclear Division, Holmes & Narver, Inc.)

[6] IAEA 1987 *Probabilistic safety assessment for research reactors (Tecdoc Series) No. 400.* (Vienna, Austria: International Atomic Energy Agency).

[7] Brayon F C *et al.* 2014 Building competence for safety assessment of nuclear installations: applying IAEA's safety guide for the development of a Level 1 probabilistic safety assessment for the TRIGA research reactor in Malaysia *Probabilistic Safety Assessment and Management Conf. 12* (PSAM12) (Sheraton Waikiki. Honolulu, Hawaii, USA).

[8] Mohamed F *et al.* 2015 Operator reliability study for probabilistic safety analysis of an operating research reactor *Ann. Nucl. Energy* **80** p 409–415,.

[9] Maskin M, Brayon F C, Hassan A, Phongsakorn P T, Ramli Z, and Mohamed F 2016 Selection of important initiating events for Level 1 probabilistic safety assessment study at Puspati TRIGA Reactor *Ann. Nucl. Energy* **92** p 198–210.

[10] Hassan A, Maskin M, Prak Tom P, Brayon F, Hlavac P and Mohamed F 2017Operator response modeling and human error probability in TRIGA Mark II research reactor probabilistic safety assessment *Ann. Nucl. Energy* 102 p 179–189.

[11] Maskin M *et al.* 2018 Quantification of initiating event frequencies and component reliability data in level 1 probabilistic safety assessment at Puspati TRIGA research reactor," *Ann. Nucl. Energy* **121** p 22–28.

[12] IAEA 2016 *Safety of Research Reactors (Specific Safety Requirements) No. SSR-3* (Vienna, Austria: International Atomic Energy Agency).

[13] IAEA 2004 *Protection against internal fires and explosions in the design of nuclear power plants- IAEA safety standards series No. NS-G-1.7* (Vienna, Austria: International Atomic Energy Agency).

[14] Fullwood R R 1998 *Probabilistic safety assessment in the chemical and nuclear industries* (United States of America: Butterworth-Heinemann Publications).

[15] Ericson C A 2015 Failure mode and effect analysis *Hazard anal. tech. for sys. safety* (Hoboken, New Jersey: John Wiley and Sons, Inc) p 235–455.

[16] Šimurka P 2014Overview of fire PSA and supporting software *23rd Int. Conf. Nucl. Energy of New Europe* p 503.1-503.8.

[17] IAEA 2010 *Development and application of level 1 probabilistic safety assessment for nuclear power plants specific safety guide, safety series no. SSG-3* (Vienna, Austria: International Atomic Energy Agency).

Acknowledgement
This paper is produced as an extension of two projects: (1) project under the Ministry of Science, Technology & Innovation (MOSTI) ScienceFund project (MOSTI 03-03-01-SF0179); and (2) IAEA – Norwegian Extra-Budgetary Program (NOKEBP) Fund, project name 'Applying PSA to Existing Facilities to Develop Transferable Skills in the use of PSA to evaluate nuclear power plant (NPP) Safety'. Both projects are aiming on developing a comprehensive technical work to master the PSA technology for a safe assessment of a nuclear facility through the practical application of the relevant recent IAEA Safety Standards on PSA to develop a PSA model for the only research reactor in Malaysia

IOP Conf. Series: Materials Science and Engineering **785** (2020) 012050 doi:10.1088/1757-899X/785/1/012050

Risk analysis and hazard identification at ALURTRON and MINTec-Sinagama

Ruzalina Baharin[1,a)], Syuhada Ramli[1], Siti Zulaiha Hairaldin[1] and Rohaizah Ahmad[2]

[1]Technical Support Division, Malaysian Nuclear Agency, Bangi, 43000 Kajang, Selangor, Malaysia.
[2]Radiation Safety Division, Malaysian Nuclear Agency, Bangi, 43000 Kajang, Selangor, Malaysia.

[a]ruzalina@nm.gov.my

Abstract. Risk Management is part of the operational element in an organization to manage their directives towards risks surrounding them. It assists the organization to identify, evaluate and prioritize the risks through risk analysis. The main objective of this work is to identify risk from hazards involved in the work place, so that control measures can be put in place to avoid or minimise the harm it can cause. The risks include hazards from radiation, chemical, biological, physical as well as ergonomic factors. Hazard identification was carried out at two radiation facilities in Malaysian Nuclear Agency that involved radiation safety. ALURTRON and MINTec-Sinagama provide radiation services to commercial industries as well as for R&D purposes. In this paper, every section in risk analysis and hazard identification are discussed.

1. Introduction

Every organization has its own risks that need to be identified at the very beginning of operation. Risks can be identified and managed accordingly prior to the probable occurrence of real incidents. It is good practice to identify the risks including hazards that might affect the operation of any organization. All levels of leadership are strongly encouraged to be involved during this period [1]. Subsequently, by evaluating and analysing the identified risks, organization can plan appropriate control or treatment to decrease the impact of the risks. At the end of the evaluating and analysing process, the organization will either accept any residual risks or continue the process of totally eliminating the risks.

1.1 Risk management

Risks need to be managed effectively for an organization in order for it to operate at its optimum level. When risk identification is correctly carried out, then the risk treatment can be put in place [2]. The process should be continuously handled and maintained at every department and level. It keeps the environment and goal-driven activities in the organization running smoothly. In turn, these activities can bring success to the organization whilst ensuring that the risks are being taken care all the time [3]. Negative and positive impacts from the identified risks will drive the organization along the right path to realise their mission and vision [4].

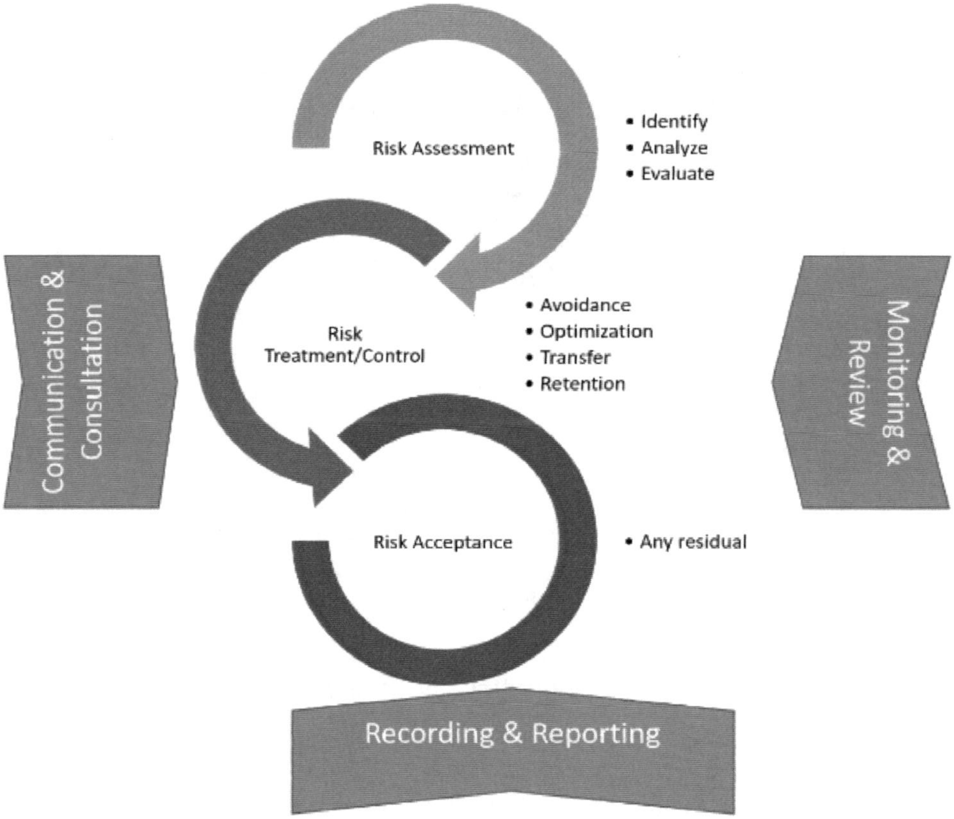

Figure 1. Process in Risk Management.

Therefore, for this purpose, this study carried out risk and hazard identification at two radiation facilities in Malaysian Nuclear Agency namely ALURTRON and MINTec-Sinagama. The main objective of this work is to identify risks from hazards involved in these workplaces. Subsequently, appropriate control measures can be applied to avoid or minimise the harm they can cause.

1.2 Radiation facility: ALURTRON [5]

ALURTRON is one of the radiation processing plants owned by Malaysian Nuclear Agency, located in Bangi, Selangor. It provides electron beam radiation services to industry, especially for wire, cable and tube (tubular product) as well as semiconductor. The purpose of this radiation process is to crosslink the polymer material of the products which can increase the application quality of the products. The most common application of these products is in the automotive industry. These are needed in the engine compartment especially for the brake line. As for treating wafer semiconductor, electron beam radiation can enhance its electrical conductivity.

For the irradiation process, ALURTRON has one electron beam machine, EPS-3000, which works with other supporting equipment and product handling system.

Figure 2. Left: Capstan drum for tubular product irradiation with scanner horn located in the bunker. Middle: Pay-off equipment for wire under beam handling system (WUBHS) which is integrated to the EPS-3000 machine. Right: Trolley for flat product irradiation.

1.3 Radiation facility: MINTec-Sinagama [6]

Located within the same compound as ALURTRON, Malaysian Nuclear Agency also has a gamma radiation processing plant, known as MINTec-Sinagama. Set-up in 1989, MINTec-Sinagama has since serve the country by treating a diversity of products: from medical devices, raw material of pharmaceuticals products, to food and non-medical products. Depending on the products, the objectives of the gamma irradiation treatment that need to be fulfilled are sterilization, decontamination, shelf-life extension and quarantine.

MINTec-Sinagama has a high radioactive Co-60 activity of approximately 0.3 MCi (1.1 x 10^{13} Bq) (as per September 2019). For product handling equipment, MINTec-Sinagama uses JS 10000 IR 219 tote irradiator system. In addition, irradiation process at MINTec-Sinagama is supported by various mechanical equipment such as compressor, cooling water system and pumps and motors.

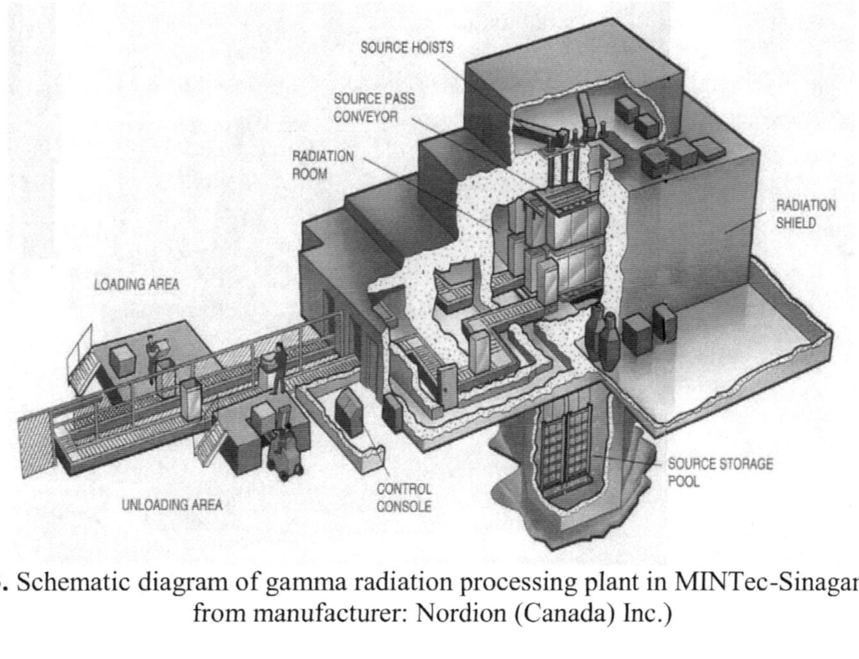

Figure 3. Schematic diagram of gamma radiation processing plant in MINTec-Sinagama (courtesy from manufacturer: Nordion (Canada) Inc.)

IOP Conf. Series: Materials Science and Engineering **785** (2020) 012050 doi:10.1088/1757-899X/785/1/012050

2. Risk analysis process

Risk analysis comprises identification, description and estimation of risks. During risk identification, ALURTRON and MINTec-Sinagama identified their main risk category with a few fitting into more than one category. Some of the risk category are strategic, compliance, financial, operational and environmental [1]. Based on the recommendation from Institute of Risk Management [2], subsequently the risk description was tabulated as shown in Table 1.

Table 1. Structured format of risk description.

Item	Description
1. Name of risk	
2. Scope of risk	Qualitative description of the events, their size, type, number and dependencies
3. Nature of risk	Eg. strategic, operational, financial, knowledge or compliance
4. Stakeholders	Stakeholders and their expectations
5. Quantification of risk	Significance and probability
6. Risk tolerance/ appetite	Loss potential and financial impact of risk Value at risk Probability and size of potential losses/gains Objective(s) for control of the risk and desired level of performance
7. Risk treatment & control mechanisms	Primary means which the risk is currently managed Levels of confidence in existing control Identification of protocols for monitoring and review
8. Potential action for improvement	Recommendations to reduce risk
9. Strategy and policy developments	Identification of function responsible for developing strategy and policy

In order to do the analysis, estimation on risk is needed. The possible impact (consequence) and likehood (occurrence) are measured depending solely on the past operational experience of the organization [2]. A mixed quantitative, semi-quantitative or qualitative approach was taken. Table 2, 3 and 4 show the table of risk estimation on likelihood and impact as well as matrix of both factors as decided by ALURTRON and MINTec-Sinagama during workshop on risk management. Subsequently, the risk is evaluated.

Table 2. Likelihood or probability of occurrence on threats.

Level	Descriptor	Description
5	Almost certain	Expected to occur in most circumstances
4	Likely	Will probably occur in most circumstances
3	Possible	Might occur in some time
2	Unlikely	Could occur at some time
1	Rare	May occur only in exceptional circumstances

Table 3. Impact Table - risk estimation on impact of risk.

Level	Description	Description
5	Very critical	Machine down/closure of facility
4	Critical	Plant closure for 1 to 3 days, high health risk
3	Moderate	Risk can cause damage, delay, customer complaint
2	Fair	Risk is real but can be averted fast
1	Risk accepted	Impact is very low

Table 4. Matrix likehood x impact table. Different colors scheme was used for easier visual inspection.

	IMPACT				
LIKELIHOOD	**1**	**2**	**3**	**4**	**5**
5	5 LOW	10 MEDIUM	15 MEDIUM	20 HIGH	25 HIGH
4	4 LOW	8 MEDIUM	12 MEDIUM	16 HIGH	20 HIGH
3	3 VERY LOW	6 LOW	9 MEDIUM	12 MEDIUM	16 HIGH
2	2 VERY LOW	4 LOW	6 LOW	8 MEDIUM	10 MEDIUM
1	1 VERY LOW	2 VERY LOW	3 VERY LOW	4 LOW	5 LOW

3. Findings and results

Possible risks involved in the work activities of ALURTRON and MINTec-SINAGAMA were explored. A few techniques were utilized – such as SWOT (Strengths, Weaknesses, Opportunities, Threat) analysis and PESTLE (Political, Economic, Social, Technical, Legal, and Environmental) analysis. Brainstorming was also applied at two workshops and in a few discussions among staffs together with the Director of both facilities. Risk analysis and findings from the above activities are tabulated in Table 5 and 6.

Table 5. Risk analysis for ALURTRON.

RISKS IDENTIFICATION			RISKS ANALYSIS & VERIFICATION					RISKS TREATMENT PLAN	
CATEGORIES	ISSUES	RISKS / OPPORTUNITIES	CURRECT CONTROL	LIKELIHOOD	IMPACT	RISKS TOTAL = LIKELIHOOD x IMPACT	RISKS LEVEL	RECOMMENDED ACTION/ ADDITIONAL CONTROL	PIC
PRODUCT RECEIVED & INSPECTION	Quantity of product not tally	Disruption during payment process	Thorough Inspection	2	2	4	LOW	Length counter (to purchase)	Admin/ OP
		Interrupted operating schedule	Thorough Inspection	2	2	4	LOW	NA	QC inspector/ executive OP
DOCUMENTATION & HANDLING PROCESS	Handling of document tagging/revision	Record not updated	Review of document	2	2	4	LOW	Document check once a year	Executive QA
IRRADIATION PROCESS	CCTV damage	Product stuck in irradiation room/ Product damage	Old CCTV	2	4	8	MEDIUM	Black and white CCTV procurement	SV OP
EQUIPMENT CALIBRATION & MAINTENANCE	Chemical hazard to worker health (fumehood)	Testing equipment performance not being checked	Service record	2	4	8	MEDIUM	Provide checklist for maintenance by BKJ	Executive QA
QA TEST - Gel Content	Employee health affected due to waste storage overload	Delayed test	Report to WASTECC for disposal	2	3	6	LOW	Disposal programme	SV QA / next disposal circle
QA TEST - Dosimetry	CTA dosimeter low in stock	Disturb & delay in work process	Procurement when left 2 batch	2	3	6	LOW	NA	Lab Assistant
PRODUCT RELEASE	Delivery	Delay release product	Target 5 days delivery (QC objective)	2	3	6	LOW	NA	Executive OP
CUSTOMER SERIVICE	Delivery	Delay & loss of customer	Target 5 day delivery objective	3	2	6	LOW	Customer service - Follow up	Admin
PUBLIC ACCEPTANCE (IP-community)	Public complaint	Termination of operation	Awareness programme (Open day)	1	2	2	LOW	NA	UKK
PAYMENT TO SERVICE/ PRODUCT PROVIDER (IP-supplier)	Delay payment	Blacklisted	Purchasing procedure	3	2	6	LOW	Follow up with BKT	PIC procurement
TRAINING AND EXPERT SERVICES (IP-NGOS)	Access to information	Complaint	UKK management	3	2	6	LOW	UKK program & PR admin	UKK
COLLABORATION & CONSULTATION (IP-investor, NGOs)	Access to information	Access to information	Website & PR tool	3	2	6	LOW	Programme by PR admin	Director
STAFF WALFARE (IP- worker)	Safety & health at work	Accident	Training & awareness program	3	2	6	LOW	Monitoring & continual training	BKS
IMPLEMENTATION OF ACT/ISO/ LEGISLATION/ GUIDELINE (IP-standard Body)	Not having enough understanding	License/ certificate revoke	Training & awareness program	3	2	6	LOW	Monitoring & continual training	BPA & ALURTRON

Table 6 Risk Analysis for MINTec-Sinagama.

RISKS IDENTIFICATION				RISKS ANALYSIS & VERIFICATION				RISKS TREATMENT PLAN	
CATEGORIES/ PROCESS	**ISSUES**	**RISKS/ OPPURTUNITIES**	**CURRENT CONTROL**	**LIKELIHOOD**	**IMPACT**	**RISKS TOTAL = LIKELIHOOD X IMPACT**	**RISKS LEVEL**	**RECOMMENDATION ACTION / ADDITIONAL CONTOL**	**PIC**
PRODUCT ARRIVAL AND RECEIVING INSPECTION	Product Damage	Product handler attitude	Use trained product handler	3	3	9	MEDIUM	Training	QA
	Integrity of product	Dangerous goods, corrosive, dirty	Memo to customer	3	3	9	MEDIUM	Reminder procedure of product figure/condition to send to MINTec-Sinagama	QA
	Customer submit false information	False identification of product in RSP1	Any false declaration under customer responsibility	2	5	10	MEDIUM	Discussion with customers / regulators	QA
PRE-IRRADIATION PROCESS	Documentation	Negligence due to the workload	Get more staff	3	2	6	LOW	Train the new part-time operator	Manager
DOSIMETER PLACEMENT	Error in dosimeter location	Negligence due to the workload	NA	2	5	10	MEDIUM	Get more staff	Manager
	Dosimeter broken during placement	Incident, more dosimeter needed	NA	3	5	15	MEDIUM	Design dosimeter hold	QA
PRODUCT LOADING	PLP configuration is wrongly managed	Irradiation is not valid	Provide updated summary PLP at loading platform	1	4	4	LOW	Recognized product that registered PLP	QA
PRODUCT ENTERS RADIATION ROOM & PRODUCT LEAVES RADIATION ROOM	Negligence of operator on handling the operating system	Disturb and delay in work process	NA	3	4	12	MEDIUM	NA	Operation
	Tote and conveyor system	Conveyor not working properly	NA	4	3	12	MEDIUM	NA	Operation
	Barcode detection	Scanner not working properly	NA	4	3	12	MEDIUM	NA	Operation
	Lack of permanent staff	Disturb and delay in work process, no replacement	Workload divided among staffs and part-time	4	4	16	HIGH	Get temporary staffs when needed	Manager
PRODUCT UNLOADING	Dosimeter broken during collection	Incident, cannot provide irradiation data reading	Put extra dosimeters and locations	3	3	9	MEDIUM	NA	QA
DOSIMETER ANALYSIS	Reduction in dosimeters stock	Disturb and delay in work process	Request to SSDL when two batches are left in stock	4	4	16	HIGH	Explore new dosimetry system	QA
	Mistake discovered in Certificate of Irradiation (CoI)	Customer complaint	Double checked by superior	4	3	12	MEDIUM	NA	QA
FINAL INSPECTION	Product Storage	Warehouse condition	Pest control monitoring and fortnightly housekeeping	3	4	12	Medium	NA	Operation
PRODUCT RELEASE	Mistake discovered in RSP 15	Customer complaint	Double checked by superior	3	4	12	LOW	NA	QA

IOP Conf. Series: Materials Science and Engineering **785** (2020) 012050 doi:10.1088/1757-899X/785/1/012050

4. Conclusion and recommendations

As a conclusion, managing risks is an important aspect at the workplace and needs to be done continuously [7]. Every factor that contributes to the risks is identified according to the activities involved in the work flow of the irradiation process. Only then can the risks be reduced to as low as possible (ALAP) even though the risk factor is low. The risk treatment is towards ALAP instead of ALARP (as low as reasonably practicable) practices.

Therefore, as a way forward, work on Risk Treatment Plan will be continued. Reviews and actions taken on recommendations in Risk Treatment Plan will ensure its continual improvement. It will be important to explore more actions as well as assessing any new risks in future analysis.

5. References

[1] Jaroslava K, Michaela D 2002 Risk management in industrial companies (SAMI 2012) doi:10.1109.

[2] Institute of Risk Management. A risk management standard (IRM) http://www.theirm.org/publications/documents/ARMS_2002_IRM.pdf

[3] Institute of Risk Management. A risk practitioners guide to ISO 31000 2018. https://www.theirm.org/media/3513119/IRM-Report-ISO-31000-2018-v3.pdf

[4] Remzi A, Besarta V 2017Analysis of international risk management standards (advantages and disadvantages) *European J. of Res. and Reflection in Management Sci.* **5**(3).

[5] Nuclear Malaysia 2019 *ALURTRON, ISO 9001:2015 Quality Management System Manual.*

[6] MINTec-Sinagama 2019 *ISO 9001:2015 and ISO 13485:2016 Quality Management System Manual.*

[7] Alok P, Meghraj S, Amit S, Rane D M, Sonawane A U 2016 Risk assessment and management of radiation safety in industrial applications of ionizing radiation: an overview *Non-Destructive Eval.* p332-340

IOP Conf. Series: Materials Science and Engineering **785** (2020) 012051 doi:10.1088/1757-899X/785/1/012051

Ageing assessment of biological shielding integrity for PUSPATI TRIGA Reactor

Nurhayati Ramli[1,a)], Julia Abdul Karim[1], Hasniyati Md Razi, [1] Muhammad Khairul Ariff Mustafa[1], Noor Azreen Masenwat[1], Suhairy Sani[1], Mohd Huzair Hussain[1]

[1]Malaysian Nuclear Agency, Bangi, 43000 Kajang, Selangor, Malaysia.

[a]nurhayati@nm.gov.my

Abstract. For more than 35 years of operation and utilization of the neutron source from the PUSPATI TRIGA Reactor (RTP), the reactor has faced ageing challenged and degradation in several structures, systems and components (SSCs). Several modifications have been made since 2010 to ensure safe operation and sustainability, however, due to recent intensively request from the user to use the neutron source, it has shown degradation behaviour in the reactor SSCs including the biological shielding of the RTP. In this paper, it described the assessment of the safety barriers integrity specifically at the biological shielding concrete structure of the RTP. In-service inspection and X-Ray diffraction methodology were used to investigate and analyses the condition. The investigation results show no irregularity except slightly degradation occurrences has been noticed and aware.

1. Introduction

PUSPATI TRIGA Reactor (RTP) is a TRIGA Mark II nuclear research reactor providing training and irradiation services to public and private institutions. It also involved and supported several research projects carried out by other research groups[1]. After serving for almost nearly forty years, the facility has faced with ageing effects and potential failures that reflect the reactor safety. In order to mitigate ageing effects during RTP operational lifetime, the facility has had to deal with several issues due to the time-dependent degradation of its structures, systems and components (SSCs).

In general, safety objectives for a nuclear research reactor are to protect human and environment by maintaining an effective defense-in-depth against radiological hazards. The defense-in-depth is achieved by a multiple barrier system where the integrity of SSCs can be affected by the failure of one or more of these barriers. However, the aging process may lead to the degradation of structural integrity and increase the probability of failures at the biological shielding concrete due to radiation field over a long period of operation time may reduce the ability to confine and contain the radioactive release from the reactor[2].

In-service Inspection (ISI) has an important role in aging management of the SSCs. ISI is incorporated with risk insights known as Risk-informed In-service Inspection programme. The main objective of ISI is to demonstrate the SSCs that important to safety will not fail during operation[3]. The ISI techniques could aide to assess the surface and volume of the structures by inspect near-surface flaws and volumetric to indicate the depth or size of cracking using radiographic, ultrasonic or eddy current techniques.

The aging mechanism is not limited to physical conditions. The ageing mechanisms such as corrosion, vibrations, temperature and radiation would be expected in the ageing process. The biological shielding concrete can be affected by radiation interactions with material and absorption of the radiation energy. The RTP biological shielding concrete is a reinforced high density concrete with approximately 3.42 g/cm^3 density and four beamport which tubular penetrates through the concrete shield[4]. Thus, it is very important to investigate and to understand the physical condition of the biological shielding concrete structure integrity to mitigate consequences (if any) for lifetime extension for the operational sustainability of the RTP.

2. Methodology

The work had conducted using three methodologies. Location mapping for ultrasonic testing has been performed at the surface of the biological shield for the most utilized beamport, i.e. beamport number 4 (BP#4). The location mapped was distributed into two segments as described in Figure 1. Prior to the mapping, the radiation field measurement was also being performed. The NDT technique was used to further assess the mapped surface and concrete specimens were taken to analyze using XRD.

2.1. Location mapping

The RTP biological shielding concrete structure is 6.6 meter in diameter and 6.5 meter height [4]. For this assessment, 8 different locations have been identified where 5 locations were located at the bottom segment and 3 locations at the top segment of the RTP concrete structure. In this study, beam port number 2 (BP#2) and beam port number 4 (BP#4) were chosen to represent the less and most utilized beam port respectively.

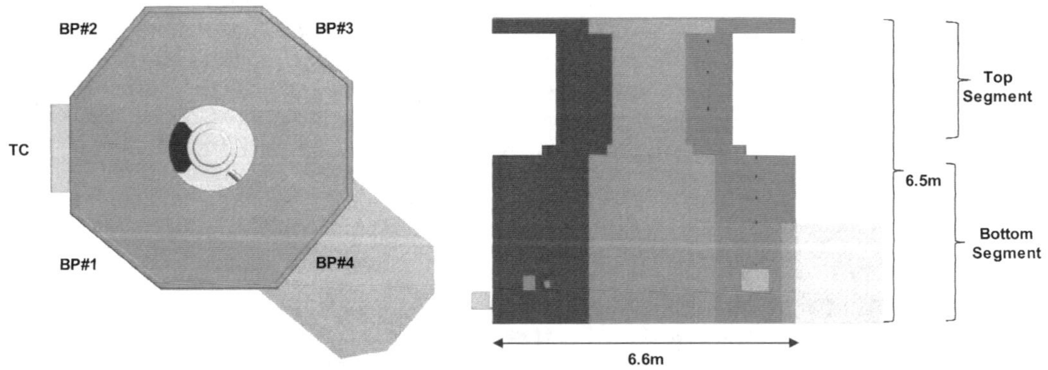

Figure 1. RTP concrete biological shielding structure.

2.2. Radiation field measurement

The Optical Stimulated Luminescence (OSL) is used to measure neutron and gamma dose at the mapped surface location at the NDT inspection region. A total of eight OSLs were exposed in eight different locations during RTP operation at 750kW for a month. Figure 2 shows the black dotted where the OSLs are located.

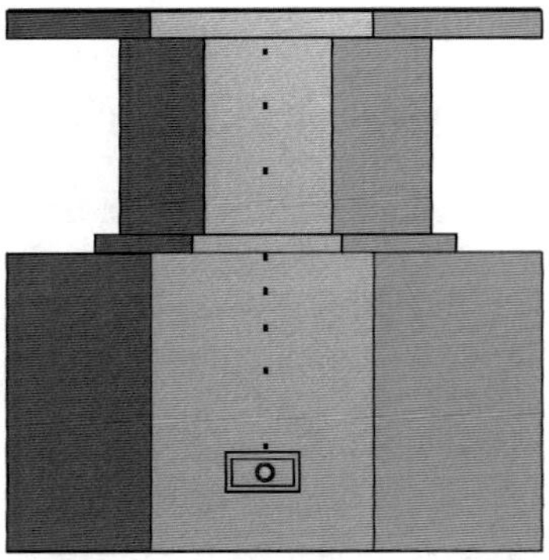

Figure 2. OSL locations.

2.3. Non-destructive testing (NDT) measurement

In-situ measurement using Ultrasonic Pulse Velocity (UPV) to determine the quality of the concrete by measuring pulse velocity. The UPV generated voltage pulses and transformed into wave bursts of mechanical energy by the transmitting transducer (which must be coupled to the specimen surface through a suitable medium). A receiving transducer is coupled to the specimen at a known distance to measure the interval between the transmission and reception of a pulse[5]. There are three practical arrangements for measuring pulse velocity, namely direct, diagonal and surface transmission arrangement. The surface transmission arrangement is used for RTP concrete as the surface concrete is accessible.

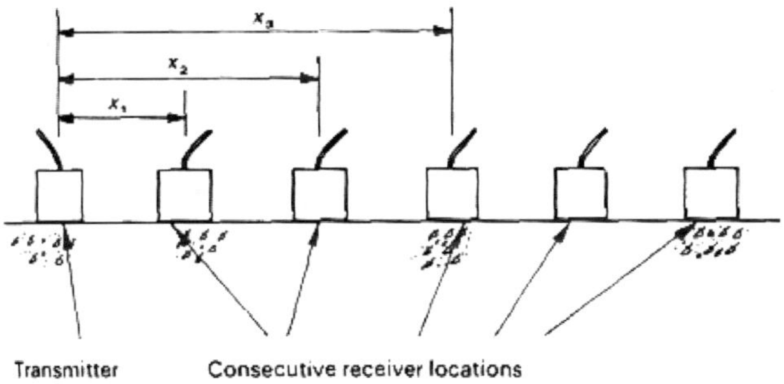

Figure 3. Surface transmission arrangement.

The estimation of the concrete quality condition is based on Table 1:

Table 1. The condition of concrete quality.

Pulse velocity (m/s)	Strength
Above 4500	Excellent
3600 – 4500	Generally good
3000 – 3600	Questionable
2100 – 3000	Generally poor
below 2100	Very poor

IOP Conf. Series: Materials Science and Engineering **785** (2020) 012051 doi:10.1088/1757-899X/785/1/012051

2.4. Quantitative Analysis using X-ray

X-ray scattering techniques reveal information about the crystallographic structure, chemical composition, and physical properties of materials and thin films. These techniques are based on observing the scattered intensity of an X-ray beam hitting sample as a function of incident and scattered angle, polarization, and wavelength or energy.

In this studies, two commonly X-ray techniques were used. X-ray Fluorescence (XRF) is used to determine elemental presents in the sample. There are two types of XRF systems, i.e. Energy Dispersive XRF (EDXRF) and Wave Length Dispersive XRF (WDXRF). In this assessment, the EDXRF is performed to determine the elements presents in the biological shielding concrete specimen.

While an X-ray diffraction (XRD) is used to obtain qualitative and quantitative analysis of crystalline compounds. In this assessment, XRD analysis was performed using Bruker AXS D8, USA and conducted in a range of Bragg angles (2θ) of 20°- 90° using Copper K-α radiation. Concrete samples were analyzed with X-ray diffraction (XRD) in powder forms. The analysis result was then plotted in a graph with intensity vs diffraction angle of 2θ.

3. Results and discussion

3.1. Neutron and gamma dose

The results of neutron and gamma dose measurement is shown in Figure 4. It can be deduced the neutron and gamma radiation of concrete exposed to irradiation related to the depth of the RTP tank. The neutron dose is slightly higher than the gamma dose due to the high penetration power and strong radiation field came from the reactor core inside concrete shielding during operation. The radiation field was observed for one month. It is found the dose was larger around 1meter depth from the bottom of RTP tank at where the core is located.

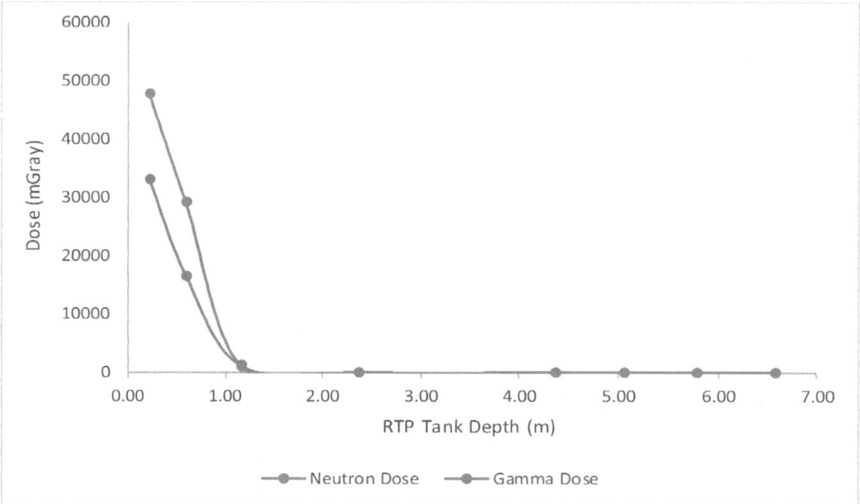

Figure 4. Relation of neutron and gamma dose with reactor tank height in meter.

3.2. Non-destructive testing (NDT) measurement

Figure 5 shows the UPV results for surface where the BP#2 and BP#4 were located. Measurement for each surfaces were taken at 8 different locations and the average velocity was calculated. It is found the velocity of BP#4 surface has higher velocity than surface BP#2. The velocity at surface BP#4 (location 6) was 3891m/s. This indicates the greater strength of the concrete according to the quality of concrete condition as described in Table 1.

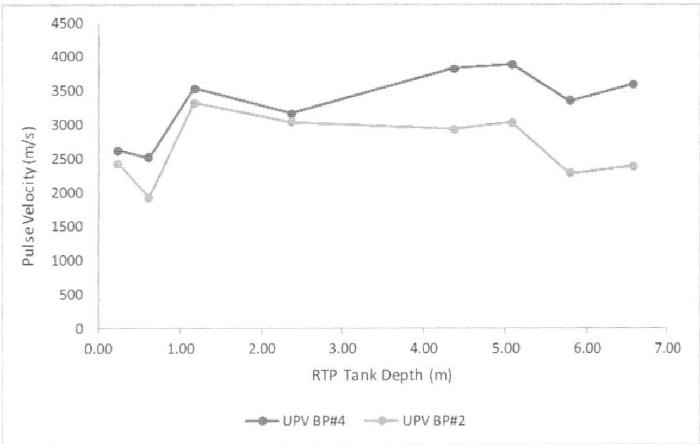

Figure 5 UPV with respect to the reactor tank height

3.3. X-ray analysis

The EDXRF and XRD analysis were performed for eight samples of the biological concrete as described in Table 2. Form the analyses, the elements in the concrete contained mainly from Silicon (Si), Aluminium (Al), Potassium (K) and Calcium (Ca). In this studies, the aggregates of the concrete were not included.

Table 2. The results of the RTP biological concrete samples using EDXRF.

Sample		Element (%)			
		Si	Al	K	Ca
Top Segment	1	80.65	6.36	3.29	7.63
	2	79.13	4.05	4.65	9.01
	3	71.10	6.31	2.76	14.32
	4	75.90	0.8	2.86	15.27
Bottom Segment	5	76.17	4.78	2.77	8.01
	6	82.78	0.83	2.21	11.37
	7	69.20	5.21	1.93	14.26
	8	68.53	5.8	2.60	11.11

The concrete samples were analyzed with X-ray diffraction (XRD) in powder forms. XRD was carried out using Bruker AXS D8, USA and conducted through a range of Bragg angles (2θ) of 20°- 90° using Copper K-α radiation. The analysis result was plotted for intensity in a function of diffraction angle at 2θ. Figure 6 and Figure 7 shows XRD patterns of the RTP biological shielding concrete compounds. The XRD patterns of the concrete compounds showed the presence of Calcium Oxide (CaO), Silicon Dioxide (SiO_2), Aluminium Oxide (Al_2O_3) and Potassium Oxide (K_2O) at all location. The XRD results are in agreement of XRF results.

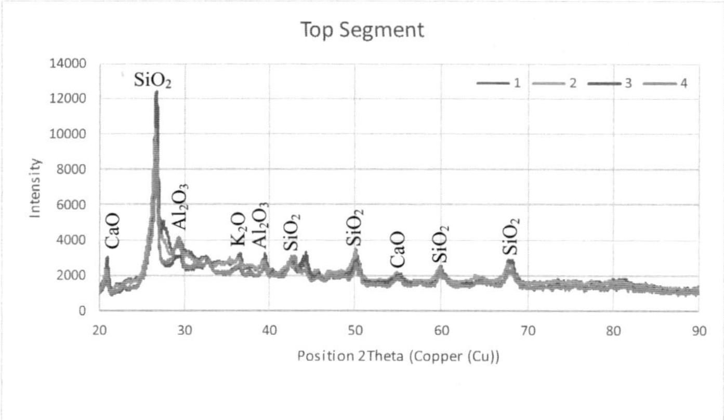

Figure 6. The XRD patterns of RTP biological shielding concrete (Top segment).

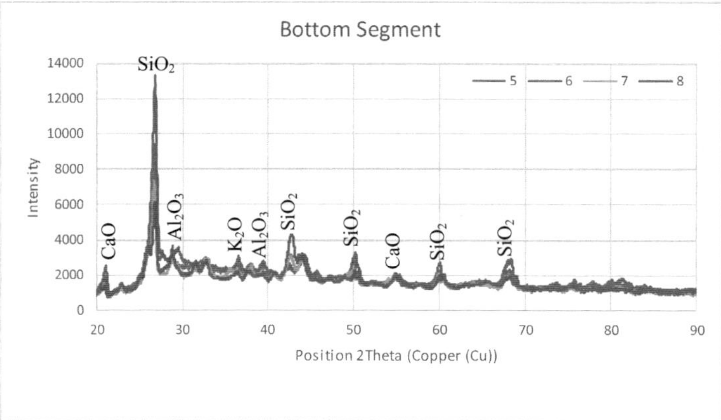

Figure 7. The XRD patterns of RTP biological shielding concrete (Bottom segment).

4. Conclusions

The ageing assessment of RTP has focused to the SSCs for lifetime extension of the sustainability in RTP operation. The biological shielding concrete has been identified as 4th barrier in defense in depth philosophy as described in the safety function of RTP. Preliminary assessment of the surface structure gives some ideas and information on the degradation of the concrete especially in the high radiation field (neutron, gamma) region. More detailed studies will be performed to obtain overall information of the degradation assessment of the reactor. Risk-informed In-service Inspection methodology are adopted for the ageing assessment of RTP and this method will be continued pursued in the next studies.

5. References

[1] Zakaria M F, Lanyau T, Saad M F and Ismail A R 2011 Ageing Management Programme for Reactor TRIGA PUSPATI (RTP) **8**(2) p 94–99.
[2] IAEA 1995 Management of Research Reactor Ageing IAEA Tecdoc-792.
[3] IAEA 2001 Application of non-destructive testing and in-service inspection to research reactors.
[4] Malaysian Nuclear Agency 2017 Safety analysis report for PUSPATI TRIGA Reactor (Malaysian

Nuclear Agency).
[5] Masenwat N A and Sani S 2019 *Concrete inspection report , Industrial Technology Division (BTI)* Nuklear Malaysia/L/2019/35(R).

IOP Conf. Series: Materials Science and Engineering **785** (2020) 012052 doi:10.1088/1757-899X/785/1/012052

ReDICS: 5 years' operation experiences and its technology

Nurfarhana Ayuni Joha[1,a], Mohd Sabri Minhat[1], Mohd Khairulezwan Abdul Manan[1], Zareen Khan Abdul Jalil Khan[1], Ridzuan Abdul Mutalib[1] and Muh. Kasim Abdul[1]

[1]Reactor Instrumentation and Control Section, Reactor Technology Centre, Malaysian Nuclear Agency, 43000 Kajang, Malaysia.

[a]nurfarhana@nuclearmalaysia.gov.my

Abstract. PUSPATI TRIGA Reactor (RTP) is the one and only 1MW research reactor in Malaysia that has been operating since 1982. RTP is used for radioisotope production, various R&D activities and operator training. After more than 3 decades of operations with an analog system, the control console is upgraded into a digital system. Reactor Digital Instrumentation and Control System (ReDICS) is designed, fabricated, manufactured, installed, tested and eventually commissioned in 2014. ReDICS mainly consists of Reactor Protection System (RPS), Operator Work Station (OWS), Data Acquisition and Control System (DACS) and Control Rod Drive Mechanism (CRDM). This paper shares the experiences and performance status throughout the 5 years of ReDICS operation after its commissioning. Besides, technical and operational data of ReDICS are presented and then compared with the old analog I&C system. This paper also describes the instrumentation and measurement technologies in ReDICS as well as the work performed to ensure ReDICS is always functioning at its best and safely.

1. Introduction

The design of ReDICS is based on the safety standard NS-R-4, Safety of Research Reactors (now known as SSR-3), published by IAEA. The safety philosophy comprising of safety concepts and principles are described in NS-R-4 to meet safety objectives such as nuclear safety, radiation protection, and technical safety [1]. It encompasses the implementation of the defense in depth concept, the establishment of legislative and regulatory infrastructure, the adoption of measures for the management and verification of safety, and the application of technical principles during the design and over the lifetime of the installation.

The defense in depth concept was applied on the design of RTP with its physical barrier protecting the fuel matrix, the cladding, the reactor pool with primary heat transport system, and the reactor building. The effectiveness of these barriers are monitored via the measurements of fuel temperature, pool water temperature, pool water level, overpower, and period signal, and the radioactivity level at various localized point within the building. The single failure criterion is applied on the design of the Reactor Protection System which will SCRAM the reactor and maintain it in a safe shutdown state during Design Basis Accidents (DBA) such as Loss of Coolant Accident (LOCA) and Reactivity Initiated Accident (RIA) [2].

The inherent safety feature is achieved by the use of TRIGA fuel which has a negative prompt coefficient of reactivity. The cooling of the fuel is achieved by the natural circulation of pool water.

The Reactor Protection System (RPS) of ReDICS has a safety function which shuts down the reactor if any of the trip parameters values exceeds the trip set point. Thus the RPS is designed to meet the single failure criterion, high reliability and means for inspection, test, and maintenance.

2. Methodology

After commissioning of ReDICS in 2014, all operation and maintenance works are done by RTP personnel. There are some replacements and developments related to RTP I&C system made to enhance the performance of ReDICS and RTP operation. The technical comparison between the analog I&C system and ReDICS are presented with respect of design criteria for general and specific design. The ReDICS system which comprises of Reactor Protection System (RPS), Operator Work Station (OWS), Data Acquisition and Control System (DACS), Control Rod Drive Mechanism (CRDM) and Wide Range Neutron Measurement System (WR-NMS) is studied in this paper including its alarms system that will notify the reactor operators on any abnormal condition or if any one reactor parameter exceeds its limit. The alarm system is the principal source of information for the detection of any specific off-normal condition. The Area Radiation Monitoring System (ARMS) and the Seismic Monitoring System (SMS) are also part of the instrumentation systems at RTP. This paper also categorizes each measurement instrumentation within RTP into neutron detector, nuclear channel, radiation monitor and non-nuclear instrumentation. The operational data of RTP are presented from 2010 until 2018 which include reactor operation (hours), total energy released (MWhours), number of samples and number of experiments. The ReDICS was tested before commissioning to ensure its correct functionality. The tests performed were Site Acceptance Test (SAT), Construction Acceptance Test (CAT), System Performance Test (SPT), Integrated System Test (IST) and Reactor Performance Test (RPT). After commissioning and full operation, the annual and semi-annual maintenance work are performed which include inspection, periodic testing, calibration and maintenance.

3. Results and discussion

RTP as a nuclear research reactor is providing training and services to private enterprises and public institutions as well as being involved in several research projects carried out by the universities and other research groups. In order to grant a safe and continuous reactor operation for the future, improving reliability and extending the system life time, many activities were carried out and successfully completed throughout 5 years of ReDICS operation to ensure RTP preparedness for such training and services which include installation of Resistance Temperature Detector (RTD) for inlet & outlet in primary cooling system, replacement of Operator Work Station (OWS), replacement of fission chamber, replacement of Area Radiation Monitoring System (ARMS) at the reactor pool top and development of user authentication for security.

Table 1 shows the comparisons of general design between the analog I&C system and ReDICS. Some design criteria need to be achieved and met in developing reliable I&C system such as redundancy and single failure criterion, diversity, independence, fail safe and ease of testing and maintenance. Table 2 shows the comparisons of specific design between the analog I&C system and ReDICS. The design criteria for the specific design of I&C system are including Reactor Protection System, Data Acquisition and Control System and control room.

Table 1. Comparisons of general design between the analog I&C system and ReDICS.

Design criteria	Analog I&C system	ReDICS
Redundancy and single failure criterion	• Uses 2 redundant safety channels • Adopts 1 out of 2 (1oo2) coincidence logic	• Same as the analog I&C system • Redundant control computer enhances system reliability
Diversity	• Uses pneumatic and electrical control rods actuators	• Same as the analog I&C system
Independence	• Electrical isolation between safety and non-safety channels	• Functional and physical separation of RPS • Isolation between RPS and DACS
Fail safe	• Fail safe design in the event of loss of power	• Same as the analog I&C system
Ease of testing and maintenance	• Considered in the design	• Considered in the design • Uses modern technology such as self-test and diagnostic capability

Table 2. Comparisons of specific design between the analog I&C system and ReDICS

Design criteria	Analog I&C system	ReDICS
Reactor Protection System	• Trip parameters: i. Fuel temperature ii. Over power iii. Reactor period iv. Loss of HV supply v. Manual trip	• Pool water level as new additional trip parameter • Manual trip button (OWS, reactor pool top, emergency exit door) • Diagnostic, display and trip event recording (using DACS and OWS)
Data Acquisition and Control System	• Analog hardwired circuit for control system • AC motors for CRs control	• Digitalized control system with dual-redundant structure • stepper motors for CRs control • Three (3) WR-NMS for better reliability • Robust control algorithm
Control room	• Accident analysis after event using limited pen recorders • Hard-wired switches	• Accident analysis after event using plotting software • Better situation awareness using Large Display Panel (LDP) • Soft switches

By the comparisons between the analog I&C system and ReDICS in Table 1 and Table 2, it can be seen that the safety philosophy of the analog I&C system has been completely preserved. Furthermore, the ReDICS has better design from the safety, reliability, operational, and maintenance point of view.

ReDICS comprises of Reactor Protection System (RPS), Operator Work Station (OWS), Data Acquisition and Control System (DACS), Control Rod Drive Mechanism (CRDM) and Wide Range Neutron Measuring System (WR-NMS) as shown in simplified block diagram in Figure 1.

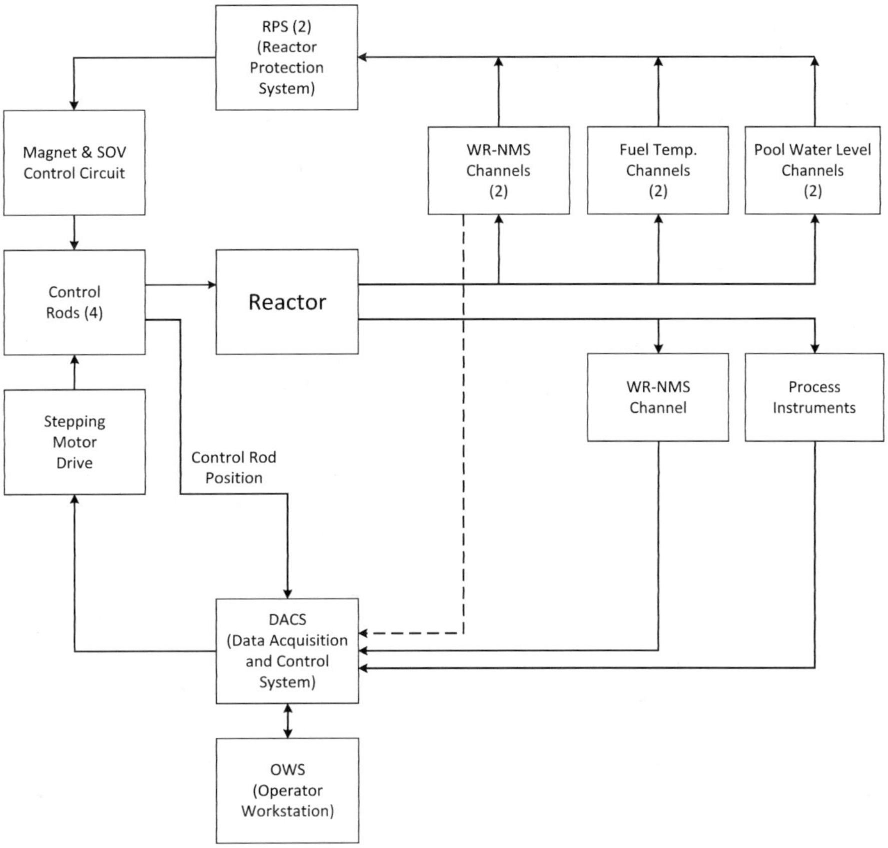

Figure 1. Block diagram of ReDICS.

The Reactor Protection System (RPS) of ReDICS provides the same function as the safety channels of the analog I&C system, that is, to shut down the reactor when any trip parameter exceeds its specified set points. The RPS trip the reactor by causing control rods to be inserted into the core. The RPS consists of 2 independent safety channels, which adopts one (1)-out-of-two (2) coincidence logic for shutdown triggering. The functions of the RPS in ReDICS are equivalent to the functions of the corresponding analog I&C system. This is to ensure that the current safety features are not jeopardized by the new design. The RPS provides the defense mechanism to prevent fuel failure which could result in the release of radioactive materials from fuel matrix. Two major accidents are postulated to happen on RTP, that is, loss of regulation and loss of coolant. Loss of regulation is described as an uncontrolled increase of power and/or increase of fuel temperature. Loss of coolant may result from large leakage of water from reactor pool due to certain events such as beam port breakage. The following trip parameters are implemented for the RPS of RTP:

a) Fuel temperature: A high value of fuel temperature indicates loss of regulation. (500°C)
b) Neutron power: A high value of neutron power indicates loss of regulation. (110%FP)
c) Reactor period: A low value of reactor period indicates loss of regulation (3sec)
d) Neutron detector amplifier high voltage: A low value indicates inability to measure a neutron power due to loss of measurement. (0.75 x Nominal HV)
e) Pool water level: A low value of pool water level indicates loss of coolant (30cm below reactor top)
f) Manual trip: The manual trip switches disconnects electrical power to all control rod magnets so that the control rods will fall freely by gravity in order to shut down the reactor. The locations of the manual trip switches are on OWS, on reactor pool top and near to reactor emergency exit door.

The RPS has displays and controls via two (2) safety panels in the control room. The displays are consisted of analog, digital or status indication as follows:

Table 3. Displays and controls of RPS in RTP control room

Analog and digital indication	• Neutron power and period • Fuel temperature • Neutron detector amplifier high voltage • Pool water level
Status indication	• Trip parameter status indicators • Reactor trip status • Channel trip status
Operator controls	• Manual trip switches • Trip Reset switches

The Operator Work Station (OWS) is the top level human machine interfacing equipment through which a reactor operator can understand the status of the reactor by monitoring the appropriate process parameters and may take appropriate action if required through control devices. The OWS desk is designed by applying ergonomic and human factor principles which consist of monitors for graphic display, keyboards, mouses, printers, and computers with the capability and capacity to implement functionalities specified on the specification. The graphic display on LCD monitor shows the current reactor condition and the operator can control the reactor power level by manipulation of the control rod up/down soft-switches or by adjustment of the demanded power. If the operating parameters exceeds the set point, alarms are displayed on a window on the monitor and audible alarms are initiated until the operator acknowledges the alarm by pressing the acknowledge button.

The Data Acquisition Control System (DACS) measures process parameters from instruments such as fission chamber, thermocouples, temperature probes, level sensors, control rods position sensors, area radiation monitors and other auxiliary systems. The signal from sensors are fed to I/O boards and processed in digital computers and transmitted to the OWS via a communication network. Interfacing between RPS and OWS is achieved through DACS. The reactor power is controlled from the source with range up to full power based on the control algorithm and operator input. The reactor operator adjusts the four control rods either manually or automatically in a stable and reliable manner. The startup of the reactor is done in manual mode and it is possible to operate in either automatic or manual operation if the reactor power is higher than 1W. During automatic operation, the four control rods are controlled by the control algorithm. The automatic reactor controller gets the reactor power level from the neutron flux monitoring system and compares with the demanded power level set by the operator and issues appropriate control rod movement signals while maintaining the reactor period longer than the trip set point. Besides the normal operation mode, square wave operation mode is also available. The square wave operation mode is used if required to raise the reactor power level quickly by firing the transient rod pneumatically. The DACS is designed not to share any hardware or software with the RPS. The DACS has no direct interface with the RPS. This is to ensure that the RPS is not prevented from performing safety function by the DACS. The DACS has two processors whose statuses are continuously and mutually monitored. If a processor fails, the failed component is disabled and excluded from the algorithm and while the working component performs all DACS functions. All of important hardware including main processor, I/O boards, and power supplies are dual redundant for system reliability and availability. The display provides accurate, complete, and timely information with appropriate engineering units and color coded designations for ease of understanding the reactor operation.

The old control rod drive mechanism of RTP consists of AC motors connected to a rack and pinion mechanism which has the function to drive the control rod up and down for power regulation. In ReDICS project, the AC motors of the old control rod drive mechanism are replaced by stepper motors.

Table 4. Comparisons of control mode between NORMAL and SQUAREWAVE operation mode.

Operation mode	NORMAL	SQUAREWAVE
Control mode	Manual & Automatic: Safety, Shim, Regulating and Transient rods are movable in both modes. Reactor operation starts with Manual mode	Manual mode: Transient rod (quick mode) Automatic mode: Safety, Shim, Regulating are movable

Table 5. Description of CRDM for all four control rods at RTP.

	Control rod drives			
	Standard control rods with electrically driven rack and pinion drives			Transient rod with pneumatic-electro-mechanical drive
	SHIM	**SAFETY**	**REGULATING**	**TRANSIENT**
Re-used (as it is)	Gear shaft (Rack & Pinion)			Ball nut & Worm gear
Old design	Induction motor	Induction motor	Servo motor	Induction motor
New design	Stepper motor Worm gear box			Stepper motor
Change	Potentiometer Limit switches			

The stepper motor has simple and robust operation and maintenance. The stepper motor is able to drive the existing rack and pinion mechanism. As such, the end shaft of the stepper motor is coupled with the pinion in order to transform rotational motion into linear motion. The control rod which is attached to the rack therefore moves up and down according to the rotation of the motor. The stepper motor is mounted inside the existing motor housing. The end shaft of the stepper motor is also coupled to the existing potentiometer to be used as the position feedback encoder. The motor drive is a set of electronic circuit that converts the number of pulses calculated by the control algorithm into a driving current for the stepper motor. The motor drive receives the input pulse from the computer system and sends out the driving current to the stepper motor of the corresponding control rod. The CRDM is connected to the DACS. The control rod movement is calculated by the computer based on the current power, demanded power and reactor period signals. The computer commands the control rod by sending signal pulses from the I/O interface to the motor drive assembly to activate the control rod drive mechanism. The signal pulses are converted into motor driving current which rotates the shaft of the stepper motor according to the calculated control rod position. The potentiometer of the control rod drive mechanism gives feedback on the actual control rod position to the computer system. Appropriate actions are taken if the difference between the actual control rod position and the calculated control rod position exceeds certain set point.

IOP Conf. Series: Materials Science and Engineering **785** (2020) 012052 doi:10.1088/1757-899X/785/1/012052

The WR-NMS consists of neutron detector (fission chamber) and electronics for signal processing. The neutron flux is measured by the fission chamber and the signal is further manipulated to give the neutron flux level by the source, wide range log or linear flux channel. The output of the WR-NMS includes log power, linear power, and reactor period. Two (2) WR-NMS are provided for RPS. From each WR-NMS, the RPS receives three (3) contact signals related to log power, period, and malfunction. These contact signals indicate if the measured value exceeds its set-point as RPS trip parameters. Another one (1) WR-NMS is provided for DACS. However, the DACS receives signals from two (2) WR-NMS on the RPS to improve its reliability. Hence, the DACS uses all three (3) WR-NMS signals. The DACS receives analog signals (log power, linear power, and reactor period) from the WR-NMS that directly indicates the current neutron power level. These values are used in the DACS control algorithm to calculate the control rods movement speed. The WR-NMS measuring range is from $10^{-8}\%$ to 200% FP for log power range, 0% to 150% for linear power range, and -30 sec to +3 seconds for reactor period signals.

ReDICS is also equipped with alarm system which is used to notify reactor operators of abnormal operating conditions, or components malfunctions. Alarm systems are meant to help operators operate the reactor safely; under both normal and abnormal operating conditions. The water temperature monitoring channel, fuel temperature monitoring channel and pool water level have alarm system in the control room which are provided with visual and audible annunciators on the RPS cabinets as well as alarms of Information & Control Display on OWS. Audible alarms are provided by using OWS for notifying the status of RTP and components to operator. The OWS supports an alarm control function. The visual and audible alarms are ergonomically designed according to style guides for ReDICS. The Area Radiation Monitoring System, Stack Monitoring System and Seismic Monitoring System are stand-alone systems in control room and independently providing alarms to operator [3].

The Area Radiation Monitoring System is to monitor the radiation level inside the reactor hall. Four area radiation monitors are positioned on the wall at strategic locations facing each of the reactor beam port of the reactor. Another radiation monitor is placed on the reactor top. A radiation monitor is also located at reactor basement. It consists of a radiation detector of GM type, electronic circuitry for signal conditioning and signal processing, a large bright LED display and an alarm siren, if the set limit is reached.

The Seismic Monitoring System is to obtain seismic information such as historical data and data on peak ground acceleration level at the reactor building. It is not related to reactor operation but is available for display of information on any earthquake event. The seismic monitoring system consists of a seismic acceleration sensor and a seismometer.

In order to operate the reactor, many variables have to be measured and monitored. Some of them are considered vital for the safety of the reactor, and their measurements have to be fast and reliable. This paper categorizes the measurement instrumentations within RTP into neutron detectors, nuclear channels, radiation monitors and non-nuclear instrumentations [4]. All the measurement instrumentation in ReDICS are described in Table 6 below.

Table 6. Instrumentation and measurement technologies in ReDICS

Category	Measurement instrumentations
Neutron detector	• Fission chambers (log percent power, linear percent power, cps for source power range, sec for wide range) • Self-powered neutron detectors (flux)
Nuclear channel	• Pre-amplifier • Log and linear amplifier • Count rate meters (source range and wide range) • Current meters (signal conditioning unit for neutron detectors operating in DC current mode) HV power supplies
Radiation monitors	• Area radiation monitoring system i. Low level gamma monitor (beamports; pre-alarm at 5 μSv/hr, alarm at 10 μSv/hr) ii. High level gamma monitor (reactor pool top; pre-alarm at 5 mSv/hr, alarm at 10 mSv/hr) iii. Area neutron monitor (Radiation Portal Monitor) • Air radiation monitoring system (Stack Monitoring System, Continuous Air Monitoring System) • Process radiation monitoring system - monitors radioactivity in the process systems (primary cooling system, heat exchanger, water purification system, effluent water monitoring)
Non-nuclear instrumentation Non-nuclear instrumentation comprises process measurements that do not involve neutron or gamma parameters. It includes both safety signals and safety related signals. Research reactors have many process variables to be measured. The sensors used to measure process variables in RTP mainly on temperature, pressure, level and flow.	• Temperature measurements i. Thermocouple (fuel temperature) ii. Resistance Temperature Detector (water temperature) iii. Thermostat (ReDICS cabinets) • Flow measurements i. Magnetic flow meter (Water flow) ii. Differential Pressure Transmitter (Air flow) • Water level measurement i. Float magnetic level transmitter • Pressure measurement i. Pressure transmitter with capacitive cells • Seismic monitoring system i. Seismic acceleration sensor ii. Seismometer

Table 7 shows the operational data of RTP from 2010 until 2018 [5]. In year 2014 which ReDICS was commissioned, there was quite significant amount of reactor operation (hours) while 2018 recorded the highest amount of reactor operation (hours). 2013 has less reactor operation (hours) due to reactor shut down for dismantling old analog console and installation of new ReDICS.

Table 7. Operational data of RTP (2010-2018).

Year	Reactor operation (hours)	Total energy released (MWhours)	No. of samples	No. of experiments
2010	522.03	335.09	3217	198
2011	348.63	241.70	2416	126
2012	413.53	278.22	3399	210
2013	291.95	149.01	3305	198
2014	544.60	365.03	5482	256
2015	453.45	235.11	3181	206
2016	414.62	218.62	1964	139
2017	497.65	299.88	2866	197
2018	675.38	404.11	1848	157

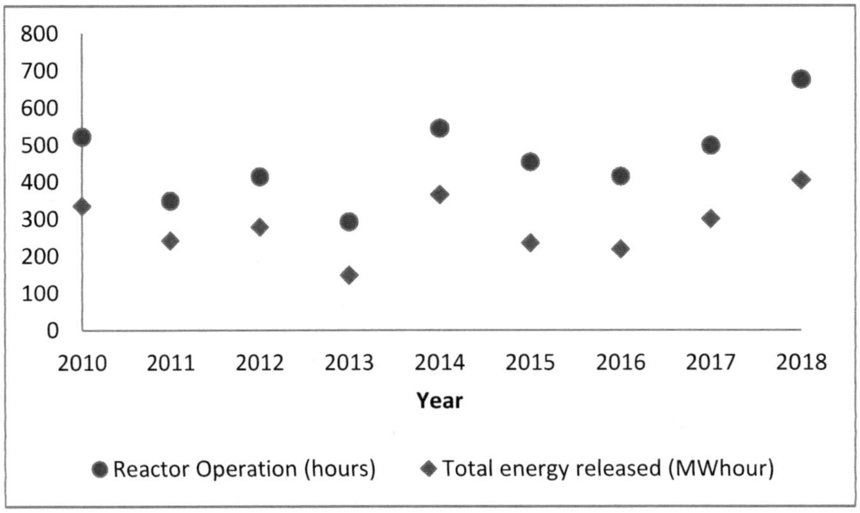

Figure 2. Graph of reactor operation time and total energy released by RTP in the year 2010-2018.

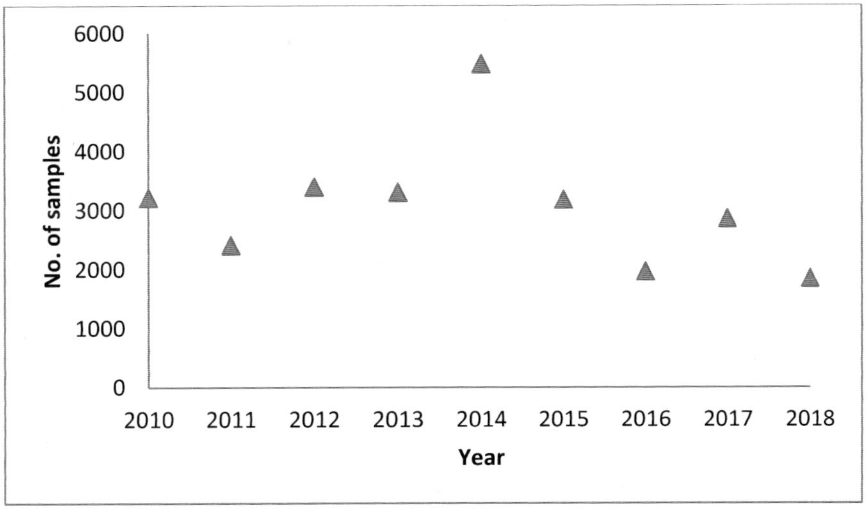

Figure 3. Graph of number of samples irradiated by RTP in the year 2010-2018.

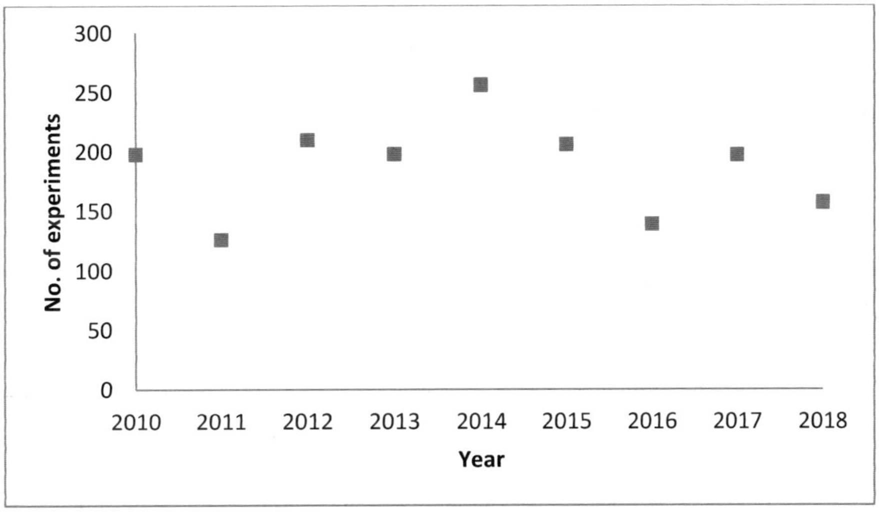

Figure 4. Graph of number of experiments performed at RTP in the year 2010-2018.

The I&C system of the RTP is very important as it connects the reactor operator and the plant. Therefore, ReDICS were tested before commissioning to verify that all components and systems in ReDICS are functioned correctly. It is also important to carry out well-planned maintenance of RTP in order to ensure the safe, efficient and reliable operation. Maintenance on I&C system of RTP is carried out twice a year to ensure the good performance of RTP operation as well as safety concern for effective monitoring and control of various process parameters [6]. Indirectly as a result, RTP shall remain as an active research reactor to continuously operate for research, training, education and training. Table 7 presented the work performed before and after commissioning of ReDICS to ensure ReDICS is always functioning at its best and safe.

Table 8. Work performed before and after commissioning of ReDICS.

	BEFORE COMMISSIONING
Site Acceptance Test (SAT)	• This test verifies that RPS, DACS, OWS, and field instruments perform their intended functions correctly after installation • The SAT includes all the required entities to ensure basic functional integrity of the systems and components.
Construction Acceptance Test (CAT)	• Differential Pressure Type Transmitter • Resistance Temperature Detector (RTD) • Level Transmitter • Panel Indicators • Area Radiation Monitoring System • Control Rod Driving Test • Rod Drop Test • Seismic Monitoring System
System Performance Test (SPT)	• Reactor Protection System • Data Acquisition and Control System • Operator Workstation
Integrated System Test (IST)	• Under Normal Operating Condition • Under Abnormal Operating Condition • Loss of Control System • Loss of Magnetic Power • Loss of Instrument Air • Response Time for RPS
Reactor Performance Test (RPT)	• Rod Worth Measurement • Power Ascension • Neutron Power Calibration • Square-Wave Operation
	AFTER COMMISSIONING
Inspection	• Standard TRIGA fuel elements • Control rods • Ventilation system • Cooling System
Periodic Testing	• Minimum safety interlock • SCRAM period • Reactivity insertion rate • Measurement of high voltage supply to console
Calibration	• Control rods, reactor power and measurement of shutdown margin • Safety channel • Monitoring system of reactor pool water temperature and level • Area radiation monitoring system • Stack monitor
Maintenance	• Control rods carrier • Rotary rack system • Water purification system

IOP Conf. Series: Materials Science and Engineering **785** (2020) 012052 doi:10.1088/1757-899X/785/1/012052

4. Conclusion

Reactor safety is an important issue with respect to employees, public, operation and utilization of the reactor. There are some activities have been planned to maintain the adequate safety and operating conditions of RTP which include the replacement of magnetic float level transducer with ultrasonic water level sensor. Other than that, ReDICS database will be developed for convenient data management and analysis as well as the integration of I&C monitoring system data for RTP. The comparisons made between the analog I&C system and ReDICS shows that the safety philosophy of the analog I&C system has been preserved. Furthermore, ReDICS has improved design with higher reliability and stability to ensure safe operation of RTP for the rest of its remaining life.

5. References

[1] IAEA 2016 Specific safety requirement *Safety of Research Reactors SSR-3*

[2] Malaysian Nuclear Agency ReDICS project technical documents

[3] Malaysian Nuclear Agency 2014 Chapter 8: Instrumentation and Control Systems *Safety Analysis Report of PUSPATI TRIGA Reactor* p 8-1 – 8-24

[4] IAEA 1995 *Research Reactor Instrumentation And Control Technology Tecdoc-973*

[5] Malaysian Nuclear Agency RTP Operating hour database (OptOur)

[6] Malaysian Nuclear Agency *RTP-W-14 Prosedur Pemeriksaan, Pengawasan Dan Penyenggaraan Tahunan.*

AUTHOR INDEX